AF867045

Experimental Brain Research Supplementum 9

Sir John Eccles discussing the supplementary motor area at Schloß Ringberg/Oberbayern 1983

Sensory-Motor Integration in the Nervous System

Edited by
O. Creutzfeldt, R. F. Schmidt and W. D. Willis

With 194 Figures

Springer-Verlag
Berlin Heidelberg New York Tokyo 1984

Prof. Dr. Otto Creutzfeldt
Department of Neurobiology,
Max-Planck-Institut für Biophysikalische Chemie, Postfach 2841,
3400 Göttingen (West-Germany)

Prof. Dr. Robert F. Schmidt
Physiologisches Institut
der Julius-Maximilians-Universität Würzburg,
Röntgenring 9, 8700 Würzburg (West-Germany)

Prof. Dr. William D. Willis
Marine Biomedical Institute,
University of Texas Medical Branch, 200 University Blvd.,
Galveston, TX 77550 USA

The cover illustration depicts the goddess of wisdom, Athene, emerging from the head of Zeus (covered with snow). This sculpture from A. Weber is at the MPI, Göttingen, where the meeting took place.

ISBN-13: 978-3-642-69933-7 e-ISBN-13: 978-3-642-69931-3
DOI: 10.1007/978-3-642-69931-3

Library of Congress Cataloging in Publication Data. Main entry under title:
Sensory-motor integration in the nervous system. (Experimental brain research supplementum; 9) "J. C. Eccles: publications from 1928–1980": p. Includes index. 1. Neurophysiology–Congresses. 2. Sensory motor integration–Congresses. 3. Eccles, John C. (John Carew), Sir, 1903. I. Creutzfeldt, O. (Otto) II. Schmidt, Robert F. III. Eccles, John C. (John Carew), Sir, 1903– . IV. Series: Experimental brain research. Supplement; 9. [DNLM: 1. Nervous System–physiology–congresses. 2. Psychomotor Performance–physiology–congresses. W1 EX485B v. 9/WL 102 S4785 1983] QP351.S45 1984 599'.0188 84-14049

This work is subject to copyright. All rights are reserved, whether the whole or part of the material is concerned, specifically those of translation, reprinting, re-use of illustrations, broadcasting, reproduction by photocopying machine or similar means, and storage in data banks. Under § 54 of the German Copyright Law where copies are made for other than private use a fee is payable to 'Verwertungs-gesellschaft Wort', Munich.

© by Springer-Verlag Berlin Heidelberg 1984

Softcover reprint of the hardcover 1st edition 1984

The use of registered names, trademarks, etc. in the publication does not imply, even in the absence of a specific statement, that such names are exempt from the relevant protective laws and regulations and therefore free for general use.

Product Liability: The publisher can give no guarantee for information about drug dosage and application thereof contained in this book. In every individual case the respective user must check its accuracy by consulting other pharmaceutical literature.

2125/3140-543210

Dedicated to Sir John Eccles
on the Occasion of his 80th Birthday

Contents

List of Contributors *

P. Andersen
Institute of Neurophysiology, Karl Johans Gate 47, Oslo 1 (Norway)

G. Andersson
Institute of Physiology, University of Lund, Sölvegatan 19, S-22362 Lund (Sweden)

M. Avoli
Institute of Neurophysiology, Karl Johans Gate 47, Oslo 1 (Norway)

G. Baumgartner
Abteilung für Neurologie der Universität Zürich, Kantonspital, Rämistr. 100, CH-8091 Zürich (Switzerland)

V.B. Brooks
Department of Physiology, University of Western Ontario, London/Ont., N6A 5C1 (Canada)

M. Brunelli
Istituto di Fisiologia Umana, Via del Giocnetto 8, 06100 Perugia (Italy)

N.C. Campbell
Institute of Physiology, University of Lund, Sölvegatan 19, S-22362 Lund (Sweden)

V. Chan-Palay
Department of Neurophysiology, Harvard Medical School, 25 Shattuck Street, Boston, Massachusetts 02115 (USA)

A.E. Cole
Department of Pharmacology and Physiology, University of California, San Francisco, CA 94143 (USA)

---

* *Only the addresses of authors who were personally present at the Symposium are listed*

J. Connor
The New York State Psychiatric Institute, 722 West 168th Street – Res. Annex, New York, NY 10032 (USA)

O. Creutzfeldt
Department of Neurobiology, Max-Planck-Institut für Biophysikalische Chemie, Postfach 2841, D-3400 Göttingen (W-Germany)

J.E. Desmedt
Brain Research Unit, University of Brussels, 115 Blvd. de Waterloo, B-1000 Bruxelles (Belgium)

J.C. Eccles
Ca "a la Gra", Contra (Locarno)/Tessin, CH-6611 (Switzerland)

C.-F. Ekerot
Institute of Physiology, University of Lund, Sölvegatan 19, S-22362 Lund (Sweden)

D.S. Faber
Division of Neurobiology, Department of Physiology, State University of New York, 313 Cary Hall, Buffalo, NY 14214 (USA)

G. Hesslow
Institute of Physiology, University of Lund, Sölvegatan 19, S-22362 Lund (Sweden)

Ø. Hvalby
Institute of Neurophysiology, Karl Johans Gate 47, Oslo 1 (Norway)

M. Ito
Department of Physiology, Faculty of Medicine, University of Tokyo, Bunkyoku, Tokyo (Japan)

E. Jankowska
Department of Physiology, University of Göteborg, Box 33031, S-40033 Göteborg (Sweden)

R. Jung
Abteilung für Neurologie und Neurophysiologie der Universität Freiburg/Br., Hansastr. 9a, D-7800 Freiburg/Br. (W-Germany)

E.R. Kandel
The New York State Psychiatric Institute, 722 West 168th Street – Res. Annex, New York, NY 10032 (USA)

M.R. Klee
Max-Planck-Institut für Hirnforschung, Deutschordenstr. 46, D-6000 Frankfurt/M 71 (W-Germany)

H. Korn
Institut Pasteur, Laboratoire de Neurobiologie Cellulaire, INSERM U261, Départment de Biologie Moléculaire, 25 Rue du Docteur Roux, 75724 Paris CEDEX 15 (France)

H.H. Kornhuber
Neurologische Klinik der Universität Ulm, D-7900 Ulm (W-Germany)

R. Kretz
Anatomisches Institut der Universität, 1, rue Gockel, CH-1700 Fribourg (Switzerland)

S. Landgren
Department of Physiology, University of Umeå, S-90187 Umeå (Sweden)

D.V. Madison
Department of Pharmacology and Physiology, University of California, San Francisco, CA94143 (USA)

F. Magni
Istituto di Fisiologia Umana, Via del Giocnetto 8, 06100 Perugia (Italy)

U. Misgeld
Max-Planck-Institut für Hirnforschung, Deutschordenstr. 46, D-6000 Frankfurt/M 71 (W-Germany)

N.R. Newberry
Department of Pharmacology and Physiology, University of California, San Francisco, CA94143 (USA)

R.A. Nicoll
Department of Pharmacology and Physiology, University of California, San Francisco, CA94143 (USA)

K.A. Olsson
Department of Physiology, University of Umeå, S-90187 Umeå (Sweden)

O. Oscarsson
Institute of Physiology, University of Lund, Sölvegatan 19, S-22362 Lund (Sweden)

T. Oshima
Department of Neurobiology, Tokyo Metropolitan Institute for Neurosciences, 2–6 Musashidai, Fuchu-shi, Tokyo, 183 (Japan)

E. Peterhans
Abteilung für Neurologie der Universität Zürich, Kantonspital, Rämistr. 100, CH-8091 Zürich (Switzerland)

K.R. Popper
Followfield, Manor Close, Manor Road, Penn, Buckinghamshire (U.K.)

K. Sasaki
Institute for Brain Research, Faculty of Medicine, Kyoto University, 606 Kyoto (Japan)

H.-G. Schaible
Physiologisches Institut der Julius Maximilians Universität zu Würzburg, Röntgenring 9, D-8700 Würzburg (W-Germany)

R.F. Schmidt
Physiologisches Institut der Julius Maximilians Universität zu Würzburg, Röntgenring 9, D-8700 Würzburg (W-Germany)

T.A. Sears
Sobell Department of Neurophysiology, Institute of Neurology, Queen Square, London WC1N 3BC (U.K.)

E. Shapiro
The New York State Psychiatric Institute, 722 West 168th Street – Res. Annex, New York, NY 10032 (USA)

H. Silfvenius
Department of Physiology, University of Umeå, S-90187 Umeå (Sweden)

P. Strata
Istituto di Fisiologia Umana dell'Universitá, Corso Raffaello 30, 10125 Torino (Italy)

J. Szentágothai
First Department of Anatomy, Semmelweis University Medical School, Tüzoltó ucta 58, H-1450 Budapest (Hungary)

N. Tsukahara
Department of Biophysical Engineering, Faculty of Engineering Science, Osaka University, Toyonaka (Japan)

R. von der Heydt
Abteilung für Neurologie der Universität Zürich, Kantonspital, Rämistr. 100, CH-8091 Zürich (Switzerland)

P.E. Voorhoeve
Department of Neurophysiology, Jan Swammerdam Institute, 1st Const. Huygensstraat 20, 1054 BW Amsterdam (Netherlands)

M. Wiesendanger
Institut de Physiologie de l'Université de Fribourg, Pérolles CH-1700 Fribourg (Switzerland)

R. Wiesendanger
Institut de Physiologie de l'Université de Fribourg, Pérolles CH-1700 Fribourg (Switzerland)

W.D. Willis
Marine Biomedical Institute, University of Texas Medical Branch, 200 University Blvd., Galveston, TX 77550 (USA)

V.J. Wilson
The Rockefeller University, New York, NY 10021 (USA)

B. Zwaagstra
Department of Neurophysiology, Jan Swammerdam Institute, 1st Const. Huygensstraat 20, 1054 BW Amsterdam (Netherlands)

Introduction

Over 200 neuroscientists met at the Max-Planck-Institut für Biophysikalische Chemie in Göttingen between April 7th–9th 1983 in order to honour a foreign member of the Institute, namely Sir John Carew Eccles, who celebrated his 80th birthday on January 27th 1983. It was a lively scientific gathering of former students, colleagues and friends of Sir John and Lady Helena. We had all come together from different parts of the world to celebrate this occasion, which gave us the wonderful chance to meet the various members of a large family, united by respect and gratitude for Sir John. The lectures were based on the many themes which have been at the centre of John Eccles' scientific zeal throughout his life. Indeed the chapter headings of this book have been taken from the book titles of Sir John's own work. We would also have liked to publish the discussions which took place after every lecture and to each of which John Eccles contributed in his usual lively manner, but editing them would not have conveyed the stimulating atmosphere, which Sir John created.

It is a wonderful tradition when scientific families meet from time to time to honour one of their patriarchs. An important function of science becomes evident at these kind of meetings, namely the unifying joy of coming together in the common endeavour to uncover the workings and wonders of nature. There is no desperate ambition to be first and smartest of all because the members of a family and their friends know each other well enough not to feel the need to attempt to impress one another.

We hope that the contents of this volume will put across the spirit of the meeting. The variety and high standard of the presentations will give some impression of the seed, which Sir John Eccles has sown in the field of Neurosciences, he himself is very aware and full of admiration for the tradition in which he grew up.

We would like to thank the Deutsche Forschungsgemeinschaft and the Max-Planck-Gesellschaft for generously funding this important symposion and all those who helped in the organization and the final production of this volume. And we wish you, Jack, many happy returns: Vivant sequentes.

O. D. Creutzfeldt
R. F. Schmidt
W. D. Willis

Introduction

Curriculum Vitae and Bibliography of Sir John Eccles

The following is an abbreviated version of the curriculum vitae and bibliography of John C. Eccles.

Present Nationality: U.S.A.

Date and Place of Birth: January 27, 1903; Melbourne, Australia

Education: University of Melbourne, M.B., B.S. (1925); Oxford, M.A., D.Phil. (1929)

Positions Held

1925–28 Rhodes Scholar, Oxford
1927–29 Christopher Welch Scholar, Oxford
1927–32 Junior Research Fellow, Exeter College, Oxford
1932–34 Staines Medical Fellow, Exeter College, Oxford
1934–37 Tutorial Fellow, Magdalen College, Oxford University
Demonstrator in Physiology, Oxford
1937–43 Director, Kanematsu Memorial Institute of Pathology, Sydney Hospital, Sydney
1944–51 Professor of Physiology, University of Otago Medical School, Dunedin, New Zealand
1951–66 Professor of Physiology, Australian National University, Canberra, A.C.T., Australia
1966–68 Member, American Medical Association, Education and Research Foundation, Institute for Biomedical Research, Chicago, Ill., USA
1968–75 Distinguished Professor of Physiology and Biophysics, State University of New York at Buffalo, N.Y., USA
1975– present Distinguished Professor of Physiology and Biophysics, State University of New York at Buffalo – Emeritus;
Address – CH 6611 Contra (TI) Switzerland

Honours:

1929 Rolleston Prize, Oxford
1929 Member, Physiological Society, U.K.

© Springer-Verlag Berlin · Heidelberg 1984

1941	Fellow, Royal Society, London
1952	Waynflete, Lecturer, Magdalen College
1952	Hon. Member, American Physiological Society
1954	Foundation Fellow, Australian Academy of Science
1955	Herter Lecturer, Johns Hopkins University
1957–61	President, Australian Academy of Science
1958	Knight Bachelor, Queen's Birthday Honours
1959	Ferrier Lecturer, Royal Society
1959	Foreign Honorary Member, American Academy of Arts and Sciences
1959	Squibb Centenary Lecturer
1960	Honorary Sc.D., Cambridge University
1961	Member, Pontifical Academy of Sciences
1961	Member, Deutsche Akademie der Naturforscher Leopoldina
1961	Baly Medal, Royal College of Physicians
1961	Honorary Fellow, Exeter College, Oxford
1962	Royal Medal, Royal Society, London
1963	Foreign Honorary Member, Accademia Nazionale dei Lincei
1963	Cothenius Medal, Deutsche Akademie der Naturforscher Leopoldina
1963	Nobel Prize in Physiology and Medicine
1963	Flinders Lecture, Australian Academy of Sciences
1964	Honorary D.Sc., University of Tasmania
1964	Honorary Fellow, Magdalen College, Oxford
1964	Honorary Fellow, American Philosophical Society
1965	Honorary LL.D., University of Melbourne
1965	Eddington Memorial Lecturer, Cambridge University
1966	Foreign Associate, National Academy of Sciences
1966	Honorary D.Sc., University of British Columbia
1966	Sherrington Lecturer, University of Liverpool
1967	Honorary D.Sc., Gustavus Adolphus College
1967	Honorary D.Sc., Marquette University
1967	Distinguished Visiting Professor, University of British Columbia
1967	Honorary Member, Academia Medica Lombarda
1968	Honorary Fellowship, Indian Academy of Sciences
1968	Honorary Member, Czechoslovak Medical Society J. E. Purkyně
1968	Alexander Forbes Lecturer, Grass Foundation
1969	Associate Member, Academie Royal de Belgique
1969	Honorary M.D., Charles University, Prague
1969	Honorary D.Sc., Loyola University, Chicago
1969	Honorary M.D., Yeshiva University, New York
1969	Foerster Lecturer, University of California at Berkeley
1971	Honorary Member, The Physiological Society, U.K.
1972	Patten Memorial Lecturer, Indiana University
1973	Foreign Member, Max-Planck-Institut für Biophysikalische Chemie, Göttingen
1973	Compton Lecturer, Washington University, St. Louis
1973–74	Phi Beta Kappa Lecturer, USA
1974	Honorary D.Sc., Oxford University

1976 Honorary Member, European Brain and Behaviour Society
1977 Honorary Member, European Neurosciences Society
1977 Phi Beta Kappa Lecturer, USA
1977 Botazzi Lecturer, Societa Italiana di Fisiologia
1978 Gifford Lecturer, University of Edinburgh
1978 Visiting Professor, Department of Biology, New York University
1979 Gifford Lecturer, University of Edinburgh
1979 Green Visiting Professor, University of Texas Medical Branch at Galveston
1980 Honorary Ph.D., University of Fribourg, Suisse
1980 Lecturer of "Werner Heisenberg Vorlesungen" at Carl Friedrich von Siemens Stiftung, München, Germany
1981 Carroll Lecturer, Georgetown University
1981 Lecturer, "100-Jahr-Feier Walter Rudolf Hess", Zürich
1982 Honorary Member, Society for Neuroscience
1983 Honorary Member, Japanese Physiological Society
1984 Honorary Member, Indian Physiological Society

Books and Major Reviews

1932 *Reflex Activity of the Spinal Cord* (with R. S. Creed, D. Denny Brown, E. G. T. Liddell and C. S. Sherrington), Clarendon Press, Oxford, 183 pages

1936 *Synaptic and Neuromuscular Transmission,* Ergebnisse der Physiologie, biologischen Chemie und experimentellen Pharmakologie 38:339–444

1953 *The Neurophysiological Basis of Mind: The Principles of Neurophysiology* (Waynflete Lectures, 1952), Clarendon Press, Oxford, 314 pages

1957 *Physiology of Nerve Cells* (Herter Lectures, 1955), Johns Hopkins Press, Baltimore, 270 pages (translated into Russian)

1961 *The Mechanism of Synaptic Action,* Ergebnisse der Physiologie, biologischen Chemie und experimentellen Pharmakologie 51:299–430

1964 *The Physiology of Synapses,* Springer, Berlin Heidelberg New York, 316 pages (translated into Japanese, Polish and Russian)

1966 The ionic mechanisms of excitatory and inhibitory synaptic action. *Ann. N.Y. Acad. Sci* 137:473–495

1966 *Brain and Conscious Experience,* Editor, Springer, Berlin Heidelberg New York (Conference of Pontifical Academy of Sciences, 1964)

1967 *The Cerebellum as a Neuronal Machine* (with M. Ito and J. Szentágothai), Springer, Berlin Heidelberg New York, 355 pages

1969 *The Inhibitory Pathways of the Central Nervous System* (Sherrington Lectures, 1966), Liverpool University Press, Liverpool, 135 pages

1970 *Facing Reality,* Heidelberg Science Series, Springer, Berlin Heidelberg New York, 210 pages (translated into German, Spanish and Italian)

1972 *Reflex Activity of the Spinal Cord,* 2nd edition (with annotations by D. P. C. Lloyd)

1973 *The Understanding of the Brain* (Patten Lectures, 1972), McGraw Hill, New York, 238 pages (translated into German, Spanish, Italian and Japanese)

1976 *The Understanding of the Brain,* 2nd edition McGraw Hill, New York (translated into German)

1977 *The Self and Its Brain* (with Sir Karl Popper), Springer International, Berlin Heidelberg New York, 597 pages (translated into Spanish, German, Italian and Portuguese)

1978 *Molecular Neurobiology of the Mammalian Brain* (with P.L. McGeer and E.G. McGeer), Plenum, New York, 644 pages

1979 *Sherrington: His Life and Thought* (with W.C. Gibson), Springer, Berlin Heidelberg New York, 269 Pages

The Human Mystery (Gifford Lectures for 1978), Springer International, Berlin Heidelberg New York, 255 pages (translated into Italian, German, and French)

1980 *The Human Psyche* (Gifford Lectures for 1979), Springer International, Berlin Heidelberg New York, 279 pages (translated into Italian and German)

Gehirn und Geist (with H. Zeier), Kindler, München, 210 pages

1981 *The modular operation of the cerebral neocortex considered as the material basis for mental events.* Neuroscience 6:1839–1856

1983 *Calcium in long-term potentiation as a model for memory.* Neuroscience 10:1071–1081

1984 *The Wonder of being Human: Our Brain and our Mind.* (with D.N. Robinson) Translated into German and Italian. The Free Press, New York, London, pp. 182

Major Research and Theoretical Papers

1928 The incidence of central inhibition on restricted fields of motor-units. *J. Physiol.* 66:109–120 (with R.S. Creed)

1929 Crossed extensor reflexes and their interaction. *J. Physiol.* 67:97–118 (with R. Granit)

Observations on electrical stimulation of the cerebellar cortex. *Proc. R. Soc.* [Biol] 104:518–536 (with D. Denny-Brown and E.G.T. Liddell)

1930 Reflex summation in the ipsilateral spinal flexion reflex. *J. Physiol.* 69:1–28 (with C.S. Sherrington)

Numbers and contraction values of individual motor-units examined in some muscles of the limb. *Proc. R. Soc.* [Biol] 106:326–357 (with C.S. Sherrington)

The isometric responses of mammalian muscles. *J. Physiol.* 69:377–385 (with S. Cooper)

1931 Studies on the flexor reflex. I. Latent period. *Proc. R. Soc.* [Biol] 107:511–534 (with C.S. Sherrington)

Studies on the flexor reflex. II. The reflex response evoked by two afferent volleys. *Proc. R. Soc.* [Biol] 107:535–556 (with C.S. Sherrington)

Studies on the flexor reflex. III. The central effects produced by an antidromic volley. *Proc. R. Soc.* [Biol] 107:557–585

Studies on the flexor reflex. IV. After-discharge. *Proc. R. Soc.* [Biol] 107:586–596 (with C.S. Sherrington)

Studies on the flexor reflex. V. General conclusions. *Proc. R. Soc.* [Biol] 107:597–605 (with C.S. Sherrington)

Studies on the flexor reflex. VI. Inhibition. *Proc. R. Soc.* [Biol] 109:91–113 (with C.S. Sherrington)

1932 The rhythmic discharge of motoneurones. *Proc. R. Soc.* [Biol] 110:485–514 (with H.E. Hoff)

1934 The rhythm of the heart beat. I. Location, action potential and electrical excitability of the pacemaker. *Proc. R. Soc.* [Biol] 115:307–327 (with H.E. Hoff)

The rhythm of the heart beat. II. Disturbance of rhythm produced by late premature beats. *Proc. R. Soc.* [Biol] 115:327–351 (with H.E. Hoff)

The rhythm of the heart beat. III. Disturbance of rhythm produced by early premature beats. *Proc. R. Soc.* [Biol] 115:352–368 (with H.E. Hoff)

The action of a single vagal volley on the rhythm of the heart beat. *J. Physiol.* 82:211–241 (with G.L. Brown)

Further experiments on vagal inhibition of the heart. *J. Physiol.* 82:242–257 (with G.L. Brown)

1935 An investigation on the visco-elastic properties of rubber. *Proc. R. Soc.* [Biol] 148:171–185 (with J.H.C. Thompson)

The action potential of the superior cervical ganglion. *J. Physiol.* 85:179–206

Facilitation and inhibition on the superior cervical ganglion. *J. Physiol.* 85:207–238

Slow potential waves in the superior cervical ganglion. *J. Physiol.* 85:464–501

1936 The actions of antidromic impulses on ganglion cells. *J. Physiol.* 88:1–39

1937 The excitation and response of smooth muscle. *J. Physiol.* 90:31–67 (with J.W. Magladery)

Rhythmic responses of smooth muscle. *J. Physiol.* 90:68–99 (with J.W. Magladery)

The discharge of impulses from ganglion cells. *J. Physiol.* 91:1–22

1939 Responses which nerve impulses evoke in mammalian striated muscles. *J. Physiol.* 97:44–102 (with W.J. O'Connor)

1941 Nature of the "endplate potential" in curarized muscle. *J. Neurophysiol.* 4:362–387 (with B. Katz and S.W. Kuffler)

Initiation of muscle impulses at neuro-muscular junction. *J. Neurophysiol.* 4:402–417 (with S.W. Kuffler)

The endplate potential during and after the muscle spike potential. *J. Neurophysiol.* 4:486–506 (with S.W. Kuffler)

Abortive impulses at the neuro-muscular junction. *J. Physiol.* 100:318–328 (with W.J. O'Connor)

1942 Effect of eserine on neuro-muscular transmission. *J. Neurophysiol.* 5:211–230 (with B. Katz and S.W. Kuffler)

1943 Synaptic potentials and transmission in sympathetic ganglion. *J. Physiol.* 101:465–483

1944 The nature of synaptic transmission in a sympathetic ganglion. *J. Physiol.* 103:27–53

Synaptic transmission in the spinal cord. *Nature* 153:432

Catelectronic potentials in the dorsal roots of the spinal cord. *Nature* 154:395–396

Investigations on muscle atrophies arising from disuse and tenotomy. *J. Physiol.* 103:253–266

1945 An electrical hypothesis of synaptic and neuro-muscular transmission. *Nature* 156:680–682

1946 Synaptic potentials of motoneurones. *J. Neurophysiol.* 9:87–120

Dorsal root potentials of the spinal cord. *J. Neurophysiol.* 9:139–160 (with J. L. Malcolm)

1947 Acetylcholine and synaptic transmission in the spinal cord. *J. Neurophysiol.* 10:197–204

An electrical hypothesis of central inhibition. *Nature* 159:760–764 (with C. McC. Brooks)

Electrical investigation of the monosynaptic pathway through the spinal cord. *J. Neurophysiol.* 10:251–274 (with C. McC. Brooks)

A study of the effects of anaesthesia and asphyxia on the mono-synaptic pathway through the spinal cord. *J. Neurophysiol.* 10:349–360 (with C. McC. Brooks)

1948 An analysis of synaptic excitatory action. *J. Neurophysiol.* 11:365–376 (with C. McC. Brooks)

Inhibitory action on a motor nucleus and focal potentials generated therein. *J. Neurophysiol.* 11:401–416 (with C. McC. Brooks)

1948 Synaptic potentials of inhibited motoneurones. *J. Neurophysiol.* 11:417–430 (with C. McC. Brooks and J. L. Malcolm)

Inhibition of antidromic responses of motoneurones. *J. Neurophysiol.* 11:431–444 (with C. McC. Brooks)

1949 Action of anti-cholinesterases on endplate potential of frog muscle. *J. Neurophysiol.* 12:59–80 (with W. V. MacFarlane)

Electric potentials generated by antidromic volleys in quadriceps and hamstring motoneurones, *J. Neurophysiol.* 12:393–424 (with T. H. Barakan and C. B. B. Downman)

1950 After-potentials and excitability of spinal motoneurones following antidromic activation. *J. Neurophysiol.* 13:9–38 (with C. McC. Brooks and C. B. B. Downman)

After-potentials and excitability of spinal motoneurones following orthodromic activation. *J. Neurophysiol.* 13:157–176 (with C. McC. Brooks and C. B. B. Downman)

Post-tetanic potentiation of responses of motoneurones. *Nature* 166:465–466 (with W. Rall)

1951 Plasticity of mammalian monosynaptic reflexes. *Nature* 167:466–468 (with A. K. McIntyre)

Experimental investigations on the afferent fibres in muscle nerves. *Proc. R. Soc.* [Biol] 138:453–475 (with L. G. Brock and W. Rall)

Repetitive monosynaptic activation of motoneurones. *Proc. R. Soc.* [Biol] 138:475–498 (with W. Rall)

Effects induced in a monosynaptic reflex path by its activation. *J. Neurophysiol.* 14:353–376 (with W. Rall)

Hypotheses relating to the brain-mind problem. *Nature* 168:53–57

Action potentials of motoneurones with intracellular electrode. *Proc. Univ. Otago Med. School.* 29:14–15 (Brock, L.G., Coombs, J.S. and Eccles, J.C.)

Interpretation of action potentials evoked in the cerebral cortex. *EEG and Clin. Neurophysiol. J.* 3:449–464

1952 The recording of potentials from motoneurones with an intracellular electrode. *J. Physiol.* 117:431–460 (with L. G. Brock and J. S. Coombs)

The nature of the monosynaptic excitatory and inhibitory processes in the spinal cord. *Proc. R. Soc.* [Biol] 140:169–176 (with L. G. Brock and J. S. Coombs)

1953 Functional changes in chromatolysed motoneurones. *J. Comp. Neurol.* 98:9–36 (with C. B. B. Downman and A. K. McIntyre)

Strychnine as a depressant of primary inhibition. *Nature* 171:1061 (with K. Bradley)

The effects of disuse and of activity of mammalian spinal reflexes. *J. Physiol.* 121:492–516 (with A. K. McIntyre)

Intracellular recording from antidromically activated motoneurones. *J. Physiol.* 122:429–461 (with L. G. Brock and J. S. Coombs)

Analysis of the fast afferent impulses from thigh muscles. *J. Physiol.* 122:462–473 (with K. Bradley)

An investigation of primary or direct inhibition. *J. Physiol.* 122:474–488 (with K. Bradley and D. M. Easton)

1954 Spinal cord potentials generated by volleys in the large muscle afferents. *J. Physiol.* 125:590–606 (with P. Fatt, S. Landgren and G. J. Winsbury)

Cholinergic and inhibitory synapses in a pathway from motor-axon collaterals to motoneurones. *J. Physiol.* 126:524–562 (with P. Fatt and K. Koketsu)

1955 Mode of action of tetanus toxin. *Nature* 175:120–121 (with V. B. Brooks and D. R. Curtis)

The central action of antidromic impulses in motor nerve fibres. *Pflügers Arch.* 260:385–415

The electrical properties of the motoneurone membrane. *J. Physiol.* 130:291–325 (with J. S. Coombs and P. Fatt)

The specific ionic conductances and the ionic movements across the motoneuronal membrane that produce the inhibitory post-synaptic potential. *J. Physiol.* 130:326–373 (with J. S. Coombs and P. Fatt)

Excitatory synaptic action in motoneurones. *J. Physiol.* 130:374–395 (with J. S. Coombs and P. Fatt)

Inhibitory suppression of reflex discharges from motoneurones. *J. Physiol.* 130:396–413 (with J. S. Coombs and P. Fatt)

1956 Central pathway for direct inhibitory action of impulses in largest afferent nerve fibres to muscle. *J. Neurophysiol.* 19:75–98 (with P. Fatt and S. Landgren)

Pharmacological investigations on a central synapse operated by acetylcholine. *J. Physiol.* 131:154–169 (with R.M. Eccles and P. Fatt)

Time courses of motoneuronal responses, *Nature* 178:1049–1050 (with J.S. Coombs and D.R. Curtis)

1957 The action of tetanus toxin on the inhibition of motoneurones. *J. Physiol.* 135:655–672 (with V.B. Brooks and D.R. Curtis)

Durations of afterhyperpolarization of motoneurones supplying fast and slow muscles. *Nature* 179:868–886 (with R.M. Eccles and A. Lundberg)

Pharmacological studies on spinal reflexes. *J. Physiol.* 136:420–434 (with D.R. Curtis and R.M. Eccles)

Synaptic actions on motoneurones in relation to the two components of the Group I muscle afferent volley. *J. Physiol.* 136:527–546 (with R.M. Eccles and A. Lundberg)

The convergence of monosynaptic excitatory afferents on to many different species of alpha motoneurones. *J. Physiol.* 137:22–50 (with R.M. Eccles and A. Lundberg)

Synaptic actions on motoneurones caused by impulses in Golgi tendon organ afferents. *J. Physiol.* 138:227–252 (with R.M. Eccles and A. Lundberg)

The interpretation of spike potentials of motoneurones. *J. Physiol.* 139:198–231 (with J.S. Coombs and D.R. Curtis)

The generation of impulses in motoneurones, *J. Physiol.* 139:232–249 (with J.S. Coombs and D.R. Curtis)

1958 The relationship between the mode of operation and the dimensions of the junctional regions at synapses and motor-end organs. *Proc. R. Soc.* [Biol] 148:38–56 (with J.C. Jaeger)

The action potentials of the alpha motoneurones supplying fast and slow muscles. *J. Physiol.* 142:275–291 (with R.M. Eccles and A. Lundberg)

The behaviour of nerve cells. In: *Neurological Basis of Behaviour* 28–47. Ciba Found, Symp., London

Problems of plasticity and organization at simplest levels of mammalian central nervous system. *Perspectives in Biology and Medicine* 1:379–396

The behaviour of chromatolysed motoneurones studied by intracellular recording. *J. Physiol.* 143:11–40 (with B. Libet and R.R. Young)

Intracellular recording from cells in Clarke's column. *Acta Physiol. Scand.* 43:303–314 (with D.R. Curtis and A. Lundberg)

1959 Delayed effects of peripheral severance of afferent nerve fibres on the efficacy of their central synapses. *J. Physiol.* 145:204–220 (with K. Krnjevic and R. Miledi)

The electrical constants of the motoneurone membrane. *J. Physiol.* 145:505–528 (with J.S. Coombs and D.R. Curtis)

The time courses of excitatory and inhibitory synaptic actions. *J. Physiol.* 154:529–546 (with D.R. Curtis)

Neuron physiology – introduction. In: *Handbook of Physiology – Neurophysiology I.* Ed. by J. Field, H.W. Magoun, and V.E. Hall, 59–73. Baltimore: Waverly Press, Inc.

Excitatory and inhibitory synaptic action. *Annals of the New York Academy of Sciences* 81:247–264

Potential changes recorded inside primary afferent fibres within the spinal cord. *J. Physiol.* 149:250–273 (with K. Krnjevic)

Presynaptic changes associated with post-tetanic potentiation in the spinal cord. *J. Physiol.* 149:274–287 (with K. Krnjevic)

1960 Synaptic action during and after repetitive stimulation. *J. Physiol.* 150:374–398 (with D. R. Curtis)

Differentiation of fast and slow muscles in the cat hind limb. *J. Physiol.* 150:399–416 (with A. J. Buller and R. M. Eccles)

Interactions between motoneurones and muscles in respect of the characteristic speeds of their responses. *J. Physiol.* 150:417–439 (with A. J. Buller and R. M. Eccles)

The properties of the dendrites. In: *Structure and Function of the Cerebellar Cortex.* Eds.: D. B. Tower and J. P. Schade. Proc. 2nd Internat. meeting of Neurobiologists, Amsterdam. Elsevier Publishing Company

Electrophysiological studies on gamma motoneurones. *Acta Physiol. Scand.* 50:32–40 (with R. M. Eccles, A. Iggo and A. Lundberg)

Monosynaptic excitatory action on motoneurones regenerated to antagonistic muscles. *J. Physiol.* 154:68–88 (with R. M. Eccles and F. Magni)

Types of neurones in and around the intermediate nucleus of the lumbosacral cord. *J. Physiol.* 154:89–114 (with R. M. Eccles and A. Lundberg)

Correlation of the inhibitory postsynaptic potential of motoneurones with the latency and time course of inhibition of monosynaptic reflexes. *J. Physiol.* 154:354–377 (with T. Araki and M. Ito)

1961 The nature of central inhibition. *Proc. R. Soc.* [Biol] 153:445–476

Membrane time constants of cat motoneurones and time courses of synaptic action. *Exp. Neurol.* 4:1–22

Intracellular recording from cells of the ventral spinocerebellar tract. *J. Physiol.* 153:486–516 (with J. I. Hubbard and O. Oscarsson)

Synaptic action of group I and II afferent fibres of muscle on the cells of the dorsal spino-cerebellar tract. *J. Physiol.* 158:517–543 (with O. Oscarsson and W. D. Willis)

The effects of use and disuse on synaptic function. In: *Brain Mechanisms and Learning,* 335–348. Ed. by J. F. Delafresnaye. 2nd. C.I.O.M.S. Conference. Oxford: Blackwells Scientific Publications

Dorsal root reflexes of muscle group I afferent fibres. *J. Physiol.* 159:128–146 (with W. Kozak and F. Magni)

Central inhibitory action attributable to presynaptic depolarization produced by muscle afferent volleys. *J. Physiol.* 149:147–166 (with R. M. Eccles and F. Magni)

Electrophysiological investigations on Renshaw cells. *J. Physiol.* 159:461–478 (with R. M. Eccles, A. Iggo and A. Lundberg)

Distribution of recurrent inhibition among motoneurones. *J. Physiol.* 159:479–499 (with R. M. Eccles, A. Iggo and M. Ito)

1962 Depolarization of central terminals of Group I afferent fibres from muscle. *J. Physiol.* 160:62–93 (with F. Magni and W. D. Willis)

Central pathways responsible for depolarization of primary afferent fibres. *J. Physiol.* 161:237–257 (with P. G. Kostyuk and R. F. Schmidt)

Presynaptic inhibition of the central actions of flexor reflex afferents. *J. Physiol.* 161:258–281 (with P. G. Kostyuk and R. F. Schmidt)

Presynaptic inhibition of the spinal monosynaptic reflex pathway. *J. Physiol.* 161:282–297 (with R. F. Schmidt and W. D. Willis)

Presynaptic inhibitory action of cerebral cortex on the spinal cord. *Nature* 194:740–741 (with P. Andersen and T. A. Sears)

Presynaptic inhibition in the cuneate nucleus. *Nature* 194:741–742 (with P. Andersen and R. F. Schmidt)

The effect of electric polarization of the spinal cord on central afferent fibres and on their excitatory synaptic action. *J. Physiol.* 162:138–150 (with P. G. Kostyuk and R. F. Schmidt)

An investigation into the effect of degenerating primary afferent fibers on the monosynaptic innervation of motoneurones. *J. Neurophysiol.* 25:544–558 (with R. M. Eccles and C. N. Shealy)

Experiments utilizing monosynaptic excitatory action on motoneurones for testing hypotheses relating to specificity of neuronal connections. *J. Neurophysiol.* 25:559–580 (with R. M. Eccles, C. N. Shealy and W. D. Willis)

Further investigations on the influence of motoneurones on the speed of muscle contractions. *J. Physiol.* 163:324–339 (with R. M. Eccles and W. Kozak)

Inhibitory phasing of neuronal discharge. *Nature* 196:645–647 (with P. Andersen)

Spinal neurones: Synaptic connexions in relation to chemical transmitters and pharmacological responses. In: Proceedings, *First Inter. Pharmacol. Meeting* 8:157–182. Ed. B. Uvnas, Oxford: Pergamon Press

1963 Depolarization of central terminals of Group Ib afferent fibers of muscle. *J. Neurophysiol.* 26:1–27 (with R. F. Schmidt and W. D. Willis)

The location and the mode of action of the presynaptic inhibitory pathways on to Group I afferent fibers from muscle. *J. Neurophysiol.* 26:506–522 (with R. F. Schmidt and W. D. Willis)

The mode of operation of the synaptic mechanism producing presynaptic inhibition. *J. Neurophysiol.* 26:523–538 (with R. F. Schmidt and W. D. Willis)

Recurrent inhibition in the hippocampus with identification of the inhibitory cell and its synapse. *Nature* 198:540–542 (with P. Andersen and Y. Loyning)

Inhibition of discharges into the dorsal and ventral spinocerebellar tracts. *J. Neurophysiol.* 26:635–645 (with R. F. Schmidt and W. D. Willis)

Depolarization of the central terminals of cutaneous afferent fibers. *J. Neurophysiol.* 26:646–661 (with R. F. Schmidt and W. D. Willis)

Inhibitory synapses on somas of Purkinje cells in the cerebellum. *Nature* 199:655–656 (with P. Andersen and P. E. Voorhoeve)

Identification of inhibitory neurones in the hippocampus. *Nature* 199:699–700 (with P. Andersen and Y. Loyning)

Pharmacological studies on presynaptic inhibition. *J. Physiol.* 168:500–530 (with R. F. Schmidt and W. D. Willis)

Strychnine-resistant central inhibition. *Nature* 200:843–845 (with P. Andersen, Y. Loyning and P. E. Voorhoeve)

1964 Cortically evoked depolarization of primary afferent fibers in the spinal cord. *J. Neurophysiol.* 27:63–77 (with P. Andersen and T. A. Sears)

Slow potential waves produced in the cuneate nucleus by cutaneous volleys and by cortical stimulation. *J. Neurophysiol.* 27:78–91 (with P. Andersen, R. F. Schmidt and T. Yokota)

Depolarization of presynaptic fibers in the cuneate nucleus. *J. Neurophysiol.* 27:92–106 (with P. Andersen, R. F. Schmidt and T. Yokota)

Effects of intracellular potassium and sodium injections on the inhibitory postsynaptic potential. *Proc. R. Soc.* [Biol] 160:181–196 (with R. M. Eccles and M. Ito)

Effects produced on inhibitory postsynaptic potentials by the coupled injections of cations and anions into motoneurones. *Proc. R. Soc.* [Biol] 160:197–210 (with R. M. Eccles and M. Ito)

Location of postsynaptic inhibitory synapses of hippocampal pyramids. *J. Neurophysiol.* 27:592–607 (with P. Andersen and Y. Loyning)

Pathway of postsynaptic inhibition in the hippocampus. *J. Neurophysiol.* 27:608–619 (with P. Andersen and Y. Loyning)

Excitation of cerebellar Purkinje cells by the climbing fibres. *Nature* 203:245–246 (with R. Llinas and K. Sasaki)

The excitatory responses of spinal neurones. In: *Progress in Brain Research, 12. Physiology of Spinal Neurons* 1–13. Ed. by J. C. Eccles and J. P. Schade. Amsterdam, Elsevier Publishing Co.

Presynaptic inhibition on the spinal cord. In: *Progress in Brain Research, 12. Physiology of Spinal Neurons* 65–89. Ed. by J. C. Eccles and J. P. Schade. Amsterdam: Elsevier Publishing Co.

The ionic mechanism of postsynaptic inhibition. *Science* 145:1140–1147

Identification of relay cells and interneurons in the cuneate nucleus. *J. Neurophysiol.* 27:1080–1095 (with P. Andersen, R. F. Schmidt und T. Yokota)

Mechanisms of synaptic transmission in the cuneate nucleus. *J. Neurophysiol.* 27:1096–1116 (with P. Andersen, T. Oshima and R. F. Schmidt)

Postsynaptic inhibition of cerebellar Purkinje cells. *J. Neurophysiol.* 27:1138–1153 (with P. Andersen and P. E. Voorhoeve)

The ventro-basal nucleus of the thalamus: types of cells, their responses and their functional organization. *J. Physiol.* 174:370–399 (with P. Andersen and T. A. Sears)

Golgi cell inhibition in the cerebellar cortex. *Nature* 204:1265–1266 (with R. Llinas and K. Sasaki)

Inhibition in thalamic and cortical neurones and its role in phasing neuronal discharges. (William Lennox Lect.) *Epilepsia* 6:89–115

Locating and identifying postsynaptic inhibitory synapses by the correlation of physiological and histological data. In: *Modern Trends in Neuromorphologica. Symp. Biol. Hung.* 5:219–214 (Andersen, P. and Eccles, J.C.)

1966 The inhibitory interneurones within the cerebellar cortex. *Exp. Brain Res.* 1:1–16 (with R. Llinas and K. Sasaki)

Parallel fibre stimulation and the responses induced thereby in the Purkinje cells of the cerebellum. *Exp. Brain Res.* 1:17–39 (with R. Llinas and K. Sasaki)

The mossy fibre-granule cell relay in the cerebellum and its inhibition by Golgi cells. *Exp. Brain Res.* 1:82–101 (with R. Llinas and K. Sasaki)

The excitatory synaptic action of climbing fibres on the Purkinje cells of the cerebellum. *J. Physiol. 182:268–296* (with R. Llinas and K. Sasaki)

Interaction experiments on the responses evoked in the purkinje cells by climbing fibres. *J. Physiol.* 182:297–315 (with R. Llinas, K. Sasaki and P. Voorhoeve)

The action of antidromic impulses on the cerebellar Purkinje cells. *J. Physiol.* 182:316–345 (with R. Llinas and K. Sasaki)

Intracellularly recorded responses of the cerebellar Purkinje cells. *Exp. Brain Res.* 1:161–183 (with R. Llinas and K. Sasaki)

Functional organization of the cerebellum in relation to its role in motor control. In: *Nobel Symposium I, Muscular afferents and motor control.* 19–36, Ed. by Ragnar Granit. Almqvist & Wiksell, Stockholm, 1966

Cerebral synaptic mechanisms. In: *Brain and Conscious Experience.* Ed. J. C. Eccles, Springer-Verlag, Heidelberg

Conscious experience and memory. In: *Brain and Conscious Experience.* Ed. J. C. Eccles, Springer-Verlag, Heidelberg

The profiles of physiological events produced by a parallel fiber volley in the cerebellar cortex. *Exp. Brain Res.* 2:18–34 (with K. Sasaki and P. Strata)

1967 Interpretation of the potential fields generated in the cerebellar cortex by a mossy fiber volley. *Exp. Brain Res.* 3:58–80 (with K. Sasaki and P. Strata)

A comparison of the inhibitory actions of Golgi cells and of basket cells. *Exp. Brain Res.* 3:81–94 (with K. Sasaki and P. Strata)

Circuits in the cerebellar control of movement. *Proc. Natl. Acad. Sci.* USA 58:336–343

Patterns of climbing fiber input to the cerebellar anterior lobe. *Brain Res.* 5:425–430 (with L. Provini, P. Strata and H. Táboříková)

1968 Responses in the dorsal accessory olive of the cat to stimulation of hind limb afferents. *J. Physiol.* 194:125–145 (with D. M. Armstrong, R. J. Harvey and P. B. C. Matthews)

The way in which the cerebellum processes sensory information from muscle. In: *Neurophysiological Basis of Normal and Abnormal Motor Activities.* Ed. M. D. Yahr and D. P. Purpura, Raven Press, Hewlett, New York. 379–406 pp

Analysis of electrical potentials evoked in the cerebellar anterior lobe by stimulation of hindlimb and forelimb nerves. *Exp. Brain Res.* 6:171–194 (with L. Provini, P. Strata and H. Táboříková)

Topographical investigations on the climbing fiber inputs from forelimb and hindlimb afferents to the cerebellar anterior lobe. *Exp. Brain Res.* 6:195–215 (with L. Provini, P. Strata and H. Táboříková)

The Distribution in the cerebellar anterior lobe of the climbing and mossy fiber inputs from the plantar and palmar cutaneous afferents. *Exp. Brain Res.* 7:1–10 (with S.T. Kitai, H. Táboříková and N. Tsukahara)

1969 Firing patterns of Purkyně cells in response to volleys from limb nerves. *Brain Res.* 14:222–226 (with D. Faber, J. Murphy, N. Sabah and H. Táboříková)

Slow potential fields generated in cerebellar cortex by mossy fiber volleys. *Brain Res.* 15:276–280 (with H. Korn, H. Táboříková and N. Tsukahara)

1970 Responses of the Purkyně cells of a selachian cerebellum (*Mustelus canis*). *Brain Res.* 17:57–86 (with H. Táboříková and N. Tsukahara)

Responses of the granule cells of the selachian cerebellum (*Mustelus canis*). *Brain Res.* 17:87–102 (with H. Táboříková and N. Tsukahara)

Inhibition of antidromic invasion of Purkyně cells in the selachian cerebellum. *Brain Res.* 17:360–365 (with D.S. Faber, S.T. Kitai, T. Shimono and H. Táboříková)

Neurogenesis and morphogenesis in the cerebellar cortex. *Proc. Natl. Acad. Sci. USA* 66:294–301

1971 The action of a parallel fiber volley on the antidromic invasion of Purkyně cells of cat cerebellum. *Brain Res.* 25:335–356 (with D.S. Faber and H. Táboříková)

Cerebellar Purkyně cell responses to cutaneous mechanoreceptors. *Brain Res.* 30:419–424 (with N.H. Sabah, R.F. Schmidt and H. Táboříková)

Afferent volleys in limb nerves influencing impulse discharges in cerebellar cortex. I. In mossy fibers and granule cells. *Exp. Brain Res.* 13:15–35 (with D.S. Faber, J.T. Murphy, N.H. Sabah and H. Táboříková)

Afferent volleys in limb nerves influencing impulse discharges in cerebellar cortex. II. In Purkyně cells. *Exp. Brain Res.* 13:36–53 (with D.S. Faber, J.T. Murphy, N.H. Sabah and H. Táboříková)

Investigations on integration of mossy fiber inputs to Purkyně cells in the anterior lobe. *Exp. Brain Res.* 13:54–77 (with D.S. Faber, J.T. Murphy, N.H. Sabah and H. Táboříková)

Responses evoked in neurones on the fastigial nucleus by cutaneous mechanoreceptors. *Brain Res.* 35:523–527 (with N.H. Sabah and H. Táboříková)

Possible synaptic mechanisms subserving learning. In: Brain and Human Behavior. Ed. by A.G. Karczmar and J.C. Eccles, 1972, Heidelberg, Springer-Verlag. pp. 39–61

1972 Mode of operation of the cerebellum in the dynamic loop control of movement. *Brain Res.* 40:73–80 (with N.H. Sabah, R.F. Schmidt and H. Táboříková)

Cutaneous afferent responses in interpositus neurones of the cat. *Brain Res.* 42:207–211 (with I. Rosen, P. Scheid and H. Táboříková)

Cutaneous mechanoreceptors influencing impulse discharges in cerebellar cortex. I. In mossy fibers. *Exp. Brain Res.* 15:245–260 (with N.H. Sabah, R.F. Schmidt and H. Táboříková)

Cutaneous mechanoreceptors influencing impulse discharges in cerebellar cortex. II. In Purkyně cells by mossy fiber input. *Exp. Brain Res.* 15:261–277 (with N.H. Sabah, R.F. Schmidt and H. Táboříková)

Cutaneous mechanoreceptors influencing impulse discharges in cerebellar cortex. III. In Purkyně cells by climbing fiber input. *Exp. Brain Res.* 15:484–497 (with N.H. Sabah, R.F. Schmidt and H. Táboříková)

Integration by Purkyně cells of mossy and climbing fiber inputs from cutaneous mechanoreceptors. *Exp. Brain Res.* 15:498–520 (with N.H. Sabah, R.F. Schmidt and H. Táboříková)

1973 The cerebellum as a computer: patterns in space and time. *J. Physiol.* 229:1–32

Responses of red nucleus neurones to cutaneous afferent input. *Brain Res.* 53:440–444 (with P. Scheid and H. Táboříková)

1974 Cerebello-spinal pathway via the fastigial nucleus and the medial reticular nucleus. *Brain Res.* 66:525–530 (with R.A. Nicoll, D.W.F. Schwarz and H. Táboříková)

Excitatory and inhibitory responses of neurones of the cerebellar fastigial nucleus. *Exp. Brain Res.* 19:61–77 (with N.H. Sabah and H. Táboříková)

The pathways responsible for excitation and inhibition of fastigial neurones. *Exp. Brain Res. 19:78–99* (with N.H. Sabah and H. Táboříková)

Somatotopic studies on cerebellar fastigial cells. *Exp. Brain Res.* 19:100–118 (with T. Rantucci, N.H. Sabah and H. Táboříková)

Temporal patterns of responses of interpositus neurons to peripheral afferent stimulation. *J. Neurophysiol.* 37:1424–1437 (with I. Rosén, P. Scheid and H. Táboříková)

Patterns of convergence onto interpositus neurons from peripheral afferents. *J. Neurophysiol.* 37:1438–1448 (with I. Rosén, P. Scheid and H. Táboříková)

Somatotopic studies on cerebellar interpositus neurons. *J. Neurophysiol.* 37:1449–1459 (with T. Rantucci, I. Rosén, P. Scheid and H. Táboříková)

The physiology and physics of the free will problem. In: *Progress in the Neurosciences and Related Fields* (Studies in the Natural Sciences, vol. 6). Ed. S.L. Mintz and S.M. Widmayer. New York and London, Plenum Press, 1–40.

Cerebral activity and consciousness. In: *Studies in the Philosophy of Biology*. Ed. F.J. Ayala and T. Dobzhansky. London: The Macmillan Press

1975 Reticulospinal neurons with and without monosynaptic inputs from cerebellar nuclei. *J. Neurophysiol.* 38:513–530 (with R.A. Nicoll, D.W.F. Schwarz, H. Tábiříková and T.J. Willey)

Medial reticular neurons projecting rostrally. *J. Neurophysiol.* 38:531–538 (with R.A. Nicoll, H. Táboříková and T.J. Willey)

The differential effect of cooling on responses of cerebellar cortex. *J. Physiol.* 249:119–138 (with I. Rosén, P. Scheid and H. Táboříková)

Responses of red nucleus neurons to antidromic and synaptic activation. *J. Neurophysiol.* 38:947–964 (with P. Scheid and H. Táboříková)

Somatotopic studies on red nucleus: spinal projection level and respective receptive fields. *J. Neurophysiol.* 38:965–980 (with Rantucci T., P. Scheid and H. Táboříková)

Prolongation of hippocampal inhibitory postsynaptic potentials by barbiturates. *Nature* 258:625–627 (with R.A. Nicoll, T. Oshima and F.J. Rubia)

1976 Medial reticular and perihypoglossal neurones projecting to cerebellum. *J. Neurophysiol.* 39:102–108 (with R.A. Nicoll, D.W.F. Schwartz, H. Táboříková and J.T. Willey)

Topographic studies on medial reticular nucleus. *J. Neurophysiol.* 39:108–118 (with R. A. Nicoll, T. Rantucci, H. Táboříková and J. T. Willey)

1977 An instruction-selection theory of learning in the cerebellar cortex. *Brain Res.* 127:327–352

The anionic permeability of the inhibitory postsynaptic membrane of hippocampal pyramidal cells. *Proc. R. Soc.* [Biol] 198:345–361 (with G. I. Allen, R. A. Nicoll, T. Oshima, F. J. Rubia)

The ionic mechanisms concerned in generating the IPSP's of hippocampal pyramidal cells. *Proc. R. Soc.* [Biol] 198:363–384 (with R. A. Nicoll, T. Oshima and F. J. Rubia)

Evolution of the brain in relation to the development of the self-conscious mind. Ann. New York Acad. Sciences 299:160–179

1978 An Instruction-Selection hypothesis of cerebral learning. In: Cerebral Correlates of Conscious Experience. Eds. Buser and Rongeul-Buser, pp. 155–175. Amsterdam: Elsevier/North-Holland

1980 Event-related potentials in the cerebral cortex. In: Nerve Cells, Transmitters and Behaviour. Ed. Rita Levi-Montalcini, pp. 551–561. Città del Vaticano: Pontificia Academia Scientiarum

1982 Cerebral mechanisms in pain perception. Panminerva Medica-Europa Medica 24:1–13.

The synapse: from electrical to chemical transmission. Ann. Rev. Neurosci. 5:325–339

The initiation of voluntary movements by the supplementary motor area. Arch. Psychiatr., Nervenkrank. 231:423–441

Language, Thought and Brain. Epistemologia IV., 97–126.

The Liaison Brain for voluntary movements: the supplementary motor area. Acta biol. Acad. Sci. Hung. 33:157–172

Some circuit operations in the mammalian brain. Acta Anat. 113:325–339

The future of studies on the cerebellum. Exp. Brain Res. Suppl. 6:607–620.

How the Self acts on the Brain. Psychoneuroendocrinology 7:271–283

Animal consciousness and human self-consciousness. Experientia 38:1384–1395

1983 The horizontal (tangential) fibres system of lamina I of the cerebral neocortex. Acta Morphologica Hungarica 31:261–284

I. The Self and Its Brain

John Eccles and Karl Popper discussing their book "The Self and Its Brain" (Contra 1978)

Critical Remarks on the Knowledge of Lower and Higher Organisms, the So-Called Sensory Motor Systems

K.R. Popper, London

I have come here to honour Jack Eccles. He has been a most wonderful friend to me for 40 years. I can only very briefly indicate how much I owe to him. I met him in Dunedin, New Zealand, and ours was a friendship at first sight. I was then in a very difficult situation, and I really needed friendship and encouragement in my work. Jack Eccles gave all that to me. He was, and he still is, generous to a fault. His readiness to share his problems with me, his quick grasp and his full appreciation of my theory of scientific knowledge, his ready use of it, all this gave me courage in a situation in which courage was much needed. I am most grateful, Jack, for this help, and for all the friendship you have shown me since, when you were in Australia and later in the United States, while I was in England.

Jack Eccles and I worked together quite a bit from the beginning of our friendship in New Zealand until the publication of our common book, *The Self and Its Brain* (Popper and Eccles 1977)

Eccles boldly chose as the topic for his Oxford Waynflete Lectures (1952) the problem of the body-mind or brain-mind interaction. The title of the book in which he published these lectures was *The Neurophysiological Basis of Mind* (Eccles 1953). I should never have had the courage to suggest to Jack 20 years later that we might write together a book on a closely related subject if he had not already published this book.

A topic central to Jack's interests and central to his book is the problem traditionally called "the problem of the freedom of the will". It was a problem to which I also had given a lot of thought. Jack was inclined to rely on our intuitive feeling that we possess freedom to make up our mind, but I wanted some stronger argument, since it was part of my theory of knowledge that intuition was very important, indeed indispensable, for suggesting hypotheses, but that it was unreliable: we ought to be highly critical of our intuitions.

It seemed to me that a critical approach to the free will problem was very important, especially in view of its involvement in one of the most violent religious disputes, the dispute between Erasmus of Rotterdam and Martin Luther, fundamentally a dispute about ethics, about the morality of war and peace. Erasmus stood for human responsibility and for peace and tolerance; Luther put the responsibility for war and cruelty and for everything else on God. "This war is the war of our Lord God", Luther wrote. "He started it, He sustains it, and it will never stop until the enemies of His word have become dung under our feet" (Luther 1525, written in reply to Erasmus' *De libero arbitrio* of 1524).

Experimental Brain Research, Suppl. 9
© Springer-Verlag Berlin · Heidelberg 1984

In order to discuss this problem in a less emotionally charged, in a less intuitive and in a more critical context, I turned to the question of determinism and indeterminism in physics, and especially to the determinism of Laplace, and in 1950 I gave a lecture in Princeton under the title "Indeterminism in Classical Physics and in Quantum Physics" (Popper 1950a,b). My lecture, I proudly remember, was attended by both Einstein and Bohr. Fifteen years later, in 1965, I gave a lecture to commemorate Arthur Holly Compton in St. Louis, under the title "Of Clouds and Clocks" (Popper 1966). In this lecture I discussed free will, yet not in the context of moral decisions but in the context of human creativeness – the creativeness of great musicians, artists, poets and scientists.

I am telling you this in order to show that Jack and I had a very central interest in common, quite apart from our great general interest in each other's work. This was the basis of our decision – and certainly a free decision – to write a book together on the mind-body problem or mind-brain problem. It became an important event in both our lives.

One of the central theses of our book is this. We assert that the physical world, the world of physical things, of physical forces, of fields of forces, is open. It is open to the world of human experiences, of joy and sorrow, of doubts and convictions, of hope and fear and of thinking. We call the world of physical objects World 1 and the world of our experiences World 2, and we assert that World 2 can be influenced by World 1 (for example, by drugs) and that World 2, in its turn, can influence World 1.

But our thesis of the openess of the physical World 1 towards the World 2 of our experiences, including thought processes, can best be shown by introducing what we call World 3; that is, the world of the products of the human mind, especially the world of problems and theories in the objective sense, and the logical relations between them, such as compatibility and incompatibility, and also of their logical consequences and the relations of these consequences to other theories.

That the physical World 1 is open can be best seen by the tremendous influence, bad and good, which quite abstract theories and their logical consequences have had and are still having on the physical world. Think of the influence of Einstein's formula $E=mc^2$; or think of a modern airliner, say the *Concord*, and of Mach numbers. The *Concord* is to a large extent based on objective, abstract, mathematically formulated theories. But it is, obviously, a physical structure. So theories, theories in the objective sense, can have a share in bringing about new physical structures. All radio and television sets are such physical structures, largely the consequence of Maxwell's equations as interpreted and tested by Heinrich Hertz. Artificial satellites were foreseen by Newton, and they utilize the consequences of Newton's theory of gravity.

Theories in the objective sense, theories as formulated in language, do not belong to the World 2 of our subjective experiences. Rather they are the objective products of World 2. Like cathedrals they are the work of many minds. Newton's theory was partly the product of the minds of Galileo and of Kepler, but also of Pythagoras, Plato, Aristotle, Euclid, Aristarchus, Ptolemy and Copernicus – to mention only the greatest of the builders and workers who contributed to the cathedral. And James Clerk Maxwell's theory was a development of theoretical ideas of Faraday, and it was, in its turn, further developed by Hertz, even before it found, in Hertz's experiments, its first physical World 1 applications. So our physical environment, the physical World 1, is not closed: it is open to the influence of the theories which are objects produced by human minds, such as cathedrals and symphonies.

Jack and I have been much criticized for holding such views, and I shall here briefly reply to one of the objections made. It asserts that our view of the influence of something non-physical upon the physical world is in conflict with the law of the conservation of energy. There are many possible replies to this criticism. Which of them is the best I just do not know.

Basic to the situation is Jack Eccles's central hypothesis: that the brain is a detector and therefore, like all detectors, an amplifier (Eccles 1953, pp. 267 ff.). Now the law of conservation of energy is a quantitative law. But detectors and amplifiers have an unlimited capacity to detect signals of almost unmeasurably small energy and amplify them so as to obtain effects that are energetically unlimited: a detector-amplifier can, in principle, detect a signal of one photon and use it to release the explosion of a hydrogen bomb.

The essential thing is that the detector-amplifier provides its own energy. The quantitative law of conservation of energy thereby becomes practically irrelevant to the issue. Moreover, the First Law holds, of course, only for closed systems, and for organisms, which are never closed, only with (more or less) good approximation.

The objector may reply that the detector-amplifier will have to detect something physical, and therefore *some* small amount of energy, even if it is very near to zero, and this small but not zero amount is at issue: it cannot arise from nothing. Maybe, but then the objection is really no longer the specific one connected with the general quantitative law of conservation of energy but, rather, it merely restates the objector's difficulty of conceiving a world in which the physical World 1 is open rather than closed, open in the sense that it is acted upon, for example, by theories. So the objector, instead of brandishing his conservation law, would have to answer *our* objection to physicalism: that the physical World 1 does not out of itself, without the guidance of minds guided in turn by theories, produce aeroplanes, airfields and the complex organization that is needed to confer upon air travel the degree of safety which it has reached. All these are, in our opinion, typical World 3 effects.

But let us go back to the conservation law. It was discovered by Joule, who established the heat equivalent of kinetic energy. Electromagnetic and chemical energy and, of course, nuclear energy were unknown forms of energy at that time, and our objectors would no doubt have argued that such forms of energy cannot exist, since their existence would clash with the law of conservation of energy as it was known then.

It is perfectly possible (although I do not think much of this possibility) that there exist purely mental forms of energy, convertible into electrochemical forms: we just do not know. (After all, the terms *ergon* (work) and *energeia* (action, life) are originally anthropomorphic terms.)

Another possibility is this. According to de Broglie, the great founder of wave mechanics and of the particle-wave (or particle-field) dualism in quantum theory, energy (and momentum) are associated with the localized object: the particle, the photon, that moves through space along a trajectory, inside the wave. The wave as such carries no energy, but it guides, it pilots the energy-carrying particle. And according to some interpretations (and even some possible and actual experiments), there seem to be empty pilot waves that can interfere with non-empty (energy-piloting particles and energy-carrying) waves. Should this most exciting theory be corroborated by the experiments going on at present, then we could say that physics itself allows for the possibility of non-energetic influences upon energetic processes. Detection and amplification of signals whose energy is precisely zero would then turn out to be a physical possibility. (See, for example, the papers of the Viennese physicist H. Rauch on neutron interferometry (Rauch 1979) and experiments suggested by A. Garuccio, J.-P. Vigier and myself (Popper et al. 1981).

Many years ago I pointed out that the Second Law of Thermodynamics is constantly violated by Brownian motion according to the Einstein-Smoluchowsky interpretation, which has long since become accepted. This is simply a consequence of the fact that there are thermodynamic fluctuations. Why should not similar considerations hold about the First Law? Several physicists have suggested this, among them Bohr, Kramers and Slater, and for different reasons Schrödinger, and for still different reasons I myself. The fundamental experiments (such as Joule's) are work with a heat equivalent, and this is obviously of a statistical character.

I should also mention a paper by Pascual Jordan (1932), the man who, together with Heisenberg and Born, wrote in 1925 the first paper on the quantum mechanics of matrices (the second paper on quantum mechanics altogether). I remember that Jordan pointed out in his paper of 1932 that an amplifier mechanism (which he conjectured might be a consequence of the protein structure) might be characteristic of biological systems in general; he did not (if I remember correctly) think of the brain in particular, but he did hope to explain free will by a combination of Heisenberg's indeterminancy with amplifiers (Popper 1966; Bartley 1982).

So much of the objectors to our book that brandish the First Law. I intend to say a little more later about some of the topics discussed in the book which Jack and I wrote together, but Jack suggested that I should first say something about my own views of science and scientific research.

There are two opposing philosophical or metaphysical views called realism and idealism. Realism may be characterized as the view that there exists a world outside of us, a world which each of us may influence a little, but a world which will remain broadly the same if we die, and will go on more or less as before. Idealism may be characterized as the view that the world may be only our dream; that we know nothing beyond our experiences, and cannot know that anything beyond our experiences exists; that what we call "time" may be unreal, an illusion, something subjective.

I must confess that I am a passionate realist. I not only believe in the reality of this wonderful world, but in its immense and immeasurable variety and beauty and in our task of trying to explore and to understand it, knowing how little we shall ever know about it. But I have known people who were passionate idealists, people who did not believe that the world in which we live is real. My greatly admired friend Erwin Schrödinger was an idealist. He believed, more or less, that the world was an illusion, and more especially that we are deceived about time. Time, he held, is unreal. Ludwig Boltzmann (whom Schrödinger did not know very closely, since Boltzmann committed suicide only a year or two after Schrödinger entered the University of Vienna), was, like myself, a passionate realist for almost all his life. However, he got into difficulties, late in his life, over the Second Law of Thermodynamics, difficulties which he met by proposing a metaphysical theory of the unreality of time, and Schrödinger believed that this theory of Boltzmann's was the deepest and most beautiful theory of physics, since physics achieved here a metaphysical result, revealing that time was an illusion and that, for this reason, death also was an illusion.

I respect this metaphysical view, but I think it is mistaken. I think that it is even morally wrong. Hiroshima and Nagasaki were not illusions but realities. Suffering and death are, I believe, very real. We have a duty to fight them and to help people who are suffering. I am, like Kant, a realist. I believe in an unknown world which it is our task to explore. And I believe – unlike Kant – in the reality of time and space which, like everything else, are beset with riddles, riddles that become more difficult the more we learn about the world. I believe in the reality of time, partly because of the evolutionary theory which entails the reality of time and of its direction – the "arrow of time". And I believe that the evolution of life on earth is very largely a cognitive adventure: that all organisms are essentially problem solvers and explorers. Knowledge is thus essentially a natural product, a product of the human organism. And like all organs, it is fallible. This is why the idea of certain knowledge, which has ruled Europe since Aristotle, is a mistake: we must turn back to the philosophy of fallible, of conjectural knowledge of the pre-Aristotelians, who said that certain knowledge is only for the gods. Man can have only uncertain, tentative knowledge: conjectural knowledge.

And so it is, especially in science: we must sharply distinguish between truth and certainty. We may find true theories, but we can never be certain of them. As Xenophanes says (the translation is mine):

But as for certain truth, no man has known it,
Nor will he know it: neither of the gods,
Nor yet of all the things of which I speak.
And even if by chance he were to utter
The perfect truth, he would himself not know it,
For all is but a woven web of guesses.

We do not know: we can only guess. Our scientific knowledge is conjectural knowledge. Many call it "probable" knowledge; and this is harmless enough as long as it is not assumed that our hypothesis are probable in the sense of the probability calculus. They are not. Few scientists care about this problem, but almost all philosophers happen to believe passionately that our hypotheses are probable in the sense of the probability calculus. They do so because they believe in induction, and because there is no other theory of induction in existence but the probabilistic theory of induction. I have produced refutations of this theory for the last 50 years.

Let me very briefly explain deduction and induction by examples. From the premise "All swans in England are white" we can deduce the conclusion "If London is in England, then all swans in London are white". From the premise "London, Manchester and Birmingham are in England and all swans in these places are white" we can inductively infer the conclusion "All swans in England are white". As you see, induction is meant to be a kind of inverse or converse of deduction.

If a deduction is valid, then, if the premise(s) is (are all) true, the conclusion is also true: in a valid deduction, truth is transmitted from the premise(s) to the conclusion. So if the conclusion can be shown to be false, (at least one of) the premise(s) must be false. Falsity is retransmitted to the premise(s). This is of immense importance in science. We test a universal theory by deriving from it deductive conclusions which we try to refute. If we succeed, then we have refuted the premise also. This is why I regard deductive logic as the *organon* (instrument) of criticism.

As long as we regard our theories as tentative conjectures, to be critically tested as severely as possible, and abandoned when we have refuted them, we do not need any induction. There is no need even to mention induction. But this means giving up the quest for certainty. It means being satisfied with unjustified, conjectural knowledge – with knowledge that falls short of what (since Aristotle) is ordinarily called "knowledge"; that is, certain, rationally justified belief. It means accepting Socrates: "I know that I do not know, and hardly that."

But (as my friend Arne Petersen says) most people are terrified by Socrates' insight that we do not know. (Perhaps this was why they killed him.) And so they reach out for a theory of induction – a theory that presents what are in fact (as I think) our guesses, our conjectures, as the conclusions of inductive inferences, derived from observational (and therefore they think secure) premises.

Now in deduction we know the difference between a valid and an invalid inference. "All swans are white, so if this bird is a swan it will be white" is valid while "All swans are white, so if this is a herring it will be red" is invalid.

Induction is, of course, never valid; or rather, in those limiting cases in which it becomes valid (in which all cases have been observed) it turns into deduction. How then can we distinguish a "good" inductive inference such as: "All observed ravens are black, therefore all ravens are black" from a "bad" inductive inference such as "All observed ravens are black, therefore all herrings are red"?

The answer to this question is, presumably, this: an inductive inference must be highly probable. If the probability reaches 1, then the induction becomes a deduction. If it is near to 1, it is a good inductive argument. If it is less than ½, it is a bad inductive argument, for this means that the negation of the conclusion is more probable than the conclusion.

Although this is only a first and a very rough approach to a probabilistic theory of induction, it indicates the need for such a theory. Since inductive inferences are not valid or secure, we must have some theory like probability theory to indicate better or worse, more nearly valid or more obviously invalid inferences.

We can write $p(x) = {}^1/_6$ for "The probability of x is ${}^1/_6$" and $p(x,y) = {}^1/_3$ for "The probability of x, given y, is ${}^1/_3$". Here x may be the proposition "The next throw with this die will be four" and y may be the information "We regard only an even number as a throw".

Now let a be "The next throw will be two" and let b be "We regard only an even prime number as a throw". Since 2 is the only even prime, a follows deductively from b, and we have therefore $p(a,b) = 1$ for a case such that a is a deductive conclusion from the "given" premise b. It has been hoped (at least since Daniel Bernoulli 1766) that for inductive inferences high probabilities could be reached, although of course short of 1.

In 1934 I published the first (German) edition of my *Logic of Scientific Discovery* (Popper 1935), in which I proposed that induction was not needed and that what was needed in science was imagination, kept under control by the severest rational criticism. Imagination was needed in order to invent bold, highly testable explanatory theories and also in order to invent bold tests of these theories.

I admitted that we often spoke of a successful theory as "probable", but I asserted that this did not mean that it was probable in the sense of the probabiliy calculus. I asserted that in a very big or infinite universe, our universal hypotheses can never reach a probability in the sense of the probability calculus which was different from zero. And I suggested that we should not prejudge the issue by continuing to talk of the probability of hypotheses but at least leave it open by calling a successful and severely tested hypothesis "*bewährt*", soon translated by Carnap as "confirmed" (Carnap 1936; Popper 1982, p. 251). If my minority views on the topic were refuted, it could, I suggested, still be said that "*Bewährung*" (I later translated it as "corroboration") had been shown to be a probability in the sense of the probability calculus.

Originally, Carnap agreed to all this. But when he was in the United States and I in New Zealand, he forgot, and under the influence of Reichenbach he identified corroboration with probability without first showing that my counter-arguments were not valid, merely on intuitive grounds. But intuition, though immensely important, is often misleading, and on page 574 of his book *Logical Foundations of Probability* (Carnap 1950), having had to admit that universal laws had the probability zero, he suggested that "universal laws are not indispensable" in science, since they "are not indispensable for making predictions". And he promised to publish all the solutions to his difficulties in a second volume. I predicted – indeed without a universal law – that the second volume would never be published. I have held on to my minority view even though all the great authorities from Daniel Bernoulli and Laplace to Bayes and Sir Harold Jeffreys, Bruno de Finetti and, literally, hundreds of other living authors were, and still are, against me.

In view of this terrifying majority, I believed that my task was far more difficult than it turned out to be. As a consequence, my first proofs of the impossibility of probabilistic induction were more difficult than my later proofs. Over the years they became simpler and simpler.

I will first explain my very simple symbolism. I am employing "$p(a)$" for the (absolute) probability of a and "$p(a,b)$" for the (relative, or conditional) probability of a, given the information b. Note that $p(a)$ may be regarded as a special case of $p(a,b)$ – the case where the information b is trivial or tautologous, like "All bees are bees", so that it can be omitted. I am employing $st(a,b) = p(a,b)$-

$p(a)$ for the (probabilistic degree of) support given to a by the information b. The degree of support is equal to the increase in the probability of a owing to the information b. While probabilities are bound by 0 and 1, $0 \leqq p(a,b) \leqq 1$, and so $0 \leqq p(a) \leqq 1$. The degree of support may vary between -1 and +1: $-1 \leqq st(a,b) \leqq +1$. If the support is negative, we speak of counter-support.

The idea of probabilistic induction (which I shall show to be untenable) can now be explained as follows. Let e be an empirical evidence statement such as "All sparrows in London are grey", and let g be a generalization of it such as "All sparrows in England are grey". (We take it for granted – as "background knowledge" – that London is in England.) Then it can be shown that:

Theorem 1 $st(g,e) \geqq 0$

which is the same as $p(g,e) \geqq p(g)$.

That is, all generalizations are positively supported by the evidence: their "prior" probability $p(g)$ is always smaller than their posterior probability $p(g,e)$. Understandably enough, this has been taken to establish probabilistic inductive support. However, this is found to be highly misleading. It turns out that theorem 1 is due entirely to the fact that e deductively follows from g, and that in all such cases without exception we have:

Theorem 2 If g deductively entails e, then e is contained in g and we have

(1) $p(g,e) = c \times p(g)$ where $c = 1/p(e) \geqq 1$
(2) $st(g,e) = k \times p(g)$ where $k = c \times (1\text{-}p(e)) \geqq 1$

Thus let g be the statement: "All swans are red and all London sparrows are grey": it will be supported by e exactly as our previous generalization for the sole reason that e is contained in g.

So the increase in probability is, for any g whatsoever, due to e being contained in or entailed by g. This leads to the proportionality constant c, and further to the proportionality constant k, on which the support of g by e depends, for any g whatsoever. Otherwise only $p(g)$ is involved, that is, the prior (or a priori) probability of the g in question; and this probability, $p(g)$, can easily be made greater simply by making g say less: for example, g_1 = "All swans are red and all London sparrows are grey" will be far less probable than "All English swans are red and all London sparrows are grey" – and still more probable will be "All swans in Regent's Park are red and all London sparrows are grey".

Now it was the idea of probabilistic induction to distinguish good inductions from bad inductions with the help of the probability calculus: "Better" generalizations were to have higher probabilities (or higher support). But our theorem 2 shows that the probability calculus is unable to do this. However, we can show more.

For a further discussion, we now define a measure of the content of a proposition a, given the information b, as a kind of inverse function of probability, as follows:

$ct(a,b) = 1\text{-}p(a,b)$ $ct(a) = 1\text{-}p(a)$

Now let h (=hypothesis) be any statement whatever, and e (=empirical evidence) be any other statement. If h is to be "induced" from e, then there will be part of its content, $ct(h)$, which can be deduced from e. This part of the content of h we call $ct(d)$, and we find that d is identical with the proposition "h or e or both", or briefly: $d=h$ or e (or both). There will be the remaining part of the content of h. We call this part $ct(f)$.

Theorem 3 If

$$ct(h)-ct(d)=ct(f)$$

so that f is the part of h that goes beyond e, then f is identical with the proposition "h if e", or briefly: $f=h$ if e. For we have indeed for this f: h is equivalent to the proposition "d and f",

(1) or briefly: $h=d$ and f

(2) $$ct(h)=ct(d)+ct(f)$$

The proof is trivial.

Taking d and f defined as before, we also have:

Theorem 4

(1) $$st(h,e)=st(d,e)+st(f,e)$$

(2) $$st(d,e)\geqq 0\geqq st(f,e)$$

The proof is again trivial.

$st(d,e)$ can be shown to be always positive simply because e entails d, that is d follows deductively from e, so that $p(d,e)=1$. Therefore, $st(d,e)$ is a purely deductive support. So by theorem 4 (1), the only non-deductive and therefore possibly "inductive" part of the support of h is $st(f,e)$, and this is $\leqq 0$, that is it is always counter-support.

[A proof to this effect by David Miller and myself has been published in *Nature* (Popper and Miller 1983).]

All this may be summed up by the corollary. Every probabilistic support either contains [according to theorem 4 (1) and (2) above] a purely deductive part that is of no interest from the point of view of induction and may easily be taken out, or else it is counter-support. So if there is probabilistic inductive support, it can only be counter-support. This corollary makes it clear that not only is the probabilistic theory of inductive support mistaken (as we have seen before), but that it clashes head on with the laws of the calculus of probability.

The theory of induction thus turns out to be untenable. It was the result of a mechanistic view of human knowledge – the view that knowledge is the result of repetition, of association. It can easily be replaced by something better: human knowledge is the result of lucky and unlucky guesses, of trials (the invention of something new) and of the elimination of the unsuccessful trials: of the elimination of error. It is the result of conjectures and refutations. "Experience" is not a collection of observations but the elimination of erroneous conjectures, an elimination which may or may not be due to the result of so-called observations, but which, I conjecture, is far more often the result of actions – trials, inventions, experiments – that did not succeed.

I have worked on these matters since 1919 and, very intensively, since 1925. I worked on them as a student of psychology, as a student of logic and as a student of probability theory. One result which I found in 1938 was published (with David Miller) in 1983 (Popper and Miller 1983) – a kind of Orwellian inversion. I have made very slow progress, but when I look back, which I rarely have time to do, I can say that I have learnt something.

I have developed a view of the growth of knowledge – of human knowledge more specifically, but also of animal knowledge – which differs greatly from nearly everybody else's. According to this view, our knowledge is not in the main derived from experience, not even from experience as I see it: the elimination of bad guesses. Most of our knowledge, and animal knowledge, and even vegetable knowledge, is rather the result of sheer invention. It could be said to be a priori, though of course not a priori valid, but a priori guesswork.

All organisms are professional problem solvers: before life, problems did not exist. Problems and life entered the world together, and with them problem solving. At first, the problems were sheer survival problems: an organism was, at first, just a professional survivor. In time this changed, and a problem today may be for some people a vastly different thing: it may be "How did life originate?" or perhaps "In which of his dialogues, if in any, does Plato attempt to portray Socrates?" – this is obviously not a survival problem.

The problems may be created by the clash of the organism with its environment But the tentative solutions, the trial and the errors all come from the organism. They are all a priori, even though the environment constantly plays its part, of course.

The view which I am trying to make attractive to you is, I am afraid, utterly different from those views which are at present held by almost everybody and which I have described, for the past 60 years, as the "bucket theory of the mind" (*die Kübeltheorie des Geistes*). But today an inventor theory of the mind and a discovery theory of the mind and a searchlight theory of the mind ought to be a little more acceptable than they were 60 years ago.

However, let us turn to a worthy opponent. Jack quotes Mountcastle in our common book, in an epilogue to his chapter on conscious perception. The passage quoted is beautiful and there is not a word in it which I could not defend myself. And yet I want to look at it quite differently, with a different emphasis. Let me first quote the passage, from "The View from Within: Pathways to the Study of Perception" (Mountcastle).

> Each of us lives within the universe – the prison – of his own brain. Projecting from it are millions of fragile sensory nerve fibres, in groups uniquely adapted to sample the energetic states of the world about us: heat, light, force, and chemical compositions. That is all we ever know of it directly: all else is logical inference.
>
> Sensory stimuli reaching us are transduced at peripheral nerve endings, and neural replicas of them dispatched brainward, to the great gray mantle of the cerebral cortex. We use them to form dynamic and continually up-dated neural maps of the external world, and of our place and orientation, and of events, within it. At the level of sensation, your images and my images are virtually the same, and readily identified one to another by verbal description, or common reaction.
>
> Beyond that, each image is conjoined with genetic and stored experiential information that makes each of us uniquely private. From that complex integral each of us constructs at a higher level of perceptual experience his own, very personal, view from within. (Popper and Eccles 1977, p. 274)

Now let me question, and criticize this beautiful quotation. In the first three lines, I certainly agree and admire the phrase "uniquely adapted to sample the energetic states of the world about us", an excellent and, as Jack Eccles says, a terse and vivid description of what the sensory system and the brain are adapted for – what they have worked hard, in millions of generations, to adapt themselves for, because there was a felt need, a problem, of knowing, of anticipating, an adventurous curiosity, a wish for knowledge and for freedom of action, but I should be inclined to look at the brain not as a prison but as an immensely powerful instrument for gaining knowledge and freedom. And as to knowing anything "directly", this I do not believe in. And as to "logical inference", this is, I suggest, only from propositions (or theories) to propositions.

But let us proceed to the next paragraph of the quotation. Except that I am made uneasy by the fact that the paragraph begins with the words "Sensory stimuli reaching us" (I shall say a little more about this later), there is nothing here to criticize, and I am especially happy about the way in which an old but silly problem is forcefully brushed aside: the problem of whether your and my images are at all comparable or even faintly similar. The problem is not brushed aside as "insoluble" or as "meaningless", but it is solved positively by saying that our images are "virtually the same". (Of course we cannot know for certain, but then we know hardly anything for certain, and Mountcastle's conjecture and his argument are reasonable.)

And yet I could never have started the paragraph with "Sensory stimuli reaching us". To me, it suggests the theory that, somehow, our knowledge starts from sensory stimuli, while in my view it starts from previous knowledge and the wish, or desire, or need, or drive, or instinct, or any terms you may choose, for more knowledge, in the form, I conjecture, of an instinct for exploration, or to use Mountcastle's own phrase, for sampling "the world around us". I find it of the greatest importance to get over the myth (for it seems to me a myth) that anything starts with a stimulus and the ability to respond to it. Everything, I suggest, starts from the organism and its eagerness, its need, which seeks for anything it can use as a message, as information or as a so-called stimulus: I am not quarrelling about words, I am insisting upon the obvious fact that it is not its capability of being stimulated which makes the organism, but the exploratory apparatus, the keenness to respond, which makes the stimulus. It is the organism which, through many trials over millions of generations, learns to respond to this kind or that kind of stimulus: it turns something into a stimulus; it invents the ability to "see" this or that, the ability to respond. It may do all this as a response to some earlier "stimulus", but it may do all this a priori, by a mutation which, in its turn, may be due to its mutationally acquired mutability.

I am a Darwinist and I am not a vitalist or a Lamarckist. I agree that the organism's aims and preferences may be conjectured to be the result of natural selection. But once they exist, they are of immense significance in evolution. The idea that environmental conditions do all the sculpturing by carving away what is less fit, and that the organism is passive, is simply false from an evolutionary point of view. Without the organism's struggle for survival there would be no evolution. And this means that it is only the organism's fight for life, its problem solving, its constant search for a better environment, for better living conditions, its active evolution of its own preferences, of its sensitivity, that makes evolution possible. It is the only active agent. It is the environment which is passive. To look at it as hostile, as so many Darwinists do, is mistaking a metaphor for a powerful reality. To think of it as a sculptor, a carver, is to make a god of it. It is always the organism which seeks a better ecological niche, better living conditions, a better life. All else is metaphor.

The biologist, observing, experimenting, begins with the stimulus. And he is in danger of believing in the famous couple, stimulus and response. But it is the organism which evolves all sorts of responses to find, interpret, exploit and learn from what we mistakenly look upon as the life-evoking, life-demonstrating stimulus of a (perhaps even passive) response.

I must close. I will close with a list of things that we should, in my view, unlearn. Associations are in my view non-existent. So are reflexes and, of course, conditioned reflexes and conditioning. Pavlov's dogs were clever animals who developed a theory, a conjecture, that the food would come when the bell rang. So they anticipated the food: anticipation of impending events is the very function of conjectural theory formation. And our sense organs are themselves theories – anticipations of what may help us to know the world better. This proves that it is impossible to explain theories – or all theories – as due to generalizations from sensory data: sensory data do not exist. Nothing is "given" to us from our senses, but our senses, which are the instruments we developed ourselves through trial and error elimination over millions of generations, are actively used by us to explore the world, and they are used by us to help us to eliminate mistaken theories, mistaken anticipations.

I wish to end with an important proposition. The difference between the growth of animal knowledge and of human knowledge is just this: human language permits us to formulate our theories outside our skin, exosomatically: this allows us to criticize them. This makes possible the evolution of human reason, of the use of imagination and of criticism in the search for truth.

References

Bartley WW III (ed) (1982) The open universe: an argument for indeterminism. Postscript to The logic of scientific discovery, vol 2, translation of Popper KR (1935) Logik der Forschung. Hutchinson, London

Carnap R (1936) Testability and meaning. Phil Sci, vol 3:

Carnap R (1950) Logical foundations of probability, 1st edn. University of Chicago Press, Chicago (2nd edn, 1962)

Eccles JC (1953) The neurophysiological basis of the mind. Clarendon, Oxford

Jordan P (1932) Die Quantenmechanik und die Grundprobleme der Biologie und der Psychologie. Naturwissenschaften 20: 815–821

Jordan P (1934) Quantenphyikalische Bemerkungen zur Biologie und Psychologie. Erkenntnis 4: 215–252

Luther Martin (1525) De servo arbitrio (The bondage of the will)

Mountcastle VB (1981) The view from within: pathways to the study of perception. Johns Hopkins Med J 136: 109–131

Popper KR (1935) Logik der Forschung. Julius Springer, Vienna (Mohr, Tübingen, 1982)

Popper KR (1950a) Indeterminism in quantum physics and in classical physics I. Br J Phil Sci 1: 117–133

Popper KR (1950b) Indeterminism in quantum physics and in classical physics II. Br J Phil Sci 1: 173–195

Popper KR (1966) Of clouds and clocks: an approach to the problems of rational and the freedom of man. Washington University Press, St. Louis (In: Popper KR (1981) Objective knowledge. An evolutionary approach. 6th edn. Oxford University Press, Oxford)

Popper KR (1982) The logic of scientific discovery. (Translation of: Logik der Forschung, Julius Springer, Vienna, 1935.) Hutchinson, London
Popper KR, Eccles JC (1977) The self and its brain. Springer, Berlin Heidelberg New York
Popper KR, Miller D (1983) A proof of the impossibility of inductive probability. Nature 302: 687–688
Popper KR, Garuccio A, Vigier JP (1981) An experiment to interpret EPR action at a distance: the possible detection of real de Broglie waves. Epistemological Letters
Rauch H (1979) In: Bonse U, Rauch H (eds) (1979) Proceedings of International Workshop at the Institute Max von Laue. Oxford University Press, Oxford

Impasses and Fallacies of the Brain-Mind Discussion

O. CREUTZFELDT, Göttingen

Summary. This lecture addresses the brain-mind question as a classical problem which brain research has been faced with since it has become a science and the discussion of which has tended to run into the same epistemological impasse since then. It is shown that a simple isomorphism between the content of our perceptions and their representation in terms of distributed neuronal activities is difficult to prove. A major difficulty is to understand the unification of the temporally and spatially distributed activities into a unified and conscious perception. On the other hand, there is no difficulty in understanding the mechanisms of the nervous system as a control system which brings the subject into a consistent relationship with his environment. But such a system does not require consciousness as we experience it. There is, in fact, no way of defining or proving the existence of consciousness and mind in biochemical, biophysical, or anatomic terms. Materialistic-monistic models of the brain-mind relation and of consciousness are only metaphors of this relationship, and so is the dualistic model of interactionism. The explanatory value of the latter for the understanding of ourselves as conscious beings is high, since duality is one of the essential aspects of conscious experience of the self and its environment, while the former may appear to be more consistent with scientific beliefs.

The Challenge

After a scientific life spent in the search for the neurophysiological basis of mind, like many other great neuroscientists, such as Fechner, Sherrington, Penfield, Albe-Fessard, and Kety, John Eccles came to the conclusion that mind cannot sufficiently be explained as a function of or by the functions of the brain. The brain is a necessary but not a sufficient condition for the expression of the mind (Popper and Eccles 1979). This conclusion found powerful support in Popper's philosophical reasoning, and is consistent with the epistemological conclusion that science never reaches sufficient explanations (Popper 1972). Consequently, Eccles' conclusions as a neuroscientist brought philosophers and neuroscientists alike into the forum, resulting in a lively revival of the brain-mind discussion. The fact that the discussion of the neuroscientists became so animated may be taken as a sign that, hidden behind the scientific interest in the details of the functional organization of the brain, is the hope of clarifying through neuroscience the mechanics of the mind. The admission that the brain might not be a suffient explanation for the expression of the mind was a challenge to this hope.

Experimental Brain Research, Suppl. 9
© Springer-Verlag Berlin · Heidelberg 1984

But philosophers were also aroused as the recently evolving analytical philosophy of mind returned the problem of mind to neurophysiology (see, e.g., Bieri 1981). Since the "ghost" was driven out of the machine by Ryle (1949, see below), the answer as to how the machine works and what it produces has been left to the neurosciences. Unfortunately, the neurosciences have as yet only expressed the hope that they may be able to solve the problem, and have invented a new word for it ("mentation"), apparently in the belief that by giving it a new name it may be made easier to solve. But since some of the leading neuroscientists have confessed that this science could not give a satisfactory answer, and since the question is now with us, the neuroscientists, the philosophers may excuse my attempt to address this question on this occasion. We are all aware of the great progress made in the neurosciences during the last 35 or 50 years, since the basic physical and biochemical laws of nerve action and interaction have been formulated. We are continually faced with new discoveries, and we may expect further surprises in the coming years with the application of methods of molecular biology to the chemical composition and the development of structures and functions of the nervous system. This is not the place to review this aspect. Instead, I shall try to say something about the difficulties with which the neuroscientist is faced in trying to answer the questions put to him by the philosophers.

The Scientific Position

Let us first remind ourselves of the state of the discussion. In trying to do this, I shall not refer to any of the most recent publications, but I wish to show that the problem has been with us since the emergence of the scientific exploration of the brain and has not changed, in principle, since then. In 1845, that is well before the time of the great neuroanatomical and neuropathological discoveries, Wilhelm Griesinger published his classic book on the pathology and therapy of mental illnesses (Griesinger 1845). In the famous first paragraph, he reasoned that mental disorders are a symptom.

> Therefore, the first step for an understanding of mental disorder should be to localize it. To which organ does the phenomenon of mental illness belong? ...If physiological and pathological facts can show that this organ can only be the brain, we must recognize in each case of mental illness above all a disease of the brain.

About 50 years later, Carl Wernicke starts his textbook of psychiatry with the categorical statement that mental illnesses are diseases of the brain, but yet – he admits – "somewhat different" (Wernicke 1906). In fact, Wernicke sees the specific location of mental diseases in the system of associations in the brain, specifically in the cortex, in contrast to diseases of projection systems, which in his classification are the causes of the neurological disorders. Griesinger goes on to argue that "although the facts force us to localize imagination (*Vorstellen*) and will (*Wollen*) in the brain, we may still have various ideas about the relation between these psychic acts to the brain, on the relation between the soul and matter in general." He alludes to various possibilities such as Descartes's and Leibniz's ideas, but comes to the conclusion that these (philosophical) hypotheses cannot be disproved empirically and ... are therefore unacceptable to the empirical scientist. Rather, an empirical view will see mental activities (*Seelentätigkeiten*) in such a unity with the body, and especially with the brain, as exists – in general – between the function and an organ. This view will regard imagination and will as the activity, the specific energy of the brain, in the same way as it regards the conduction in the nerves, the reflex action of the spinal cord, etc. as the functions of these parts, and thus the soul will be, in this view, the sum of all brain states.

So much for the empirical view. But Griesinger does not leave it at that. He goes on to write:

In accepting this view, one cannot warn enough against the abstract and shallow materialism which would like to throw overboard the most general facts about human consciousness because one cannot grasp them in the brain with one's hands. While the empirical view ascribes the phenomena of perception, imagination (*Vorstellen*), and action (*Wollen*) to the brain as its activities, it leaves untouched the actual content of the human mental life (*Seelenleben*) in all its richness, and retains especially the fact of free self-determination. It leaves open the metaphysical question of what it might be that enters as mental substance (*Seelensubstanz*) into these relations of sensation, imagination (*Vorstellen*), and action (*Wollen*) and takes the form of mental existence. [The empirical view] may quietly wait for the time when the questions on the connection between the contents of the human mental life (*Seelenleben*) and its form become physiological instead of metaphysical questions.

While the empirical view spreads out the soul into a manifold and a multitude of mental processes, it does not lose the unity of the mental life (*Seelenleben*). It appears first as that unity which also exists in the soul of animals, comparable to an accord or a harmony, which are also a one composed of many [tones]. Beyond that, it appears as a higher and conscious unity, in which the alterations of mental states collect themselves into a One, an apparently lasting (*Bleibendes*), a Self (*Ich*). Of course, this resulting unity is not resting and rigid, but with the Self [I, ego] the soul confronts itself with itself, a Thou (*Du*), and the progress of this duality (*Entzweiung*) to increasingly wider and higher unifications is the driving moment of human mental life (*Seelenleben*).

The controversy on the materiality or immateriality of mental processes cannot be decided with our present concepts. All comparisons with the imponderables [such as temperature, force, and acceleration] which are in a similar relation to matter are of little help, since, although they are also immaterial, they are elicited and modified in their effects by material changes and themselves cause changes in matter. The mental agent has nothing analogous in the remaining world. Any theory would find, as Locke said, the same difficulties, whether it lets matter think, or whether it tries to understand the action of something immaterial on the matter.

Attempts at Scientific Explanations

This is, you may argue, all history, since it was written nearly 150 years ago. But I take the position that this text – like many other intelligent writings on that problem during the following years, such as those in Wundt's (1896), James' (1890), or Ribot's (1899) textbooks of psychology – represents the program of the present-day discussion. These sentences clearly address the various positions, such as materialistic monism, emergentism, and metaphysical dualism, and critically define their position in relation to the empirical approach. And for those who can listen, these statements (in fact, one should read all the first 100 pages of Griesingers book) contain the various ideas present in those days which later on became programs of psychiatry, with the neuropathologically oriented biological psychiatry of Nissl and Alzheimer at the dawn of our century, the neurophysiology (*Hirnpathologie*) of Wernicke, Liepmann, and Kleist at the end of the nineteenth and the first decennia of this century, the evolution of psychoanalysis around 1900 (the belief in a Self, the ego, and its counterpart), of psychophysics (Fechner's *Elements of Psychophysics* appeared in 1859, the first volume of Helmholtz' *Handbook of Physiological Optics* in 1856 and the second in 1866), and of physiological psychology (Wundt 1874). In fact, the program expressed in these thoughts by Griesinger in the mid-nineteenth century is that of empirical reductionism, but represents – at the same time – the insight that the "mental," especially in its appearance as "a higher conscious unity, its unification in a conscious self," is something unique in our (physical) world, not accessible to empirical science.

Before we turn to some of the modern considerations, let us remain with history for a moment. The subsequent reductionism to the elements of mind, i.e., sensation, cognition, consciousness, will, and action, became of course the program of psychophysics, the most famous exponents of which were Fechner and Mach, who both tried to explain perceptions by a combination of such elements, which should be explainable in physical terms. Mach claimed, in his *Analyse der Empfindungen* (1901), that "the contrast between the Self (*Ich*) and the world, sensation or appearance, and object should be described as a physical relation between the object, the body, and its sensations." In other words, Mach and the psychophysical positivists tried to introduce a "language" in which mental processes could be described and analyzed within one (physical) system, through which the Self, sensations, and the world of objects were related to each other, so that the gulf between objects and sensations, between inside and outside, between the material and the mental world would disappear. Mach's critique of the usage of mental terms reminds one of Ryle's linguistic critique of the misuse of mental terms by giving them ontological meaning. Ryle pointed out in *The Concept of Mind* (1949) that this confusion of mental categories and actual mechanisms resulted in a sort of pseudo- or paramechanics of mental mechanisms, which – as he saw it – necessarily resulted in the model of a (mental) ghost in the (physical) machine. (We may leave aside, here, a critical analysis of Ryle's argumentation: to some extent he first put the ghost into the machine by his argumentation, in order to be able to drive it out afterward). Such and other logical and linguistic analyses, including the *Philosophical Investigations* of Wittgenstein, opened the way for the apparently new "analytical philosophy of mind" (for a short outline of this school of philosophy and some representative texts see Feigl 1967; Bieri 1981 or Dennett 1979), which sees itself as a mediator between the old mentalistic descriptions and explanations of persons and new empirical theories, waiting for the time when "the connections between the contacts of human mental life and its form become physiological questions," as Griesinger put it 150 years ago.

The central state theories, for which we may also use Griesinger's formulation that "the mind (*Seele*) is the sum of all brain states", has found some formalistic expressions in the modern discussion (e.g., Armstrong 1968; Bunge 1980; Hebb 1980), which have been criticized on epistemological grounds by Popper (Popper and Eccles 1977) and Seifert (1979). In such theories, a mental state becomes a formula which is assumed to be isomorphic with a physiological, i.e., a physical state. Such theories are rather unsatisfactory, as these formulas are empty. They state – essentially – a correspondence of words, while the correspondence between the two levels is in no way explained. Especially, the dimensions of these states are not defined and thus such theories cannot be of great heuristic value.

More fruitful appear to be ideas which introduce evolutionary aspects and try to trace conditions and constraints of the human mind back to its evolutionary routes. Such an attempt was introduced by Konrad Lorenz (1941, 1973), and has developed into what is now called "evolutionary epistemology" (Campbell 1974; Vollmer 1975). According to this school of thought, the elements of mind are not so much psychophysical elements as in Mach's and Fechner's positivistic psychophysical theories, but elements of behavior. This approach recognizes the evolutionary history of mind; it states that mind is an emergent property of matter once it has assumed a certain configuration, but the explanatory value for the mental phenomena as we experience them as humans is small (Creutzfeldt 1979).

An important aspect which had been neglected in most discussions is the realization that mind is not an activity in itself, but that an essential constituent of it is that it rests in the brain as well as in the world of mind, in worlds 2 and 3 in Karl Popper's terms (Popper 1972). Here the story of interactionism begins, which has raised so much controversy and brought new color into the picture of the more and more abstract discussion. But we should realize that even the postulate of a separate entity beyond the brain, i. e., a self-conscious mind, does not explain the mechanisms of consciousness and conscious experience but only transposes the problem onto a higher level and leaves it there to be explained by still further mechanisms. Thus a mystery remains even in such a model.

Limitations of the Neurophysiological Model

Let us now try to see whether, and if so how, the neurosciences are able to rise to the challenge put to them. We have to ask ourselves first what to look for. The term "mind" is used for a variety of functions of the brain and its organism. Most of these functions can indeed be defined in terms either of perception or of action. And the neurosciences have gone far indeed to demonstrate how the physical and chemical stimuli coming from the objects of our environment are transformed in the sense organs by receptors into a code by which the nervous system communicates with its environment, its organism, and even with itself. This code is, of course, the modulation of the discharge frequency of electrical pulses along nerve fibers. We understand, down to a molecular level, how this transformation takes place, how the nerve cells are excited, and how they transmit this information. And we know, and can make visible by modern radiochemical methods that brain areas in which such information is being processed increase their metabolism. Thus this physical aspect of mind needs energy.

We know the code, can decipher and simulate it with the help of machines. This neuronal algorithm is, in fact, largely isomorphic with what we actually perceive, and it would have satisfied Fechner as the neuronal correlate for at least some of the elements of psychophysics. But already at the simple level we are faced with a serious problem: while the receptors faithfully measure the energy of the stimulus they are tuned to and transfer this information to the nervous system, the brain tends to reduce its signalling to changes of stimulus energies. Therefore, at the highest level, in the cortex cerebri, the real stimulus environment is represented not faithfully and completely but as a processed image, which largely corresponds to a first derivative of the original. Thus what we consciously perceive, or what the nervous system attributes to the environment, is not fully represented by the activities of the nervous system and, for perception of a complete image, the nervous system must fill in gaps. This filling in is not necessarily coded as action potentials.

On the other hand, such processing of the raw sensory material makes sense if we regard the central representations not as the coded percept but only as signals to enable the subject to respond appropriately to objects of his environment – so as to fixate the gaze on the edge of an object, to superimpose (or fuse) the images from the two eyes onto the same point in the physical environment, or to represent the temporal structure of auditory stimuli, to respond to a change in the physical environment during motor performance, etc. (for further details see Creutzfeldt 1981b, 1983). In fact, we have no great difficulty in understanding and describing in mathematical terms the neuronal machine for perception as the input stage of a servoloop system to elicit appropriate responses or to control action within a simple or complex environment. But within this servoloop, the true image, the representation of the object, fades away and is reduced to lines, edges, and transients.

Let us go one step further. An important principle of the genetically determined organization of the nervous system is what is referred to as the topographical or topological principle. This means that relations of neighborhood between neurons are largely maintained from the periphery to the highest centers, and from there down to the motor periphery. Sensory structures and motor systems are layed out like maps along the surface of the brain. Therefore, not only the form of energy to which a receptor is specifically tuned but also its location is essential. This, of course, also makes sense if we consider the receptors and their central representations as the afferent arm of a servoloop system: The system should not only appropriately respond to the energy of the stimulus, but also to its place. But conscious perception now has the formidable task of integrating these layed out maps into an undivided, unified gestalt.

In addition, since perception is an active process, any perceived object does not enter, except in some experimental situations (e.g., using stroboscopic stimuli), at once, but is spread out over time. If we touch an object, we actively explore it over time; if we look at an object, we go over it with our eyes (or rather let our eyes jump from one corner or edge in the picture to another), and an auditory stimulus like a word or a tune evolves over time into a semantic signal. Although our system represents only momentary aspects by its code, perception is able to combine these temporally spread out signals into coherent percepts. Thus it needs what the engineer would call a "sample-and-hold system," the code of which is – once again – not ongoing discharge activity, and we have as yet no indication that such a sample-and-hold system is represented in specific neurons or discharge patterns. But here again, we have no difficulty in understanding this in terms of a servosystem which is designed actively to explore the environment and to respond to every change when it appears (Creutzfeldt 1981b).

An important discovery about the organization of the nervous system is that sensory surfaces and the motor system are not only once but multiply layed out in the nervous system (for review and detailed aspects of this see Creutzfeldt 1983). These multiple representations or maps differ not only in that responses to different stimulus aspects are represented, such as the movement of an object, the quality of a somatosensory stimulus or auditory space, but they also differ in their respective connections with the efferent motor control systems. Thus they represent various active, behavioral relationships between the receptors and through them with the environment and the body. Thus these multiple representations of sensory surfaces represent, at the same time, various dispositions to act in response to a stimulus. All these dispositions are united into a single and appropriate response or action at lower motor control levels.

For our servoloop system, this is a fine device, since various object-subject-object relations are represented in parallel and can be calculated simultaneously as in a distributed, parallel computer system (this term was introduced, in a different context, by Mountcastle, in Edelman and Mountcastle 1978), taking into account various parameters for evaluation and for action. But for perception, the distributed representations of sensory inputs give each representation a different meaning according to the connections of each of these representations. Thus in each representation the percept receives a specific label of meaning as defined by the output connection of the respective representation field. This poses an additional problem for a unified conscious perception, as these sensorimotor dispositions, these labels of meaning, and the exact image representation in the primary sensory areas have to be combined into a unified percept of the object and its meaning, of the subject and its action.

The projections of sensory inputs to the limbic system, either directly or through cortical connections mainly via the temporal association cortex, bring the neuronal substrates for drives and internal needs under the control of some aspects of these inputs and their cortical representations. On the other hand, through the projections of limbic structures to the frontal cortex, these limbic activities have access to and control over the cortical circuits. These activities give an emotional label to the conscious perception being represented in specific cortical areas, which also has to be integrated into the unified percept.

The next complication is that the sensory channels contain various filters, filters for tone frequencies, colors, odors, etc. All these filters are broadly tuned and overlapping. Thus any pure color, pure tone, or somatosensory stimulus such as touch, heat, or pain, is not represented by excitation of just one type of filter (or receptor) but is coded by a certain relationship of various relative response amplitudes of different filters. Unequivocal representation in conscious perception is therefore not only dependent on integration of activities over cortical space and time, but also on a comparison of relative amplitudes of neuronal responses distributed across the brain. We have not yet found a neuronal algorithm or code which extracts from such relative excitation profiles an unequivocal representation of percepts.

In sum, the neurosciences provide ample material which allows us to develop concepts and ideas on how this system controls the subject in his environment, and how these sensorimotor servoloops control and come under the control of internal needs and drives. All these network control systems are genetically built into the brain, but are adaptable in youth and able to learn later within the frame of a fixed functional *Bauplan* (building scheme). In principle, for all these elements of mind we may find isomorphic representations in the anatomy and the activity of the brain, but no such mechanism is sufficient to explain how these distributed activities are united into a conscious experience of a conscious self.

It is, in fact, at this point that classical materialism as well as the modern neurosciences come to an end. For the materialistic view, consciousness and conscious experience are only an epiphenomenon, and all that the neurophysiologist finds in the brain are either representations of processed images, of dispositions for sensori-motor responses or – if he relates his measurements closer to the motor end of the system – of driving signals for motor responses. The analysis of the nervous system by physical, biochemical, or anatomic methods will reveal physical, biochemical, or anatomic structures, but nobody investigating the nervous system in such a way would come to the conclusion that this system has consciousness at all (Creutzfeldt 1979, 1983). In fact, for the control system aspect of the brain, consciousness is not a necessary condition, and if one wants to have it in there, it is merely a mental projection from ourself to that machine. The nervous system is a system through which information flows, which leads to action and thus produces new information. This flow of information can be translated into behavioral responses and behavior, and an isomorphism between neural activities of the nervous system with behaving machines should indeed be possible in principle, notwithstanding the fact that it might be too complicated to be realized. But it is clearly not possible to tell from neurophysiological investigations why certain neuronal activities enter consciousness and others do not, why and how consciousness is able to fill in the gaps between certain trains of activity and not between others.

The Mental and the Physical Impasse

Therefore, in spite of all their progress the neurosciences have not been able to rise to the challenge put to them by philosophy as far as the mechanisms and explanations of conscious experience are concerned. Conscious experience must involve the ability of the brain selectively to perceive its own activities and to reconstruct from these distributed and selected activities a coherent picture of the Self, its environment, and its relation to it. This is what one might call the "duality aspect" of consciousness, which Griesinger alluded to when he said "With the self the soul confronts itself with itself, a Thou." It is this aspect of consciousness with which we are confronted as humans, and which probably does not exist in lower organisms. There is no doubt that this ability for conscious self-representation is closely connected with the linguistic abilities of the human brain (for review and references see Sombart 1938; Creutzfeldt 1981a). These enable it to collect its activities to semantic entities and to present them back to itself, thus adding consciousness to its sensations and actions. We may call this the reflective mode of the brain's activities, or the reflective loop (Creutzfeldt 1979). Although the content of such reflection may still be coded inside the brain in the usual way by neuronal discharges, the content itself is a symbol of the reality present in the brain and presented to it. But by being re-presented to the brain and represented in the brain, these symbols actually become a neurophysiological, that is, a physical reality and through this the symbols develop into an independent world of symbols with ontological significance. This world of symbols and signals we call the world of mind, a metastructure built upon the real world already present in the brain.

But we realize that we are now using a mental language rather than describing physiological processes in physical terms. This is the impasse in this discussion: we can, by replacing mental descriptions with physiological or physical terms, get rid of Ryle's ghost in the machine, but in order to describe and understand ourselves and our conscious experience, we have to use the language and the algorithm of this specific experience, the essence of which is, due to the very brain mechanisms on which it is based, duality as we heard from Griesinger. But a description of this duality aspect of the working of our self, whether in mental or in physical terms, must remain metaphorical. One such metaphor is, of course, the dualistic model of brain and mind, and another the materialistic-monistic model. Through such mental metaphors we understand ourselves in this world and make ourselves understood to others. The neurosciences may contribute, like the analytical philosophy of mind, to keep these metaphors as close as possible to reality and to the logic of science. But it is impossible and impractical that the neurosciences with their biophysical and biochemical analysis should ever be able to replace the mental concepts of mind, since they are reached by an experience different from that which the biophysical and biochemical methods give us. Biophysical, biochemical, and anatomic techniques may reveal biophysical mechanisms, biochemical structures, and anatomic connections, but not mind. It may turn out, in fact, that the metaphor of the brain-mind dualism may be better suited to describe and to understand the reality of mind and of ourselves than the materialistic-monistic theory, which is, in essence, like all our models of this world, also only a metaphor, a Platonic reflection of the reality in our mind.

References

Armstrong DM (1968) A materialist theory of mind. Routledge, and Kegan Paul, London
Bieri P (ed) (1981) Analytische Philosophie des Geistes. Hain, Königstein
Bunge M (1980) The mind-body problem. Pergamon, New York
Campbell R (1974) Evolutionary epistemology. In: Schillpp A (ed) The philosophy of Karl Popper. Open Court, La Solle, pp 413–463
Creutzfeldt O (1979) Neurophysiological mechanisms and consiousness. In: Brain and mind. Ciba Foundation Symposion 69 (new series). Excerpta medica, Amsterdam, pp 217–233
Creutzfeldt O (1981a) Philosophische Probleme der Neurophysiologie. In: Roesner H (ed) Rückblick in die Zukunft. Severin and Siedler, Berlin, pp 256–278
Creutzfeldt O (1981b) Diversification and synthesis of sensory systems across the cortical link. In: Pompeiano O, Ajmone-Marsan C (eds) Brain mechanisms of perceptual awareness and purposeful behavior. Raven, New York, pp 153–166
Creutzfeldt O (1983) Cortex cerebri. Springer, Berlin Heidelberg New York
Dennett DC (1979) Brainstorms. Harvester, Hassocks
Edelmann GM, Mountcastle V (1978) The mindful brain. MIT Press, Cambridge

Feigl H (1967) The "mental" and the "physical." University of Minnesota Press, Minneapolis
Griesinger W (1845) Die Pathologie und Therapie der psychischen Krankheiten. Krabbe, Stuttgart
Hebb DO (1980) Essay on mind. Erlbaum, Hillsdale
James W (1890) The principles of psychology. Holt, New York
Lorenz K (1941) Kants Lehre vom Apriorischen im Lichte gegenwärtiger Biologie. Blätter für deutsche Philosophie 15: 94–125
Lorenz K (1973) Die Rückseite des Spiegels. Piper, München
Mach E (1901) Die Analyse der Empfindungen und das Verhältnis des Physischen zum Psychischen. Fischer, Jena
Popper KR (1972) Objective knowledge: an evolutionary approach. Clarendon, Oxford
Popper KR, Eccles JC (1977) The self and its brain. An argument for interactionism. Springer, Berlin Heidelberg New York
Ribot T (1899) La psychologie des sentiments. Paris
Ryle S (1949) The concept of mind. Hutchinson, London
Seifert J (1979) Das Leib-Seele-Problem in der gegenwärtigen philosophischen Diskussion. Wissenschaftliche Buchgesellschaft, Stuttgart
Sombart W (1938) Vom Menschen. Buchholz and Weißwange, Berlin
Sperry R (1980) Mind-brain interaction: mentalism, yes; dualism, no. Neuroscience 5: 195–206
Vollmer G (1975) Evolutionäre Erkenntnistheorie. Hirzel, Stuttgart
Wernicke C (1906) Grundriß der Psychiatrie, 2nd edn, (Liepmann H ed). Thieme, Leipzig
Wundt W (1896) Grundriß der Psychologie. Kröner, Leipzig

II. Reflex Activity of the Spinal Cord

John Eccles and Charles S. Sherrington
at Ipswich 1936

Non-invasive Analysis of the Spinal Cord Generators Activated by Somatosensory Input in Man: Near Field and Far Field Potentials*

JOHN E. DESMEDT, Brussels

Summary. Electronic averaging makes it possible to analyze somatosensory evoked potential (SEP) components recorded non-invasively from the body surface in man. With a non-cephalic reference electrode, the scalp SEP discloses several potentials that are volume conducted from distant neural generators. Among the open field generators with a geometry adequate to produce an external potential gradient at the skin surface, it is useful to distinguish the near field and the far field potentials. Far fields have a widespread distribution over the head and present stationary onset and peak latencies throughout. Near fields are influenced to a greater extent by the position of the recording electrodes; their latencies vary if the corresponding neural generator is propagated and the electrodes are placed along congruent directions. Mapping of these potentials over the skin indeed provides pertinent data about the corresponding neural generators.

The earliest SEP component to stimulation of the median nerve or fingers, P_9, is a far field generated by the peripheral nerve volley at its entry into the brachial plexus. The scalp far field P_{11} reflects the afferent volley propagated up the dorsal column, and it corresponds to the near field N_{11}, which can be recorded at the posterior neck over the Cv7 vertebra. The near field N_{13} at the posterior neck reflects a horizontal dipole generated by postsynaptic activities in the dorsal horn of the spinal cord. N_{13} phase reverses into a "P_{13}" near field when recorded from either oesophageal or anterior neck sites. Because of its orientation at right angles to the axis of the spinal cord, N_{13} does not produce any far field at the scalp. It must be distinguished from the scalp P_{14} far field, which is generated by the medial lemniscus volley. Consideration of conduction times and axonal diameters in various sections of the afferent pathway results in a coherent picture of the somatosensory system in man. The data are consistent with the alterations produced in SEPs by focal lesions of the brain in neurological patients.

Key words: *Spinal cord generators – Clinical neurophysiology – Somatosensory evoked potentials – Far field potentials – Near field potentials – man*

Introduction

Spinal cord potentials activated by somatosensory input were first demonstrated in the anaesthetized cat by Gasser and Graham (1933). These results have stood

* Supported by grants from the Fonds de la Recherche Scientifique Médicale, Belgium

© Springer-Verlag Berlin · Heidelberg 1984

the test of time (Bernhard 1952; Eccles et al. 1962) and have recently received renewed attention in conjunction with non-invasive analysis of spinal cord potentials in intact man (Desmedt and Cheron 1981a). Briefly put, electrophysiological recordings from the exposed cord dorsum in cat or monkey disclosed, after a single-afferent stimulus, a spike (currently related to the volley ascending the dorsal column) followed by "intermediary cord potentials" of which the first was a negative wave. Gasser and Graham (1933) considered several interpretations for this wave, but concluded that it was "produced beyond the primary neuron" and that it presumably involved "internuncial neurons" (p. 316). Recent studies have indeed related the initial wave to a focal negativity of the spinal dorsal horn that was elicited by lowest-threshold A-alpha-beta afferents (Beall et al. 1977).

Current interest for spinal and brain somatosensory evoked potentials (SEPs) in man results from the feasibility of non-invasive studies involving electronic averaging. The method allows the (small) evoked potentials to be recorded from the scalp or body surface instead of directly from neural tissue. SEPs provide invaluable data about the normal or pathological function of the nervous system. They are used for patient monitoring, as a diagnostic aid to localize neurological lesions and also to study electrophysiological correlates of psychological processes such as attention or memory (see Callaway et al. 1978; Desmedt 1979, 1980). The present paper considers recent evidence on the neural generators of early SEP components in man.

Volume Conduction of Potentials in the Brain

Spinal and brain potentials recorded from the body surface are small (1 μV or less) and present complex distributions. They must be interpreted in conjunction with the properties of potential fields in conductive media.

Phasic changes of extracellular potential fields are produced by volume conduction either of a synchronized volley of action potentials in nerve trunk or corticipetal tract (Lorente de No 1947), or postsynaptic potentials generated in the geometrically coherent dendrites of cortical neurons (Eccles 1951).

Lorente de No (1947) distinguished several pertinent sets of geometric parameters. A spike volley in a tract of parallel nerve fibers can be viewed as a propagated dipole and it represents a good example of an "open field" system that can generate in the volume conductor a coherent extracellular field at a distance. Beyond the termination of such a tract, volume conduction of the dipole is recorded as a positive-going approach wave without subsequent negativity ("killed end" recording). Coherent geometrical orientation of the individual nerve fibers in the tract is required for observing sizeable volume conduction at a distance. Cancellation of individual potential fields would occur if the individual fibers were oriented in different directions. Non-coherent geometry is actually a situation prevailing in many central nuclei and it corresponds to the "closed field" system of Lorente de No (1947). For example, depolarization of neurons in a nuclear structure with dendrites radiating in different directions from the center would produce inward flows of current towards the center of the nucleus, but virtually no recordable potential difference outside its anatomical boundaries (Klee and Rall 1977).

Early interpretations of brain waves in electroencephalography (EEG) considered summation of action potentials as a possibility. Eccles (1951) made an essential contribution when he proposed that cortical EEG potentials were rather related to the postsynaptic potentials generated in the geometrically coherent apical dendrites of cortical pyramidal neurons. These indeed represent an open field system in the sense of Lorente de No and can generate volume-conducted potentials at a distance (see Towe 1966; Humphrey 1968; Creutzfeldt et al. 1969; Klee and Rall 1977).

Near Field and Far Field Potentials

Human SEP studies have only recently identified the true relevance of volume conduction of potentials generated at a distance from the recording electrodes. This important issue has been clarified by the use of non-cephalic reference recording (Cracco and Cracco 1976; Desmedt and Cheron 1980).

As a rule, electrophysiological recordings use differential amplifiers to reject in-phase (common mode) interference. Therefore, all recordings are in fact bipolar and they measure the potential difference between the "active" electrode connected to grid 1 and the "reference" electrode connected to grid 2 of the amplifier. If the reference is placed on the head, it picks up potentials generated by the brain under the electrode site and subtracts them from the potentials derived by the "active" electrode. Moreover, both electrodes are influenced to a similar extent by the widespread potential fields that are volume-conducted from distant neural generators, so that these potentials cancel in this montage and will thus be ignored. These severe distortions are avoided when using a non-cephalic reference placed on the hand or shoulder on the non-stimulated side. Interference from the electrocardiogram (ECG) can then be a problem, but this can be solved with appropriate electronic circuits triggering the duty cycle of the averaging computer between the T and P waves of the ECG.

Non-cephalic reference recording ensures that any open field generators recorded by a scalp electrode appear in the recorded trace. Accurate interpretation must then distinguish between near field and far field potentials (Jewett and Williston 1971). The voltage of the recorded potential is large close to a generator, but it attenuates sharply as the electrode is moved away from the dipole. The cortical SEP generators are located near the brain convexity right under the skull (this is not the case for auditory or visual receiving areas in man) and they can be regarded as relatively near field for a postcentral electrode. The latter is nevertheless 12–30 mm away from the active cortex, which results in attenuation by about 10–80 for scalp as compared to direct cortical recordings (Domino et al. 1964). The SEP voltage also decreases with horizontal distance from the focus along the scalp, and the potential gradients are somewhat smoothed out because of the distance from cortex to scalp (see Lorente de No 1947). This does not prevent distinct cortical generators being mapped out at the scalp (Desmedt and Cheron 1981b; Mauguière et al. 1983b).

Subcortical SEP generators are even more distant from the scalp. Their volume-conducted field potentials are small and change relatively little even for large changes in electrode position on the scalp (far field recording condition). However, small far fields can be detected by electronic averaging, which increases the signal-to-noise ratio. Familiar examples are the brain stem auditory responses, which have about 1 mV in focal brain stem recordings in cats, but attenuate about 1000 times when averaged from the scalp (Jewett and Williston 1981). The terms "near field" and "far field" should not imply a strict dichotomy, since they relate in fact to a continuum of voltage distributions in the volume conductor of the head.

Thus scalp-recorded brain potentials relate to open field generators, and the extent to which any generator will be represented in the averaged response depends on several factors: geometry of the active neural units (Klee and Rall 1977); temporal synchronization, duration and spatial distribution of the transmembrane currents in these neural units ("size" of resulting dipole) (Humphrey 1968); and amount of summation or cancellation of field potentials volume-conducted from multiple concomitant sources.

Methods

Experiments were carried out on 22 normal subjects of either sex aged 20–30 years, mostly medical or dental students. They had given informed consent and most of them volunteered for repeat sessions. They were in good health and free from neurological disease. The subjects were selected from a larger group on the basis of good yields for SEP averaging and of ability to relax fully so as to minimize muscle and eyeblink interference. They lay comfortably on a couch in a sound-proofed, electrically shielded and air-conditioned room at 24°C. The procedure was noninvasive and carried no risk. Electrical square pulses of 0.2-ms duration and 3–10 mA (intensity monitored throughout) were delivered either to the left median nerve at the wrist (about thumb twitch threshold) or to fingers I-II-III. Stimulation was random (mean rate less than 4/s). In some experiments each stimulus was triggered via a special delay circuit from the ECG. The upper limb temperature was 35°–37°C.

Scalp, earlobes and neck sites were recorded from with sterile unvarnished stainless steel needles of 0.2 mm diameter inserted subcutaneously. An oesophageal probe of flexible plastic tubing terminated by a smooth perspex rod was inserted through the nostril and adjusted under X-ray control for prevertebral recordings (Desmedt and Cheron 1981a). The reference electrode was placed on the right hand or shoulder. Differential amplifiers with bandpass from 0.5 to 2500 Hz were used in conjunction with a Nicolet 1074 or a PDP 11-34 averaging computer, using bin width of 80 or 200 μs (details in Desmedt 1977; Desmedt and Cheron 1980; Desmedt et al. 1983a). SEP components were labeled from their positive (P) or negative (N) polarity and modal peak latency in normal adults of average body size (Donchin et al. 1977).

Fibre diameters were measured in the sensory nerves of fingers II and III removed at autopsy less than 35 h after death in eight subjects with no neurological disease. The nerves were stained with Flemming's fluid, embedded in paraffin, and cut in thin transverse sections which were photographically enlarged 1000 times to enable measurement of the total diameters of all myelinated axons.

Results

Non-cephalic reference recording of SEPs to median nerve stimulation on the mid-line of the posterior neck disclosed an early positivity, P_9, followed by the negative components N_{11} and N_{13} (Fig. 1, E,F,G). The onset latency of N_{11} shifted by nearly 1 ms from cervical levels Cv7 up to Cv3 (black triangle). N_{13} was largest at about Cv5 level. Simultaneous records from the mid-line frontal (Fz in A) or occipital (Oz in B) scalp and from right earlobe (C) or inion (D) disclosed far field positive components P_9, P_{11} and P_{14}, which had a fixed-onset latency over the entire head. The latency of P_9 is indicated by the vertical interrupted line with black diamond in Fig. 1. The latency of the scalp P_{11} far field corresponded to that of the neck N_{11} "near field" at level Cv7 (Fig. 1, G). The P_{14} scalp far field was not seen at the neck as it is generated above foramen magnum in medial lemniscus (Desmedt and Cheron 1981b; Mauguière et al. 1983c).

Figure 2 identifies the spinal entry time of the peripheral volley. The stimulus was not delivered to the median nerve (to exclude antidromic motor axons and group-I afferent potentials), but to fingers I-II-III (thumb shock delayed by 0.5 ms to make up for different conduction distance; Debecker and Desmedt 1964). Sensory nerve action potentials were recorded with subcutaneous needles from the median nerve up to Erb's point. The averaged traces A–E were separated proportionally to the conduction distances along the nerve. The calculated linear regression gave a conduction velocity (CV) of 72 m/s from wrist to shoulder in this subject. Extrapolation to the level of spinal segments Cv6–Cv7 (Fig. 3) consistently coincided with both N_{11} onset at the lower neck (Fig. 2, F) and onset of the P_{11} scalp far field (Fig. 2, H). The central transit time from spinal entry to arrival at the cortex was taken to the onset of the parietal N_{20} at the contralateral postcentral scalp (Desmedt and Cheron 1981b; Mauguière et al. 1983b). Figure 2 also shows that the P_9 far field occurred before spinal entry and reflected the volume-conducted peripheral volley as it entered the brachial plexus (between axilla and Erb's point, under the lateral part of the clavicula) (Desmedt et al. 1983b).

The spinal N_{11} must be related to the dorsal column in view of its latency shift (Fig. 1, E–F), whose mean value along the posterior neck was 0.94 ± 0.24 (SD) ms in 18 normal adults (Desmedt and Cheron 1980, 1981a). Anatomical data in human cadavers clarified its interpretation. Since the distance between recording electrodes over the posterior processes of vertebrae Cv7–Cv1 (just below the inion) was about 100 mm (Fig. 3), this would suggest a CV of 100:0.94 = 105 m/s, but this is improbably high as compared to the fastest mean afferent CV of 71.1 ± 4.0 m/s in the median nerve. The sensory innervation of finger I depends on spinal segment Cv6 and that of fingers II and III depends on Cv7 (Foerster 1936). Anatomy showed that the Cv6–Cv7 rootlets ran a downwards oblique course

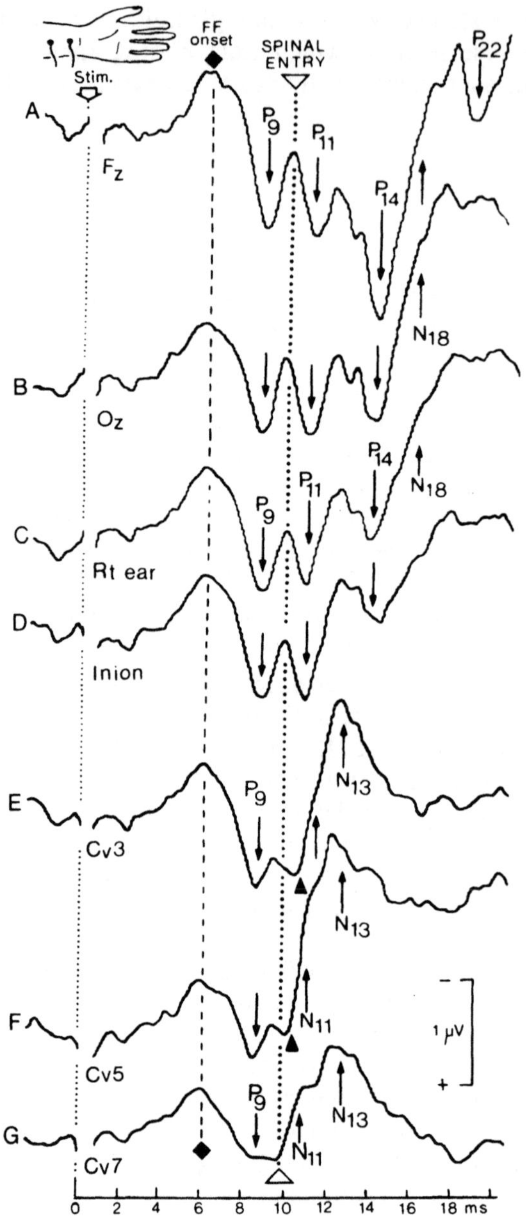

Fig. 1. Early somatosensory evoked potentials to stimulation of left median nerve at the wrist (*vertical dotted line on the left side*). Non-cephalic reference on the right hand. Active electrodes on the mid-line front (*Fz*) in *A*, mid-line occipital region (*Oz*) in *B*, right earlobe (*C*), 2 cm above the inion (*D*), and above the spinous processes of vertebrae Cv3 (*E*), Cv5 (*F*) and Cv7 (*G*). Negativity of the active electrode records upwards in this and all figures. Same calibration for all traces. *Vertical interrupted line with diamonds* shows onset of P_9 far field (*FF*). *Vertical dotted line with white triangles on the right side* shows spinal entry time. The onset of N_{11} shifts to the right from *G* to *E* (*black triangles*). (Modified from Desmedt and Cheron 1980)

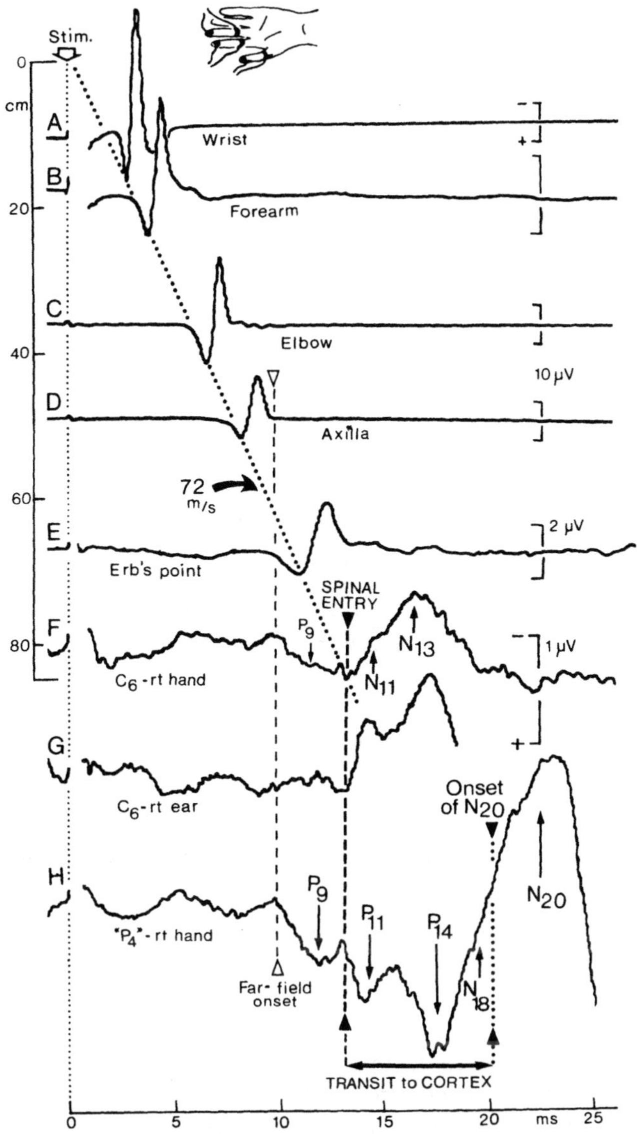

Fig. 2. Estimation of spinal entry time of an afferent volley elicited in median nerve by stimulation (three times sensory threshold) of left fingers I-II-III. Thumb stimulus delayed by 0.5 ms to achieve a synchronized sensory nerve volley which is recorded by subcutaneous steel needles (active close to the nerve; reference inserted 3 cm at right angles to the nerve course) at the wrist (*A*), mid-forearm (*B*), elbow (*C*), axilla (*D*) and Erb's point (*E*). The *vertical separation of the averaged traces* in proportional to distances along the nerve. Calculated linear regression traced obliquely through onsets of the first negative phase of the sensory nerve potentials. The neck somatosensory evoked potential (SEP) recorded with non-cephalic (*F*) or earlobe (*G*) reference is picked up over the spinous process of the Cv7 vertebra. The contralateral parietal scalp SEP is shown in *H* (non-cephalic reference). *Vertical interrupted line with white triangles* shows onset of P_9 far field. *Vertical interrupted line with black triangles* shows spinal entry time extrapolated from the peripheral conduction. The onset of the cortical parietal N_{20} is indicated. (Desmedt and Cheron 1980)

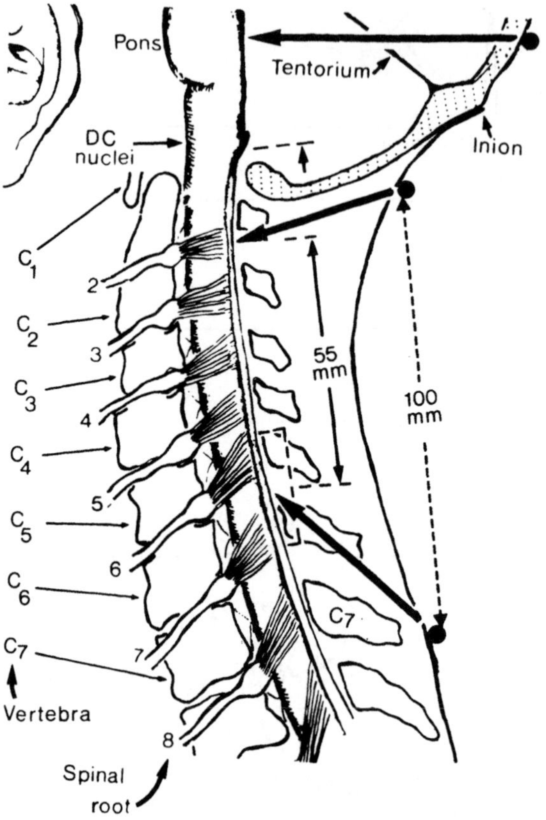

Fig. 3. Sketched lateral view of a dissection of the medulla and cervical spinal cord with roots (*numbered*) and vertebrae (*arrows*) in a formalin-fixed adult man. Skull cut sagittally on the midline. Skin contours indicated, with usual positions of electrodes over the Cv7 spinous process, 2 cm below and 2 cm above the inion. *Large arrows* point to presumed sites of actual recording. (Desmedt and Cheron 1980)

from the cord to the upper surface of the posterior arch of the corresponding vertebrae (Fig. 3). The spinal segments Cv6–Cv7 (identified from roots entry) were about 20 mm more cephalad than vertebra Cv7. Therefore, the actual distance travelled along the dorsal column approximated a mean of 55 mm, rather than 100 mm (Fig. 3), and the true mean maximum CV was recalculated as 55:0.94 = 58.5 m/s. That the proximal fibres of the spinal root ganglion conducted slower (not faster) than their peripheral part was in line with data on cat (Rindos et al. 1984). Previous erroneous interpretations of N_{11} (based on scalp reference recordings) were clarified through non-cephalic reference recordings (see Desmedt and Cheron 1980).

The fact that an electrode slightly above the inion recorded a P_{14} far field (Fig. 1, D) that was not seen by the upper neck electrode (E) may have to do with inhomogeneities in the volume conductor, namely in relation to the basal occipital bone (Fig. 3, heavy horizontal arrow on top).

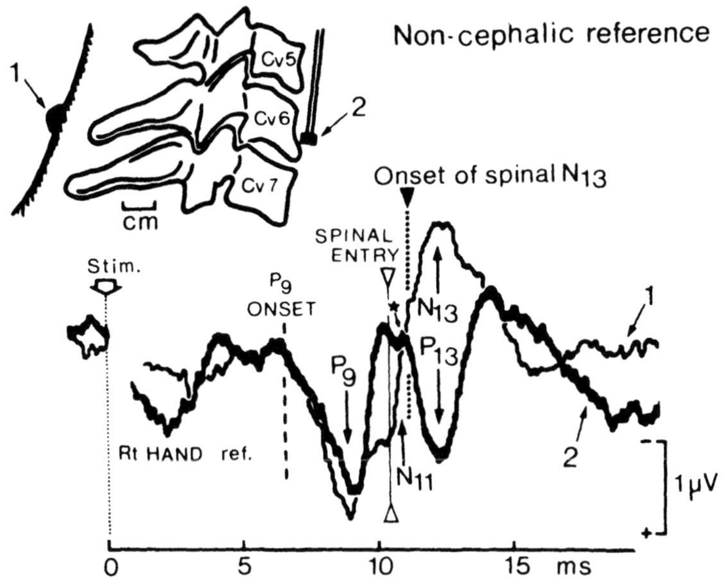

Fig. 4. Phase reversal of N_{13} in prevertebral recording. Drawing of a lateral X-ray of this subject (*upper left*) showing the oesophageal electrode (*2*) in front of Cv6 vertebra and the posterior neck electrode (*1*) over Cv6 spinous process. Non-cephalic reference on right hand. Stimulation of left median nerve at the wrist. Records are superimposed, with oesophageal trace *thicker* (several writings with a slight vertical displacement by the X-Y plotter). The *black star* points to the (small) N_{11} peak seen by the oesophageal electrode, after the negative root potential which precedes spinal entry time (*vertical line with white triangles*). (Modified from Desmedt and Cheron 1981a)

Non-cephalic reference recordings from posterior neck revealed another component, N_{13}, whose latency did not seem to shift like that of N_{11}. N_{13} and N_{11} were better dissociated at the lower neck (Fig. 1, A) than above (E,F), as if N_{11} progressively moved under N_{13}. This was clarified by oesophageal (prevertebral) recordings (Fig. 4). Such traces showed a typical P_9 far field followed by a root potential which preceded spinal entry time. The oesophagus is indeed closer to the spinal roots at the anterolateral aspect of the vertebrae (Desmedt and Cheron 1981a), but farther from the dorsal column, hence the small N_{11} (star in Fig. 4).

A major finding was the unexpected divergence of the oesophageal trace into a positive component, "P_{13}", whose profile was a phase reversal of the N_{13} simultaneously recorded at the posterior neck (Fig. 4). The mean onset and peak latencies of the N_{13} and P_{13} did not differ significantly in ten subjects, but P_{13} had a larger peak voltage (Table 1). Pooled data from 62 oesophageal recorded sites clearly showed that, in any subject, the P_{13} voltage was largest between vertebrae Cv3 and Cv7 and sharply dropped either above or below (Fig. 5B). The P_{13} onset latency was remarkably stable along the cord in any given subject, irrespective of its voltage changes (Fig. 5C).

The P_{13} effect can also be recorded by a surface skin electrode over the hyoid bone, at the anterior neck (Fig. 6A). Spatiotemporal mapping (cf. Rémond 1969)

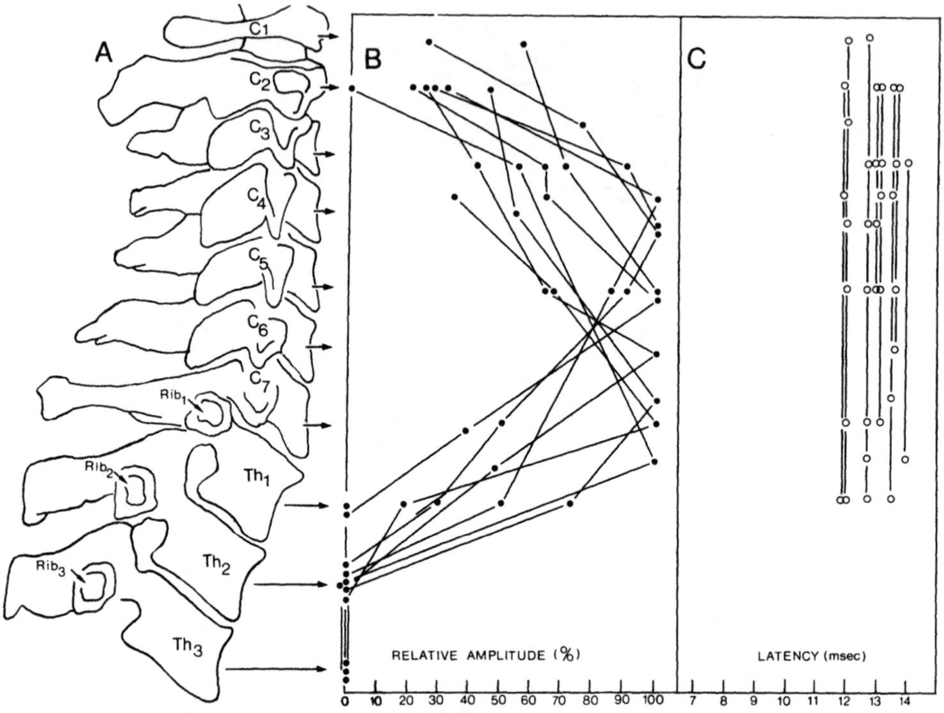

Fig. 5A–C. Longitudinal distribution of P_{13} near field in front of vertebrae which are redrawn from a lateral X-ray in *A*. This drawing provides the *ordinate for B* and *C*. Pooled data from oesophageal sites recorded in ten subjects (usually six sites per subject). *B* Voltage of P_{13} measured from onset to peak, in percent of its maximum voltage in the same subject (*abscissa*). *C* Peak latency of P_{13} (ms, *abscissa*). Data for any given subject are joined by a line. (Desmedt and Cheron 1981a)

Table 1. Data from ten normal adults (see Fig. 5)

	Posterior neck N_{13}	Oesophageal P_{13}	*t* test
Onset latency (ms)	11.96 ± 0.94	11.95 ± 0.94	$p > 0.6$
Peak latency (ms)	13.15 ± 0.91	13.10 ± 0.80	$p > 0.7$
Duration (ms)	–	4.19 ± 0.64	–
Voltage (μV)	0.47 ± 0.35	0.93 ± 0.45	$p < 0.06$

of the surface distribution of potential fields around the neck disclosed the steep gradients of the widespread P_9, followed by the phase reversal (at about electrode 3 in B) between the anterior P_{13} positivity and the posterior N_{13} negativity (Fig. 6C). Thus the orientation of the N_{13} dipole is roughly at right angles to that of the N_{11} dipole (Fig. 7B).

In the cat, the first negative cord dorsum potential was generated in the dorsal horn of the lumbosacral spinal cord (Lindblom and Ottoson 1953; Coombs et al. 1956). Intraspinal microelectrode mapping of the response to low-threshold cuta-

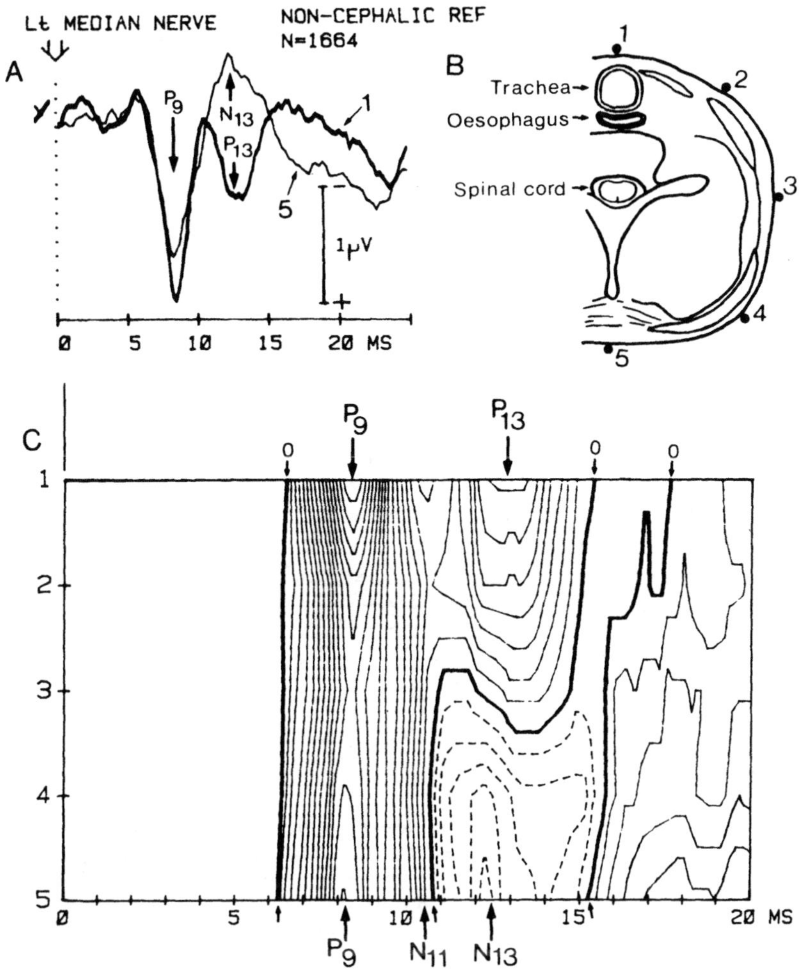

Fig. 6A–C. Mapping of early somatosensory evoked potentials with skin electrodes placed around the neck, as shown in the sketched horizontal section in *B*. Stimulation of left median nerve at the wrist. Non-cephalic reference on right shoulder. *A* Superimposed traces from posterior (*electrode 5*) and anterior (*1, thicker trace*) neck mid-line. *C* Spatiotemporal mapping of potential gradients at the five electrodes (*ordinate*) as they change over time (ms, *abscissa*). *Thicker lines (small arrows)* indicate zero potential. Isopotential contours determined using a PDP 11-34 computer. Negative gradients in *interrupted lines*.

neous afferents in the monkey disclosed a negative field focus in the lateral part of laminae IV and V of Rexed (1954) and a positive field more ventrally (Fig. 7A).

Figure 8 shows a typical histogram of axonal diameters in a ventral digital nerve of finger II in a female of 38 years (autopsy material). The largest cutaneous afferents have a diameter of 14 μm. Since the mean maximum CV in median nerve from fingers to wrist is 63.2 ± 3.3 m/s in normal adults at 35°–36°C tissue temperature (Nielsen 1973; Desmedt and Cheron 1980), the conversion factor was calculated as 63:14 = 4.5 m/s/μm (Table 2).

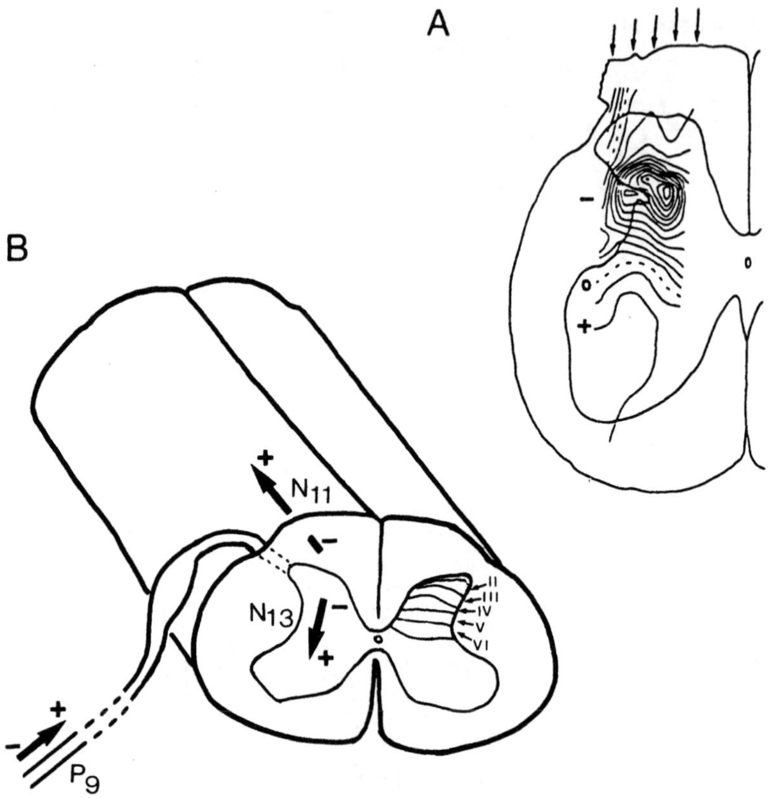

Fig. 7A. Isopotential contour map of the early (lumbosacral) cord dorsum negativity evoked by stimulation of the sural nerve in anesthetized monkey. Recordings made every 200 μm along five microelectrode tracks (*arrows*). Zero isopotential line *dashed*. Focal negativity in dorsal horn and ventral positivity. (Beall et al. 1977) *B* Cartoon of the cervical spinal cord (Rexed layers II–VI sketched *on the right*) with proposed longitudinal propagated N_{11} dipole, horizontal fixed N_{13} dipole and peripheral P_9 dipole.

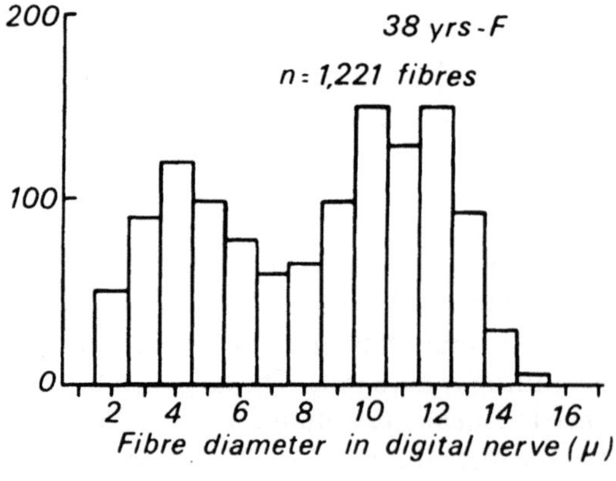

Fig. 8. Histogram of diameters (μm, *abscissa*) of myelinated axons in a ventral digital nerve of finger II (female of 38 years, autopsy material).

Table 2. Data from normal adults

	Maximum diameter of axons (μm)	Conduction distance (mm)	Maximum CV (m/s)	Conduction time (ms)	Cumulated time (ms)	Measured time from spinal entry (ms)
Finger-to-wrist	14 (Fig. 8)		63.2±3.2			
Wrist-to-cord	16 (calculated)		71.1±4.0			
Dorsal column	13 (Bok 1928)	70	58.5	1.2	1.2	
Synaptic delay in cuneate				0.8	2.0	2.03±0.28 (onset of P_{14})
Medial lemniscus	8 (Verhaart)	70	36	1.94	3.94	

CV, conduction velocity

Discussion

The data illustrate how non-invasive recording from the body surface combined with electronic averaging can be used to analyze SEP neural generators in man. This is important because data obtained in other mammals cannot be safely extrapolated to the human SEPs, whose detailed features must receive robust interpretations before their changes in disease can be understood and put to clinical use.

It is remarkable that SEPs recorded from the scalp with non-cephalic reference can so clearly disclose far field potentials that relate to the peripheral nerve and subcortical afferent pathway (Fig. 2, H). The far fields have about 1 μV amplitude and an identical latency throughout their widespread scalp distribution (Fig. 1, A–D). Their duration of 1–2 ms is too brief for postsynaptic activity, but is consistent with volume conduction of synchronized action potential volleys (Desmedt and Cheron 1981b, Figs. 10 and 11; see also Arezzo et al. 1979; Stockard and Sharbrough 1980).

The P_9 Far Field Potential

P_9 is recorded over all the scalp and neck (Figs. 1, 4, 6) and is the first event in SEPs to median nerve or finger stimulation. It is volume-conducted from a dipole generated by the peripheral nerve volley (Cracco and Cracco 1976; Kritchevsky and Wiederholt 1978; Kimura et al. 1978). The P_9 onset precedes the arrival of the volley at Erb's point, but it is later than its arrival at the axilla (Fig. 2) (Desmedt and Cheron 1980). Thus, the P_9 far field with stationary onset and peak throughout the head is generated as the nerve volley arrives at a particular site, as discussed by Desmedt et al. (1983b), who consider the role of inhomogeneities in the surrounding tissue (Lorente de No 1947) and of changes in the axial orientation of the propagated dipole. P_9 persisted in patients with traction injuries of the brachial plexus when the spinal roots had been severed and subsequent SEP components were lost (Anziska and Cracco 1980).

The P_{11} Far Field and N_{11} Near Field Potentials

The second scalp deflection P_{11} is also a brief far field with stationary onset and peak latencies. It represents volume conduction of the dipole generated by the volley of action potentials in the dorsal column, at the level of the Cv6–Cv7 spinal segments (Desmedt and Cheron 1980). Strong evidence is provided by the latency shift of N_{11} along the posterior neck (Fig. 1, G–E), which corresponds to a mean CV of 58.5 m/s in the dorsal column (Table 2). The onset latency of P_{11} coincides with the time of arrival of the afferent volley at the spinal cord (Fig. 2). P_{11} onset also coincides with the onset of the negative potential N_{11} recorded over the Cv7 vertebra, and N_{11} is regarded as the "near field" counterpart of the P_{11} "far field".

It is indeed appropriate for the near field potential of a propagated volley to have a negative polarity and to shift latency when recording further up along the nerve tract (Fig. 7B). What is remarkable is that the P_{11} far field appears stationary at the scalp (Fig. 1, A–D) and corresponds in time with the initiation of the dorsal column volley at the Cv6–Cv7 spinal segments by the activated spinal roots.

The N_{13} Near Field Potential

At posterior neck SEP recordings, N_{13} superimposed on N_{11} and its features failed to be clearly delineated until it was found that N_{13} phase-reversed into a P_{13} at prevertebral (oesophageal) sites (Fig. 4, Table 1) or at the anterior neck (Fig. 6). Both N_{13} and P_{13} must be regarded as near field potentials corresponding in man to the two sides of the dorsal horn dipole analyzed through intraspinal microelectrode mapping in the monkey (Fig. 7A).

This dipole involves the activation of interneurons with dendrites and axon in the horizontal plane (Wall 1973) with a current sink (extracellular negativity) in the dorsal horn and a current source (extracellular positivity) in the ventral region (Fig. 7B). The horizontal orientation of this dipole explains why it is not associated with any concomitant far field potential at the scalp (the P_{14} scalp far field has a longer latency and shorter time course) (Desmedt and Cheron 1981a, b). While the brief scalp far fields represent volume-conducted action potential volleys, the longer N_{13} near field must be related to postsynaptic activities in the dorsal horn interneurons.

The longitudinal distribution of P_{13} has been accurately plotted with oesophageal electrodes (X-ray control of position) and found to extend from Cv3 to Cv7 vertebrae, hence from Cv4 to Th1 spinal cord levels (Figs. 3 and 5). This is in line with the cord distribution of primary cutaneous afferents. After spinal entry, the large afferents bifurcate into an ascending branch that reaches the cuneate nucleus, and a shorter descending branch. Collaterals of either branch penetrate into the dorsal horn through its mediodorsal aspects and form a dense plexus in Rexed layers IV and V (Cajal 1909; Sprague and Hang 1964). The density of such collaterals is highest near root level (here Cv6–Cv7; Fig. 3) and drops in the neighboring segments (Sterling and Kuypers 1967; Rethelyi and Szentágothai 1973). From the autopsy findings of Dejérine and Thomas (1896) in brachial plexus lesions, collaterals should be given to about two segments above and two segments below root entry in man.

The data are consistent with a fixed-latency (Fig. 5C) horizontal dipole generator that produces N_{13}-P_{13} near fields around the neck (Fig. 6) but no far field at the scalp, in marked contrast with N_{11} (Fig. 7B). This generator is not a serial link in the somatosensory pathway to the cortex and it can undergo dissociated changes: for example, it is reduced or delayed in certain patients with multiple sclerosis, while the cortical N_{20} SEP component can remain within normal range (Mastaglia et al. 1978; Small et al. 1980). On the other hand, in a patient with unilateral traumatic section at lower medulla, N_{11} and N_{13} were preserved while the P_{14} scalp far field and subsequent SEPs were lost on the lesion side (Mauguière et al. 1983a), as expected if N_{13} is generated below foramen magnum.

Conduction in the Somatosensory Pathway

The calculated conversion factor of 4.5 m/s/μm for CV in human sensory axons (Fig. 8, Table 2) will be used as first approximation. The mean maximum afferent CV in median nerve from wrist to shoulder is 71.1 ± 4.0 m/s, which suggests a maximum fiber diameter of about $71{:}4.5 = 16\,\mu$m. An excess of 2 μm with respect to the same axons in the digital nerves (Fig. 8) is acceptable in view of the known distal tapering of long axons. In the dorsal column, the primary axons originating from fingers have a maximum CV of 58.5 m/s (see above), which fits with the maximum axonal diameter of 13 μm reported by Bok (1928), since $58.5{:}4.5 = 13\,\mu$m. The conduction distance in dorsal column up to the cuneate nucleus was measured as 70 mm in cadavers (Fig. 3), hence the calculated conduction time of $70{:}58.5 = 1.2$ ms (Table 2).

The cuneate synapses present large synaptic boutons around clusters of neurons and dendrites, and they have a high safety factor transmission (Krnjevic and Morris 1976). From Coombs et al. (1956) and Beall et al. (1977) the synaptic delay at the cuneate is taken as 0.8 ms, which implies that the afferent volley starts involving the medical lemniscus $1.2 + 0.8 = 2.0$ ms after spinal entry. This fits with the mean value of 2.03 ± 0.28 ms measured in our adult group for the transit time between P_{11} onset (spinal entry time) and the onset of the scalp-recorded P_{14} (Desmedt and Nguyen 1984). One implication is that the onset of the stationary scalp far field P_{14} reflects activation of the caudal part of medical lemniscus, a situation not unlike the one between P_{11} and the dorsal column volley (see above).

Lemniscal CV can be roughly estimated from the diameter of 8 μm for the largest second-order axons in the subgroup of arcuate fibres (ml-3) that relay the input from the human hand (W. Verhaart, personal communication). The maximum CV is calculated as $8 \times 4.5 = 36$ m/s. Since dissection of cadavers suggests a mean conduction distance of 70 mm from cuneate to posterior thalamus, the conduction time in medial lemniscus is estimated as $70{:}36 = 1.94$ ms. This excludes the possibility that P_{14} could be generated at or above the thalamus, in agreement with data in patients with a thalamic lesion in whom P_{14} persists while the cortical SEP components are lost (Nakanishi et al. 1978; Mauguière et al. 1983c).

References

Anziska B, Cracco RQ (1980) Short-latency somatosensory evoked potentials: patients with focal neurological disease. Electroencephalogr Clin Neurophysiol 49: 227–239

Arezzo J, Legatt AD, Vaughan HG (1979) Topography and intracranial sources of somatosensory evoked potentials in the monkey: early components. Electroencephalogr Clin Neurophysiol 46: 155–172

Beall JE, Applebaum AE, Foreman RD, Willis WD (1977) Spinal cord potentials evoked by cutaneous afferents in the monkey. J Neurophysiol 40: 199–211

Bernhard CG (1952) The cord dorsum potentials in relation to peripheral source of afferent stimulation. Cold Spring Harbor Symp Quant Biol 17: 221–232

Bok ST (1928) Das Rückenmark. In: v Möllendorff W (ed) Handbuch der mikroskopischen Anatomie der Menschen 4: 478–573

Cajal SR (1909) Histologie du système nerveux de l'homme et des vertébrés. Maloine, Paris

Callaway E, Tueting P, Koslow S (eds) (1978) Event-related brain potentials in man. Academic, New York, pp 1–631

Coombs JS, Curtis DR, Landgren S (1956) Spinal cord potentials generated by impulses in muscle and cutaneous afferent fibers. J Neurophysiol 19: 452–467

Cracco RQ, Cracco JB (1976) Somatosensory evoked potentials in man: far-field potentials. Electroencephalogr Clin Neurophysiol 41: 460–466

Creutzfeldt O, Watanabe S, Lux HD (1969) Relations between EEG phenomena and potentials in single cells. Electroencephalogr Clin Neurophysiol 20: 1–18

Debecker J, Desmedt JE (1964) Les potentiels évoqués cérébraux et potentiels de nerf sensible chez l'homme. Acta Neurol Psychiatr Belg 64: 1212–1248

Dejérine J, Thomas A (1896) Contribution à l'étude du trajet intramédullaire des racines postérieures dans la région cervicale et dorsale supérieure de la moelle épinière. CR Soc Biol (Paris) 48: 675–679

Desmedt JE (1977) Some observations on the methodology of cerebral evoked potentials in man. In: Desmedt JE (ed) Attention, voluntary contraction and event-related cerebral potentials. Karger, Basel, pp 12–29 (Prog Clin Neurophysiol, vol 1)

Desmedt JE (ed.) (1979) Cognitive components in cerebral event-related potentials and selective attention. Karger, Basel, pp 1–319 (Prog Clin Neurophysiol, vol 6)

Desmedt JE (ed) (1980) Clinical uses of cortical, brainstem and spinal somatosensory evoked potentials. Karger, Basel, pp 1–352 (Prog Clin Neurophysiol, vol 7)

Desmedt JE, Cheron G (1980) Central somatosensory conduction in man: neural generators and interpeak latencies of the far-field components recorded from neck and right or left scalp and earlobes. Electroencephalogr Clin Neurophysiol 50: 382–403

Desmedt JE, Cheron G (1981a) Prevertebral (oesophageal) recording of subcortical somatosensory evoked potentials in man: the spinal P13 component and the dual nature of the spinal generators. Electroencephalogr Clin Neurophysiol 52: 257–275

Desmedt JE, Cheron G (1981b) Non-cephalic reference recording of early somatosensory potentials to finger stimulation in adult or aging man: differentation of widespread N18 and contralateral N20 from the prerolandic P22 and N30 components. Electroencephalogr Clin Neurophysiol 52: 553–570

Desmedt JE, Nguyen TH, Bourguet M (1983a) The cognitive P40, N60 and P100 components of somatosensory evoked potentials and the earliest electrical signs of sensory processing in man. Electroencephalogr Clin Neurophysiol 56: 272–282

Desmedt JE, Nguyen TH, Carmeliet J (1983b) Unexpected latency shifts of the stationary P9 somatosensory evoked potentials far field with changes in shoulder position. Electroencephalogr Clin Neurophysiol 56: 628–634

Domino EF, Matsuoka S, Waltz J, Cooper I (1964) Simultaneous recordings from scalp and epidural somatosensory evoked response in man. Science 145: 1199–1200

Donchin E, Callaway E, Cooper R, Desmedt JE, Goff WR, Hillyard SA, Sutton S (1977) Publication criteria for studies of evoked potentials in man. In: Desmedt JE (ed) Attention, voluntary contraction and event-related cerebral potentials. Karger, Basel, pp 1–11 (Prog Clin Neurophysiol, vol 1)

Eccles JC (1951) Interpretation of action potentials evoked in the cerebral cortex. Electroencephalogr Clin Neurophysiol 3: 449–464

Eccles JC, Kostyuk PG, Schmidt RF (1962) Central pathways responsible for depolarization of primary afferent fibres. J Physiol (Lond) 161: 237–257

Foerster O (1936) Symptomatologie der Erkrankungen des Rückenmarks und seiner Wurzeln. In: Bumke O, Foerster O (eds) Handbuch der Neurologie, vol 5. Springer, Berlin Heidelberg New York, pp 1–403

Gasser HS, Graham HT (1933) Potentials produced in the spinal cord by stimulation of dorsal roots. Am J Physiol 103: 303–320

Humphrey DR (1968) Re-analysis of the antidromic cortical response: contribution of cell discharge and PSPs to evoked potentials. Electroencephalogr Clin Neurophysiol 5: 421–442

Jewett DL, Williston JS (1971) Auditory evoked far-fields averaged from the scalp in humans. Brain 94: 681–696

Kimura J, Yamada T, Kawamura H (1978) Central latencies of somatosensory cerebral evoked potentials. Arch Neurol 35: 683–688

Klee M, Rall W (1977) Computed potentials of cortically arranged populations of neurons. J Neurophysiol 40: 647–666

Kritchevsky M, Wiederholdt WC (1978) Short-latency somatosensory evoked responses in man. Arch Neurol 35: 706–711

Krnjevic K, Morris ME (1976) Input-output relation of transmission through cuneate nucleus. J Physiol (Lond) 257: 791–815

Lesser RP, Lueders H, Hahn J, Klem G (1981) Early somatosensory potentials evoked by median nerve stimulation: intraoperative monitoring. Neurology 31: 1519–1523

Lindblom UF, Ottoson JO (1953) Localization of the structure generating the negative cord dorsum potential evoked by stimulation of low-threshold cutaneous fibres. Acta Physiol Scand 29: (Suppl. 106) 180–190

Lorente de No R (1947) A study of nerve physiology, vol 132. Studies from the Rockefeller Institute, New York, pp 1–548

Mastaglia FL, Black JL, Edis R, Collins DWK (1978) The contribution of evoked potentials in the functional assessment of the somatosensory pathway. In: Tyrer JH, Eadie MJ (eds) Clinical and Experimental Neurology, Adis Press, Sydney 15: 279–298

Mauguière F, Courjon J, Schott B (1983a) Dissociation of early SEP components in unilateral traumatic section of the lower medulla. Ann Neurol 13: 309–313

Mauguière F, Desmedt JE, Courjon J (1983b) Astereognosis and dissociated loss of frontal or parietal components of somatosensory evoked potentials in hemispheric lesions: detailed correlations with clinical signs and CT scanning. Brain 106: 271–311

Mauguière F, Desmedt JE, Courjon J (1983c) Neural generators of N18 and P14 far field somatosensory evoked potentials: patients with lesion of thalamus or thalamo-cortical radiations. Electroencephalogr Clin Neurophysiol 56: 283–292

Nakanishi T, Shimada Y, Sakuta M, Toyokura Y (1978) The initial positive component of scalp-recorded somatosensory evoked potentials in normal subjects and in patients with neurological disorders. Electroencephalogr Clin Neurophysiol 45: 26–34

Nielsen VK (1973) Sensory and motor conduction in the median nerve in normal subjects. Acta Med Scand 194: 435–443

Rémond A (1969) Topographic data in EEG phenomena and an electrical model to reproduce them. Electroencephalogr Clin Neurophysiol [Suppl 27]: 29–49

Rethelyi M, Szentágothai J (1973) Distribution and connexions of afferent fibres in the spinal cord. In: Iggo A (ed) Handbook of sensory physiology, vol 2. Springer, Berlin Heidelberg New York, pp 207–252

Rexed B (1954) A cytoarchitectonic atlas of the spinal cord in the cat. J Comp Neurol 100: 297–379

Rindos AJ, Loeb GE, Levitan H (1984) Conduction velocity changes along lumbar primary afferents in cats. Brain Res (to be published)

Small DG, Beauchamp M, Matthews WB (1980) Subcortical somatosensory evoked potentials in normal man and patients with CNS lesions. In: Desmedt JE (ed) Clinical uses of cerebral, brainstem and spinal somatosensory evoked potentials. Karger, Basel, pp 190–204 (Progr Clin Neurophysiol, vol 7)

Sprague JM, Hang CH (1964) The terminal fields of dorsal root fibers in the lumbosacral spinal cord of the cat, and the dendrite organization of the motor nuclei. Prog Brain Res 11: 120–154
Sterling P, Kuypers H (1967) Anatomical organization of the brachial spinal cord of the cat: the distribution of dorsal root fibers. Brain Res 4: 1–15
Stockard JJ, Sharbrough FW (1980) Unique contributions of short-latency auditory and somatosensory evoked potentials to neurologic diagnosis. In: Desmedt JE (ed) Clinical uses of cerebral, brainstem and spinal somatosensory evoked potentials. Karger, Basel, pp 231–263 (Progr Clin Neurophysiol, vol 7)
Towe AL (1966) Nature of the primary evoked response. Exp Neurol 15: 113–139
Wall PD (1973) Dorsal horn electrophysiology. In: Iggo A (ed) Handbook of sensory physiology, vol 2. Springer, Berlin Heidelberg New York, pp 253–270

Organization of Reflexes Evoked by Stimulation of Neck Receptors*

VICTOR J. WILSON, New York

Summary. Spinal reflexes originating with receptors in the neck present a variety of opportunities for investigation. The cervicocollic reflex produced in the splenius muscle by yaw rotation of the head of labyrinthectomized cats around a C1–C2 axis has a rostrocaudal gain gradient. The rostral part of the muscle may be excited selectively by this stimulus, an excellent example of reflex localization. To a considerable extent, the distribution of the cervicocollic reflex evoked by head rotation parallels the distribution of the Ia afferents, which have an asymmetric rostrocaudal distribution within the motor nucleus. This is one line of indirect evidence suggesting that the reflex is, at least in part, due to these afferents. If so, then stretch reflexes in the neck have a higher gain than has been thought, at least when evoked by high-frequency stimuli. We still have much to learn about another reflex due to neck receptors, the tonic neck reflex acting on the limbs. Despite its name the reflex is not purely tonic: gain increases with frequency and phase sometimes advances. The receptors giving rise to the reflex have not been identified and very little is known about the polysynaptic pathways linking neck afferents to limb motoneurons. We have identified interneurons whose dynamics resemble those of the reflex, but their relation to the reflex pathway remains to be determined.

Introduction

It is a great privilege to be asked to contribute to a symposium honoring Sir John Eccles by participating in a session on the Reflex Activity of the Spinal Cord. When I first met Sir John I was just being introduced to the spinal cord. After a long diversion to the vestibular system I am back almost where I started, having, however, moved rostrally from the lumbosacral enlargement. Our focus is now on neck reflexes, discovered by Magnus, some aspects of which were discussed 50 years ago in the book that gives this session its name (Creed et al. 1932).

Many of the reflexes involved in regulation of posture have been examined in great detail. This is certainly true for vestibulospinal reflexes, which have been studied by methods ranging from electroanatomic investigation of single neurons to systems analysis in alert animals. Knowledge about these reflexes is extensive, though far from complete (Wilson and Melvill Jones 1979; Wilson and Peterson 1981). Vestibulospinal reflexes are only half of a more complex control system,

* Supported by grants from NIH (NSO 2619) and NASA (NSG 2380)

Experimental Brain Research, Suppl. 9
© Springer-Verlag Berlin · Heidelberg 1984

the other half of which is reflexes arising in neck receptors. These consist of the tonic neck reflex acting on the limbs (von Holst and Mittelstaedt 1950; Lindsay et al. 1976), and the cervicocollic reflex, acting on the neck itself (e.g., Peterson et al. 1981a, b). There have been a number of experiments in various laboratories on both sets of neck reflexes in recent years, but much remains to be learned about their properties and about the mechanism of their interaction with vestibular reflexes. In this paper I will mainly review some of our recent experiments on neck reflexes and will try to answer some of the many questions that need to be addressed.

The Cervicocollic Reflex

Stimulation of receptors in the neck by rotation of the body with respect to a fixed head or by rotation of the head in labyrinthectomized cats evokes prominent reflexes in neck muscles (Peterson et al. 1981a, b; Ezure et al. 1983a). It would appear a priori that these reflexes are at least in part a response to stretch because the stimulus stretches the muscles, which are particularly rich in spindles (e.g., Richmond and Abrahams 1975b) and whose motoneurons receive typical Ia connections (Rapoport 1979; Brink et al. 1981). It has been thought that stretch reflexes in neck muscles are weak, but this belief was based on experiments with electrical stimulation (Abrahams et al. 1975) or on specific behavioral paradigms (Bizzi et al. 1978). More recently, Peterson et al. (1981a) have suggested that the cervicocollic reflexes which they observed in their experiments could be due to spindle afferents.

Our interest in cervicocollic reflexes had two origins. First, the wealth of spindles coupled with the apparent weakness of the stretch reflex raised the question of the spindles' function. Secondly, the unusual division of muscles such as the splenius and biventer cervicis into "compartments" partly separated anatomically by tendinous inscriptions (Richmond and Abrahams 1975a) raised the question of how afferents from spindles located in different compartments were distributed to the motor nucleus, which is spread out rostrocaudally over three to four segments. We began with the second question (Brink et al. 1981). By stimulating the peripheral end of cut muscle nerves and recording the electromyogram (EMG) with a roving electrode, we showed that the splenius is indeed often subdivided into four areas, each innervated by a different segment (Fig. 1); when there is no C1 branch to the splenius, the most rostral, medial area, which is not delimited by an inscription, is innervated by a rostral branch of the C2 nerve. Our results suggest that, as also shown by Frances Richmond (personal communication), muscle fibers are short and do not appear to cross the inscriptions. Next, we recorded intracellularly from splenius motoneurons while stimulating the three or four nerves innervating the muscle. The results we obtained are summarized in Fig. 1. In brief, Ia afferents from the area of the muscle innervated by a particular segment make monosynaptic connections with motoneurons of that and more rostral, but not more caudal, segments. The results were the same whether stimuli near threshold or as strong as five times threshold were used, implying that any contribution from group II afferents

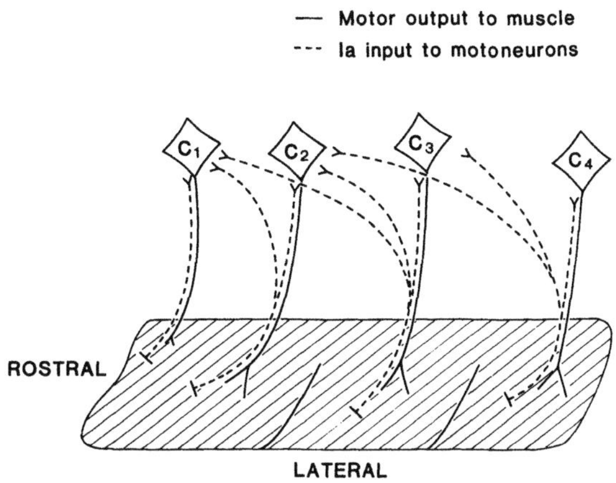

Fig. 1. Schematic representation of the innervation of the splenius muscle of the cat. The *two dark lines* represent tendinous inscriptions, which partly subdivide the muscle into three regions. *Diagonal lines* indicate the approximate direction of muscle fibers. The muscle shown is innervated by C1–C4. In some cats there is no C1 nerve to the splenius.

has a similar distribution. This result suggests that reflexes elicited by stretch of the rostral part of the muscle will be localized to that part, but that there will not be such discrete localization of reflexes elicited by stretch of caudal muscle fibers. We have tested this idea, first qualitatively, then quantitatively, by rotating the heads of decerebrate, labyrinthectomized cats in the yaw plane (Bilotto et al. 1982; Ezure et al. 1983a). EMG was recorded by electrodes placed on the surface of the four splenius compartments. Stimulation consisted of computer-driven head rotations, usually sums of eight or ten sinusoids, up to 5° per component. As had been shown previously by Peterson et al. (1981a, b), the cervicocollic reflex is sufficiently linear for a sinusoidal analysis. Their experiments also showed that at low frequencies of stimulation reflex gain is low and the response is in phase with head position; at around 0.2–0.5 Hz phase advances and gain begins to increase. We have confirmend these observations (Fig. 2A). The point I want to stress, however, is that the structure of the muscle is such that when the head is rotated around a vertical axis between C1 and C2, with movement of the vertebral column mainly at the joints between the skull and C1 and between C1 and C2, it is mainly the rostral part of the splenius (and other muscles with similar attachments) that is stimulated. As the Bode plot of Fig. 2A shows, with such a stimulus, modulation of the splenius EMG is strongest in the most rostral, medial part of the muscle (compartment 1) and weakest in the caudal part (compartment 4). In some instances only compartments 1 and 2 are modulated at low frequencies, where reflex gain is low. Phase of modulation is the same throughout the muscle. Note, by the way, that at high frequencies reflex gain is quite high: in compartment 1 average EMG modulation is about 30% per degree of head rotation at 4 Hz. If the axis of rotation is changed so that it is over C4–C5, there is still some movement at the C1-skull and C1–C2 joints, but there is additional deformation at C5–C6, i. e., there is more activation of caudally located receptors. Figure 2B shows that in this situation the strength of modulation is now

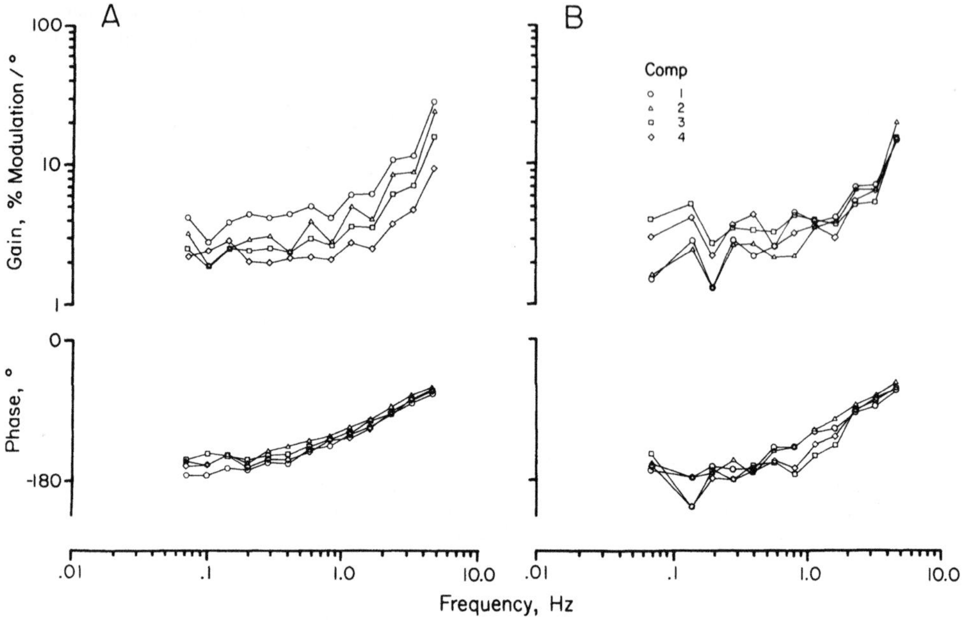

Fig. 2A,B. Modulation of the four compartments of the splenius in response to multiple sine waves. Gain and phase expressed with respect to head position; -180° corresponds to maximal contralateral flexion of the neck. *A* average response of six experiments with rotation around a C1–C2 axis; *B* average response of four experiments with rotation around a C4–C5 axis. (Ezure et al. 1983a)

comparable in all parts of the muscle at higher frequencies; at low frequencies there is more modulation in the caudal region.

Two other aspects of our results need to be mentioned. First, it seems that not all the receptors giving rise to the reflex are in the splenius: there is a much weaker but otherwise normal reflex when the muscle, its innervation intact, is completely detached from its origins and insertions. Secondly, the reflex is a spinal one because it survives transection of the spinal cord just rostral to C1. A supraspinal contribution is, of course, not ruled out.

Some properties of the cervicocollic reflex, namely the similarity of its dynamics to those of the vestibulocollic reflex and its interaction with the latter, have been discussed elsewhere (Peterson et al. 1981a, b). I will restrict myself to a consideration of two other issues, the probable origin of the reflex and its localization. There is no direct evidence to identify the receptors responsible for the reflex we studied, but there is reasonable indirect evidence to suggest an important role for muscle spindles. First, head rotation visibly stretches splenius. Secondly, as pointed out by Peterson et al. (1981a), the dynamics of the cervicocollic reflex have a strong resemblance to the dynamics of spindles in their linear range (Matthews and Stein 1969; Poppele and Bowman 1970). Thirdly, the distribution of activity within the splenius when different axes of rotation are used (Fig. 2) is consistent with the distribution of Ia afferents (Fig. 1), i. e., stimulation of rostral muscle spindles would be expected to influence the rostral part of the muscle, while stimulation of caudal receptors would exert reflex effects on both caudal and rostral regions. It is a reasonable hypothesis that the cervicocollic reflex evoked by head rotation is due at least in part to activation of spindle afferents in the splenius and its synergists.

It is, of course, not news that stretch reflexes are localized, and recently there has been renewed emphasis on the fact that Ia afferents may cause particularly effective reflex activation of the part of the muscle in which their spindles lie in addition to their broader actions (e.g., Botterman et al. 1978; Binder and Stuart 1980; Lüscher et al. 1980; Brink et al. 1981; Botterman et al. 1981). Neck muscles, subdivided into compartments, provide a particularly good example of localization.

The Tonic Neck Reflex

The basic description of this reflex has been with us for a long time: when the body is rotated on a stationary head, either in roll or in yaw, there is reciprocal activation of the limbs such that the one nearest the chin is extended, the other flexed (Magnus and de Kleijn 1912). Two important advances were made about 40 years later. First, the reflex was shown to be due to receptors in the neck, located near the vertebral column (McCouch et al. 1951). Secondly, it was suggested that there is cancellation of tonic neck reflexes and tonic vestibular reflexes, which have the opposite polarity, allowing the head to move without a change in position of the platform that supports it (von Holst and Mittelstaedt 1950). This hypothesis has received support from experiments in which the C1 and C2 nerves were cut so that vestibular reflexes could be evoked by rotating the head, and weakened neck reflexes subserved by only the C3 and C4 nerves by rotating the C2 vertebra (Lindsay et al. 1976). This work aside, it is easier to list what we do not know about the tonic neck reflex than what we do know. The important receptors are near the vertebral column, but their exact nature and location remain uncertain. It was customary to ascribe the reflex to joint receptors, but as of this time no slowly adapting receptors have been found in or near neck joints (Richmond and Bakker 1982). Perhaps the reflex originates from muscle spindles which are found in great numbers in perivertebral muscles (Richmond and Abrahams 1979; Bakker and Richmond 1982; Richmond and Bakker 1982). There are reasonably short polysynaptic pathways linking neck afferents to limb motoneurons (Kenins et al. 1978; Nakajima et al. 1981), but we do not know whether these pathways are related to the tonic neck reflex. We also do not know whether the reflex is purely tonic, and over how wide a frequency range it can cancel vestibular reflexes. Finally, it is not even clear that the tonic neck reflex is a spinal one (Magnus 1926; Wenzel and Thoden 1977; Nakajima et al. 1981).

We have recently begun a study of neck reflexes acting on the forelimbs (Wilson and Ezure 1983). So far we have emphasized two approaches, investigation of reflex dynamics and a search for spinal interneurons that may be in the reflex pathway. As in our work on the cervicocollic reflex, the stimulus consisted of head rotation in decerebrate, acutely labyrinthecomized cats; rotation, however, was in roll rather than in yaw, and individual rather than multiple sine waves were

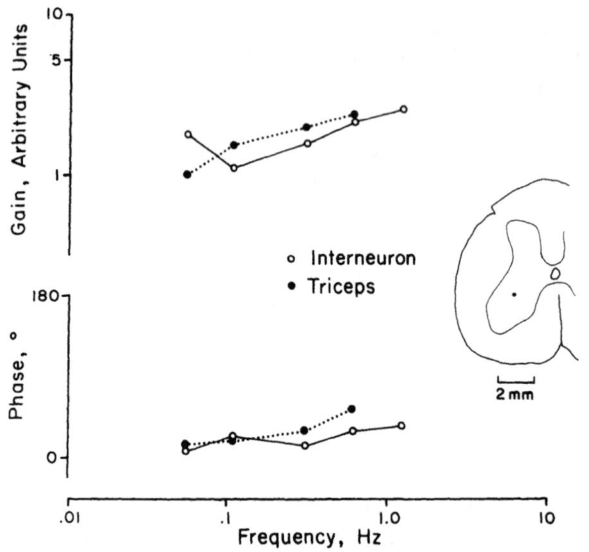

Fig. 3. Response of the left medial head of the triceps branchii and a spinal interneuron on the left side of the spinal cord to head rotation. Both phase and gain are expressed with respect to head position; phase is 0 when chin is maximally rotated to the left. Recording from interneuron was after cat was paralyzed. *Inset* shows location of neuron in C7.

used. EMG was recorded from the long and medial heads of the triceps, bilaterally. Control experiments showed that when the stimuli are large enough (typically 20°–40° at low frequencies up to 0.2 Hz) the reflex is sufficiently linear for a sinusoidal analysis.

Not all cats had a reflex, but in all those that did reflex properties were very similar. As expected, the behavior of the muscles in the two limbs was reciprocal and the "chin limb" extended; we also found that at low frequencies peak reflex activity is in phase with position. The reflex is not purely tonic, however. As illustrated in Fig. 3 (open circles), reflex gain increases and phase sometimes advances with frequency.

When the behavior of the triceps in several cats is averaged, gain is seen to increase with a slope of about 10 dB per decade and phase at 4 Hz leads position by about 30°. I will continue to use the expression "tonic neck reflex" for convenience, but this reflex obviously has some sensitivity to the velocity of the stimulus.

At this point it is interesting to compare the dynamics of the cervicocollic and tonic neck reflexes. Whether the stimulus is rotation in roll or in yaw, the cervicocollic reflex has a steeper gain slope and develops a much greater phase advance with respect to head position (on average 115° in the splenius vs 34° in the triceps at 4 Hz). Do the dynamics differ because the receptors have different properties, or because the two reflexes start with a similar signal but have different central processing? We cannot answer this question until the receptors for the two reflexes are identified with certainty. As I stated earlier, the receptors for the cervicocollic reflex may consist of spindle primary receptors, particularly but not only in the muscle being reflexly activated. Spindles in perivertebral muscles are a reasonable candidate for the receptors of the tonic neck reflex. Their localization would fit with the original finding that the receptors are near the vertebral column (McCouch et al. 1951), and some of these spindles at least seem to have appropriate dynamics, a pronounced tonic discharge with a small phasic component (Fig. 12 in Richmond and Abrahams 1979).

How effective is cancellation of neck and vestibular reflexes? The latter have recently been studied by Schor and Miller (1981) in preparations similar to ours. At low frequencies of roll tilt, 0.2 Hz and below, the reflexes are approximately 180° out of phase and have comparable gain, 1% modulation per degree or less. This means that there is potentially complete cancellation of otolith and neck reflexes. In our recent experiments performed on cats with intact labyrinths it has been possible to rotate the whole animal, the body with respect to a fixed head, or the head, and we have been able to test this prediction. Over the frequency range studied, up to 0.5 Hz, the dynamics and absolute gain of tonic neck reflexes are similar to those in labyrinthectomized cats, and there is often complete cancellation of neck and vestibular reflexes. The platform supporting the head will indeed be stable. At higher frequencies, which we have not been able to test in this fashion, this should no longer be the case because vestibular (now largely canal) reflexes develop a considerable phase advance over the neck reflex and the slopes of the gains are different. It appears that during head movements with a significant content of high-frequency components, and such movements are not uncommon (Donaghy 1980), the platform may no longer be so stable.

The next question is, can neurons in the tonic neck reflex pathway be identified? As a first step we have tried to find spinal neurons responding to head rotation with dynamics similar to those of the tonic neck reflex (Ezure et al. 1983b). Cats were prepared as for muscle recording but a laminectomy exposing about three cervical segments was performed in addition. Once we were satisfied that a tonic neck reflex was present, the animals were paralyzed and maintained on artificial respiration and the cervical spinal column was stabilized for recording from neurons. In successful preparations motion was restricted to the skull-C1 and C1–C2 joints, and there was no motion of the spinal cord itself. It was possible to record extracellularly from spinal interneurons, sometimes for an extended period. So far we have searched from C4 to C8 and found a population of neurons responding to head rotation. Cells were mainly located in laminae 7–8, although some were found more dorsally.

Figure 3 (filled circles) illustrates the behavior of a cell in C7–C8 whose firing increased when the chin was rotated to the ipsilateral side, behavior similar to that of ipsilateral extensor muscles. We have called such cells "type I neurons"; type II neurons, firing when contralateral extensor muscles were excited, were also seen. It is apparent from Fig. 3 that not only are neuron and muscle approximately in phase but the slopes of the gains are also quite similar. Behavior like this was often observed, although on average there was a tendency for the phase of neurons to lead that of muscle.

Neurons that behave like the one in Fig. 3 may be in the pathway to motoneurons. Such a pathway could include propriospinal neurons in the upper cervical segments, or interneurons in the enlargement, or both. Of course, some of these neurons are probably not in motor pathways at all. For example, they may be spinoreticular neurons similar to those that transmit signals of utricular origin to the lateral reticular nucleus (Coulter et al. 1976; Corvaja et al. 1977); those located rostrally may be in the central cervical nucleus (Hirai et al. 1978). Still another possibility is that, like some propriospinal neurons, neurons may project to both forelimb motoneurons and the lateral reticular nucleus (Alstermark et al. 1981). Further experiments are required to provide direct evidence about the projections of our population of neurons.

Acknowledgment. I thank my colleague K. Ezure for allowing me to discuss some unpublished work.

References

Abrahams VC, Richmond FJR, Rose PK (1975) Absence of monosynaptic reflex in dorsal neck muscles of the cat. Brain Res 92: 130–131

Alstermark B, Lindström S, Lundberg A, Sybirska E (1981) Integration in descending motor pathways controlling the forelimb in the cat. 8. Ascending projections to the lateral reticular nucleus from C3–C4 propriospinal neurones also projecting to forelimb motoneurones. Exp Brain Res 42: 282–298

Bakker DA, Richmond FJR (1982) Muscle spindle complexes in muscles around upper cervical vertebrae in the cat. J Neurophysiol 48: 62–74

Bilotto G, Schor RH, Uchino Y, Wilson VJ (1982) Localization of proprioceptive reflexes in the splenius muscle of the cat. Brain Res 238: 217–221

Binder MD, Stuart DG (1980) Motor unit-muscle receptors interactions: design features in the neuromuscular control system. In: Desmedt JE (ed) Spinal and supraspinal mechanisms of voluntary motor control and locomotion. Karger, Basel, pp 72–98

Bizzi E, Dev P, Morasso P, Polit A (1978) Effect of load disturbances during centrally initiated movements. J Neurophysiol 41: 542–556

Botterman BR, Binder MD, Stuart DG (1978) Functional anatomy of the association between motor units and muscle receptors. Am Zool 18: 49–66

Botterman BR, Hamm TM, Reinking RM, Stuart DG (1981) Intramuscular reflex localization in relation to muscle design and action. Soc Neurosci Abst 7: 557

Brink EE, Jinnai K, Wilson VJ (1981) Pattern of segmental monosynaptic input to cat dorsal neck motoneurons. J Neurophysiol 46: 496–505

Corvaja N, Grofova I, Pompeiano O, Walberg F (1977) The lateral reticular nucleus in the cat – 1. An experimental anatomical study of its spinal and supraspinal afferent connections. Neuroscience 2: 537–553

Coulter JD, Mergner T, Pompeiano O (1976) Effects of static tilt on cervical spinoreticular tract neurons. J Neurophysiol 39: 45–62

Creed RS, Denny-Brown D, Eccles JC, Liddell EGT, Sherrington CS (1932) Reflex activity of the spinal cord. Oxford University Press, London

Donaghy M (1980) The cat's vestibulo-ocular reflex. J Physiol 300: 337–351

Ezure K, Fukushima K, Schor RH, Wilson VJ (1983a) Compartmentalization of the cervicocollic reflex in cat splenius muscle. Exp Brain Res 51: 397–404

Ezure K, Schor RH, Wilson VJ (1983b) Response of spinal interneurons to natural stimulation of neck receptors. Soc Neurosci Abst 9: 468

Hirai N, Hongo T, Sasaki S (1978) Cerebellar projection and input organizations of the spinocerebellar tract arising from the central cervical nucleus in the cat. Brain Res 157: 341–345

Kenins K, Kikillus H, Schomburg ED (1978) Short and long latency reflex pathways from neck afferents to hindlimb motoneurons in the cat. Brain Res 149: 235–238

Lindsay KW, Roberts TDM, Rosenberg JR (1976) Asymmetric tonic labyrinth reflexes and their interaction with neck reflexes in the decerebrate cat. J Physiol 261: 583–601

Lüscher HR, Ruenzel P, Henneman E (1980) Topographic distribution of terminals of Ia and group II fibers in spinal cord, as revealed by postsynaptic population potentials. J Neurophysiol 43: 968–985
Magnus R (1926) Some results of studies in the physiology of posture, part I. Lancet 211: 531–536
Magnus R, de Kleijn A (1912) Die Abhängigkeit des Tonus der Extremitätenmuskeln von der Kopfstellung. Pflügers Arch 145: 455–548
Matthews PBC, Stein RB (1969) The sensitivity of muscle spindle afferents to small sinusoidal changes of length. J Physiol 200: 723–743
McCouch GP, Deering ID, Ling TH (1951) Location of receptors for tonic neck reflexes. J Neurophysiol 14: 191–195
Nakajima K, Maeda M, Ishii S, Miyazaki M (1981) Neuronal organization of the tonic neck reflex. Equilibrium res 40: 195–201
Peterson BW, Bilotto G, Fuller JH, Goldberg J, Leeman B (1981a) Interaction of vestibular and neck reflexes in the control of gaze. In: Fuchs AF, Becker W (eds) Progress in oculomotor research. Elsevier, Amsterdam, pp 335–342
Peterson BW, Bilotto G, Goldberg J, Wilson VJ (1981b) Dynamics of vestibulo-ocular, vestibulocollic and cervicocollic reflexes. In: Cohen B (ed) Ann NY Acad Sci 374: 395–402
Poppele RE, Bowman RJ (1970) Quantitative description of linear behavior of mammalian muscle spindles. J Neurophysiol 33: 59–72
Rapoport S (1979) Reflex connections of motoneurons of muscles involved in head movement in the cat. J Physiol 289: 311–327
Richmond FJR, Abrahams VC (1975a) Morphology and enzyme histochemistry of dorsal muscles of the cat neck. J Neurophysiol 38: 1312–1321
Richmond FJR, Abrahams VC (1975b) Morphology and distribution of muscle spindles in dorsal muscles of the cat neck. J Neurophysiol 38: 1322–1339
Richmond FJR, Abrahams VC (1979) Physiological properties of muscle spindles in dorsal neck muscles of the cat. J Neurophysiol 42: 604–617
Richmond FJR, Bakker DA (1982) Anatomical organization and sensory receptor content of soft tisues surrounding upper cervical vertebrae in cat. J Neurophysiol 48: 49–61
Schor RH, Miller AD (1981) Vestibular reflexes in neck and forelimb muscles evoked by roll tilt. J Neurophysiol 46: 167–187
von Holst E, Mittelstaedt H (1950) Das Reafferenzprinzip. Naturwissenschaften 37: 464–476
Wenzel D, Thoden U (1977) Modulation of hindlimb reflexes by tonic neck positions in cats. Pflügers Arch 370: 277–282
Wilson VJ, Ezure K (1983) Dynamics of the tonic neck reflex. Soc Neurosci Abst 9: 468
Wilson VJ, Melvill Jones G (1979) Mammalian vestibular physiology. Plenum, New York
Wilson VJ, Peterson BW (1981) Vestibulospinal and reticulospinal systems. In: Brooks VB (ed) The nervous system II. American Physiological Society, Bethesda, pp 667–702 (Handbook of physiology)

On Neuronal Pathways of Presynaptic Depolarization of Group I Muscle Afferents

E. Jankowska, Göteborg

Introduction

Our knowledge of the central pathways responsible for the depolarization of primary afferent fibers (PAD), the basis of the presynaptic inhibition, has not progressed much since the first studies in which it was analyzed. By comparing the successive reviews of Eccles (1961, 1964), Lundberg (1964), Wall (1964), Schmidt (1971, 1973), Burke and Rudomin (1977), Nicoll and Alger (1979), Levy (1980) and Baldissera et al. (1981), one can see that interests in presynaptic inhibition have been primarily focused on other aspects than its pathways.

Studies of PAD evoked in various groups of afferents, in particular those of Eccles et al. (1962a, b), led to the conclusion that two distinct neuronal systems are responsible for PAD: (a) neuronal systems mediating PAD from group I afferents and (b) those mediating it from cutaneous and other flexor reflex afferents. With regard to the first of these systems, and the only one to be dealt with in this chapter, it was proposed that it operates via at least two interneurons in series and that the first synaptic relay is on cells of the intermediate nucleus of the spinal cord (see below). Both the first-order interneurons and any later-order interneurons of this system were expected to be excited by group I but not cutaneous or group II or III muscle afferents, and to be particularly strongly affected by group I volleys from flexor nerves. This was indicated by the relative potency of PAD evoked from different subgroups of afferents in group I afferents, as summarized in the well known diagrams of Eccles et al. (1963a) and Schmidt (1971, 1973), which are shown in Fig. 1 with some modifications. The cells were further expected to exhibit a clear temporal facilitation and convergence from group I afferents, of the same or different flexor nerves, in view of a pronounced temporal and spatial facilitation of PAD. Figure 2 shows a plot and records of Eccles et al. (1962b) illustrating facilitation of these two kinds.

The postulated interneuronal pathways of PAD of group I afferents are shown in Fig. 3. It was first considered that the same cells (represented by the black D cell in Fig. 3A) would have synaptic endings on both Ia and Ib afferent fibers (Eccles et al. 1962a), but a separate set of neurons due to depolarize Ib afferents was introduced with the diagram shown in Fig. 3B (Eccles 1964). A reservation was, however, made that "much more investigation is required before any conclusion can be reached on the properties and on locations of the interneurons of the two systems that are concerned in producing PADs" (Eccles et al. 1962a).

Experimental Brain Research, Suppl. 9
© Springer-Verlag Berlin · Heidelberg 1984

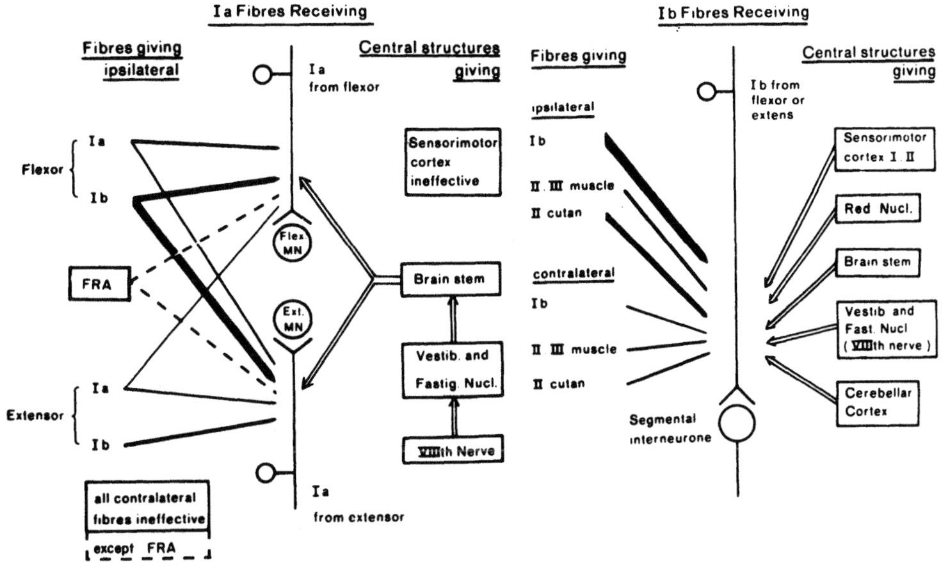

Fig. 1. Origin of primary afferent depolarization (PAD) of Ia and Ib afferents. The origin of PAD from primary afferents is to the *left* and from the supraspinal structures to the *right*. The *thickness of the lines* from the primary afferents indicates the approximate relative contribution of these afferents. The diagrams are reproduced from Figs. 9 and 10 of Schmidt (1973). They are supplemented with data of Hongo et al. (1972) with respect to rubral actions on Ib afferents, and with data of Anden et al. (1966) and Jankowska et al. (1966) with respect to flexor reflex afferent (FRA) (Eccles and Lundberg 1959a) actions on Ia afferents. The FRA effects are indicated by *dashed lines*, as they have been seen only in preparations treated with monoamine precursors. A line is also added to account for PAD evoked from extensor Ia afferents upon flexor Ia afferents. The line was present in the original diagram of Eccles et al. (1963a). It later disappeared (Schmidt 1971), but recent observations based on intra-axonal recording from flexor afferents and on a selective activation of extensor Ia afferents (Brink et al. 1984) show that it should be restored.

Subsequent studies have shown that PAD from group I afferents is supplemented by PAD from other afferents, not only in Ib (Eccles at al. 1962b, 1963a) but also in Ia afferents. Thus Anden et al. (1966) have demonstrated the existence of pathways from flexor reflex afferents (FRA; Eccles and Lundberg 1959a) to Ia afferents in lower lumbar segments. These pathways are now added to Schmidt's diagram in Fig. 1; they are drawn in dashed lines as operating only under conditions when transmission through them is released. It has further been demonstrated that both Ia and Ib afferents are depolarized by descending systems activated in the brain stem and the cerebellum, and that Ib afferents are also depolarized by the sensorimotor cortex and the red nucleus (for references see Schmidt 1971, 1973 and Hongo et al. 1972). Another and particularly interesting group of data concerned the control – both inhibitory and facilitatory – of transmission via interneurons in different PAD pathways (see Table 1). Inhibition of transmission has been found from four fiber systems: from the FRA (Lund et al. 1965), from the brain stem reticular formation (Lundberg and Vyklicky 1966), from the sensorimotor cortex (cf. Lund et al. 1965) and from the red nucleus

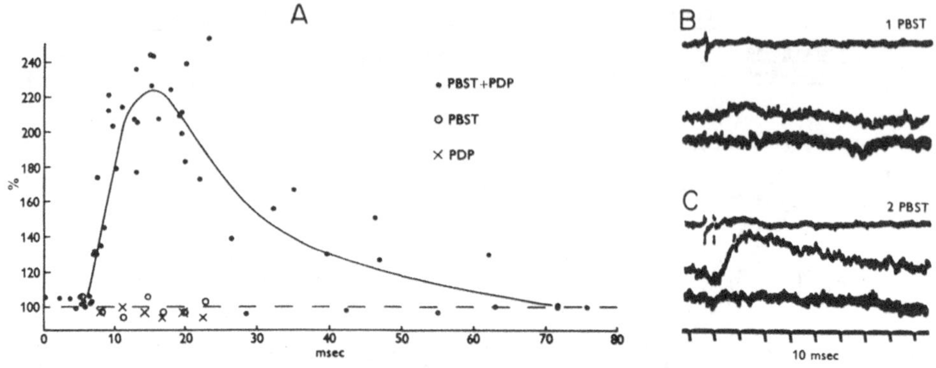

Fig. 2A–C. Illustration of *A* spatial and *B,C* temporal facilitation in primary afferent depolarization pathways. Reproduction of Fig. 12 and of a part of Fig. 14 of Eccles et al. (1962b). The *filled circles* in the diagram in *A* show time course of the increase in the excitability of afferents of flexor digitorum longus nerve stimulated near their terminals in the motor nucleus following single stimuli applied simultaneously to the posterior biceps-semitendinosus (PBSt) and pretibial flexors (PDP) nerves. The *open circles and crosses* show that separate PBSt and PDP volleys produced no increase in the excitability. The records *to the right* are from a quadriceps Ia afferent (*middle traces*), from just outside this afferent (*bottom traces*) and from the surface of the spinal cord (*top traces*). They show that two stimuli applied to PBSt nerves evoked a considerable depolarization of the fiber, while single stimuli were practically without effect.

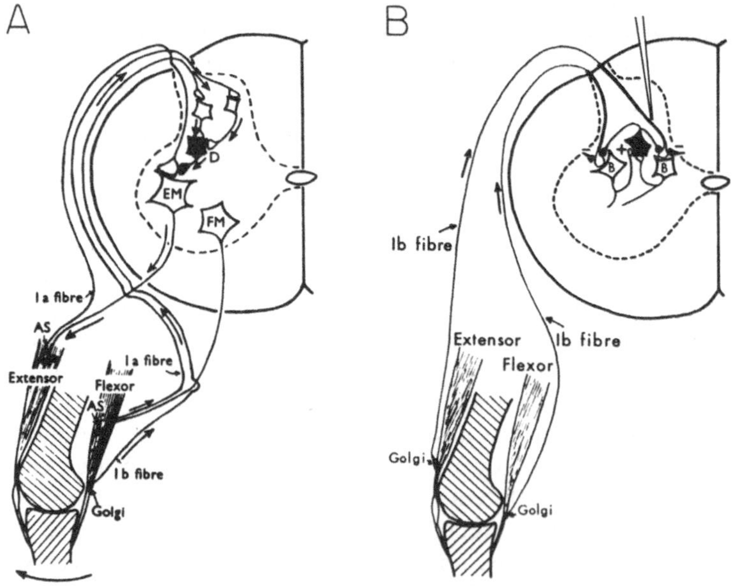

Fig. 3A,B. Schematic diagrams from *A* Eccles et al. (1962a) and *B* Eccles (1964) of the suggested pathways for presynaptic inhibition of Ia and Ib afferents. *EM* and *FM* symbolize extensor and flexor motoneurons. The *white* and *black* interneurons represent the first-(*B*) and last-(*D*)order interneurons in primary afferent depolarization pathways. Note a postulated convergence of flexor Ia and Ib pathways in *A* and of flexor and extensor Ib pathways in *B*. *AS*, muscle spindle.

Table 1. Comparison of patterns of convergence and of axonal projection of interneurons mediating postsynaptic (non-reciprocal) inhibition of motoneurons and of interneurons mediating primary afferent depolarization. The convergence was most often indicated by a mutual facilitation, +, or inhibition, −, of actions evoked by the indicated fiber systems at the interneuronal level. In parantheses are actions present only under certain conditions.

	Interneurones in pathways to:		
Convergence from:	Motoneurones	Ia afferents	Ib afferents
Ia and Ib afferents of extensors	+	0	0
Ia and Ib afferents of flexors	?	+	0
Various Ia afferents	+	+	0
Various Ib afferents	+	+	+
Group Ib and cutaneous afferents	+	0	+
Group Ib and FRA	+	0 (−)	+
Group Ib and cortico-spinal fibres	+	0 (−)	+
Group Ib and rubro-spinal	+	0 (±)	+
Group Ib and reticulo-spinal	0 (−)	+ (−)	+
Group Ib and vestibulo-spinal	0	+	+
Projection to Clarke's column	+	0	0

(Hongo et al. 1972); in all four cases it involved only interneurons in PAD pathways from Ia to Ia afferents. The facilitatory control appeared to be more differentiated, with volleys in rubrospinal tract fibers facilitating transmission from both Ia and Ib afferents (Hongo et al. 1972), while facilitation from the corticospinal tract fibers only involved interneurons in PAD pathways from Ib afferents to Ib afferents (Carpenter et al. 1963). However, new observations pertinent to the original hypotheses of Eccles and his colleagues on neuronal organization of PAD pathways from group I to group I afferents were made only very recently, and this is a particularly good opportunity to summarize them.

Relative Contribution of Specific Neuronal Pathways and of Other Factors to PAD of Group I Origin

Several factors have been regarded as adding to the depolarization evoked by specialized neuronal pathways (for review and references see Burke and Rudomin 1977; Levy 1980; Rudomin et al. 1981). However, the relative contribution of these factors and of neuronal pathways to PAD of group I origin has been investigated only recently (Rudomin et al. 1981; Rudomin and Jankowska 1981; Rudomin et al. 1983; Jimenez et al. 1983). In one approach to this problem, Rudomin and his colleagues compared PAD evoked in group I afferents to PAD in other fibers which terminate nearby. Since these fibers should be similarly affected by such non-specific changes in their surroundings as changes in K^+ concentration or electric field potentials, any major differences in PAD induced in group I fibers, as compared to other fibers, were taken to indicate actions of specific neuronal pathways. In one series of experiments single group I and rubrospinal tract fibers were simultaneously stimulated in laminae V–VI of Rexed, in the area where they both have terminal branches (Rudomin and Jankowska 1981; Rudomin et al. 1981, 1983). Volleys in flexor Ia afferents increased the excitability of Ia afferents with virtually no effect on the rubrospinal tract fibers, while volleys in cutaneous afferents which increased the excitability of the rubrospinal tract fibers decreased the excitability of Ia afferents. Similar observations were made with respect to presynaptic inhibition of transmission from group I afferents and descending tract fibers to the same motoneurons (Rudomin et al. 1975). These observations thus add to the list of previously established cases of PAD which may be attributed to specific neuronal pathways (see Fig. 1). The second approach of Rudomin and his colleagues was more direct. They examined the relationship between PAD of single fibers and changes in K^+ concentration, as well as some other factors (Jimenez et al. 1983; Rudomin et al. 1981). The study has shown a correlation between PAD of group I afferents and changes in K^+ concentration only in cases of a very weak PAD evoked by volleys in mixed or cutaneous nerves, but not when strong PAD was induced by volleys in group I afferents. The incompatibility of the time course of PAD with the time course of effects induced by much slower developing changes in K^+ concentration, or by shorter-lasting changes in electric field potentials, has been indicated previously (see Burke and Rudomin 1977). All observations, including previous observations that the same conditioning stimuli (applied to peripheral nerves or to supraspinal structures) have different effects upon Ia and Ib afferents, or upon Ia afferents of flexors and extensors terminating in the same area, thus point to neuronal pathways as being the main source of PAD of group I origin.

Number of Neurons in Series in the Specific Neuronal Pathways of PAD Evoked from Group I Afferents

The shortest reported latencies of presynaptic inhibition evoked from group I afferents were 2.5 ms (see Fig. 1 of Eccles et al. 1961a) and the shortest segmental latency of intra-axonally recorded PAD measured from the first effective stimulus was similarly 2.5 ms (see Fig. 14K of Eccles et al. 1962b, which is reproduced in Fig. 2C of this chapter). The minimal latencies of PAD of group I origin were, nevertheless, estimated at about 4 ms, taking into account the earliest clear changes in the excitability of fibers tested according to Wall's (1958) technique. Since excitability measurements may not be the most sensitive means of detecting PAD, we made a number of measurements of the segmental latencies of PAD from intra-axonal records and from dorsal root potentials, DRPs (Jankowska et al. 1981d; Brink et al. 1984). The latencies were measured between the time of arrival of the afferent volleys at the spinal cord and the onset of the particular components of the PAD which were evoked by them. Since large potentials with as distinct an onset as in Fig. 2C were rarely seen, the records were routinely averaged. Figure 4 shows examples of the DRPs and the distribution of three samples of DRP and PAD latencies. To correct the latency measurements in records being made at a certain distance from the terminals, we subtracted 0.2 ms from their onset to account for the electrotonic delay. The minimal values of the actual latencies of PAD thus obtained were 1.5–1.8 ms, fully compatible with the trisynaptic

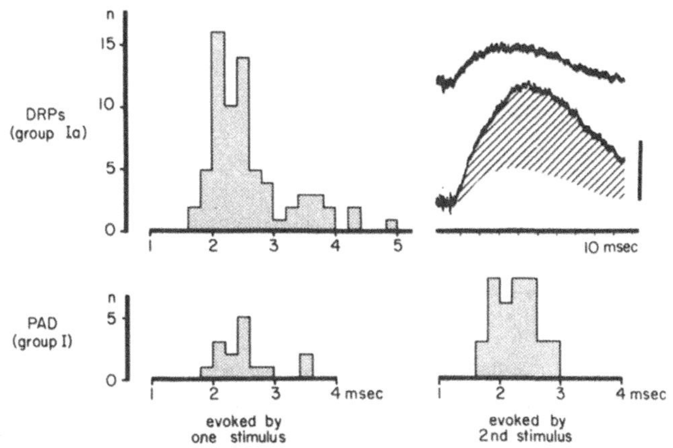

Fig. 4. Distribution of minimal estimates of latencies of DRPs and of primary afferent depolarization recorded intra-axonally. They were evoked by electrical stimuli subthreshold for Ib afferents of quadriceps and posterior biceps and semitendinosus. Sample records of DRPs evoked by single and double stimuli applied to group I afferents of posterior biceps and semitendinosus are at the *top right*. The additional depolarization due to the second stimulus is indicated by *hatching*. The voltage calibration is 2 mV. (Modified from Jankowska et al. 1981d and Brink et al. 1984)

linkage in the pathways and with the mediation of PAD of group I origin by the minimal number of two interposed interneurons in series, as postulated by Eccles et al. (1962a, b).

Location of Interneurons Mediating PAD of Group I Afferents

The intermediate nucleus was originally proposed as the first relay in PAD pathways since it was, at the time, the only known relay of both group Ia and Ib impulses on the way to interneurons (Eccles et al. 1962a, b). Since then neurons with group I input have also been found at other locations: in lamina VII (Hultborn et al. 1971), lamina IV (Jankowska et al. 1979) and lamina VIII (P.J. Harrison, E. Jankowska and D. Zytnicki, unpublished observations). However, support for the original proposal has come from experiments in which we compared the effectiveness of weak electrical stimuli applied at different locations in the spinal cord in evoking PAD of group I afferents to define the most likely location of the responsible interneurons (Jankowska et al. 1981d). The most effective were stimuli applied within the intermediate nucleus (Fig. 5C), at depths 2–3 mm from the surface (Fig. 5A), where the largest focal group I potentials were recorded. The minimal latencies of PAD evoked from this area were 0.5–1.0 ms shorter than the minimal segmental latencies of PAD evoked by peripheral afferent volleys, and as short as 0.8 ms (Fig. 5B). The shortest latencies are clearly compatible with the monosynaptically evoked effects and the longer ones with disynaptic effects. The PAD evoked with these latencies was therefore attributed to directly activated last- and first-order interneurons of the trisynaptic PAD pathways, and led to the conclusion that the two groups of interneurons might be located within the same area. On the basis of such observations it is, of course, impossible to exclude that

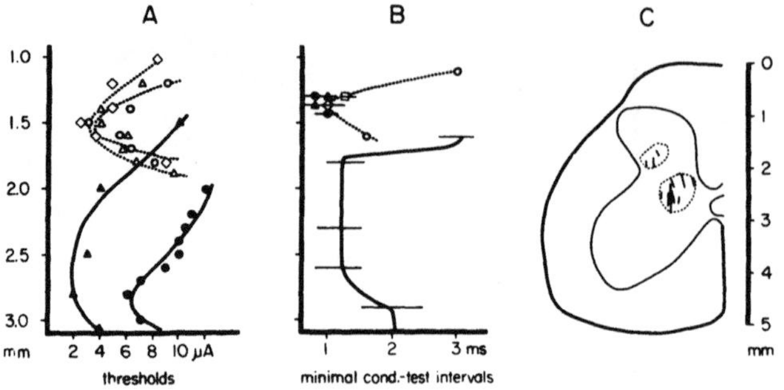

Fig. 5A–C. Optimal spinal cord regions for evoking primary afferent depolarization (PAD) by intraspinal stimuli. *A* Thresholds for an increase in excitability of cutaneous (*open symbols*) and group I (*filled symbols*) afferents produced by intraspinal stimuli as a function of depth. *B* Minimal intervals between conditioning and test stimuli at which an excitability increase was detected; *horizontal lines* connect the shortest effective and the longest ineffective intervals. *C* Spinal cord regions from which PAD was evoked at lowest thresholds and with shortest intervals between the conditioning and test stimuli. *Upper and lower areas enclosed by dashed lines* indicate such regions for cutaneous and group I afferents respectively. *Vertical and diagonal lines* within these areas indicate the most effective parts of the individual electrode tracks. (Modified from Jankowska et al. 1981d)

the somata of either the first- or the last-order interneurons are located outside the intermediate nucleus and that only their axons cross it. However, if this were the case, the somata of these neurons would be likely to be located close to the intermediate nucleus, since the thresholds for their direct activation by stimuli applied in laminae V-VI could be easily increased, or decreased, by peripheral nerve stimulation, which could evoke postsynaptic inhibition or excitation (Rudomin et al. 1983). In contrast, thresholds of PAD evoked by intraspinal stimuli applied in the motor nucleus, only about 1–2 mm ventrolaterally from the intermediate nucleus, were unaffected by the same conditioning stimuli.

Comparison of Integrative Functions of Interneurons Mediating Presynaptic and Postsynaptic Actions from Group I Afferents

In one of their first articles on spinal interneurons, Eccles et al. (1954) raised the question of the raison d'être of these neurons. They considered two main functions of interneurons: (a) a co-ordinative function and (b) the function of providing a change in the type of neuron, or transmitter, along the pathway. The pathway primarily under consideration was that of the reciprocal inhibition, which clearly required an interneuron capable of hyperpolarizing motoneurons, and a co-ordinative function of these neurons appeared at first rather unlikely. Subsequent studies, from that of Eccles and Lundberg (1959b) to the most recent ones, have stressed, however, that the co-ordinative function is the main function of both these and any other known interneurons in pathways to motoneurons (see Lundberg 1979). With respect to the interneurons mediating PAD, very little is known about their specific co-ordinative functions. Only scarce indications of such functions have been reported (Jänig et al. 1968; Hongo et al. 1972), and PAD has most often been regarded in terms of a rather non-specific feedback, or damping of effects of rather large classes of afferents (see Eccles 1964). The most specific co-ordinative functions of neuronal systems of PAD of group I afferents so far considered (Hongo et al. 1972) have been the following. One concerned PAD of descending origin. It was proposed to assist the selection of one of two uses of spinal interneurons, either in segmental reflexes or in descending control, strong PAD of descending origin favoring the use of particular spinal interneurons by descending rather than segmental systems. Another was to control transmission in reflex paths from small groups of muscles in order to assist the interneuronal control systems in selecting neurons subserving different movements. It has therefore become of interest to compare properties of interneurons in pathways of postsynaptic inhibition or excitation of motoneurons, with clearly specific co-ordinative functions, to the properties of interneurons in pathways of PAD of group I origin.

Convergence upon Interneurons in PAD Pathways

The depolarization of Ia and Ib afferents is evoked from several fiber systems (see Fig. 1), but only occasional observations have previously been made to investigate whether the involved neuronal pathways are separate or shared. The first observations were that simultaneous stimulation of various Ia flexor afferents (Eccles et al. 1962b; see Fig. 2) greatly enhances their actions by spatial facilitation – the evidence that these actions are evoked by some common interneurons. Similar observations were made on actions of Ib afferents and cortico- or rubrospinal tract fibers (Carpenter et al. 1963; Hongo et al. 1972). The postulated convergence of Ia and Ib afferents of flexor (Fig. 3A) or of Ib afferents of flexors and extensors (Fig. 3B) on the same interneurons of PAD pathways did, on the other hand, anticipate the results of a later study. In this study (Brink et al. 1984), we made a survey of shared PAD pathways from and to different subgroups of group I afferents. Using spatial facilitation of transmission as an indication of interneurons used in common, we found shared pathways from various Ia afferents (whether of flexors or extensors) and from various Ib afferents (similarly of either flexors or extensors). We found them also from any tested Ia afferents and flexor Ib afferents, but not from Ia afferents and extensor Ib afferents. These data are summarized in Table 2 and illustrated in Fig. 6A–C.

The common use of the same interneurons in PAD pathways from different group I afferents, in particular from flexors and extensors, and from muscle spindles and tendon organs, is in sharp contrast to the much more selective use of interneurons in pathways of Ia reciprocal inhibition by afferents from only one functional group of muscles (Laporte and Lloyd 1952; Eccles and Lundberg 1958; Hultborn et al. 1971). It is, on the other hand, comparable to the situation in the pathways of excitation and of non-reciprocal inhibition from Ia and Ib afferents (Eccles et al. 1957; Jankowska et al. 1981b, c; Harrison et al. 1983). However, different groups of afferents co-excite neurons in the pathways to motoneurons and those mediating PAD. Interneurons in PAD pathways are generally much more extensively used by flexor afferents, but the most striking difference is with respect

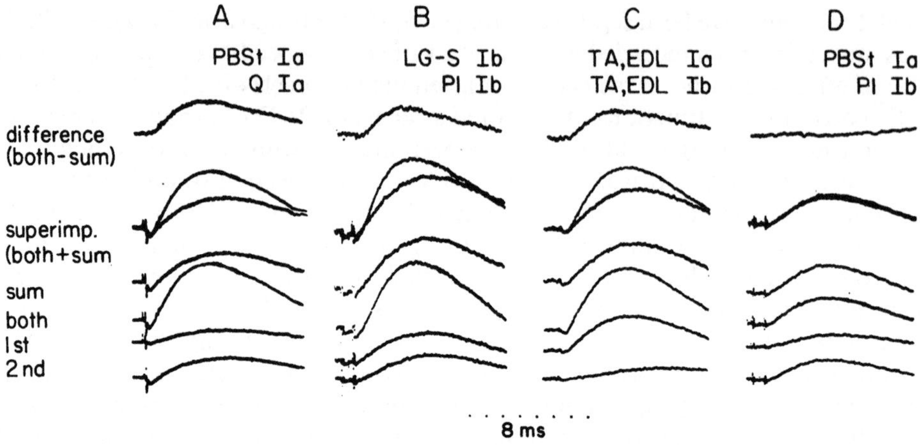

Fig. 6A–D. Examples of spatial facilitation in primary afferent depolarization pathways from different Ia afferents (*A*), different Ib afferents (*B*), and Ib afferents of flexors (*C*) but not of extensors (*D*) and Ia afferents. *From the lowest traces upwards* the records are: DRPs evoked by selective stimulation of Ia or Ib afferents of the first or the second of two nerves, as indicated above, alone, and by simultaneous stimulation of both of them. The next records are: the sum of the first and second traces, the superimposed records of *both* and *sum* traces and the result of subtracting the *sum* trace from the *both* trace. *PBSt*, posterior biceps-semitendinosus; *Q*, quadriceps; *LG-S*, lateral gastrochemius-soleus; *PL*, plantaris; TA, anterior tibial; *EDL*, extensor digitorum longus. (Modified from Figs. 4, 7 and 8 of Brink et al. 1984)

Table 2. Occurrence (+) and lack of occurrence (o) of spatial facilitation in primary afferent depolarization (PAD) pathways from Ia and Ib afferents of the listed nerves. The spatial facilitation was tested on DRPs or on PAD recorded from identified Ia and Ib afferents. The selective stimulation of Ia afferents of Q and PBSt was by weak electrical stimuli and that of other muscles by their stretches. The technique of Coppin et al. (1970) was adopted for the selective activation of Ib afferents. (Modified from Brink et al. 1984)

With	Ia PBSt	Ia GS, Pl stretch	Ia TA, EDL stretch	Ib GS, Pl	Ib TA, EDL
Ia Q	+	+	+	o	+
Ia PBSt		+	+	o	+
Ia GS, Pl stretch			+	o	+
Ia TA, EDL stretch				o	+
Ib GS, Pl				+	+

to the pathways from extensor Ia and Ib afferents. They are separate in the case of PAD (Brink et al. 1983b) and commonly shared on the way to motoneurons (Jankowska and McCrea 1983; Harrison et al. 1983).

The differences in the combinations of afferents which have access to these two classes of neurons may be particularly meaningful from the point of view of co-ordinative functions of PAD interneurons, and therefore will be worth further studies. The dominating presynaptic actions of flexor group I afferents might, for instance, be used in one of two ways: either to assist the extensor-dominated pattern of Ib postsynaptic actions (via PAD of flexor afferents) or else to damp the extensor actions (via PAD of extensor afferents). It would therefore be of interest to find out whether or not the same interneurons mediate PAD of flexor and extensor afferents. There may exist specialized subgroups of interneurons mediating PAD of either flexor or extensor afferents; these interneurons could be recruited by some particular combinations of afferents and descending fibers, assisting only particular kinds of movement.

Presynaptic actions evoked by extensor afferents could also be of importance for specific co-ordinative PAD function. Separate pathways of PAD of Ia and of Ib origin might, for instance, be used to change the balance between Ia and Ib input to interneurons with postsynaptic actions upon motoneurons. Some of the laminae V–VI interneurons have been recently identified as interneurons mediating postsynaptic inhibition of motoneurons from Ia and Ib afferents (Hongo et al. 1983a, b; Brink et al. 1983) which allowed us to investigate their input in some detail. About half of their population were found to be excited by both Ia and Ib afferents and the other half by only Ib afferents (Harrison and Jankowska in press). One could, therefore, expect that various subpopulations of these neurons may be activated under conditions when either Ia or Ib actions are depressed by presynaptic inhibition. For instance, under conditions of strong activation of tendon organs and weak activation of muscle spindles of extensors (as in fast isotonic movements, cf. Prochazka 1981; Appenteng et al. 1982), the Ib excited PAD interneurons would evoke presynaptic depolarization of both Ia and Ib afferents of these muscles (see Fig. 1). However, since Ia actions on laminae V–VI interneurons tend to be weaker than Ib actions, a reduction in Ia input by PAD of Ia afferents could leave the Ib input as the only one of importance. With weak or more restricted activation of tendon organs, combined with a stronger activation of muscle spindle afferents (as in slow isometric movements), the contribution of Ia input to the activation of the laminae V–VI interneurons would be greately enhanced. This would occur in particular because PAD of extensor Ia afferents evoked from extensor Ia afferents is only marginal (cf. Brink et al. 1984 and Fig. 1), and also because in the situation of reduced Ib input the PAD from Ib afferents would be smaller. Furthermore, since the Ia afferents are in addition a potent source of inhibition of laminae V–VI interneurons (Jankowska et al. 1981a), different subpopulations of interneurons could be selected. This arrangement would, of course, be meaningful only if individual laminae V–VI interneurons were functionally specialized and subserved different movements. It is clear that we must know much more about the organization of both these neurons, and of neurons mediating PAD, to understand their interactions.

The selective PAD actions from and to different afferents (summarized in Table 1) should aid in the investigation of which particular PAD actions are evoked by different and which by the same subgroups of PAD interneurons. For instance, the depression by cutaneous and cortico- and rubrospinal volleys of PAD of Ia afferents, with a concomitant facilitation of PAD of Ib afferents, led Rudomin et al. (1983) to conclude that different neurons are interposed in PAD pathways to these afferents. Convergence of Ia and of flexor Ib afferents but not of extensor Ib afferents would further indicate that at least two sets of neurons mediate PAD of extensor Ia afferents (see Brink et al. 1984).

It should also be noted that some information is already available as to which of the interneurons in PAD pathways are excited or inhibited by other than group I fiber systems (Rudomin et al. 1983). Stimulation within the brain stem has effects on the last-order interneurons, whereas cutaneous, rubrospinal, and corticospinal effects are on the first-order interneurons. The advantages of such a differentiation of the input to these interneurons remain, however, to be investigated.

Axonal Projections

All the spinal interneurons that have been identified so far are funicular cells: Renshaw cells, lamina VII interneurons mediating Ia reciprocal inhibition and laminae V–VI interneurons mediating Ib and Ia non-reciprocal inhibition. The latter have been found to project rostrally as far as to L2–L3 segments, and to terminate in Clarke's column (Hongo et al. 1983a, b), as indicated in the upper part of Fig. 7. It has therefore become of interest to determine whether interneurons mediating PAD of group I afferents are also funicular cells, and whether their co-ordinative functions involve afferents at different spinal levels. In particular PAD of group I afferents in Clarke's column (Eccles et al. 1961b, 1963b; Jankowska and Padel 1984) might be evoked by the same interneurons which mediate PAD of group I terminals in lower lumbar segments. To test this hypothesis we used two different approaches. In one series of experiments, changes in excitability of terminals of group I afferents stimulated in Clarke's column in L3 (following conditioning stimulation of group I afferents of hind limb muscles) were used as a measure of PAD (Jankowska and Padel 1984). Excitability of afferents was tested before and after a lesion of the spinal cord at L4 which spared only the dorsal columns, and was made rostral to the level of entry of the stimulated hind limb afferents. Since the effectiveness of PAD found after these lesions did not show any marked deterioration, it appeared to be evoked primarily by interneurons located in the same segments as Clarke's column, interneurons activated by axon collaterals of group I afferents given off at this level.

In the second series of experiments the dorsal columns (at L4) were transected just caudal to Clarke's column and attempts were made to evoke DRPs in fibers of L6 or L7 dorsal roots by intraspinal stimuli applied above the level of the transection (Harrison and Jankowska 1984). Any DRPs then obtained would be attributable to funicular neurons with axons running in the lateral or ventral funiculi. Under these conditions we failed to detect any DRPs following stimuli applied within Clarke's column, or its surroundings, and DRPs evoked from other regions rostral to the lesion appeared to be mediated by neurons in PAD pathways from high-threshold muscle and cutaneous afferents but not from group I afferents. The two series of experiments thus lead to the conclusion that PAD of group I origin is evoked by interneurons with fairly local actions, as indicated in the lower diagram of Fig. 7.

Future studies will undoubtedly reveal further differences in the input and in the axonal projections of interneurons in reflex pathways to motoneurons and in PAD pathways and help in a better understanding of their role in motor control. However, it appears that the already disclosed differences are so marked that it

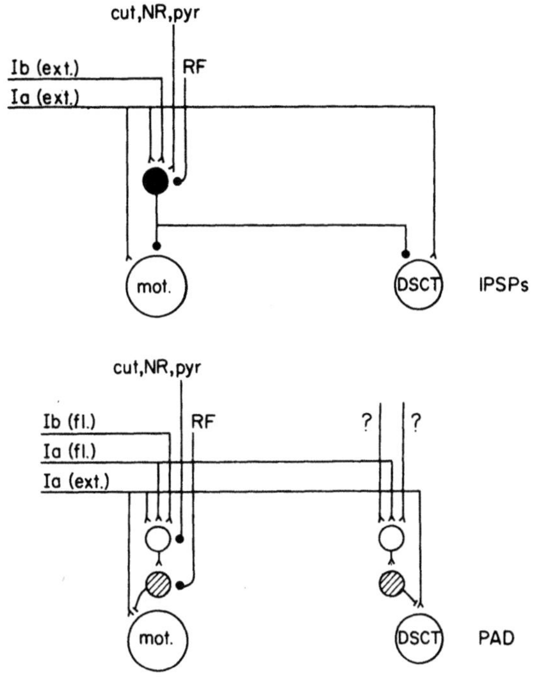

Fig. 7. Patterns of input and types of axonal projection which may be used to differentiate interneurons mediating postsynaptic inhibition of motoneurons of Ia-Ib origin from interneurons mediating primary afferent depolarization of Ia-Ib origin. Interneurons in excitatory pathways to motoneurons would have the same input as those in inhibitory pathways but lack the ascending axonal projection upon dorsal spinocerebellar tract (*DSCT*) cells in Clarke's column (see text). *cut,* cutaneous; *pyr,* pyramidal tract; *NR,* red nucleus; *RF,* reticular formation; ⅄ excitatory synapses; ↓ inhibitory synapses; ⊥ axoaxonic synapses

may now only be a matter of time before interneurons in PAD pathways from group I afferents are finally identified, and a more detailed analysis of these neurons at a cellular level becomes possible.

Acknowledgments. I wish to thank Drs. E. Brink, P.J. Harrison, D.A. McCrea, P. Rudomin and B. Skoog for their permission to use some of the unpublished material.

References

Anden NE, Jukes MGM, Lundberg A, Vyklicky L (1966) The effect of Dopa on the spinal cord. 3. Depolarization evoked in the central terminals of ipsilateral Ia afferents by volleys in the flexor reflex afferents. Acta Physiol Scand 68: 322–336

Appenteng K, Prochazka A, Proske U, Wand P (1982) Effect of fusimotor stimulation on Ia discharge during shortening of cat soleus muscle at different speeds. J Physiol (Lond) 329: 509–526

Baldissera F, Hultborn H, Illert M (1981) Integration in spinal neuronal systems. In: Brooks VB (ed.) The nervous system. Motor control. American Physiology Society, Bethesda, pp 509–595 (Handbook of physiology, section 1, vol II.)

Brink E, Harrison PJ, Jankowska E, McCrea DA, Skoog B (1983) Postsynaptic potentials in a population of motoneurones following activity of single interneurones. J Physiol (Lond) 343: 341–359
Brink E, Jankowska E, Skoog B (1984) Convergence onto interneurones subserving primary afferent depolarization of group I afferents. J Neurophysiol 51: 432–449
Burke RE, Rudomin P (1977) Spinal neurons and synapses. In: Brooks VB (ed) The nervous system. American Physiology Society, Bethesda, pp 877–944 (Handbook of Physiology, section 1, vol 1, part 2)
Carpenter D, Lundberg A, Norrsell U (1963) Primary afferent depolarization evoked from the sensorimotor cortex. Acta Physiol Scand 59: 126–142
Coppin CML, Jack JJB, MacLennan CR (1970) A method for the selective activation of tendon organ afferent fibres from the cat soleus muscle. J Physiol (Lond) 210:18–20P
Eccles JC (1961) The mechanism of synaptic transmission. Ergebnisse der Physiologie, biologischen Chemie und experimentellen Pharmakologie 51:299–430
Eccles JC (1964) Presynaptic inhibition in the spinal cord. In: Eccles JC, Schade PJ (eds) Physiology of spinal neurons. Elsevier, Amsterdam, pp 65–91 (Progress in brain research, vol 12)
Eccles JC, Fatt P, Landgren S, Winsbury GJ (1954) Spinal cord potentials generated by volleys in the large muscle afferents. J Physiol (Lond) 125: 590–606
Eccles JC, Eccles RM, Lundberg A (1957) Synaptic action of motoneurones caused by impulses in Golgi tendon organ afferents. J Physiol (Lond) 138: 227–252
Eccles JC, Eccles RM, Magni F (1961a) Central inhibitory action attributable to presynaptic depolarization produced by muscle afferent volleys. J Physiol (Lond) 159: 147–166
Eccles JC, Oscarsson O, Willis WD (1961b) Synaptic action of group I and II afferent fibres of muscle on the cells of the dorsal spinocerebellar tract. J Physiol (Lond) 158: 517–543
Eccles JC, Kostyuk PG, Schmidt RF (1962a) Central pathways responsible for depolarization of primary afferent fibres. J Physiol (Lond) 161: 237–257
Eccles JC, Magni F, Willis WD (1962b) Depolarization of central terminals of group I afferent fibres from muscle. J Physiol (Lond) 160: 62–93
Eccles JC, Schmidt RF, Willis WD (1963a) Depolarization of central terminals of group Ib afferent fibres of muscle. J Neurophysiol 26: 1–27
Eccles JC, Schmidt RF, Willis WD (1963b) Inhibition of discharge into the dorsal and ventral spino-cerebellar tract. J Neurophysiol 26: 635–645
Eccles RM, Lundberg A (1958) Integrative pattern of IA synaptic actions on motoneurones of hip and knee muscles. J Physiol (Lond) 144: 271–298
Eccles RM, Lundberg A (1959a) Synaptic actions in motoneurones by afferents which may evoke the flexion reflex. Arch Ital Biol 97: 199–221
Eccles RM, Lundberg A (1959b) Supraspinal control of interneurones mediating spinal reflexes. J Physiol (Lond) 147: 565–584
Harrison P, Jankowska E (1984) Do interneurones in lower lumbar segments contribute to the presynaptic depolarization of group I muscle afferents in Clarke's column? Brain Res 295: 203–210
Harrison P, Jankowska E, Johannisson T (1983) Shared reflex pathways of group I afferents of different cat hind limb muscles. J Physiol (Lond) 338: 113–127
Hongo T, Jankowska E, Lundberg A (1972) The rubrospinal tract. III. Effects on primary afferent terminals. Exp Brain Res 15: 39–53
Hongo T, Jankowska E, Ohno T, Sasaki S, Yamashita M, Yoshida K (1983a) Inhibition of dorsal spino-cerebellar tract cells by interneurones in upper and lower lumbar segments in the cat. J Physiol (Lond) 342: 145–160
Hongo T, Jankowska E, Ohno T, Sasaki S, Yamashita M, Yoshida K (1983b) The same interneurones inhibit dorsal spinocerebellar tract cells and lumbar motoneurones in the cat. J Physiol (Lond) 342: 161–180
Hultborn H, Jankowska E, Lindstrom S (1971) Recurrent inhibition of interneurones monosynaptically activated from group Ia afferents. J Physiol (Lond) 215: 316–363
Jänig W, Schmidt RF, Zimmermann M (1968) Two specific feedback pathways to the central afferent terminals of phasic and tonic mechanoreceptors. Exp Brain Res 6: 116–129
Jankowska E, McCrea D (1983) Shared neuronal pathways of reflex actions from Ib tendon organ afferents and Ib muscle spindle afferents in the cat. J Physiol (Lond) 338: 99–111

Jankowska E, Padel Y (1984) On the origin of presynaptic depolarization of group I muscle afferents in Clarke's column in the cat. Brain Res 295: 203–210
Jankowska E, Lund S, Lundberg A (1966) The effect of Dopa on the spinal cord. 4. Depolarization evoked in the central terminals of contralateral Ia afferent terminals by volleys in the flexor afferents. Acta Physiol Scand 68: 337–341
Jankowska E, Rastad J, Zarzecki P (1979) Segmental and supraspinal input to cells of origin of non-primary fibres in the feline dorsal columns. J Physiol (Lond) 290: 185–200
Jankowska E, Johannisson T, Lipski J (1981a) Common interneurones in reflex pathways from group Ia and Ib afferents of ankle extensors in the cat. J Physiol (Lond) 310: 381–402
Jankowska E, McCrea D, Mackel R (1981b) Pattern of "non-reciprocal" inhibition of motoneurones by impulses in group Ia muscle spindle afferents. J Physiol (Lond) 316: 393–409
Jankowska E, McCrea D, Mackel R (1981c) Oligosynaptic excitation of motoneurones by impulses in group Ia muscle spindle afferents. J Physiol (Lond) 316: 411–425
Jankowska E, McCrea D, Rudomin P, Sykova E (1981d) Observations on neuronal pathways subserving primary afferent depolarization. J Neurophysiol 46: 505–516
Jimenez I, Rudomin P, Solodkin M, Vyklicky L (1983) Specific and potassium components in the depolarization of the Ia afferents in the spinal cord of the cat. Brain Res 272: 179–184
Laporte Y, Lloyd DPC (1952) Nature and significance of the reflex connections established by large afferent fibres of muscles. Am J Physiol 169: 609–621
Levy RA (1980) Presynaptic control of input to the central nervous system. Can J Physiol Pharmacol 58: 751–766
Lund S, Lundberg A, Vyklicky L (1965) Inhibitory action from the flexor reflex afferents on transmission to Ia afferents. Acta Physiol Scand 64: 345–355
Lundberg A (1964) Supraspinal control of transmission in reflex paths to motoneurones and primary afferents. In: Eccles JC, Schade JP (eds) Prog Brain Res 12: 197–221
Lundberg A (1979) Multisensory control of spinal reflex pathways. Prog Brain Res 50: 11–28
Lundberg A, Vyklicky L (1966) Inhibition of transmission to primary afferents by electrical stimulation of the brain stem. Arch Ital Biol 104: 86–97
Nicoll RA, Alger BE (1979) Presynaptic inhibition: transmitter and ionic mechanisms. Int Rev Neurobiol 21: 217–258
Prochazka A (1981) Muscle spindle function during normal movement. Int Rev Physiol 25: 47–90
Rudomin P, Jankowska E (1981) Presynaptic depolarization of terminals of rubrospinal tract fibres in intermediate nucleus of cat spinal cord. J Neurophysiol 46: 517–531
Rudomin P, Nunez R, Madrid J (1975) Modulation of synaptic effectiveness of Ia and descending fibers in the spinal cord. J Neurophysiol 39: 1181–1195
Rudomin P, Engberg I, Jimenez I (1981) Mechanisms involved in presynaptic depolarization of group I and rubrospinal fibres in cat spinal cord. J Neurophysiol 46: 532–548
Rudomin P, Jimenez I, Solodkin M, Duenas S (1983) Site of action of segmental and descending control of transmission on pathways mediating PAD of Ia and Ib afferent fibers in cat spinal cord. J. Neurophysiol 50: 743–769
Schmidt RF (1971) Presynaptic inhibition in the vertebrate central nervous system. Ergebnisse der Physiologie, biologischen Chemie und experimentellen Pharmakologie. Springer, Berlin Heidelberg New York, pp 20–101
Schmidt RF (1973) Control of the access of afferent activity to somatosensory pathways. In: Iggo A (ed) Somatosensory systems. Springer, Berlin Heidelberg New York, pp 151–206 (Handbook of sensory physiology, vol II)
Wall PD (1958) Excitability changes in afferent fibre terminations and their relation to slow potentials. J Physiol 142: 1–21
Wall PD (1964) Presynaptic control of impulses at the first central synapses in the cutaneous pathways. Prog Brain Res 12: 92–114

Spinal Integration of the Command for Respiratory Movements

T.A. Sears, London

We each experience a sense of privilege at being present on this occasion to honour Sir John as scientist, colleague and friend. Mindful of the need to do justice to the occasion, I shall simply present two aspects of respiratory motoneurone performance which encapsulate the influence which he has had on my own principal field of research, the physiology of the movements of the rib cage; this influence starts with the background of ideas that prompted me to seek the opportunity of joining the "Canberra School".

Introduction

The neurophysiology of respiratory movements is a field of research which has only in recent years found affinity with that of the rest of the somatic musculature, but the reasons for this are as much historic as scientific. Thus concepts which are absolutely central to work on the limbs, such as the integrative action of the motoneuron itself and the spinal integration of segmental and suprasegmental mechanisms, had been paid scant attention at the time I started my own work on the thoracic spinal cord, a region referred to as a "no man's land" by one distinguished clinical neurologist! These concepts have epistemological roots planted by Sherrington at the turn of the century, and through the work particularly of the Oxford school were brought into recent times by ,notably, Eccles, Lloyd and Lundberg. In contrast, work on respiratory movements has been, and continues to be, dominated by the search for a "respiratory centre". This is deemed to exist to coordinate the activity of motoneurons at widely differing levels of the neuraxis, to permit integration of the diverse excitatory and inhibitory pathways which influence respiratory movements, and to match them to the metabolic and behavioural demands of the moment (cf. Pitts 1946). It would seem that most authors find the use of the term "respiratory centre" a convenient shorthand for the preceding sentence, but I have always doubted whether such a system is "localizable" from a neurophysiological point of view, as many claim. The dominance of this concept stems very much from its age, since the idea that in the medulla there exists a region *"dans lequel reside le premier mobile de la respiration"* (in which resides the origin of the respiration) was advanced by Le Gallois (1812) one century before publication of *The Integrative Action of the Nervous System* by Sherrington, before Marshall Hall had conceived the hierarchal interaction of segmental, intersegmental and suprasegmental reflexes, and even antedating the proof of the sensory function of the posterior spinal roots. Again, it preceded the

Experimental Brain Research, Suppl. 9
© Springer-Verlag Berlin · Heidelberg 1984

development of ideas about the "involuntary" or "autonomic" nervous system, the control of which was similarly thought to proceed without concious intervention like that of automatic breathing. But even this idea of breathing as wholly automatic is to some extent overstated, at least for the awake state. Thus the slightest mechanical hindrance to one's breathing movements instantly registers into consciousness. This suggests a quite remarkable mechanism of sensorimotor adaptation to these ongoing movements, which in principle must differ little from that which is involved in one's abrupt perception of a step missed in walking, to instance only one of the immense repertoire of motor skills which, once initiated by volition, are highly automatic in their execution (Sears 1974). Nevertheless, one obvious way in which the central "demand" for respiratory movement differs from the demand for limb movement is the wholly automatic manner in which the respiratory rhythm is initiated by the chemical drives from peripheral and central chemoreceptors, provided that irritant or obstructive stimuli do not antecede them. Even here a new perspective can be introduced. This arises from the observation, as discussed below, that for central and peripheral chemoreceptor drives the demand for respiratory motoneuron excitation can persist in the absence of rhythm generation as a tonic discharge of either inspiratory or expiratory motoneurons, or both, depending on the balance of excitation between these chemical drives. The sensory manifestation of these activities is unknown. Finally, the discovery of the systems of powerful respiratory and cardiovascular reflexes from the lungs and the carotid body and carotid sinus were to determine thereafter that the study of breathing movements was to have much closer links with the study of autonomic function than with the control of the somatic musculature, while a further consequence was that the biophysics of synaptic transmission was to figure less in the investigation of the nervous mechanism of breathing.

Thus greatly impressed as I was by the importance of the motoneuron as the "final common path", stimulated by Lloyd's elegant analysis of connectivity of afferent, propriospinal, and suprasegmental pathways to motoneurons, but above all excited by the dimension which Eccles had added to the investigation of the central nervous system at the cellular level, it was perhaps not surprising that I did not find the "respiratory centre" approach greatly appealing, the published work having virtually ignored the very mechanisms which had excited my curiosity. This was the background that brought me to Canberra, and within a relatively short time I had obtained an answer to one question that had preoccupied me: what role does inhibition play in the generation of respiratory motoneuron activity?

Central Respiratory Drive Potentials

One idea common to the then current concepts about the respiratory centres was the belief that the alternating bursts of inspiratory and expiratory motoneuron activity were due simply to the periodic bursts of excitation arising from the "centres". Clearly, the most direct way of examining this idea was to record intracellularly from these motoneurons during spontaneous breathing. Because of my previous interest in intercostal motoneurons (Sears 1958) and the fact that the

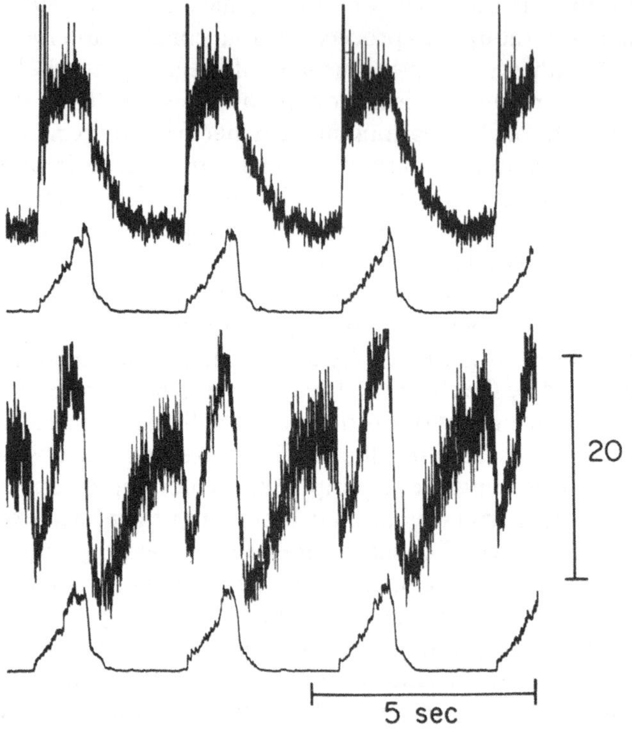

Fig. 1. Effect of hyperpolarizing current on the central respiratory drive potential of a phrenic motoneuron. *Upper trace,* DC recording of membrane potential; *vertical calibration bar* = 20 mV. *Lower trace,* integrated phrenic nerve discharge (time constant, 50 ms). Note apparent monophasic wave of depolarization with respect to each respiratory cycle. *Lower pair of traces* recorded while passing 10 nA hyperpolarizing current through microelectrode via bridge circuit. Note two phases of depolarization in each respiratory cycle due to reversal of inhibition. (Berger 1979)

antagonistic nature and multisegmental organization of the intercostal and abdominal muscles offered a better prospect of interesting reflex interactions, the work was done on the thoracic spinal cord. Such recordings in anaesthetized cats showed that during spontaneous breathing thoracic inspiratory and expiratory motoneurons displayed rhythmic changes in membrane potential which could reach threshold in some motoneurons to generate the typical burst pattern of activity recorded as single-unit discharges from filaments of the internal (expiratory) or external (inspiratory) intercostal nerves (Sears 1963). These slow variations in potential were named "central respiratory drive potentials" (CRDPs) to link them with the suprasegmental demand for breathing movements. The CRDPs could be up to 12–15 mV peak-to-peak amplitude in individual motoneurons, but were usually much smaller than this in eupnoea. Analogous recordings were made from phrenic motoneurons by Gill and Kuno (1963; Webber and Pleschka 1976). A beautiful example of a phrenic motoneuron CRDP from the more recent work of Berger (1979) is illustrated in Fig. 1, chosen because it illus-

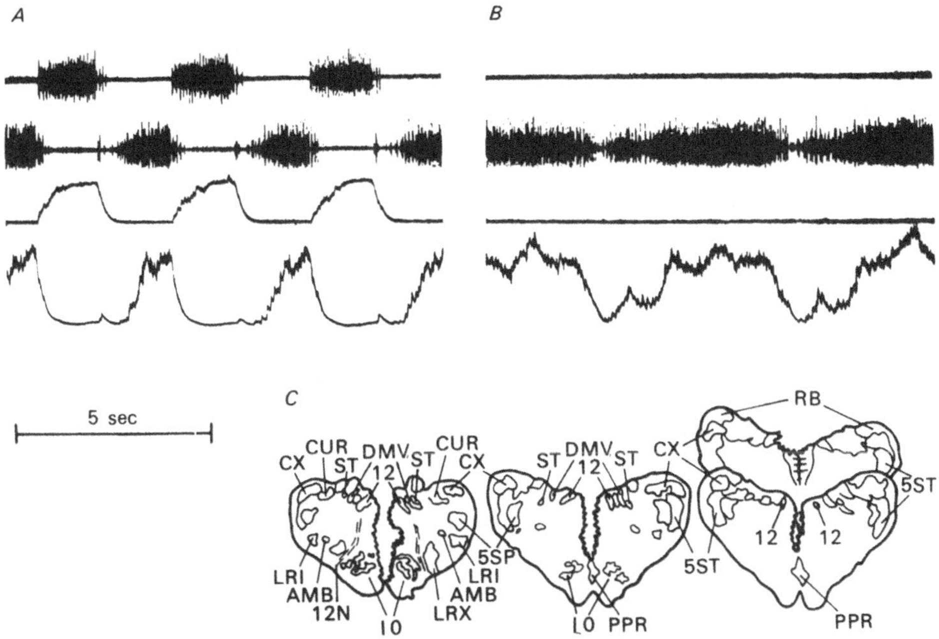

Fig. 2A–C. Effect of sagittal section of the medulla starting rostral to obex. *Upper two traces*, diaphragm and expiratory intercostal electromyograms respectively; *lower two traces*, integrated versions of the upper two. *A* is the control recording and *B* is after the section was made from 3 mm rostral to 2 mm caudal to obex. Extending the lesion more caudally abolished all activity. Artificial respiration, $P_{A\ CO_2}$ constant at 31.5 torr. *C* Transverse sections of the medulla from this cat to show rostral extent of lesion (1.5, 1.0 and 0.5 mm between sections, caudal to rostral). Notation following Berman (1968): *5SP*, spinal trigeminal nucleus; *5ST*, spinal trigeminal tract; *12*, hypoglossal nucleus; *12N*, hypoglossal nerve; *AMB*, nucleus ambiguus; *CUR*, cuneate nucleus, rostral division; *CX*, external cuneate nucleus; *DMV*, dorsal motor nucleus of vagus; *IO*, inferior olive; *LRI*, lateral reticular nucleus, internal division; *LRX*, lateral reticular nucleus, external division; *PPR*, postpyramidal nucleus of the raphe; *RB*, restiform body; *ST*, solitary tract. (Bainton et al. 1978)

trates very well a later point of discussion. With the phrenic neurogram used as the reference phase, the CRDP could be interpreted as a monophasic wave of synaptically evoked depolarization for each respiratory cycle. However, as Berger demonstrates with a hyperpolarizing current, and I demonstrated originally by chloride ion injection into thoracic motoneurons (Sears 1964a, b), the occurrence of two phases of depolarization in each respiratory cycle provided direct evidence by "reversal" (cf. Coombs et al. 1955) of a phase of postsynaptic inhibition during the antagonistic respiratory phase. It can be seen that there is temporal overlap of these excitatory and inhibitory synaptic drives which, in alternating sequence, determine the breath-by-breath shape of the CRDP and therefore of the resulting discharge patterns (Sears 1964b). The relative amplitudes of the inhibitory and excitatory phases of the CRDPs varies between individual motoneurons and also breath by breath, reflecting changes in the breathing pattern. Thus in deep anaesthesia, when there is a marked inspiratory shift in this pattern, the CRDPs of expiratory motoneurons can be formed almost entirely by inhibition (Sears 1964a). The intensity of the inhibition shows a close relationship to the intensity of the excitatory drive to the antagonistic motoneurons and there is some evidence, mainly indirect, that the relevant interneurons are in the spinal cord (Sears 1966; Aminoff and Sears 1971). Additional evidence is offered in Fig. 2 from the work of Bainton et al. (1978), who studied the effects of hypocapnic apnoea on respiratory motoneurons. In the control electromyograms (Fig. 2A) recorded from a decerebrate cat, unparalysed but artificially ventilated to maintain alveolar CO_2 constant, inspiratory activity of the diaphragm (upper trace) alternated with that of the expiratory activity of the internal intercostal muscle. A discrete mid-line incision, aimed at interrupting the inspiratory bulbospinal axons where they cross rostral to the obex (Merrill 1974), selectively abolished the inspiratory motoneuron activity (Fig. 2B). In contrast, the expiratory activity now persisted throughout the respiratory cycle, which was only slightly slowed by the lesion. Thus with the loss of the inspiratory bulbospinal drive, the expiratory motoneurons were disinhibited. This also allowed the segmental chest wall inflation reflex of the expiratory intercostal motoneurons (Sears 1958, 1963; Euler and Fritts 1963) to be expressed in the inspiratory phase, whereas before the lesion it was only visible as a modulation during the expiratory phase of the central respiratory drive.

Eccles has always insisted that new physiological facts need to form part of an overall story. For this I proposed that the normal functional operation of the inhibitory phase of the CRDP is to prevent reflex excitation of the antagonistic motoneurons due to stretch of their homonymous muscle spindles by agonist muscle contraction. The need for this can be understood because these muscles, like those of the hip flexors (Eccles and Lundberg 1959), lack reciprocal inhibition. The result described above fully supports my "story". However, our most recent work, in which we have studied the interaction between central and peripheral chemoreceptor drives under hypocapnic conditions, suggests that these inhibitions have a wider significance than was previously recognized.

Reciprocal Tonic Activation of Inspiratory and Expiratory Motoneurons in Response to Chemical Drives

When CO_2 is lowered to abolish spontaneous respiratory movements (hypocapnic apnoea), expiratory motoneurons show a tonic discharge which is modulated by a segmental stretch reflex due to the pump-induced inflation of the chest wall (Sears 1958, 1964a, b). Later studies showed that the tonic discharge shows a graded increase with graded increases in CO_2 and dependence on a brain-stem-mediated activity conveyed in axons which decussate below the obex (Sears et al. 1975; Bainton et al. 1978). We postulated that this activity was dependent on the caudal group of expiratory bulbospinal neurons (Bianchi 1971; Merrill 1970), which monosynaptically excite expiratory motoneurons (Kirkwood and Sears 1973), and this hypothesis was verified by Bainton and Kirkwood (1979), who demonstrated the predicted behaviour of the relevant neurons in anaesthetized animals.

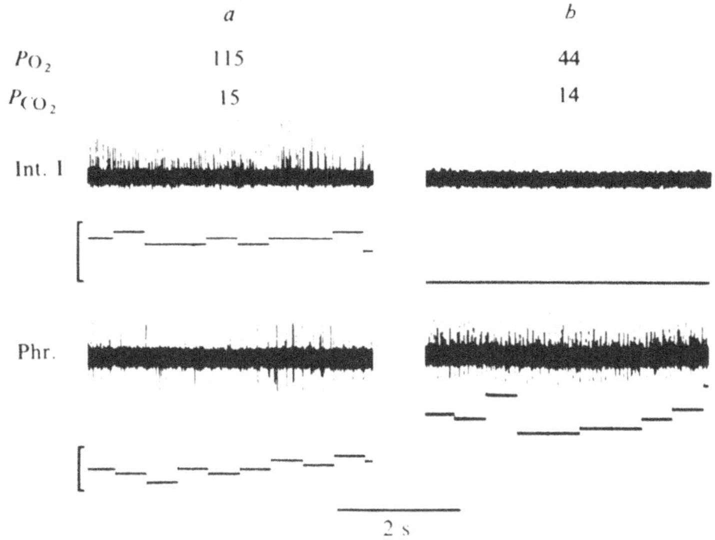

Fig. 3a,b. Recordings of inspiratory and expiratory motoneuron discharges from the phrenic nerve (*Phr*) and internal intercostal nerve filaments (*Int. 1*) respectively, with their corresponding integrated records to provide an estimate of rate below. Anaesthetized, paralysed cat, artificially ventilated with gas mixtures to produce the steady-state arterial blood gas tensions of CO_2 and O_2 given above.

Using mid-sagittal incisions of the medulla, which both elevated the threshold for rhythm generation and interrupted the axons of inspiratory bulbospinal neurons which decussate above the obex (Merrill 1974), we showed that this graded, tonic action of CO_2 extends over a wide range, up to 60 torr (Bainton et al. 1978). A few decerebrate cats were inspiratory "biassed", and in these graded CO_2 in hypocapnic apnoea caused a tonic discharge of inspiratory motoneurons, with reciprocal inhibition of any tonic expiratory activity present. All of these effects were attributed to the central chemosensitivity to CO_2, since the hyperoxia we also used would have abolished any reflex drive from peripheral chemoreceptors.

We have now studied the effects of peripheral chemoreceptor excitation by adding controlled levels of hypoxic stimulation to anaesthetized cats in hypocapnic apnoea (Sears et al. 1982). This always caused a tonic discharge of inspiratory motoneurons with reciprocal inhibition of the expiratory motoneuron discharge due to the hypocapnia (Fig. 3). A further increase in either stimulus caused rhythm generation, and this was seen to be sculpted by inhibition from the underlying tonic activities (cf. Bainton et al. 1978). These and related phenomena (Sears et al. 1982) are brought together in the diagram of Fig. 4, which illustrates how, through mutual reciprocal inhibition, the tonic chemical drives from peripheral and central chemoreceptors are integrated at segmental level.

In hyperoxic, hypocapnic apnoea, there is tonic firing of expiratory motoneurons over the expiratory bulbospinal pathway, with reciprocal inhibition of the inspiratory motoneurons. At low CO_2 inspiratory bulbospinal neurons are in-

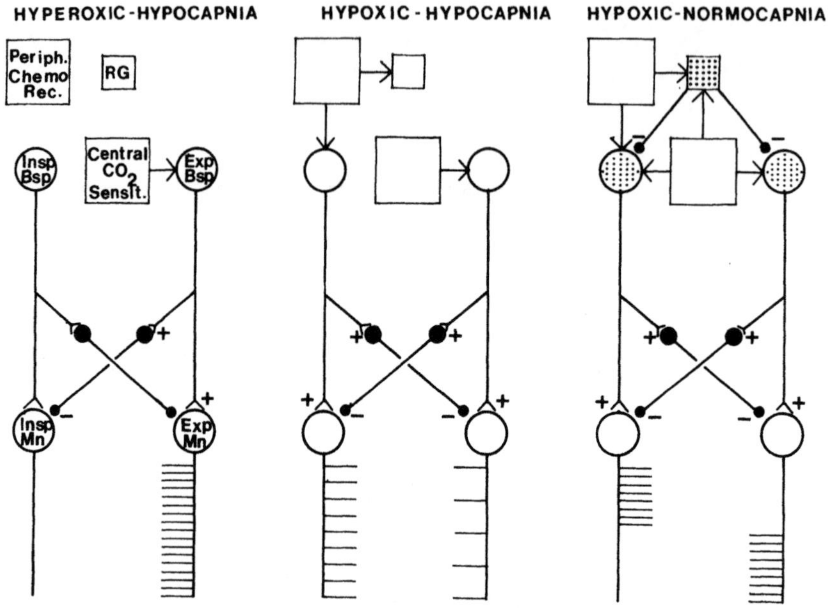

Fig. 4. Schematic representation of response of inspiratory and expiratory motoneurons to brainstem-mediated central and peripheral chemoreceptor drives. *RG*, rhythm generator which alternately inhibits inspiratory and expiratory bulbospinal neurons. *Arrows* indicate effective inputs. Peripheral chemoreceptor input to rhythm generator is included to represent the effect it has on lowering the threshold for rhythm generation in response to CO_2 (T.A. Sears, A.J. Berger and E.A. Phillipson, unpublished observations)

active (Nesland and Plum 1965; Cohen 1968). The addition of a peripheral chemoreceptor drive excites inspiratory bulbospinal neurons [species yet to be identified (Berger et al. 1983)] and their corresponding inspiratory motoneurons; this inspiratory drive gives graded reciprocal inhibition to the expiratory motoneurons. With balanced stimulation there can be coactivation of inspiratory and expiratory motoneurons, but mutual inhibition depresses the effectiveness of either drive.

The addition of a capnic stimulus (or increasing hypoxia) now causes rhythm generation. With this model the "rhythm generator" or "central pattern generator" can be understood as a mechanism which alternately inhibits the inspiratory and expiratory bulbospinal neurons, thus allowing the disinhibited pathway to express the prevailing balance of peripheral and central chemoreceptor stimulation. This model thus explains the disinhibition illustrated in Fig. 2 when the inspiratory bulbospinal axons are interrupted. It can equally be taken as a model of the bulbospinal neurons themselves, since these display the requisite inhibition (Mitchell and Herbert 1974; Richter et al. 1975; Richter 1982), thus providing a testable hypothesis for further studies. Certainly it can be suggested that the dominance of the expiratory system of neurons at low levels of CO_2, coupled with the increasing dominance of the inspiratory half of the system at higher levels of CO_2, represents an asymmetry of action providing the instability necessary for the promotion of rhythm generation in a neural network. This basic model differs considerably from that advanced by von Euler and collaborators (Bradley et al. 1975) by including expiratory bulbospinal neurons and their inputs into the organization of the central mechanisms for rhythm generation and by emphasizing the important role segmental integration plays in determining the final output pattern of respiratory motoneuron discharge (cf. Sears 1977).

The renewal of interest in spinal generators of the respiratory rhythm (Langendorff and Nitschmann 1980) through the recent work of Viala et al. (1979, 1983) and Aoki et al. (1980) provides additional possibilities of testing the hypotheses outlined above. Whatever merit these proposals should prove to have for future developments in respiratory neurophysiology, they will owe considerably to the "model" provided by Eccles through his work, inspiration, and encouragement.

References

Aminoff MJ, Sears TA (1971) Spinal integration of segmental, cortical and breathing inputs to thoracic respiratory motoneurones. J Physiol (Lond) 215: 557–575

Aoki M, Mori S, Kawahara K, Watanabe H, Ebata N (1980) Generation of spontaneous respiratory rhythm in high spinal cats. Brain Res 202: 51–63

Bainton CR, Kirkwood PA (1979) The effect of carbon dioxide on the tonic and the rhythmic discharges of expiratory bulbospinal neurones. J Physiol (Lond) 296: 291–314

Bainton CR, Kirkwood PA, Sears TA (1978) On the transmission of the stimulating effects of carbon dioxide to the muscles of respiration. J Physiol (Lond) 280: 249–272

Berger AJ (1979) Phrenic motoneurones in the cat: subpopulations and nature of respiratory drive potentials. J Neurophysiol 42: 76–90

Berger AJ, Phillipson EA, Sears TA (1983) Time signatures of peripheral and central chemoreceptor drives to feline respiratory motoneurones. J Physiol (Lond) 334: 41P

Berman AL (1968) The brain stem of the cat. The University of Wisconsin Press. Madison Milwaukee London

Bianchi AL (1971) Localization et étude des neurones respiratoire bulbaires. J Physiol (Paris) 36: 5–40

Bradley GW, Euler C von, Martilla I, Roos B (1975) A model of the central and reflex inhibition of inspiration in the cat. Biol Cybern 19: 105–116

Cohen MI (1968) Discharge patterns of brain-stem respiratory neurons in relation to carbon dioxide tension. J Neurophysiol 31: 142–165

Coombs JS, Eccles JC, Fatt P (1955) The specific ion conductances and the ionic movements that produce the inhibitory post-synaptic potential. J Physiol (Lond) 130: 326–373

Eccles RM, Lundberg A (1959) Supra-spinal control of interneurones mediating spinal reflexes. J Physiol (Lond) 147: 565–584

Euler C von, Fritts HW (1963) Quantitative aspects of respiratory reflexes from the lungs and chest walls of cats. Acta Physiol Scand 57: 284–300

Gill PK, Kuno M (1963) Properties of phrenic motoneurones. J Physiol (Lond) 168: 258–273

Kirkwood PA, Sears TA (1973) Monosynaptic excitation of thoracic expiratory motoneurones from lateral respiratory neurones in the medulla of the cat. J Physiol (Lond) 234: 87–89P

Langendorff O, Nitschmann R (1980) Studien über die Innervation der Atembewegungen. I.Über die spinalen Zentren der Atmung. Arch Physiol 4: 518–549

Le Gallois M (1812) Experiences sur le principe de la vie. D'Hautel, Paris

Merrill EG (1970) The lateral respiratory neurones of the medulla, their associations with nucleus ambigualis, the spinal accessory nucleus and the spinal cord. Brain Res 24: 11–28

Merrill EG (1974) Finding a respiratory function for the medullary respiratory neurones. In: Bellairs R, Gray EG (eds) Essays on the nervous system. Clarendon, Oxford, pp 451–486

Mitchell RA, Herbert DA (1974) The effect of carbon dioxide on the membrane potential of medullary respiratory neurons. Brain Res 75: 345–349
Nesland R, Plum F (1965) Sub-types of medullary respiratory neurons. Exp Neurol 12: 337–348
Pitts RF (1946) Organization of the respiratory centre. Physiol Rev 26: 609–630
Richter DW (1982) Generation and maintenance of the respiratory rhythm. J Exp Biol 100: 93–108
Richter DW, Heyde F, Gabriel M (1975) Intracellular recordings from different types of medullary respiratory neurones of the cat. J Neurophysiol 38: 1162–1171
Sears TA (1958) Electrical activity in expiratory muscles of the cat during inflation of the chest. J Physiol (Lond) 152: 35P
Sears TA (1963) Activity of fusimotor fibres innervating muscle spindles in the intercostal muscles of the cat. Nature 197: 1013–1014
Sears TA (1964a) The slow potentials of thoracic respiratory motoneurones and their relation to breathing. J Physiol (Lond) 175: 404–424
Sears TA (1964b) Investigations on respiratory motoneurones of the cat's thoracic spinal cord. In: Eccles JC, Schade JP (eds) Prog Brain Res 12: pp 259–272
Sears TA (1966) Pathways of supra-spinal origin regulating the activity of respiratory motoneurones. In: Granit R (ed) Nobel symposium I. Muscular afferents and motor control. Wiksell, Stockholm
Sears TA (1974) The afferent regulation of learnt movements. Brain Res 71: 465–473
Sears TA (1977) The respiratory motoneurone and apneusis. Fed Proc 35: 2013–2023
Sears TA, Kirkwood PA, Bainton CR (1975) Medullary-spinal neural connectivity and breathing. Bull Physio-Path Resp 11: 82–83P
Sears TA, Berger AJ, Phillipson EA (1982) Reciprocal tonic activation of inspiratory and expiratory motoneurones by chemical drives. Nature 299: 728–730
Viala D, Freton E (1983) Evidence for respiratory and locomotor pattern generators in the rabbit cervico-thoracic cord and for their interactions. Exp Brain Res 49: 247–256
Viala D, Vidal C, Freton E (1979) Coordinated rhythmic bursting in respiratory and locomotor muscle nerves in the spinal rabbit. Neurosci Lett 11: 155–159
Webber CL Jr, Pleschka K (1976) Structural and functional characteristics of individual phrenic motoneurones. Pflügers Arch 364: 113–121

Segmental Control of Pain

W.D. WILLIS, Galveston

Introduction

For the past decade, my laboratory has concentrated on two major lines of investigation, one dependent upon the other. The first has been the demonstration that a specific somatosensory pathway, the spinothalamic tract, is capable of conveying to the brain information about noxious stimuli that may underlie, or at least contribute to, pain sensation. The second series of studies has made use of the nociceptive responses of spinothalamic tract cells as an assey system for investigations of segmental mechanisms of analgesia. The rationales of this approach are that: (a) inhibition of the discharges of spinothalamic cells may result from the activation of those neural pathways that produce analgesia by an action at the spinal cord level, and (b) this inhibition may contribute to the analgesia.

"Analgesia pathways" have been proposed to operate at a spinal cord level in recent years, on the basis of behavioral observations using spinal nociceptive reflexes, such as the tail flick in the rat (Basbaum et al. 1977; Mayer and Liebeskind 1974; cf. Reynolds 1969; see review by Willis 1982). However, changes in a flexor withdrawal reflex may not always reflect changes in pain sensation (McGrath et al. 1981; Willer et al. 1979) and so our objective has been to develop a way to examine the action of "analgesia systems" on a nociceptive sensory pathway. Many other laboratories have done similar studies to those of our laboratory, but have been content to record from nociceptive spinal cord interneurons. Unfortunately, there is an ambiguity in such studies in that the neural systems to which particular nociceptive dorsal horn interneurons belong have not yet been identified. For example, such cells may belong to reflex circuits, such as the flexion reflex pathway, or they may contribute to the activation (or inhibition) of sensory tracts (see Willis 1982). In our opinion, many of these problems are best circumvented by focusing attention on nociceptive neurons that belong to an identified population of ascending tract cells.

Historical Aside

The motivation for our studies on the spinothalamic tract came not from an initial focus on pain mechanisms, but rather from an interest in the study of identified spinal cord neurons. After spending 2 years learning spinal cord neurophysiology from John C. Eccles in Canberra, I was impressed by the elegance of experiments centered around well-characterized entities, such as motoneurons in-

Experimental Brain Research, Suppl. 9
© Springer-Verlag Berlin · Heidelberg 1984

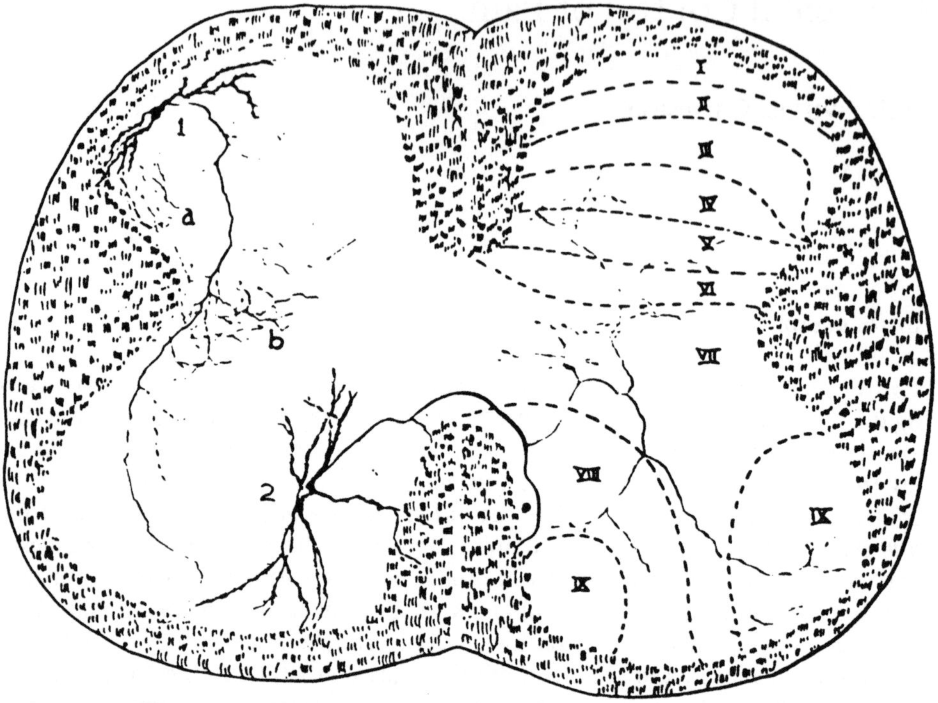

Fig. 1. Two "proprioneurons" in the spinal cord of a 4 to 6-day-old mouse. Neuron *1* is in the superficial dorsal horn and projects its axon ventrally, giving off collaterals before entering the ventral white matter. Neuron *2* is said to be located lateral to the "Renshaw area," and its main axonal projection is to the contralateral side. Rapid Golgi method. *a* and *b* are collaterals of neuron *1*. *I–IX* are Rexed's laminae. (Scheibel and Scheibel 1966)

nervating particular muscles and tract cells connecting the spinal cord with the cerebellum (see Eccles 1957, 1964). Recordings could be made from single neurons identified according to the projection targets of their axons, and the role of these identified cells in particular reflex pathways could be examined by determining what synaptic inputs affected the cells. The identification of synaptic inputs become crucial for neurons whose axonal projections are difficult to define, such as many classes of spinal cord interneurons, whose functional role in most instances is still uncertain.

My attention was directed to tract cells as a result of an issue raised at one of the meetings of the American Association of Anatomists by Arnold Scheibel (Scheibel and Scheibel 1964). The Scheibels had been studying spinal cord interneurons using the Golgi technique. They found that most interneurons in the medial part of the ventral horn projected their axons to the contralateral side of the cord (Fig. 1; Scheibel and Scheibel 1966, 1971). Since Renshaw cells were thought to be ipsilaterally projecting interneurons located in the medial ventral horn (Eccles et al. 1954), the Scheibels were struck by what they regarded as a discrepancy between their anatomic findings and the electrophysiological data concerning this class of interneuron. Other questions were also raised about the capability of recurrent collaterals to mediate recurrent inhibition (Scheibel and Scheibel

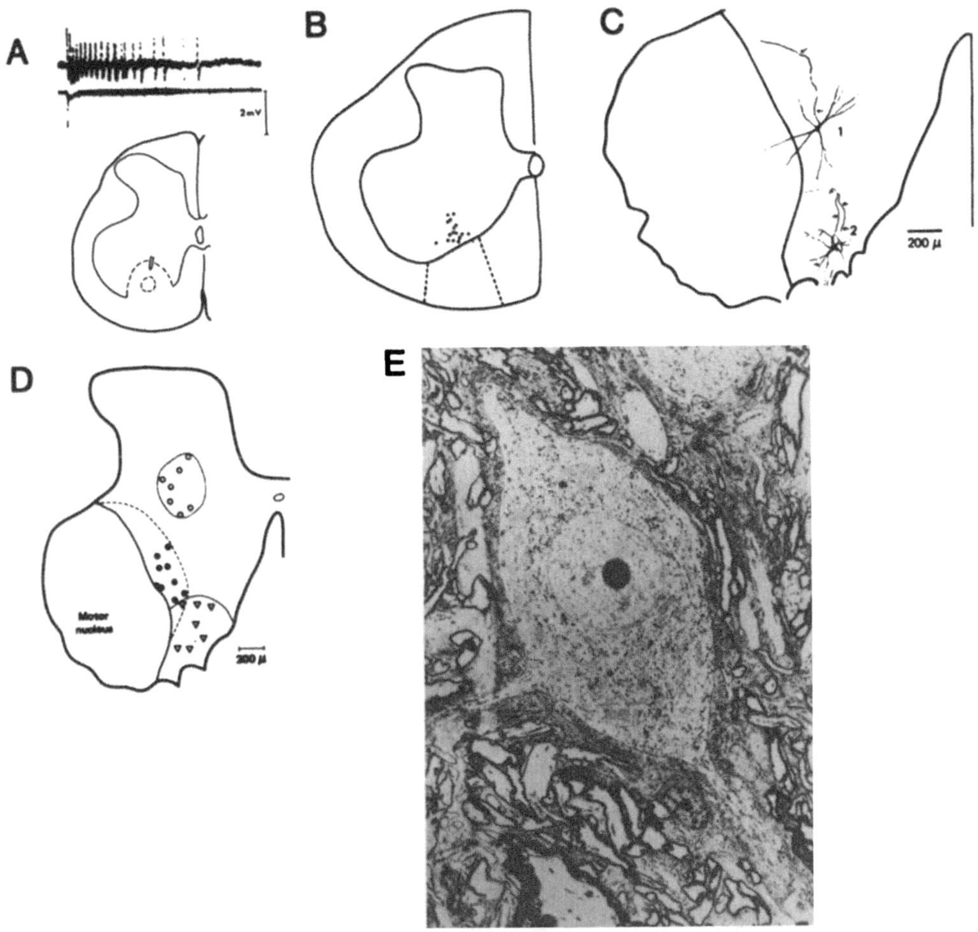

Fig. 2A–E. Progressive refinement in the anatomic definition of the Renshaw cell. In *A* is shown the discharge of a Renshaw cell in response to stimulation of the L7 and S1 ventral roots and also the location of the electrode tip from which the recording was made. The mark was made by extrusion from the electrode by iontophoretic current of a solution containing a copper complex that caused a large lesion. (Willis and Willis 1966) *B* shows the locations of a series of Renshaw cells using ejection of fast green from the recording microelectrodes. (Thomas and Wilson 1965) In *C* are drawings of two interneurons marked by intracellular injection of procion yellow dye. Cell 1 is a Ia inhibitory interneuron and cell 2 a Renshaw cell; these neurons are identified by electrophysiological criteria. *D* shows the distributions of a series of intracellularly marked interneurons, including Renshaw cells (*triangles*), Ia inhibitory interneurons (*filled circles*), and other Ia interneurons (*open circles*). (*C* and *D*, Jankowska and Lindström 1973). The electron micrograph in *E* is of a thin section through a cell identified as a Renshaw cell by following a labeled motor axon collateral to its synaptic terminal on this neuron. (Lagerbäck and Ronnevi 1982)

1971). At any rate, we thought it would be useful to mark the locations of Renshaw cells to determine if they were in the same region as that examined anatomically by the Scheibels (Willis and Willis 1964, 1966; see also Willis 1971). Our marks included several sites in the dorsal and lateral parts of lamina VII, but most of the marks were in the ventral part of lamina VII in the region of exodus of mo-

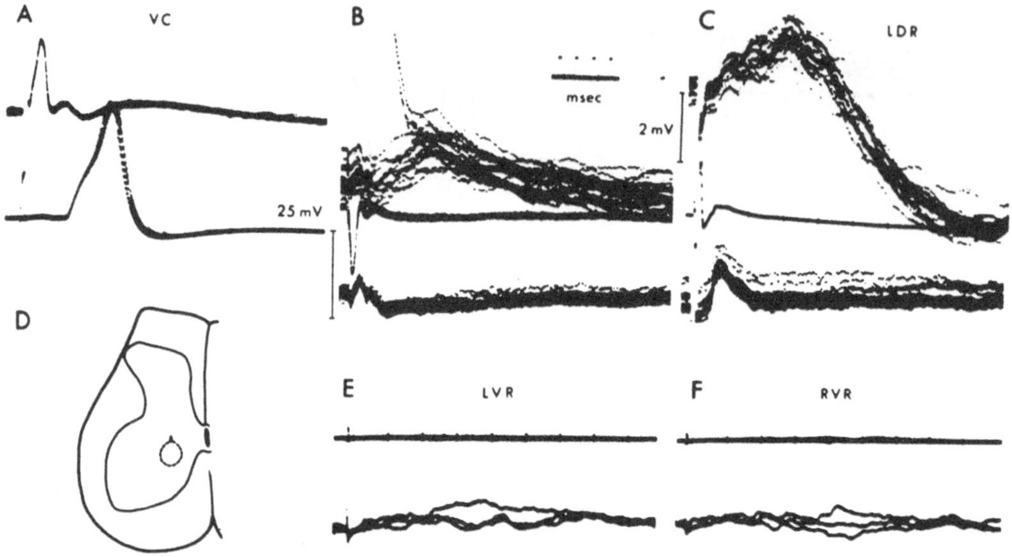

Fig. 3A–F. Evidence that some of the neurons in the medial ventral horn are tract cells. In *A* is shown the antidromic action potential recorded intracellularly (*lower trace*) from a cell in the medial ventral horn following stimulation of the ventral white matter of the cord at L1. The *upper trace* shows the cord dorsum potential produced by the descending volley. *B* and *C* are recordings of synaptic potentials evoked by stimulation of axons in the ventral white matter of the cord and of the ipsilateral L7 and S1 dorsal roots (*upper traces*). Cord dorsum potentials are shown in the *middle traces* and extracellular field potentials in the *lower traces*. The drawing in *D* shows the position of the tip of the recording electrode. The *lower traces* in *E* and *F* show that no synaptic potentials resulted in this cell following stimulation of the ipsilateral or contralateral L7 and S1 ventral roots. (Willis and Willis 1966). Abbreviations: *LDR* left dorsal root; *LVR* left ventral root; *RVR* right ventral root; *VC* ventral cord

tor axons from the ventral horn (Fig. 2A). Unfortunately, the only marking technique that was available to us for use with glass microelectrodes was the technique of Galifret and Szabo (1960), which produced rather large spherical lesions. Thomas and Wilson (1965), who had been developing a marking technique using deposits of fast green dye, replicated our study and found all of their Renshaw cells in the ventral part of lamina VII (Fig. 2B). Jankowska and Lindström (1971, 1973) and Van Keulen (1971) carried the identification further by marking Renshaw cells intracellularly with Procion yellow dye. One of the cells of Jankowska and Lindström is shown in Fig. 2C (cell 2), and a comparison of the distributions of Renshaw cells with two other classes of spinal cord interneurons is shown in Fig. 2D. Recently, Lagerbäck and Ronnevi (1982) have traced labeled motor axon collaterals to synaptic endings on interneurons in ventral lamina VII in the electron microscope. Figure 2E shows an electron micrograph of a section through an interneuron in ventral lamina VII identified in this way as a Renshaw cell.

By contrast to Renshaw cells, which are excited following stimulation of motor axons, interneurons in the medial part of the ventral horn (lamina VIII;

Fig. 3D) were unaffected by volleys in ventral root axons (Fig. 3E,F; Willis and Wilis, 1966). However, a clue as to the identification of some of these cells came from the observation that they could be antidromically activated by stimulation of the ventral white matter of the spinal cord at an upper lumbar level (Fig. 3A). Consistent with the observations of the Scheibels (Scheibel and Scheibel 1966, 1971), at least some of these axons were shown to decussate (Willis and Willis, 1966).

We decided to pursue the question of the identity of the contralaterally projecting neurons of the medial ventral horn. Two candidate pathways to which these neurons might contribute had been described by Lundberg and Oscarsson (1962). These were termed the "bilateral ventral flexion reflex tract" (bVFRT) and the "contralateral ventral flexion reflex tract" (cVFRT). Lundberg and Oscarsson (1962) were able to demonstrate possible termination zones for the bVFRT in the lateral reticular nucleus and midbrain. They suggested that the cVFRT might include axons belonging to the spinothalamic tract.

We thought that the best approach for us to take was to utilize antidromic activation of axons to provide a means of identifying tract cells. Furthermore, we judged that the least ambiguous results should come from studies on the spinothalamic tract, since this is the most rostrally projecting tract originating in the spinal cord. That is to say, if we found neurons in the spinal cord that could be activated antidromically by stimulating in the thalamus, such cells had to be spinothalamic neurons, since no spinal neurons are known that project to sites rostral to the thalamus, such as the basal ganglia or the cerebral cortex. Fortunately, later studies using sensitive anatomic techniques for tracing central pathways have not contradicted this assumption.

Our plan, then, was to study the spinothalamic tract first. When we knew the characteristics of this pathway, we thought we could move our antidromic stimulating electrode caudally and examine other pathways as well. Although we recognized at that time (1970) that little was known about the spinothalamic tract, despite its obvious clinical importance in humans, we would not have predicted that practically all of our efforts for more than a decade would be concentrated on this one system.

Spinothalamic Tract in Cat

We began our investigation of the spinothalamic tract in the animal model to which we were accustomed, the cat. It proved possible to activate cat spinothalamic tract cells antidromically (cf. Dilly et al. 1968), and we mapped their distribution in the lumbosacral enlargement (Trevino et al. 1972). Unfortunately, the cells were hard to find, and the spikes were small relative to those of neighboring cells, and so it was difficult to isolate units in extracellular recordings in order to map receptive fields. Curiously, we had better luck at this in a study of spinothalamic cells in the thoracic spinal cord of the cat (Hancock et al. 1975).

Despite these difficulties, we did show that many cat spinothalamic cells are in laminae VII and VIII (Fig. 4A,B) and that these cells usually project their axons to the contralateral thalamus, thus identifying at least some of the neurons stud-

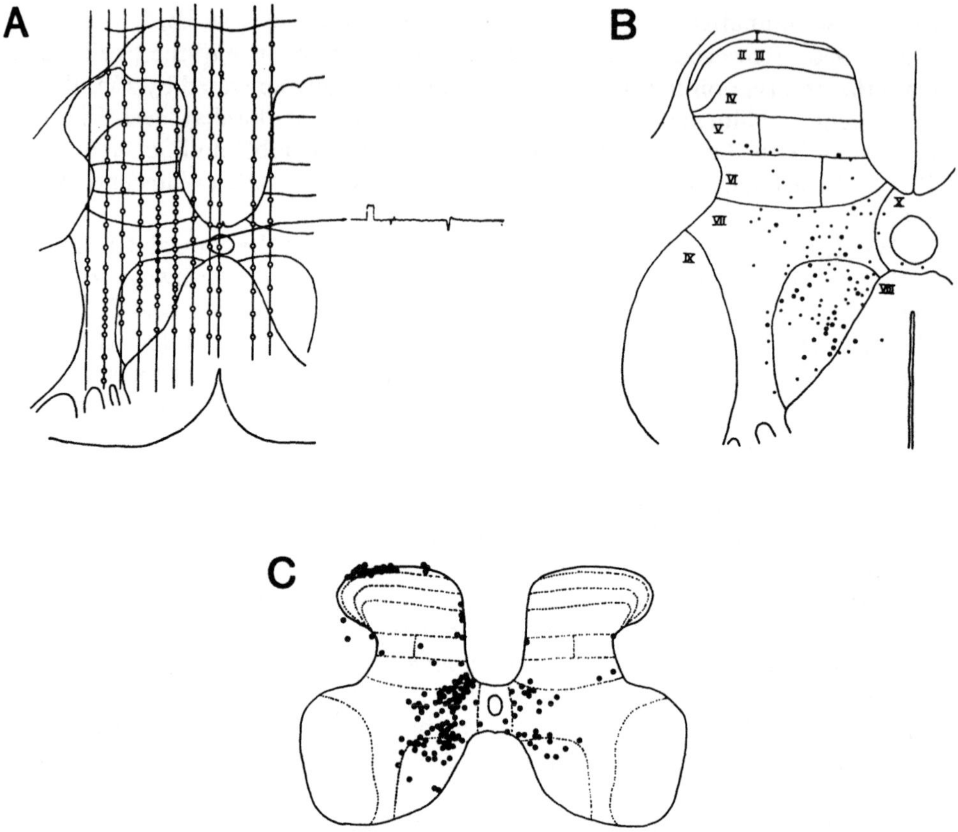

Fig. 4A–C. Locations of the spinothalamic tract cells in the cat. The signal-averaged recording in *A* is from the point indicated in one of a series of tracks across the spinal cord. The antidromic action potential resulted from stimulation in the contralateral thalamus. The potential could be observed only when recordings were made at the points indicated by the *filled circles*. Calibration 100 μV, 1 ms. In *B* are plotted the locations of recording points from which antidromic action potentials could be recorded from 105 spinothalamic tract cells in 25 cats. *Large dots* indicate antidromic spikes of 100 μV or greater in signal-averaged records. (*A* and *B*, Trevino et al. 1972) The drawing in *C* shows the locations of spinothalamic neurons marked in the cat by retrogradely transported horseradish peroxidase (HRP). The HRP had been injected into the contralateral thalamus. *I–X* are Rexed's laminae. (Trevino and Carstens 1975)

ied by the Scheibels (1966, 1973) in the medial ventral horn as neurons of origin of the spinothalamic tract. This observation was later confirmed in horseradish peroxidase experiments in the cat (Fig. 4C; Trevino and Carstens 1975). However, we missed an important set of spinothalamic neurons in our electrophysiological study of the cat, a population located in lamina I, the marginal zone (Fig. 4C). Our failure to record from these was no doubt a sampling error, because these cells have been demonstrated not only by horseradish peroxidase, but also recently by antidromic activation (Craig and Kniffki 1982). The significance of these cells is that many of them are nociceptive or thermoreceptive (Christensen and Perl 1970; Kumazawa et al. 1975; Craig and Kniffki 1982).

Spinothalamic Tract in Monkey

Recognizing the possibility that spinothalamic tract cells might be more abundant in primates (Mehler 1969), we decided to try our luck in the monkey. A particular issue was whether or not there were large numbers of spinothalamic tract cells in the dorsal horn in the monkey, since this is the traditionally assumed location of the cells of origin of this tract. In fact, it proved easy to record from spinothalamic cells in the monkey dorsal horn, including many in laminae I and V of Rexed

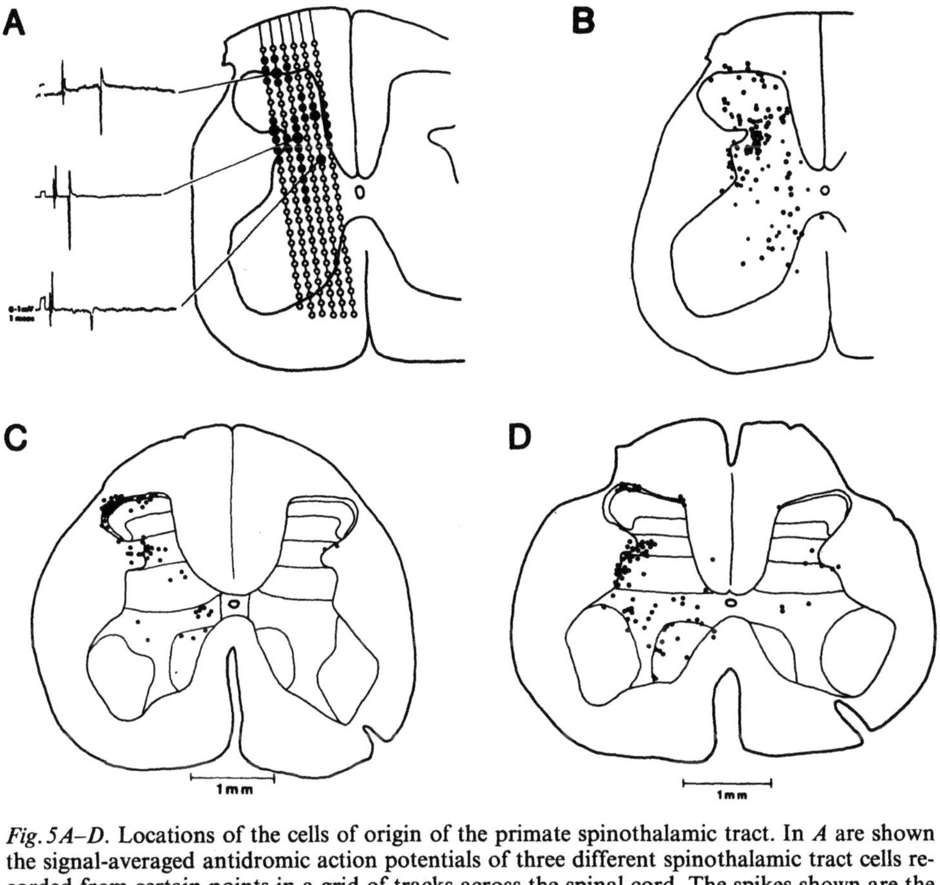

Fig. 5A–D. Locations of the cells of origin of the primate spinothalamic tract. In *A* are shown the signal-averaged antidromic action potentials of three different spinothalamic tract cells recorded from certain points in a grid of tracks across the spinal cord. The spikes shown are the largest seen from these particular units, being recorded at the points indicated by the *large filled circles*. The antidromic spikes of two other units were also observed at this level (not illustrated), as shown by the *large filled circles without leads*. Note that in this grid of tracks, records could be made from five different spinothalamic tract cells, of which two appeared to be in lamina V and one in lamina I. In *B* are summarized the locations of 118 spinothalamic tract cells in 15 monkeys. The highest concentrations of cells were in lamina I and V. (Trevino et al. 1973) In *C* and *D* are shown the distributions of spinothalamic tract cells labeled by retrogradely transported horseradish peroxidase (HRP). The HRP in the experiment illustrated in *C* was injected into the region of the contralateral ventral posterior lateral thalamic nucleus, while in the experiment illustrated in *D* the HRP was introduced into the region of the contralateral central lateral intralaminar nucleus. (Willis et al. 1979)

(Fig. 5A,B; Rexed 1952; Trevino et al. 1973; Willis et al. 1974). In a later horseradish peroxidase study, we showed that cells projecting from the spinal cord to the region of the ventral posterior lateral nucleus of the contralateral thalamus were concentrated in laminae I and V, although some of these cells were in other laminae, including VII and VIII (Fig. 5C; Willis et al. 1979; cf. Trevino and Carstens 1975). However, other spinothalamic cells projecting to the region of the central lateral nucleus in the intralaminar complex of the contralateral thalamus were more concentrated in the ventral horn (Fig. 5D; Willis et al. 1979). Thus we were able to verify that in the monkey, as in the cat, some of the tract cells in laminae VII and VIII project to the thalamus. Similar findings apply also to the rat (Giesler et al. 1979).

Other Tract Cells in Medial Ventral Horn

Several investigations, using either electrophysiological or anatomic mapping techniques, have demonstrated that many of the neurons in laminae VII and VIII project to the reticular formation (Corvaja et al. 1977; Fields et al. 1977a; Haber et al. 1982; Kevetter et al. 1982). Some of these neurons also project to the thalamus and so are both spinoreticular and spinothalamic tract cells, as shown by antidromic activation from both levels of the brain or by an anatomic double-labeling technique (Giesler et al. 1981; Haber et al. 1982; Kevetter and Willis 1982).

Neurons in laminae VII and VIII have also been found to contribute to propriospinal tracts (Molenaar and Kuypers 1978), and it would not be surprising if they projected to other targets as well.

Sensory Role of Spinothalamic Tract

Having found it relatively easy to obtain good extracellular recordings from at least the larger cells of the primate spinothalamic tract, we decided to investigate the response properties of these neurons. From clinical evidence (Foerster and Gagel 1931; Kuru 1949; White and Sweet 1955), we anticipated that many spinothalamic tract cells would be nociceptive. Others we expected to be thermoreceptive, and still others tactile.

In our initial survey of the response properties of spinothalamic tract cells (Willis et al. 1974), we restricted our recordings to cells projecting to the ventral posterior lateral nucleus. Of a series of 186 cells, 30% had high thresholds for stimuli applied to their cutaneous receptive fields. These spinothalamic cells were similar to nociceptive dorsal horn interneurons in studies done in other laboratories (Christensen and Perl 1970; Kumazawa et al. 1975; Cervero et al. 1976; Handwerker et al. 1975; Price and Mayer 1974). Recordings from high-threshold spinothalamic cells could be most commonly made when the electrode was in lamina I, but similar responses were found in lamina V and other deep laminae of the dorsal horn.

Figure 6B shows the excitatory receptive field of a high-threshold spinothalamic tract cell on a limited portion of the left foot (blackened area, plus signs). Stimulation of some areas of the body (minus signs) produced an inhibition of

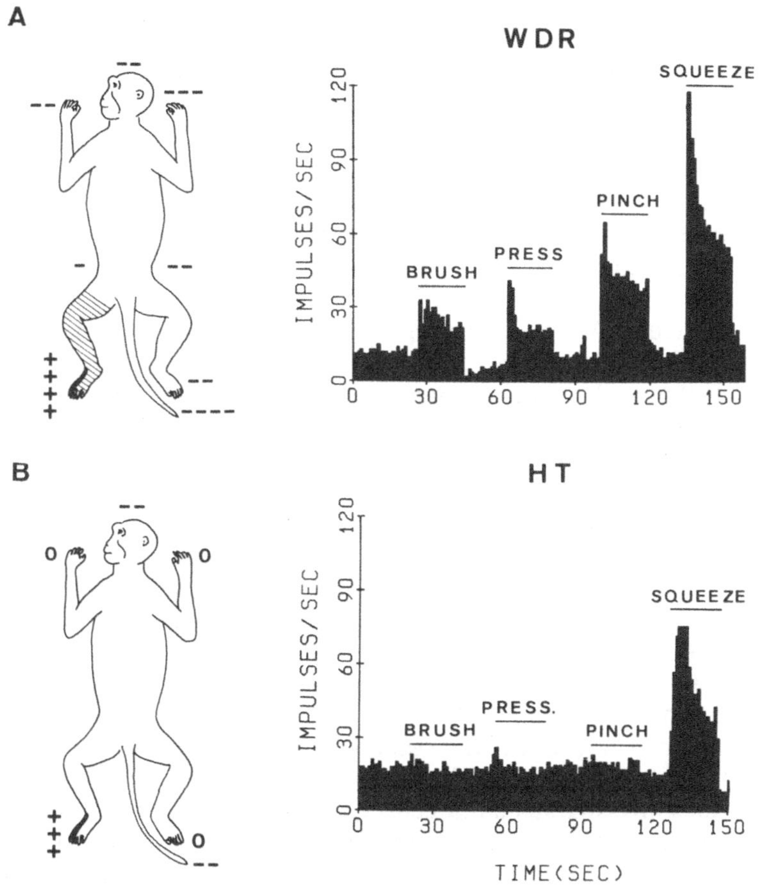

Fig. 6A,B. Receptive field properties of spinothalamic tract cells. In *A* is shown the receptive field of a wide-dynamic-range (*WDR*) spinothalamic tract cell. The excitatory field is indicated by the *plus signs* and the *hatched* and *blackened* areas, and the inhibitory field by the *minus signs*. The signal-pass peristimulus time *histogram* shows the responses of the cell to a graded series of mechanical stimuli. In *B* are shown the receptive field of a high-threshold (*HT*) spinothalamic tract cell (*plus signs* and *blackened area* for excitatory field, *minus signs* for inhibitory field) and a *histogram* of the responses to graded mechanical stimuli. (Willis 1981)

the cell, and so such areas were part of an inhibitory receptive field. The histogram in Figure 6B shows that the neuron failed to respond to brushing the hair, an innocuous mechanical stimulus, or to the application of pressure or pinch stimuli, using two gradations of stimulus intensity that were marginally or clearly painful when applied to human skin. However, squeezing the skin with forceps, a strong noxious stimulus, elicited a vigorous discharge. This particular cell had an unusually high threshold to mechanical stimuli. Others responded well to pressure or pinch stimuli, but not to brushing the skin.

The most common type of spinothalamic tract cell in our initial study (59%; Willis et al. 1974) could be activated by weak mechanical stimuli, such as hair movement or gentle indentation of the glabrous skin. Although it was tempting to ascribe a tactile function to these cells, the fact that most of the cells were more powerfully excited by noxious stimuli suggested an additional or alternative role in nociception. We have tended to call these cells "wide-dynamic-range neurons," following the terminology introduced by Mendell (1966) for comparable cells of the dorsal horn. However, a better term would be desirable. Similar neurons have been described by many other laboratories (Hillman and Wall 1969; Handwerker et al. 1975; Price and Mayer 1974). Wide-dynamic-range spinothalamic tract cells appeared to be most abundant in the neck of the dorsal horn, especially in the lateral part of lamina V. However, similar responses can be recorded from cells in lamina I and in other laminae.

Figure 6A shows the receptive field of a typical wide-dynamic-range spinothalamic tract cell (figurine). The excitatory receptive field occupied much of the left leg (hatched area), although there was a lower threshold zone on the foot (blackened area). The inhibitory receptive field was very large, occupying the rest of the surface of the body and face. The histogram shows that the cell responded to brushing the hair, but that it responded more vigorously to progressively more intense noxious stimuli.

Some of the spinothalamic tract cells responding to tactile stimuli appeared not to have a convergent input from nociceptors. We did not make an accurate determination of the proportion of these cells in our initial survey (Willis et al. 1974), but our impression now is that such cells are rare (Chung et al. 1979).

Another kind of spinothalamic tract cell that we found had a prominent input from proprioceptors of the hind limb (Willis et al. 1974). Such cells were fairly common (11% of the initial sample), and they were located both in the deep layers of the dorsal horn and in the ventral horn. The axons of many of these proprioceptive spinothalamic cells were later observed to lie along the ventrolateral margin of the spinal cord (Applebaum et al. 1975). We have recently studied a special population of such cells in Stilling's nucleus in the sacral spinal cord (Milne et al. 1982). These cells of Stilling's nucleus appear to signal tail position and to receive an input from stretch receptors in tail muscles.

Curiously, we have so far been unable to demonstrate spinothalamic tract cells with an input from specific thermoreceptors. We are inclined to attribute this to technical difficulties. Perhaps such cells are small and have therefore been overlooked in our sample. Alternatively, such cells may project to other regions of the thalamus than the ventral posterior lateral nucleus. Both of these hypotheses may prove to be correct, since Craig and Kniffki (1982) have recently recorded from a thermoreceptive spinothalamic tract cell in the cat that projected to the medial thalamus. On the other hand, most spinothalamic tract cells that have been tested respond to noxious heat (Kenshalo et al. 1979).

Although most of our recordings have been made from spinothalamic tract cells that can be activated antidromically from the ventral posterior lateral (VPL) nucleus of the thalamus, we have done one study to compare the responses of these cells to the responses of spinothalamic cells projecting to the central lateral (CL) nucleus of the intralaminar complex (Giesler et al. 1981). One observation that we were able to make was that there are several populations of spinothalamic tract cells according to the destinations of their thalamic projections. One set of spinothalamic tract cells appears to project just to the VPL nucleus; another sends

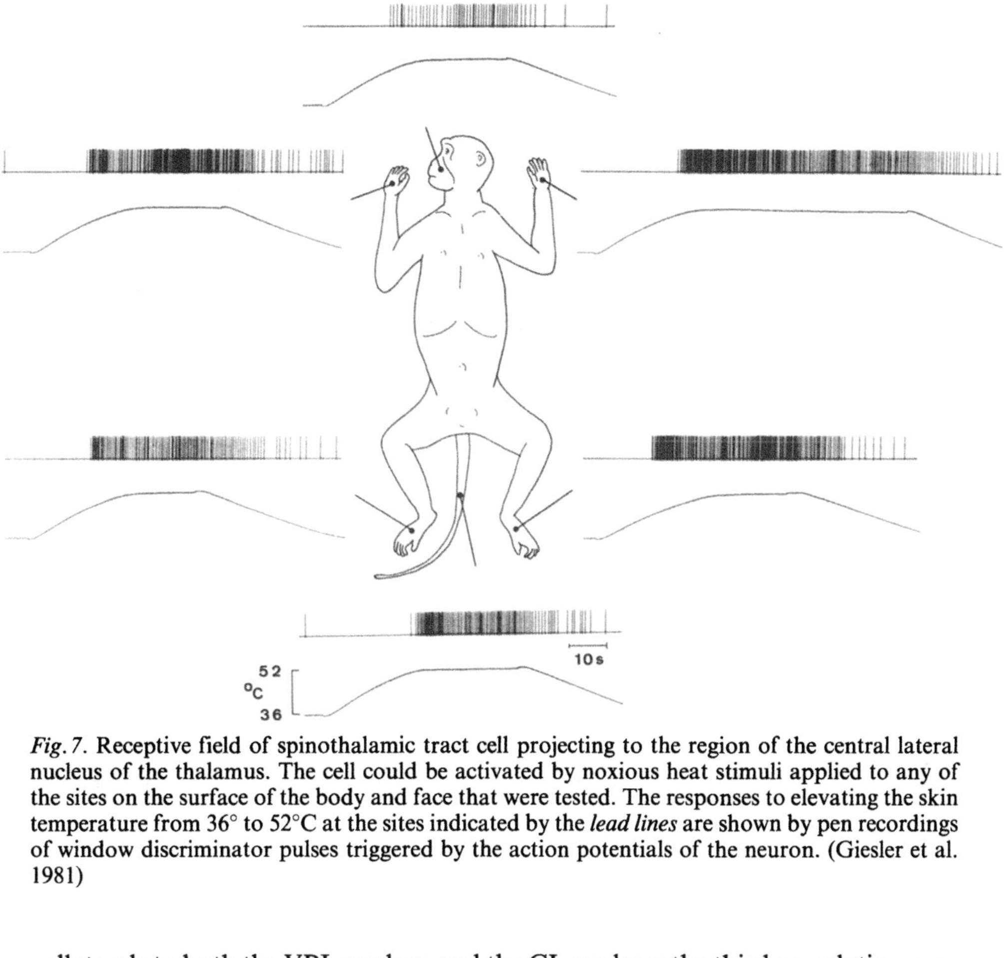

Fig. 7. Receptive field of spinothalamic tract cell projecting to the region of the central lateral nucleus of the thalamus. The cell could be activated by noxious heat stimuli applied to any of the sites on the surface of the body and face that were tested. The responses to elevating the skin temperature from 36° to 52°C at the sites indicated by the *lead lines* are shown by pen recordings of window discriminator pulses triggered by the action potentials of the neuron. (Giesler et al. 1981)

collaterals to both the VPL nucleus and the CL nucleus; the third population projects exclusively to the CL nucleus (see also Craig and Kniffki 1982).

The response properties of spinothalamic tract cells projecting just to the VPL nucleus or to both the VPL and CL nuclei were identical. These cells were usually classified as high-threshold or wide-dynamic-range cells. An important observation was that their cutaneous receptive fields were usually resticted to a portion of the contralateral hind limb. However, spinothalamic cells projecting exclusively to the CL nucleus often had quite different properties. These cells were most commonly classified as high-threshold cells, but what made them unique was that their receptive fields often extended over the entire body surface, so far as could be tested (Fig. 7). The neural connections responsible for these extraordinarily large receptive fields appeared to include a supraspinal loop, since transection of the spinal cord at an upper cervical level reduced the receptive field to only the hind limb on the side ipsilateral to the neuron under investigation. It was easy to excite such neurons by stimulation in the pontomedullary reticular formation, and so we hypothesized that the supraspinal loop involved connections with reticulospinal neurons.

Descending Control of Spinothalamic Tract Cells

A number of studies have shown the locations of sites within the brain stem that, when stimulated, result in behavioral signs of analgesia (reviewed by Willis 1982). For example, a powerful stimulation-produced analgesia results from electrical stimulation within the raphe nuclei of the brain stem, especially the nucleus raphe magnus (NRM) (Oliveras et al. 1975, 1979), or within the midbrain periaqueductal gray (PAG) (Reynolds 1969; Oliveras et al. 1974; Mayer and Liebeskind 1974). For this reason, we have examined the effects of stimulation in the NRM and PAG on neurons of the primate spinothalamic tract.

Stimulation in the NRM produces a powerful inhibition of spinothalamic tract cells (Willis et al. 1977). An example of such an action is shown in Fig. 8.

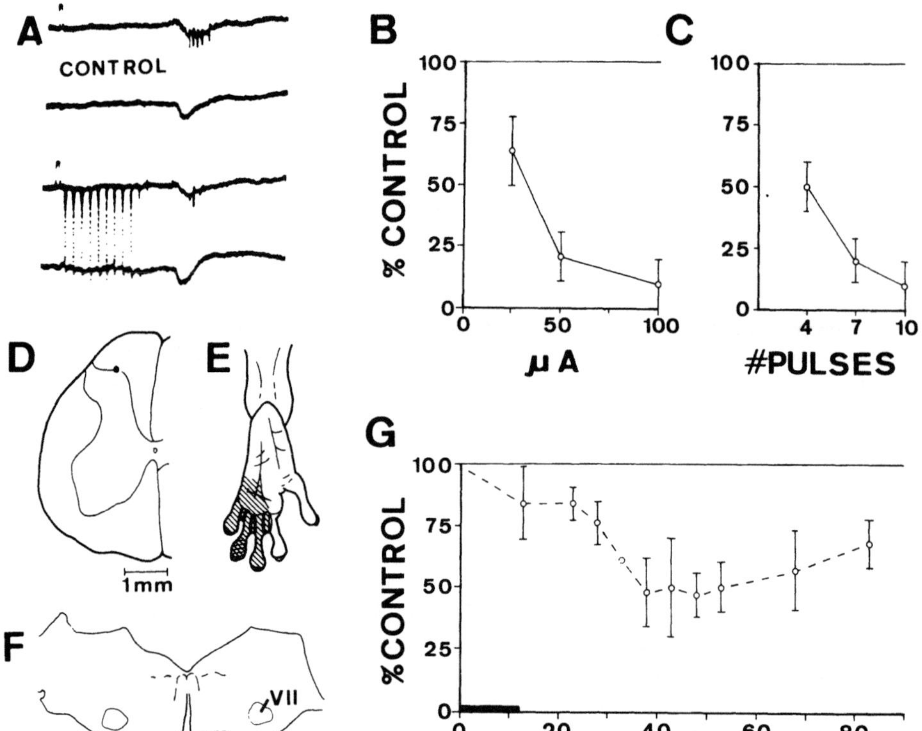

Fig. 8A–G. Inhibition of spinothalamic tract cell following stimulation in the nucleus raphe magnus (NRM). The *upper trace* in *A* shows an extracellular recording of the discharges of a spinothalamic tract cell in response to electrical stimulation of the skin. The *second trace* is a cord dorsum recording taken at the same time. The *lower pair of traces* were recorded in the same fashion, but in this case the cutaneous stimulus was preceded by brief train of stimuli applied within the NRM. *B* and *C* show the amount of inhibition, expressed in terms of the number of discharges as a percentage of control, when the stimulus strength or the number of stimuli were varied. *D* shows the location of the cell, *E* the excitatory receptive field, and *F* the stimulus site in the NRM (*RM* is raphe magnus, *VII* the facial nucleus). *G* shows the time course of inhibition (conditioning train duration indicated by *dark line at the start of the abscissa*). (Willis et al. 1977)

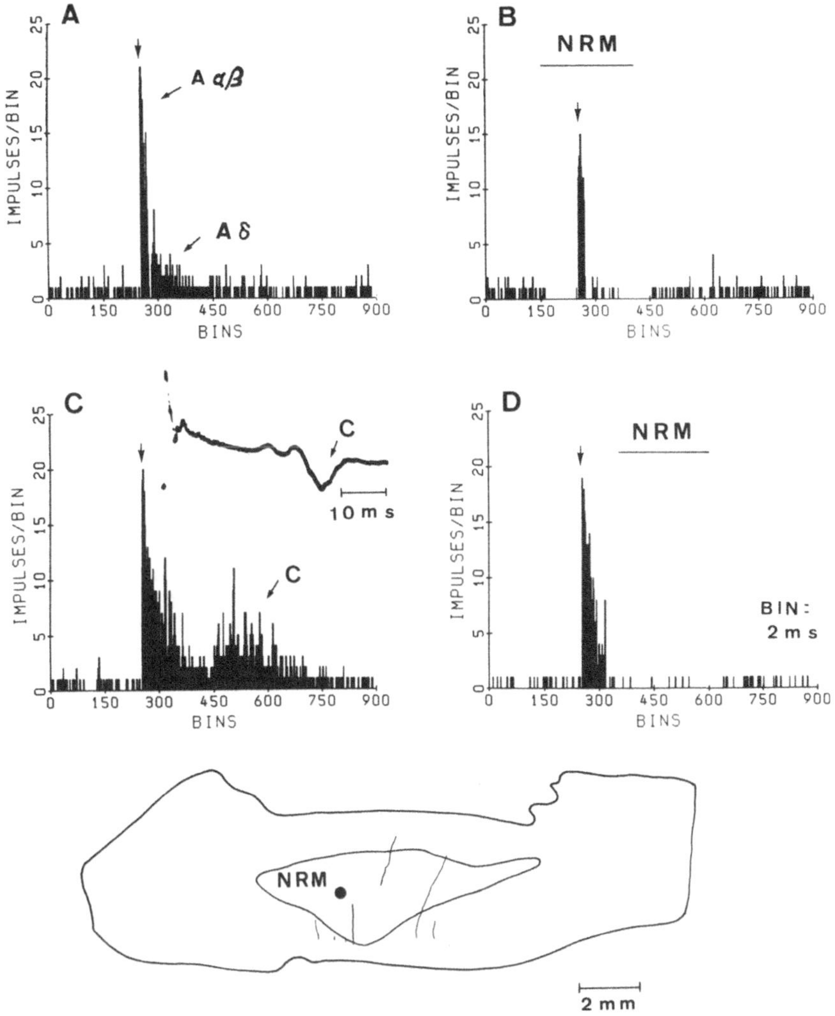

Fig. 9A–E. Inhibition of the responses of a spinothalamic tract cell to the A and C components of an afferent volley following stimulation in the nucleus raphe magnus (*NRM*). The peristimulus time *histograms* in *A* and *C* show the control responses to ten successive volleys in the A fibers or in the A and C fibers of the sural nerve. The C fiber volley is shown in the *inset* in *C*. In *B* and *D*, a 500-ms train of stimuli was applied in the NRM (0.1-ms, 150-μA pulses at 333 Hz) during the discharges evoked, respectively, by the A or C fiber volleys. In *E* is shown the location of the tip of the stimulating electrode. (Gerhart et al. 1981)

The spinothalamic tract cell in this case was a high-threshold cell located in lamina I of the dorsal horn (Fig. 8D). The receptive field was on a restricted portion of the foot (Fig. 8E). Stimulation of the skin within the receptive field, using electric shocks, caused the cell to discharge a brief burst of spikes (Fig. 8A, upper trace). When a brief train of stimuli was applied within the NRM at the site shown in Fig. 8F just prior to the discharge of the cell, only a single spike was evoked

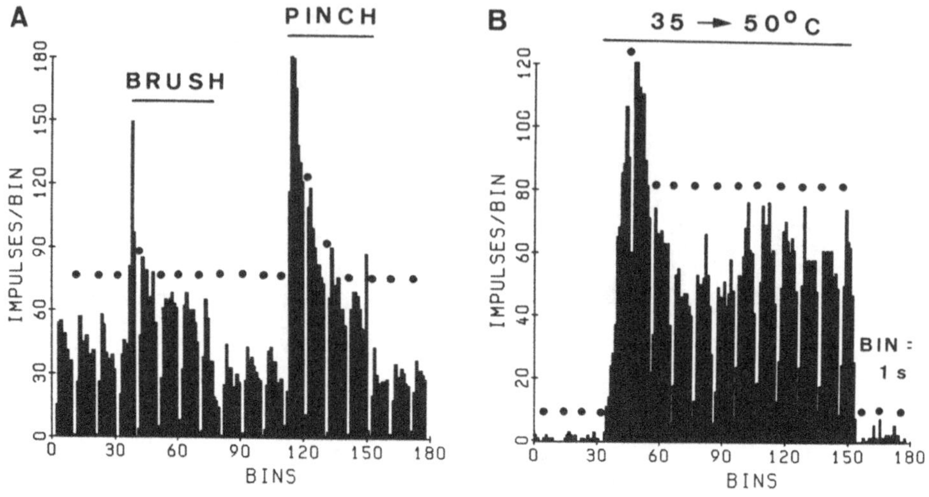

Fig. 10A,B. Inhibition of the responses of spinothalamic tract cells to innocuous and noxious stimuli following stimulation in the nucleus raphe magnus (NRM). The single-pass peristimulus time *histograms* in *A* and *B* were recorded from two different wide-dynamic-range spinothalamic tract cells. In each case, stimuli were applied in the NRM at the times *indicated by the dots* (2-s trains, 0.1-ms, 100-μA pulses at 333 Hz, repeated every 10 s). In *A*, NRM stimulation resulted in inhibition of the background activity and the responses to brushing and pinching the skin. In *B*, the response to heating the skin from 35° to 50°C was inhibited. (Gerhart et al. 1981)

by skin stimulation (Fig. 8A, third trace). The stimulus strength applied in the NRM was varied, and it could be shown that inhibition was produced by stimuli as weak as 25 μA (Fig. 8B). The inhibition was increased by temporal summation (Fig. 8C), and the time course of inhibition was prolonged (Fig. 8G).

The inhibitory action of stimulation in the NRM is somewhat selective in that the responses of spinothalamic tract cells to volleys in Aδ fibers are more effectively counteracted than are the responses to volleys in Aαβ fibers (Willis et al. 1977). Furthermore, the responses of spinothalamic tract cells to volleys in C fibers are more effectively inhibited than are responses to volleys in A fibers (Gerhart et al. 1981). An illustration of the very powerful inhibition that resulted from stimulation in the NRM of the responses of a spinothalamic tract cell to volleys in A and C fibers is shown in Fig. 9. Note that the activity due to the A and C components of the afferent volley was essentially eliminated, whereas the initial response to the Aαβ volley was largely intact during NRM stimulation.

The inhibition of the response of a spinothalamic tract cell to tactile stimulation and to noxious mechanical and thermal stimulation is shown in Fig. 10. The NRM was stimulated intermittently at the times indicated by the dots. There was a powerful inhibition of the background activity of the cell, as well as of the responses of the cell to brushing or pinching the skin (Fig. 10A) or elevation of the skin temperature from 35° to 50°C (Fig. 10B). Usually, the inhibition produced by NRM stimulation was nonselective, as in this instance, affecting the responses to innocuous and noxious stimulation alike. However, in some instances NRM inhibition was selective, causing a reduction in the responses to noxious, but not innocuous, stimuli (Gerhart et al. 1981).

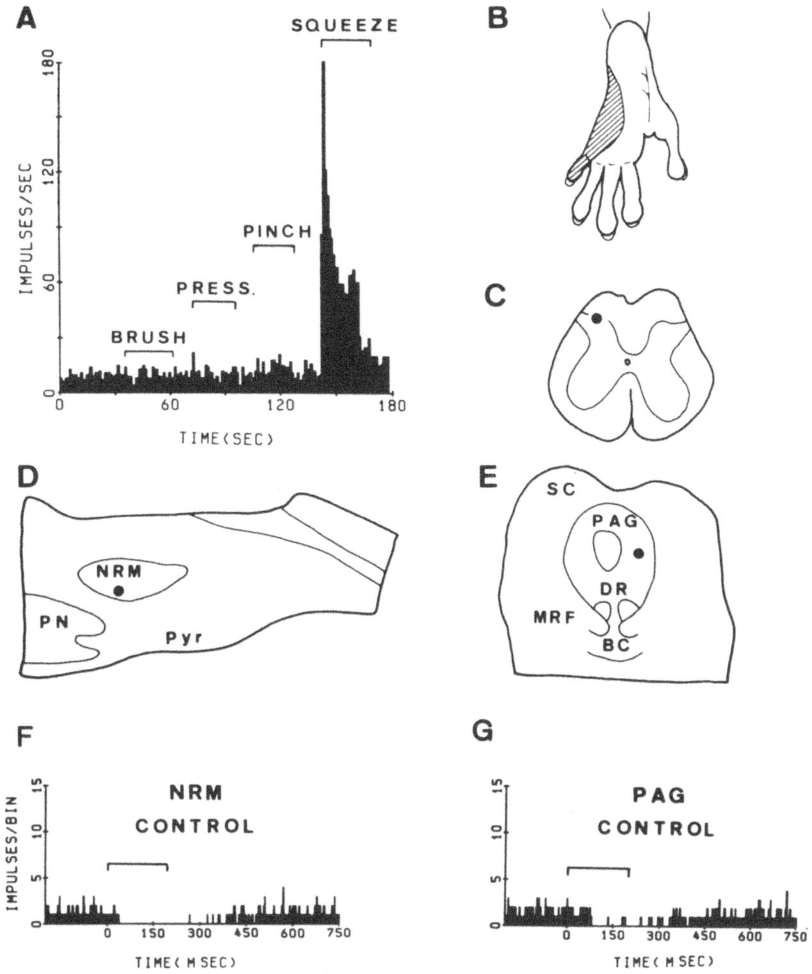

Fig. 11A–G. Inhibition of a high-threshold spinothalamic tract cell by stimulation in either the nucleus raphe magnus (*NRM*) or the periaqueductal gray (*PAG*). The single-pass peristimulus *histogram* in *A* shows that the cell responded only to the most intense mechanical stimulus tried, squeezing the skin with forceps. The receptive field is shown in *B*, the location of the cell in lamina I in *C*, and the stimulus sites in the NRM and PAG in *D* and *E*. The peristimulus time *histograms* in *F* and *G* resulted from ten successive sweeps, and the activity of the cell was enhanced during this time by squeezing the receptive field with forceps. Stimuli were applied in the NRM or PAG at the times indicated by the brackets (200-ms trains, 0.1-ms, 200-μA pulses at 333 Hz). Abbreviations: *PN* pontine nuclei; *Pyr* pyramids; *SC* superior colliculus; *DR* dorsal raphe nucleus; *MRF* mesencephalic reticular formation; *BC* brachium conjunctivum. (Yezierski et al. 1982).

In some experiments, we stimulated in both the NRM and in the PAG to determine if spinothalamic tract cells can be inhibited from both these sites, since both are associated with stimulation-produced analgesia. We found that stimulation in both the NRM and the PAG could inhibit a given spinothalamic tract cell (Yezierski et al. 1982). For instance, the cell in Fig. 11 was a high-threshold spi-

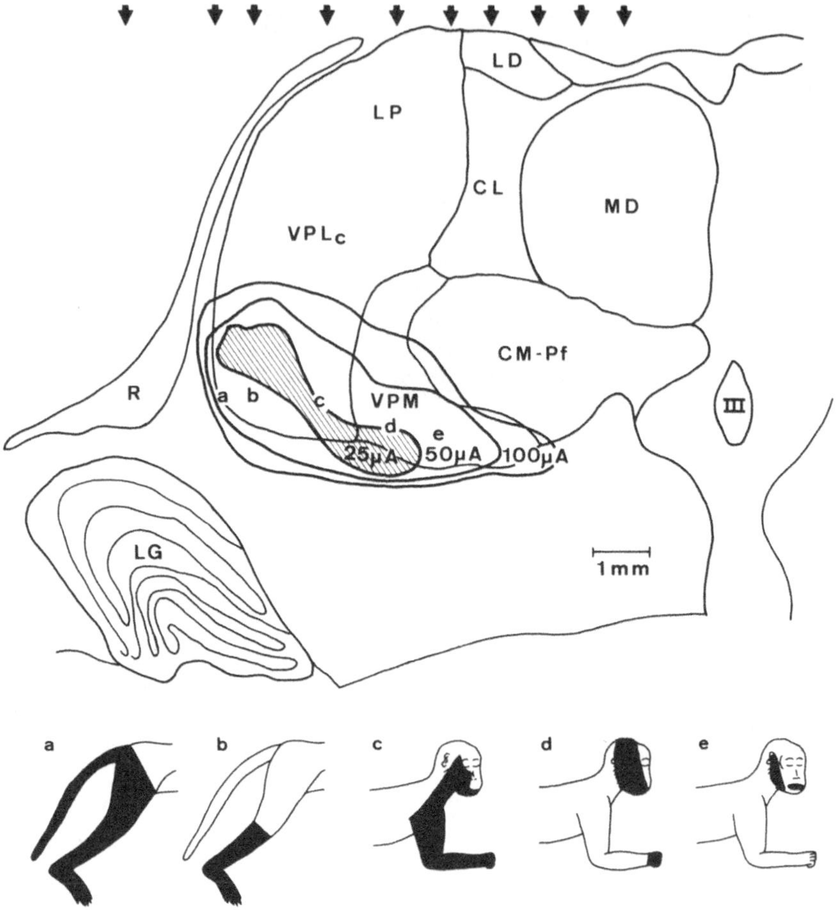

Fig. 12. Map of thalamic sites on the side of the body ipsilateral to the spinothalamic tract cell under study that when stimulated caused inhibition at low stimulus strengths. The *arrows* show the direction of a series of parallel tracks made through the thalamus by the stimulating electrode. The *contour plot* over the ventrobasal complex shows the areas in which stimuli at 100, 50, or 25 μA could be shown to result in inhibition of the spinothalamic tract cell (several-second trains at 333 Hz). The location of the electrode tip was verified anatomically after the experiment and also by recording multiunit receptive fields; these are shown for the sites indicated by the corresponding letters on the map by the *blackened areas on figurines a–e*. Abbreviations of thalamic nuclei: *CL* central lateral; *CM-Pf* centre median-parafascicular; *LD* lateral dorsal; *LG* lateral geniculate; *LP* lateral posterior; *MD* medial dorsal; *R* reticular; *VPL_c* ventral posterior lateral, pars caudalis; *VPM* ventral posterior medial; *III* third ventricle. (Gerhart et al. 1983)

nothalamic tract cell, as judged by its responses to a graded series of mechanical stimuli (Fig. 11A). The excitatory receptive field was in a limited area of the foot (Fig. 11B), and the cell was located in lamina I of the dorsal horn (Fig. 11C). Stimulation in the NRM (Fig. 11D) or PAG (Fig. 11E) caused an inhibition of the activity of the cell (Fig. 11F,G). As in the case of stimulation in the NRM, the inhibition produced by stimulation in the PAG tended to be more powerful with

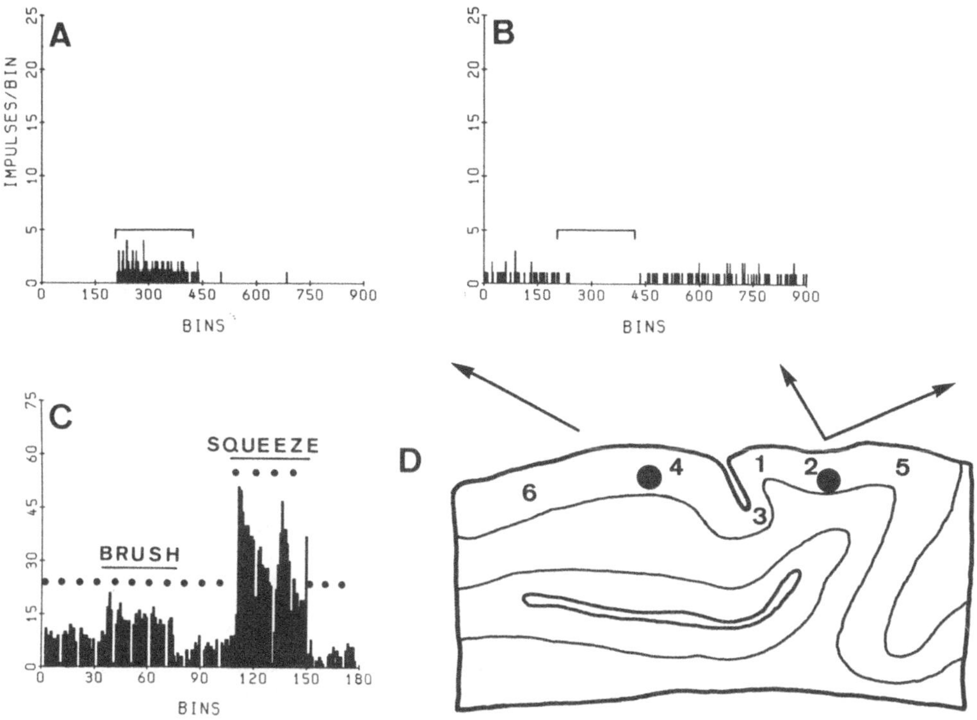

Fig. 13A–D. Excitation or inhibition of a spinothalamic tract cell following stimulation in the cerebral cortex. The peristimulus time *histograms* in *A* and *B* represent activity summed over ten successive stimulus trials, whereas the *histogram* in *C* is a single-pass one. In *A*, the cell is seen to be excited when stimuli are applied at the point in area *4* shown in the drawing below (200-ms trains, 0.1-ms, 200-μA pulses, 333 Hz). Comparable stimulation in area *2* of the SI sensory cortex produced the inhibition shown in *B* (the activity of the cell was enhanced by squeezing the skin with forceps). Inhibition of activity evoked by brushing or squeezing the skin is shown in *C* (2-s trains of stimuli repeated at 10-s intervals). (Yezierski et al. 1983)

respect to the responses of spinothalamic tract cells to volleys in C fibers than in A fibers, yet there was usually little evidence of selectivity with respect to responses to innocuous versus noxious forms of cutaneous stimulation. However, in some cases responses to noxious stimuli were inhibited, while responses to tactile stimuli were not.

Analgesia can be produced by stimulation not only in the brain stem but also in the thalamus (Hosobuchi et al. 1973; Mazars 1975). We have examined the possibility that thalamic stimulation will inhibit spinothalamic tract cells and found that the answer is affirmative (Gerhart et al. 1983). Figure 12 provides an example of an experiment in which a grid of tracks was made in order to stimulate in various parts of the thalamus to see what region provided the best inhibition with a limited stimulus strength. In this case, it was possible to demonstrate inhibition with stimuli of 25 μA or less when the stimulating electrode was in the ventrobasal complex of the thalamus. It seems to us likely that the inhibition in this case was due to the antidromic activation of pathways ascending to the ventrobasal complex, rather than by way of a thalamocortical loop, although we cannot rule out the latter. However, it is of interest that stimulation in the ventrobasal complex has been shown to excite raphe spinal neurons (Tsubokawa et al. 1981), which in turn could mediate the inhibition.

In a preliminary study, we found that stimulation of the sensorimotor cerebral cortex produces either inhibition or excitation and inhibition of spinothalamic tract cells (Coulter et al. 1974). However, the inhibition appeared to be selective, in that the responses to tactile stimulation were inhibited, whereas responses to noxious stimuli were not. More recently, although we have confirmed that such a selective inhibition can occur, we have also observed a nonselective inhibition of the responses of spinothalamic tract cells to both tactile and noxious stimuli (Yezierski et al. 1983). The inhibition was most commonly observed when stimuli were applied within the first somatosensory or SI cortex (Fig. 13). When stimulation was applied in the motor cortex, we often found that spinothalamic tract cells were excited (Fig. 13). This excitation could also be produced following stimulation in the medullary pyramid, and it was eliminated by transection of the region of the spinal cord that contains the lateral corticospinal tract.

Conclusions

These findings suggest that brain stem pathways that produce analgesia in behavioral tests also cause a powerful inhibition of spinothalamic tract cells. Work in other laboratories shows that comparable stimulation will result in the inhibition of nociceptive dorsal horn interneurons (Engberg et al. 1968; Liebeskind et al. 1973; Duggan and Griersmith 1979; Carstens et al. 1979; Fields et al. 1977b; Guilbaud et al. 1977). However, in our opinion, the observation of an inhibition of nociceptive sensory tract cells, such as spinothalamic cells, should provide a more direct correlation with changes in the sensation of pain than will alterations of the discharges of unidentified dorsal horn interneurons. The latter may show changes that reflect alterations in reflex transmission or in sensory tract discharges, depending upon the particular pathways to which they contribute. Our laboratory, of course, is only one of those using the responses of identified tract cells for such studies (cf. McCreery et al. 1979; Hayes et al. 1979).

The observation that stimulation of the cerebral cortex can produce either an inhibition or an excitation of spinothalamic tract cells is intriguing, since it suggests that the cerebral cortex may play a role not only in analgesia (cf. Fields and Adams 1974), but also in the activation of neural pathways producing pain. The latter possibility may be important for the eventual understanding of the mechanism of central pain states (Cassinari and Pagni 1969).

Acknowledgments. The author expresses his appreciation for the collaborative efforts of the following individuals in this work: A.E. Applebaum, J.E. Beall, R.N. Bryan, J.M. Chung, J.D. Coulter, Z.R. Fang, R.D. Foreman, K.D. Gerhart, G.J. Giesler, L.H. Haber, M.B. Hancock, D.R. Kenshalo Jr, G.A. Kevetter, R.B. Leonard, R.F. Martin, R.A. Maunz, R.J. Milne, B.D. Moore, B.J. Schrock, D.L. Trevino, T.K. Wilcox, Jean C. Willis, R.P. Yezierski. He also thanks Gail Silver, Helen Willcockson, and Griselda Gonzalez for their expert technical assistance and Phyllis Waldrop and Margie Watson for typing the manuscript. The work was supported by NIH Research Grants NS 09743 and 11255, and a grant from the Moody Foundation.

References

Applebaum AE, Beall JE, Foreman RD, Willis WD (1975) Organization and receptive fields of primate spinothalamic tract neurons. J Neurophysiol 38: 572–586

Basbaum AI, Marley NJE, O'Keefe JO, Clanton CH (1977) Reversal of morphine and stimulus-produced analgesia by subtotal spinal cord lesions. Pain 3: 43–56

Carstens E, Yokota T, Zimmermann M (1979) Inhibition of spinal neuronal responses to noxious skin heating by stimulation of mesencephalic periaqueductal gray in the cat. J Neurophysiol 42: 558–568

Cassinari V, Pagni CA (1969) Central Pain. A neurosurgical survey. Harvard University Press, Cambridge

Christensen BN, Perl ER (1970) Spinal neurons specifically excited by noxious or thermal stimuli: marginal zone of the dorsal horn. J Neurophysiol 33: 293–307

Chung JM, Kenshalo DR Jr, Gerhart KD, Willis WD (1979) Excitation of primate spinothalamic neurons by cutaneous C-fiber volleys. J Neurophysiol 42: 1354–1369

Cervero F, Iggo A, Ogawa H (1976) Nociceptor-driven dorsal horn neurones in the lumbar spinal cord of the cat. Pain 2: 5–24

Corvaja N, Grofová I, Pompeiano O, Walberg F (1977) The lateral reticular nucleus in the cat – I. An experimental anatomical study of its spinal and supraspinal afferent connections. Neuroscience 2: 537–553

Coulter JD, Maunz RA, Willis WD (1974) Effects of stimulation of sensorimotor cortex on primate spinothalamic neurons. Brain Res 65: 351–356

Craig AD, Kniffki KD (1982) Lumbosacral lamina I cells projecting to medial and/or lateral thalamus in the cat. Soc. Neurosci Abstr 8: 95

Dilly PN, Wall PD, Webster KE (1968) Cells of origin of the spinothalamic tract in the cat and rat. Exp Neurol 21: 550–562

Duggan AW, Griersmith BT (1979) Inhibition of the spinal transmission of nociceptive information by supraspinal stimulation in the cat. Pain 6: 149–161

Eccles JC (1957) The physiology of nerve cells. Hopkins, Baltimore

Eccles JC (1964) The physiology of synapses. Springer, Berlin Heidelberg New York

Eccles JC, Fatt P, Koketsu K (1954) Cholinergic and inhibitory synapses in a pathway from motor-axon collaterals to motoneurones. J Physiol 126: 524–562

Engberg I, Lundberg A, Ryall RW (1968) Reticulospinal inhibition of interneurones. J Physiol 194: 225–236

Fields HL, Adams JE (1974) Pain after cortical injury relieved by electrical stimulation of the internal capsule. Brain 97: 169–178

Fields HL, Clanton CH, Anderson SD (1977a) Somatosensory properties of spinoreticular neurons in the cat. Brain Res 120: 49–66

Fields HL, Basbaum AI, Clanton CH, Anderson SD (1977b) Nucleus raphe magnus inhibition of spinal cord dorsal horn neurons. Brain Res 126: 441–453

Foerster O, Gagel O (1931) Die Vorderseitenstrangdurchschneidung beim Menschen. Eine klinisch-patho-physiologisch-anatomische Studie. Z gesamte Neurol Psychiat 138: 1–92

Galifret Y, Szabo T (1960) Locating capillary microelectrode tips within nervous tissue. Nature 188: 1033–1034

Gerhart KD, Wilcox TK, Chung JM, Willis WD (1981) Inhibition of nociceptive and non-nociceptive responses of primate spinothalamic cells by stimulation in medial brain stem. J Neurophysiol 45: 121–136

Gerhart KD, Yezierski RP, Fang ZR, Willis WD (1983) Inhibition of primate spinothalamic tract neurons by stimulation in ventral posterior lateral (VPL_c) thalamic nucleus: possible mechanisms. J Neurophysiol 49: 406–423

Giesler GJ, Menétrey D, Basbaum AI (1979) Differential origins of spinothalamic tract projections to medial and lateral thalamus in the rat. J Comp Neurol 184: 107–126

Giesler GJ, Yezierski RP, Gerhart KD, Willis WD (1981) Spinothalamic tract neurons that project to medial and/or lateral thalamic nuclei: evidence for a physiologically novel population of spinal cord neurons. J Neurophysiol 46: 1285–1308
Guilbaud G, Oliveras JL, Giesler G, Besson JM (1977) Effects induced by stimulation of the centralis inferior nucleus of the raphe on dorsal horn interneurons in cat's spinal cord. Brain Res 126: 355–360
Haber LH, Moore BD, Willis WD (1982) Electrophysiological response properties of spinoreticular neurons in the monkey. J Comp Neurol 207: 75–84
Hancock MB, Foreman RD, Willis WD (1975) Convergence of visceral and cutaneous input onto spinothalamic tract cells in the thoracic spinal cord of the cat. Exp Neurol 47: 240–248
Handwerker HO, Iggo A, Zimmermann M (1975) Segmental and supraspinal actions on dorsal horn neurons responding to noxious and non-noxious skin stimuli. Pain 1: 147–165
Hayes RL, Price DD, Ruda M, Dubner R (1979) Suppression of nociceptive responses in the primate by electrical stimulation of the brain or morphine administration: behavioral and electrophysiological comparisons. Brain Res 167: 417–421
Hillman P, Wall PD (1969) Inhibitory and excitatory factors influencing the receptive fields of lamina V spinal cord cells. Exp Brain Res 9: 284–306
Hosobuchi Y, Adams JE, Rutkin B (1973) Chronic thalamic stimulation for the control of facial anaesthesia dolorosa. Arch Neurol 29: 158–161
Jankowska E, Lindström S (1971) Morphological identification of Renshaw cells. Acta physiol Scand 81: 428–430
Jankowska E, Lindström S (1973) Procion yellow staining of functionally identified interneurons in the spinal cord of the cat. In: Kater SB, Nicholson C (eds) Intracellular staining in neurobiology. Springer, Berlin Heidelberg New York, pp 199–207
Kenshalo DR Jr, Leonard RB, Chung JM, Willis WD (1979) Responses of primate spinothalamic neurons to graded and to repeated noxious heat stimuli. J Neurophysiol 42: 1370–1389
Kevetter GA, Willis WD (1982) Spinothalamic cells in the rat lumbar cord with collaterals to the medullary reticular formation. Brain Res 238: 181–185
Kevetter GA, Haber LH, Yezierski RP, Chung JM, Martin RF, Willis WD (1982) Cells of origin of the spinoreticular tract in the monkey. J Comp Neurol 207: 61–74
Kumazawa T, Perl ER, Burgess PR, Whitehorn D (1975) Ascending projections from marginal zone (lamina I) neurons of the spinal dorsal horn. J Comp Neurol 162: 1–12
Kuru M (1949) Sensory paths in the spinal cord and brain stem of man. Sogensya, Tokyo
Lagerbäck PA, Ronnevi LO (1982) An ultrastructural study of serially sectioned Renshaw cells. I. Architecture of the cell body, axon hillock, initial axon segment and proximal dendrites. Brain Res 235: 1–15
Liebeskind JC, Guilbaud G, Besson JM, Oliveras JL (1983) Analgesia from electrical stimulation of the periaqueductal gray matter in the cat: behavioral observations and inhibitory effects on spinal cord interneurons. Brain Res 50: 441–446
Lundberg A, Oscarsson O (1962) Two ascending spinal pathways in the ventral part of the cord. Acta Physiol Scand 54: 270–286
Mayer DJ, Liebeskind JC (1974) Pain reduction by focal electrical stimulation of the brain: an anatomical and behavioral analysis. Brain Res 68: 73–93
Mazars GJ (1975) Intermittent stimulation of nucleus ventralis posterolateralis for intractable pain. Surg Neurol 4: 93–95
McCreery DB, Bloedel JR, Hames EG (1979) Effects of stimulating in raphe nuclei and in reticular formation on response of spinothalamic neurons to mechanical stimuli. J Neurophysiol 42: 166–182
McGrath PA, Sharav Y, Dubner R, Gracely RH (1981) Masseter inhibitory periods and sensations evoked by electrical tooth pulp stimulation. Pain 10: 1–17
Mehler WR (1969) Some neurological species differences – a posteriori. Ann NY Acad Sci 167: 424–468
Mendell LM (1966) Physiological properties of unmyelinated fiber projection to the spinal cord. Exp Neurol 16: 316–332
Milne RJ, Foreman RD, Willis WD (1982) Responses of primate spinothalamic neurons located in the sacral intermediomedial gray (Stilling's nucleus) to proprioceptive input from the tail. Brain Res 234: 227–236

Molenaar I, Kuypers HGJM (1978) Cells of origin of propriospinal fibers and of fibers ascending to supraspinal levels. A HRP study in the cat and rhesus monkey. Brain Res 152: 429–450
Oliveras JL, Woda A, Guilbaud G, Besson JM (1974) Inhibition of the jaw opening reflex by electrical stimulation of the periaqueductal gray matter in the awake, unrestrained cat. Brain Res 72: 328–331
Oliveras JL, Redjemi F, Guilbaud G, Besson JM (1975) Analgesia induced by electrical stimulation of the inferior centralis nucleus of the raphe in the cat. Pain 1: 139–145
Oliveras JL, Guilbaud G, Besson JM (1979) A map of serotoninergic structures involved in stimulation producing analgesia in unrestrained freely moving cats. Brain Res 164: 317–322
Price DD, Mayer DJ (1974) Physiological laminar organization of the dorsal horn of *M. mulatta*. Brain Res 79: 321–325
Reynolds DV (1969) Surgery in the rat during electrical analgesia induced by focal brain stimulation. Science 164: 444–445
Rexed B (1952) The cytoarchitectonic organization of the spinal cord in the cat. J Comp Neurol 96: 415–495
Scheibel ME, Scheibel AB (1964) Are there Renshaw cells? Anat Rec 148: 332
Scheibel ME, Scheibel AB (1966) Spinal motoneurons, interneurons and Renshaw cells. A Golgi study. Arch Ital Biol 104: 328–353
Scheibel ME, Scheibel AB (1971) Inhibition and the Renshaw cell. A structural critique. Brain Behav Evol 4: 53–93
Thomas RC, Wilson VJ (1965) Precise localization of Renshaw cells with a new marking technique. Nature 206: 211–213
Trevino DL, Carstens E (1975) Confirmation of the location of spinothalamic neurons in the cat and monkey by the retrograde transport of horseradish peroxidase. Brain Res 98: 177–182
Trevino DL, Maunz RA, Bryan RN, Willis WD (1972) Location of cells of origin of the spinothalamic tract in the lumbar enlargement of cat. Exp Neurol 34: 64–77
Trevino DL, Coulter JD, Willis WD (1973) Location of cells of origin of spinothalamic tract in lumbar enlargement of the monkey. J Neurophysiol 36: 750–761
Tsubokawa T, Yamamoto T, Katayama Y, Moriyasu N (1981) Diencephalic modulation of activities of raphe-spinal neurons in the cat. Exp Neurol 74: 561–572
Van Keulen LCM (1971) Morphology of Renshaw cells. Pflügers Arch 328: 235–236
White JC, Sweet WH (1955) Pain, its mechanisms and neurosurgical control. Thomas, Springfield
Willer JC, Boureau F, Albe-Fessard D (1979) Supraspinal influences on nociceptive flexion reflex and pain sensation in man. Brain Res 179: 61–68
Willis WD (1971) The case for the Renshaw cell. Brain Behav Evol 4: 5–52
Willis WD (1981) Ascending pathways from the dorsal horn. In: Brown AG, Rethelyi R (eds) Spinal cord sensation. Sensory processing in the dorsal horn. Scottish Academic, Edinburgh, pp 169–178
Willis WD (1982) Control of nociceptive transmission in the spinal cord. In: Ottoson D (ed) Progress in sensory physiology 3. Springer, Berlin Heidelberg New York
Willis WD, Willis JC (1964) Location of Renshaw cells. Nature 204: 1214–1215
Willis WD, Willis JC (1966) Properties of interneurons in the ventral spinal cord. Arch Ital Biol 104: 354–386
Willis WD, Trevino DL, Coulter JD, Maunz RA (1974) Responses of primate spinothalamic tract neurons to natural stimulation of hindlimb. J Neurophysiol 37: 358–372
Willis WD, Haber LH, Martin RF (1977) Inhibition of spinothalamic tract cells and interneurons by brain stem stimulation in the monkey. J Neurophysiol 40: 968–981
Willis WD, Kenshalo DR Jr, Leonard RB (1979) The cells of origin of the primate spinothalamic tract. J Comp Neurol 188: 543–574
Yezierski RP, Wilcox TK, Willis WD (1982) The effects of serotonin antagonists on the inhibition of primate spinothalamic tract cells produced by stimulation in nucleus raphe magnus or periaqueductal gray. J Pharmacol Exp Ther 220: 266–277
Yezierski RP, Gerhart KD, Schrock BJ, Willis WD (1983) A further examination of effects of cortical stimulation on primate spinothalamic tract cells. J Neurophysiol 49: 424–441

Central Effects by Ventral Root Nociceptive Afferents

P.E. VOORHOEVE and B. ZWAAGSTRA, Amsterdam

Summary. Close-arterial injection of bradykinin in a cat low-spinal preparation with bilateral transection of all dorsal roots evoked a number of responses that are characteristic for a flexor reflex. There was an increase in motoneuronal discharge rate to the flexor muscle tenuissimus that could reversibly be blocked by morphine. There was a decrease in threshold for intraspinal microstimulation of afferent fibers and a decrease in size of evoked dorsal root potentials. Both phenomena are indicative of primary afferent depolarization. Ten out of 36 class 2 wide-dynamic-range cells in the dorsal horn were activated by the injection. These effects are attributed to activation of peripheral nociceptors that enter the spinal cord via unmyelinated afferent fibers in the ventral roots.

Key words: *nociception – ventral root afferents – flexor reflex*

Introduction

It is by now a well (re)established fact that lumbosacral ventral roots in frog, rat, cat, and man contain a large number of unmyelinated afferent fibers (Applebaum et al. 1976; Coggeshall 1980; Coggeshall et al. 1974, 1975). The receptive fields and adequate stimuli for these afferents have been worked out in detail by the Galveston group (Clifton et al. 1976; Coggeshall and Ito 1977). These results may be summarized as follows:

All lumbosacral ventral roots from L_7 downward contain some 30% unmyelinated axons. In the segments with an autonomic outflow, S_3 and Co_1 in particular, half of these are preganglionic fibers, but the remaining 15% are afferents having their cell bodies in the corresponding spinal ganglia. In L_7–S_1 all C fibers in the ventral roots are afferents, amounting to some 1000–2000 fibers per root.

In L_7 and S_1 70% of the nociceptive afferents are of somatic origin, and 30% innervate pelvic viscera. In S_3 and Co_1 this proportion is reversed and some 30% are of somatic origin, i.e., 5% of the total number of unmyelinated afferent fibers in that root.

The ratio is less clear in S_2. A higher proportion of somatic afferents was found in this root compared to the more caudal ones, and it has been suggested that cutaneous receptive fields might be more common in the spinal enlargement (Coggeshall und Ito 1977). That would fit with the observation by De Groat and Ryall (1968) that the distribution of autonomic fibers over S_2 and S_3 differed in

Experimental Brain Research, Suppl. 9
© Springer-Verlag Berlin · Heidelberg 1984

different cat. In 25% of their preparations S_2 did not contain any preganglionic fibers at al.

Almost all somatic ventral root afferent fibers are nociceptors of some kind. Not all visceral afferents, however, could be classified as nociceptive. Some units supplying bladder, urethra, and lower bowel reacted to moderate extension as has been previously been described by other authors of dorsal root afferents. The majority of mucosal units, on the other hand, reacted similarly to cutaneous polymodal nociceptors.

Reports on the central terminations of these fibers, however, are scant. In one study horseradish peroxidase (HRP) was applied to proximal ventral roots of the coccygeal and sacral spinal cord of cats (Light and Metz 1978). Some stained fibers assumed to be primary afferents of small caliber (0.1–1.0 μm) went right up to the dorsal horn with terminal enlargements in the substantia gelatinosa. In two other studies by Maynard et al. (1977) and by Yamamoto et al. (1977), respectively, HRP was injected into the dorsal horn of the lumbosacral spinal cord of cats with severed dorsal roots. A number of labeled cells were then found in the spinal ganglia. All reports are suggestive, but none of them is conclusive with regard to the central termination, and doubts are still expressed as to whether they do enter the cord at all (Risling 1981). An extension of this approach is certainly needed.

The physiological role ascribed to this afferent system is also a matter of uncertainty, as is its significance in man, e. g., in the failure of dorsal rhizotomy for relief of intractable pain (Coggeshall et al. 1975; Hosobuchi 1980). If we limit ourselves to experimental physiology, only three studies report a positive effect of ventral root stimulation. Longhurst et al. (1980) noted changes in blood pressure in the cat on stimulation of the sciatic nerve after sectioning of the corresponding dorsal roots. These results, however, are not generally accepted by circulatory physiologists. In another study by Niechaj et al. (1977) the occurrence of dorsal root potentials (DRP) is described in all neighboring dorsal roots on both sides of the cord on stimulation of a ventral root in the lower sacral cord in the cat at C fiber strength. A dorsal root reflex (DRR), however, could only be evoked in the ipsilateral root of the same segment. Since both these responses are expressions of the same phenomenon of primary afferent depolarization (PAD), it seemed unlikely that such specificity in evoking DRRs would exist (Schmidt 1971). Moreover, the latency of the DRRs did not match the necessary conduction time for a C fiber volley. These results have subsequently been refuted in a previous report from our laboratory, and are attributed to spread of current due to too strong electrical stimulation and to inadequate grounding of the stimulated ventral root (Voorhoeve and Nauta 1983).

Since one would expect at first hand a general flexor reflex pattern to be evoked also by nociceptive ventral root afferents, we have in the same study applied a number of standard tests for this response. The results obtained on electrical stimulation of ventral roots with the appropriate precautions against current spread, and monitoring the C fiber action potential, turned out to be negative, however. Activation of all nociceptors in the lower part of the body by intraarterial injection of bradykinin (BK) in the aortic bifurcation (Beck and Handwerker 1974; Besson et al. 1972; Juan and Lembeck 1974; Mense 1977), on the other hand, evoked in a low-spinal preparation with all dorsal roots severed bilaterally a number of responses that are in accordance with the flexor reflex. The present study reports on such effects.

Methods

Cats were used throughout. They were anesthetized with 40 mg chloralose per kilogram and additional urethane when needed. They were spinalized at the L_3–L_4 level and in the crucial phase of the experiment all dorsal roots on both sides of the cord had been transected below that level. Hence the preparation was, in the classical sense, completely deafferentiated. In order to obtain a higher degree of spontaneous unit activity three animals were aseptically spinalized at that level 1 week before the experiment for microelectrode recording in the dorsal horn. The acute experiment could then be performed with only 20 mg chloralose per kilogram, supplemented by minute doses of urethane or the short-acting barbiturate methitural.

Body temperature, blood pressure, blood gases, and acid-base balance were monitored during the experiment and kept within physiological limits by a heating pad and appropriate infusions. The correction of pH with 5% $NaHCO_3$ in particular is considered to be essential for keeping the preparation in optimal condition. Mean arterial blood pressure was usually around 120 mmHg, and was never allowed to fall below 100 mmHg.

All electrical stimulation and recording was done by conventional means on distally severed roots or nerves. Pulse duration for C fiber stimulation was at least 0.5 ms. Single-cell discharges in the dorsal horn were recorded with low-resistance glass microelectrodes filled with 4M NaCl. Poststimulus-time histograms were obtained after passing the recorded potentials through a window discriminator (Neurolog NL 200) and counted on an Ortec system (4620/21). Doses of 10–100 μg BK were dissolved in 1 ml Locke's solution and rapidly injected via a cannula in the right femoral artery with its tip at the aortic bifurcation.

Results

Motoneuronal Discharges

"Spontaneous" discharges in fusimotor (γ) and occasionally in skeletomotor (α) neurons of the flexor muscle tenuissimus do occur in the acute spinal preparation, even after bilateral total dorsal rhizotomy (Voorhoeve 1960). This preparation provides a sensitive tool in establishing flexor reflex effects (Laporte et al. 1958). No increase in the rate of discharge could be obtained by repetitive stimulation of sacral ventral roots at several times maximal C fiber strength (Voorhoeve and Nauta 1983). The preparation responded vividly, however, to close-arterial injection of BK. Varying doses of 10–100 μg in 1 ml bolus injection were applied, flooding the pelvis and the left hind limb with this algesic agent. After a latency of some 10–20 s the overall discharge rate to tenuissimus increased in a dose de-

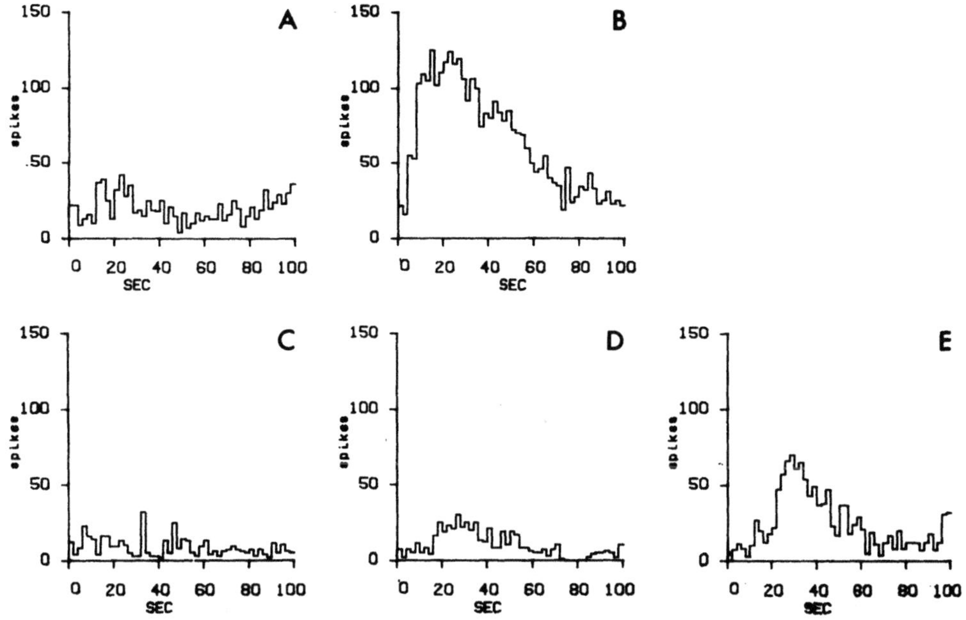

Fig. 1A–E. Poststimulus-time histograms of motor discharges to tenuissimus muscle in the same preparation. Bin width 2 s. *A* spontaneous; *B* after 25 μg bradykinin (BK) intraarterially at zero time; *C* spontaneous after 2 mg/kg morphine; *D* 25 μg BK; *E* 25 μg BK after 0.07 mg/kg naloxone.

pendent way. At higher doses alpha discharges occurred also. The period of increased activity lasted for about a minute, and the response could be evoked several times in the same preparation, although some diminished response (tachyphylaxia) occurred on repeated injections. Systemically applied morphine blocked the response, which could be reobtained after naloxone (Fig. 1). No discharges were evoked in a previously silent branch to the gastrocnemius muscle, so the response appeared to be an expression of the flexor reflex. The concurrent fall in blood pressure and the bradycardia some 20–30 s after the injection did not seem to be the cause of the increased firing rate. These circulatory effects were moderate and had a longer latency, and hypotension by thoracic compression did not evoke an increase in discharge to tenuissimus.

Primary Afferent Depolarization

Impulses in flexor reflex afferents (FRA) evoke PAD in other FRA, as can be shown in several ways (Schmidt 1971). Selective electrical activation of C fibers, or differential blocking of large-diameter A fibers, have also been employed to demonstrate that in an intact preparation C fiber activation evokes PAD in myelinated fibers (Burke et al. 1971; Franz and Iggo 1968; Jänig and Zimmermann 1971; Zimmermann 1968). It is technically far more difficult to investigate terminal depolarization in C fibers, but some studies reporting on this phenomenon have appeared in recent years (Calvillo 1978; Fitzgerald and Woolf 1981; Hentall and Fields 1979).

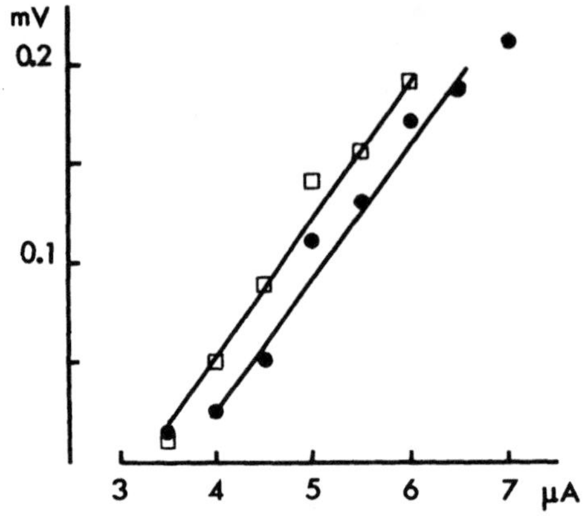

Fig. 2. Excitability curves for dorsal root terminals on intraspinal microstimulation at 1.8 mm depth. Size of antidromically recorded spike potential in mV plotted against stimulus strength in μA. ● test; □ during injection of 50 μg bradykinin intraarterially.

The amount of PAD can be assessed by stimulating the afferent terminals through a microelectrode and recording the antidromically conducted action potential in the corresponding nerve (Wall 1958). In our experiments this was done in a dorsal root filament, so no identification of the fibers receiving PAD was possible. Because of the short period available following the injection of BK, and the limited number of times this procedure could be repeated, only a few points on the excitability curve have been established. Nevertheless, the results obtained were fairly consistent. The curve relating amplitude of the antidromically recorded action potential as a function of the intraspinal stimulus intensity shifted to the left (Fig. 2), disclosing an increase in excitability of some 10% (Eccles et al. 1962).

Another, more indirect way of demonstrating PAD is either by recording the electrotonic spread of the depolarization along a distally severed dorsal root filament, the DRP, or by recording a slow potential on the surface of the cord, the so-called P wave. The latter is considered to be less discriminative.

Injection of BK as such did not evoke any detectable sign of PAD, which is not so surprising in view of the asynchronous and prolonged character of the input. We did not, however, record intracellularly from primary afferent fibers during the injection. In the present experiments DRPs were evoked in a dorsal root filament of L_6 or L_7 by stimulating electrically adjacent filaments every 2 s at low intensity. Following close-arterial injection of BK the amplitude of the DRPs decreased for a period of about 1 min (Fig. 3). This duration corresponded in general with the known increase in discharge rate in tenuissimus motoneurons.

Since it is assumed that both of the fibers giving and those receiving PAD in the testing situation are already depolarized by the conditioning nociceptor activation by BK, this might theoretically result in summation, occlusion, and/or presynaptic inhibition by diminishing the spike size of the test volley. Apparently, the decreasing influences dominated.

Both observations, however, are a clear indication for a moderate degree of PAD in coarse afferent fibers, presumably FRA (Jänig and Zimmermann 1971),

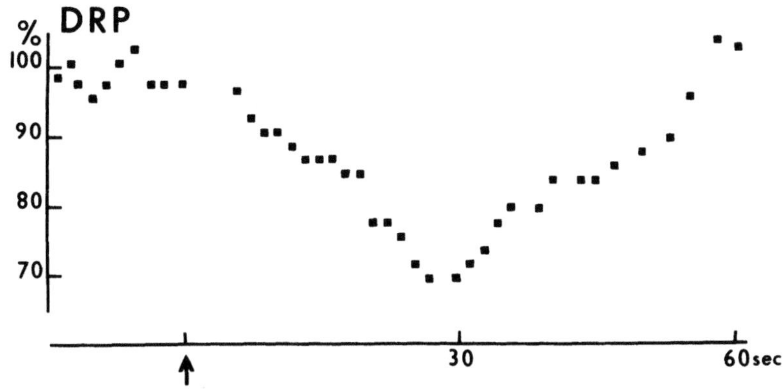

Fig. 3. Amplitude of dorsal root potential (DRP) in percentage of test in L_6 filament evoked by stimulation of L_7 dorsal root filament, after injection of 100 μg bradykinin intraarterially at *arrow*.

evoked by asynchronous impulse activity in the ventral root nociceptive afferent C fibers. No attempts have been made to establish whether C fiber terminals were also depolarized.

Dorsal Horn Neurons

In the deeper layers of the dorsal horn, especially in lamina V, one meets a class of neurons that is activated by afferents conveying innocuous as well as noxious information. Such cells are designated by a variety of names, "class 2 cells", "wide-dynamic-range cells", "polymodal cells" and "convergent neurons" being the ones most commonly employed (Cervero et al. 1976; Fitzgerald and Wall 1980; Gregor and Zimmermann 1972; Handwerker et al. 1975; Menétrey et al. 1977; Pomeranz et al. 1968; Price and Dubner 1977).

Cells of this type were identified by a prolonged or a second discharge on increasing a dorsal root stimulus to C fiber strength. In our initial series of experiments on 11 such cells, none could be fired by ventral root stimulation at supramaximal C fiber strength (Voorhoeve and Nauta 1983). Although intraarterial injection of BK is a commonly applied technique to study the effect of nociceptive input to the cord by the classical entry via the dorsal roots (Beck and Handwerker 1974; Besson et al. 1972; Hong et al. 1979; Mense 1977; Mense and Schmidt 1974), even then the yield is meager (Soja and Sinclair 1980). The response of dorsal horn cells to radiant heat seems to be more consistent, but failed in our hands when only the ventral roots were intact. This may be due to the fact that one can activate only limited receptive fields in this way, whereas a whole area can be activated by BK.

Responding cells in our experiments reacted after a short latency of some 15 s either with an increase in spontaneous irregular bursting activity (Fig. 4), or with a fairly even enhanced firing rate, matching in time the already known phenomenon of tenuissimus discharges (Fig. 5). In order to start from a maximum spontaneous background activity, three experiments were performed in chronic spinal animals. In the acute experiment these also needed less anesthetic than prepara-

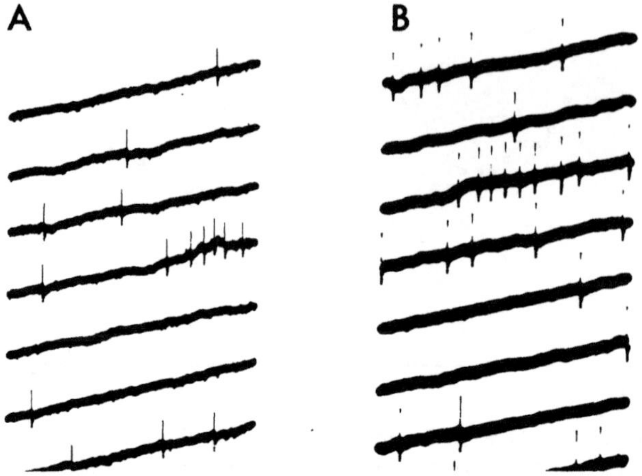

Fig. 4. Class 2 cell in dorsal horn, continuous recording on running film, sweep duration 200 ms. *A* spontaneous firing (spikes slightly retouched); *B* after injection of 100 μg bradykinin intraarterially, note increased spike size.

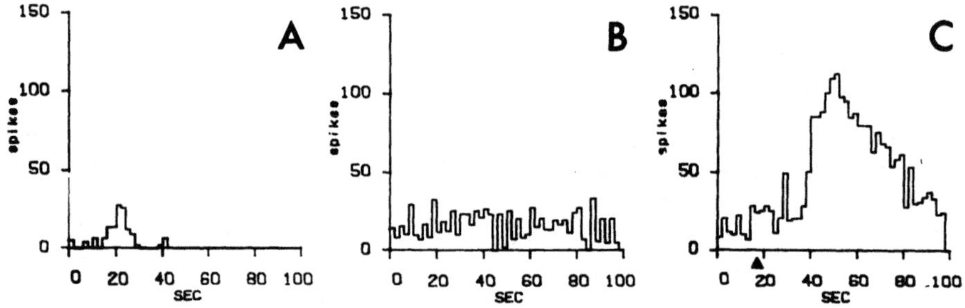

Fig. 5. Poststimulus-time histograms of firing rate of two different class 2 cells. Bin width 2 s. *A* no spontaneous firing, small response to injection of 50 μg bradykinin (BK) intraarterially at zero time; *B* chronically spinalized preparation, spontaneously firing unit; *C* the same after injection of 100 μg BK at *arrow*.

tions on which the whole operation had to be performed. The results did not differ essentially from those obtained in the acutely spinalized cord, although the "best" cells were encountered in these preparations. Proper recordings were obtained from ten cells only out of 37 units encountered. Of the remainder, 13 cells showed dubious responses and 14 did not respond at all. All units recorded from were located at a depth of about 1.5 mm when entering the cord just medial from the dorsal root entry zone, pointing 10° medially.

When BK is injected, even intraarterially, however, circulatory perturbations are unavoidable that may, and often do, influence microelectrode recordings in the cord. This is probably due to temporary shrinking and swelling of the tissue, disrupting the contact between the tip of the electrode and the unit under study, especially since higher doses of 100 μg had to be applied to produce any response at all. A cell could also start firing, or increase its firing rate 10–15 s after the injection, and then be injured by the tip of the electrode after 20-odd s. So it was obviously activated by the nociceptive input, but could not be properly analyzed.

All such responses, however, had a longer latency, occurred after a drop in blood pressure, were often accompanied by an increase in spike size, and usually ended with an injury discharge. Even if one cannot always be quite sure, we have the impression that one can distinguish both patterns and that nociceptively evoked responses were obtained only in about one out of four cells encountered. This figure corresponds with the one mentioned by Soja and Sinclair on the difference in cell responses evoked by BK and radiant heat respectively (Soja and Sinclair 1980).

Discussion

Our search for flexor responses has not been exhaustive, but the results obtained are consistent. Increased motoneuronal firing to flexor muscles and PAD in large myelinated afferent fibers are characteristic for the flexor reflex. The results obtained on dorsal horn neurons, on the other hand, are more ambiguous. There is a certain discrepancy between the ease with which flexor motor discharges are produced and the paucity of responses in the dorsal horn, but this is not a unique property of the ventral root input system.

At first hand one would expect all incoming information to be processed in the dorsal horn before being funneled toward the flexor final common path, and that there would therefore be a large amount of unit activity in the dorsal parts of the cord. There is, however, undoubtedly a bias toward larger cells in our sampling method. We did not record from the substantia gelatinosa, where most of the C fiber input establishes its first synaptic contacts, at least when entering via the dorsal roots (Cervero and Iggo 1980).

Scrutinizing the literature, one also gets the impression that the dorsal horn is not so packed with detectable unit activity as one would hope for, and all sorts of responses, excitatory and inhibitory, can be obtained (Cervero et al. 1976, 1977; Wall et al. 1979).

Another puzzling observation, although not emphasized in the present study, is the lack of response upon electrical stimulation of the ventral roots. In an analysis by Coggeshall and Ito (1977) of 95 ventral root nociceptive C fibers in L_7 and S_1, the mean conduction velocity was reported to be 0.8 m/s (0.2–2.4). In that study 19 fibers were conducted at a lower speed, 23 at the mean velocity and 53 at higher speed, the median value being also 0.8 m/s. The conduction velocity of the C fiber action potential in our previous study was in the same range (Voorhoeve and Nauta 1983). So it seemed unlikely that electrical stimulation at several times supramaximal intensity for evoking a maximum C potentials would not activate a large number of afferents. On the other hand, one cannot be sure, since the C fiber compound action potential is best obtained in the longer sacral roots, where the afferents are only 15% of the total number of fibers and where it may be masked by the antidromic preganglionic efferent potential. We still do not know whether there was a proximal block in C fiber conduction or a failure to activate electrically a sufficient number of the finest C fiber afferents, or whether it is a matter of a different synaptic organization. With the present data available we do not suppose the ventrally entering nociceptive system to be functionally different from the dorsal one, even if this possibility cannot be excluded.

Unfortunately, no exact figures are available on the proportion of unmyelinated nociceptive fibers in the dorsal roots, even if the total percentage of C fibers is estimated at around 50%.

If there is any difference, it seems to be quantitative rather than functional, but why does such an elusive afferent system exist?

References

Applebaum ML, Clifton GL, Coggeshall RE, Coulter JD, Vance WH, Willis WD (1976) Unmyelinated fibres in the sacral 3 and caudal 1 ventral roots of the cat. J Physiol (Lond) 256: 557–572

Beck PW, Handwerker HO (1974) Bradykinin and serotonin effects on various types of cutaneous nerve fibres. Pflügers Arch 347: 209–222

Besson JM, Conseiller C, Haman KF, Maillard MC (1972) Modifications of dorsal horn cell activities in the spinal cord after intra-arterial injection of bradykinin. J Physiol (Lond) 221: 189–205

Burke RE, Rudomin P, Vyklický L, Zajac FE (1971) Primary afferent depolarization and flexion reflexes produced by radiant heat stimulation of the skin. J Physiol (Lond) 213: 185–214

Calvillo O (1978) Primary afferent depolarization of C fibres in the spinal cord of the cat. Can J Physiol Pharmacol 56: 154–157

Cervero F, Iggo A (1980) The substantia gelatinosa of the spinal cord. A critical review. Brain 103: 717–772

Cervero F, Iggo A, Ogawa H (1976) Nociceptor-driven dorsal horn neurones in the lumbar spinal cord of the cat. Pain 2: 5–24

Cervero F, Moloney VZ, Iggo A (1977) Extracellular and intracellular recordings from neurons in the substantia gelatinosa Rolandi. Brain Res 136: 565–569

Clifton GL, Coggeshall RE, Vance WH, Willis WD (1976) Receptive fields of unmyelinated ventral root afferent fibres in the cat. J Physiol (Lond) 256: 573–600

Coggeshall RE (1980) Law of separation of function of the spinal roots. Physiol Rev 60: 716–755

Coggeshall RE, Ito H (1977) Sensory fibres in ventral roots L7 and S1 in the cat. J Physiol 267: 215–235

Coggeshall RE, Coulter JD, Willis WD (1974) Unmyelinated axons in the ventral roots of the cat lumbosacral enlargement. J Comp Neurol 153: 39–58

Coggeshall RE, Applebaum ML, Frazen M, Stubbs TB, Sykes MT (1975) Unmyelinated axons in human ventral roots, a possible explanation for the failure of dorsal rhizotomy to relieve pain. Brain 98: 157–166

De Groat WC, Ryall RW (1968) The indentification and characteristics of sacral parasympathetic preganglionic neurons. J Physiol (Lond) 196: 563–577

Eccles JC, Magni F, Willis WD (1962) Depolarization of central terminals of group I afferent fibres from muscle. J Physiol (Lond) 160: 62–93

Fitzgerald M, Wall PD (1980) The laminar organization of dorsal horn cells responding to peripheral C fibre stimulation. Exp Brain Res 41: 36–44

Fitzgerald M, Woolf CJ (1981) Effects of cutaneous nerve and intraspinal conditioning on C-fibre afferent terminal excitability in decerebrate spinal rats. J Physiol 318: 25–39

Franz DN, Iggo A (1968) Dorsal root potentials and ventral root reflexes evoked by nonmyelinated fibers. Science 162: 1140–1142

Gregor M, Zimmermann M (1972) Characteristics of spinal neurones responding to cutaneous myelinated and unmyelinated fibres. J Physiol (Lond) 221: 555–576

Handwerker Ho, Iggo A, Zimmermann M (1975) Segmental and supraspinal actions on dorsal horn neurons responding to noxious and non-noxious skin. Pain 1: 147–165

Hentall ID, Fields HL (1979) Segmental and descending influences on intraspinal thresholds of single C-fibers. J Neurophysiol 42: 1527–1537
Hong SK, Kniffki KD, Mense S, Schmidt RF, Wendisch M (1979) Descending influences on the responses of spinocervical tract neurons to chemical stimulation of fine muscle afferents. J Physiol (Lond) 290: 129–140
Hosobuchi Y (1980) The majority of unmyelinated afferent axons in human ventral roots probably conduct pain. Pain 8: 167–180
Jänig W, Zimmermann M (1971) Presynaptic depolarization of myelinated afferent fibres evoked by stimulation of cutaneous C fibres. J Physiol (Lond) 214: 29–50
Juan H, Lembeck F (1974) Action of peptides and other algesic agents on paravascular pain receptors of the isolated perfused rabbit ear. Naunyn Schmiedebergs Arch Pharmacol 283: 151–164
Laporte Y, Bessou P, Voorhoeve PE (1958) Action de volées amyéliniques d'origine cutanée sur l'activité tonique de motoneurones de muscles fléchisseurs ipsilatéraux. CR Soc Biol 152: 663–665
Light AR, Metz CB (1978) The morphology of the spinal cord efferent and afferent neurons contributing to the ventral roots of the cat. J Comp Neurol 179: 501–516
Longhurst JC, Mitchell JH, Moore MB (1980) The spinal cord ventral root: an afferent pathway of the hind-limb pressor reflex in cats. J Physiol (Lond) 301: 467–476
Maynard CW, Leonard RB, Coulter JD, Coggeshall RE (1977) Central connections of ventral root afferents as demonstrated by the HRP method. J Comp Neurol 172: 601–608
Menétrey D, Giesler GJ, Besson JM (1977) An analysis of response properties of spinal cord dorsal horn neurons to non-noxious and noxious stimuli in the spinal rat. Exp Brain Res 27: 15–33
Mense S (1977) Nervous outflow from skeletal muscle following chemical noxious stimulation. J Physiol (Lond) 267: 75–88
Mense S, Schmidt RF (1974) Activation of group IV afferents from muscle by algesic agents. Brain Res 72: 305–310
Niechaj A, Lupa K, Ozóg M (1977) Dorsal root potentials evoked by stimulation of ventral roots in the lower sacral cord of the cat. Brain Res 137: 356–360
Pomeranz P, Wall PD, Weber WV (1968) Cord cells responding to fine myelinated afferents from viscera, muscle and skin. J Physiol (Lond) 199: 511–532
Price DD, Dubner R (1977) Neurons that subserve the sensory-discriminative aspects of pain. Pain 3: 307–338
Risling M (1981) Oral communication at the 3rd IASP congress on pain, Edinburgh (no abstract)
Schmidt RF (1971) Presynaptic inhibition in the vertebrate central nervous system. Rev Physiol 63: 20–101
Soja PJ, Sinclair JG (1980) The response of dorsal horn neurones of the cat intra-arterial bradykinin and noxious radiant heat. Neurosci Lett 20: 183–188
Voorhoeve PE (1960) Autochthonous activity of fusimotor neurones in the cat. Acta Physiol Pharmacol Neerlandica 9: 1–43
Voorhoeve PE, Nauta J (1983) Do nociceptive ventral root afferents exert central somatic effects? In: Bonica JJ, Lindblom U, Iggo A (eds) Advances in pain research and therapy, vol 5. Raven, New York, pp 105–110
Wall PD (1958) Excitability changes in afferent fibre terminations and their relation to slow potentials. J Physiol (Lond) 142: 1–21
Wall PD, Merrill EC, Yaksh TL (1979) Responses of single units in laminae 2 and 3 of cat spinal cord. Brain Res 160: 245–260
Yamamota T, Takahashi K, Satomi H (1977) Origins of primary afferent fibers in the spinal ventral root in the cat as demonstrated by the horseradish peroxidase method. Brain Res 126: 350–354
Zimmermann M (1968) Dorsal root potentials after C-fiber stimulation. Science 160: 896–898

III. The Cerebellum as a Neuronal Machine

Frederic Bremer, John Eccles, and Ragnar Granit at a symposium in Groningen 1955

Purkinje Cells of the Cerebellum: Localization and Function of Multiple Neuroactive Substances*

V. CHAN-PALAY, Boston

Coexistence denotes the situation in which more than one neuroactive substance can be localized by cytochemical means in an individual neuron. Usually, this has meant the combination of a traditional neurotransmitter, such as acetylcholine or norepinephrine or an amino acid, together with a peptide, such as substance P or enkephalin. But this definition does not exclude the presence of more than one traditional neurotransmitter in the same cell or more than one neuroactive substance, such as peptides. The issue of the presence of multiple neuroactive substance in neurons is not limited to the mammalian central nervous system (CNS) – the peripheral nervous system of vertebrates and identified neurons in a number of invertebrates also have examples of this phenomenon. True coexistence of neuroactive substances in single neurons as defined here must be distinguished from the looser definitions used by many investigators to denote the simultaneous existence of multiple neuroactive substances in different nerve cells in a particular geographic area of the brain. This can be illustrated by specific examples: single serotonin neurons in the medulla and raphe nuclei have been shown also to contain substance P (Chan-Palay et al. 1978; Chan-Palay 1979; Hökfelt et al. 1978); Purkinje cells in the cerebellum have been found to have motilin and glutamic acid decarboxylase (Chan-Palay et al. 1981). However, enkephalin, substance P, and glutamic acid decarboxylase immunoreactivity have been demonstrated in separate cell types in the neostriatum (Panula et al. 1980); somatostatin and substance P have been distinguished in separate cell groups in primary afferent neurons (Hökfelt et al. 1976); and serotonin, tyrosine hydroxylase, and enkephalin exist simultaneously in separate cells elsewhere in the rat CNS (Barber et al. 1979). Since it is clear that one geographic area or nucleus in the brain consists of many nerve cells that may be heterogeneous, it would be expected that these neurons could utilize a number of different chemicals in neural communication. Thus the simultaneous existence of a number of neuroactive compounds in a nerve cell group is not surprising. The phenomenon of true coexistence of multiple neuroactive substances in a single cell is, however, more interesting, and calls for careful investigation of the parameters of such coexistence, its function, the cytological substrates that allow for orderly synthesis, transport and release of the compounds, and, of course, its significance in the complexity of nerve circuits.

Gamma-aminobutyric acid (GABA) is a major inhibitory neurotransmitter in the mammalian CNS and deficits of GABA have been implicated in certain neu-

---

* Supported in part by USPHS grants NS 14740 and NS 18539 and AFOSR 82-0328

Experimental Brain Research, Suppl. 9
© Springer-Verlag Berlin · Heidelberg 1984

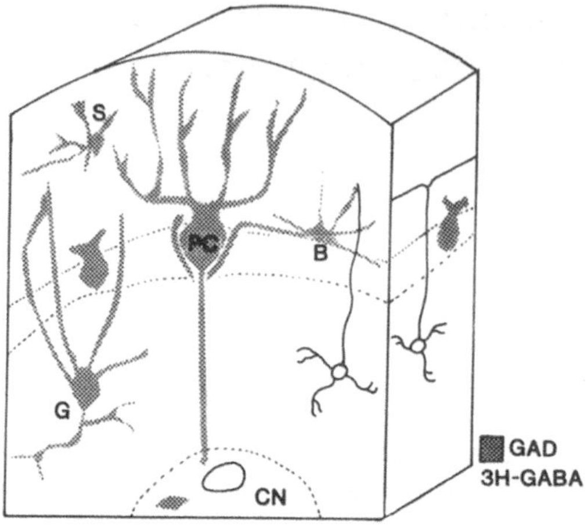

Fig. 1. Cerebellar cortical and nuclear neurons with L-glutamic acid decarboxylase (*GAD*) immunoreactivity and uptake capacities for gamma-aminobutyric acid (^{3}H-GABA) are indicated by *stippling*. These are stellate (*S*), basket (*B*), Purkinje (*PC*), and Golgi cells (*G*) of the cortex, and some small neurons in the deep cerebellar nuclei (*CN*).

rological and psychiatric disorders, such as Huntington's chorea, Parkinson's disease, and schizophrenia. Considerable interest has been attached in recent years to the identification of GABA-containing neurons, their possible inhibitory function, the identification of receptors, and the pharmacologic manipulations of GABA-activated receptor sites. The cerebellum remains a major area in the CNS where the amino acid transmitters figure largely, and a considerable literature supports the identification of GABA as the inhibitory transmitter for many cerebellar neurons. Among the approaches that have been used to support the localization and role of GABA are: (a) autoradiography following the uptake of ^{3}H-GABA (Hökfelt and Ljungdahl 1970); (b) immunocytochemistry on tissue sections with antibodies directed against the enzyme L-glutamic acid decarboxylase (GAD) responsible for the synthesis of GABA (Barber and Saito 1976; McLaughlin et al. 1974, 1975; Wood et al. 1976), or with anti-GAD antibody injected directly into the live tissue to trace GABA-specific pathways (Chan-Palay et al. 1979a); (c) immunocytochemistry with antibodies directed against the enzyme GABA-transaminase (GABA-T), the major metabolic emzyme for GABA (Barber and Saito 1976; Chan-Palay et al. 1979b); and (d) autoradiographic demonstrations of various receptor-GABA binding sites with ^{3}H-GABA analogs such as ^{3}H-muscimol (Chan-Palay 1978; Chan-Palay and Palay 1978). Of the cerebellar neurons in the cortex, a majority (although not all) of the small interneurons – e.g., stellate, basket, and Golgi cells – are labeled by uptake of ^{3}H-GABA, anti-GAD antibody, GABA-T, and ^{3}H-muscimol binding, thus indicating that GABA is a reasonably consistent constituent. In the deep cerebellar nuclei, the largest projection neurons are never marked by these GABA methods, whereas a population of smaller neurons is (Chan-Palay 1977; Mugnaini and Oertel 1981) (see Fig. 1). The present studies, however, are concerned with the Purkinje cells, the intracortical projection neuron responsible for the major relay of information from the complex cerebellar cortical network to the deep nuclear cells and to the lateral vestibular nucleus. Since the basic organization, cytology, and connections of these cells are well known and have been the subject of major works (Eccles et al. 1967; Palay and Chan-Palay 1974), the details need not be repeated here. Only the pertinent facts will be drawn out to substantiate the case that coexistence of neuroactive substances occurs in Purkinje cells, and that this opens up a frontier for important advances in our understanding of the cerebellum.

Purkinje cells were demonstrated by a combination of pharmacologic and physiological approaches to contain GABA and to have a postsynaptic inhibitory effect on Deiters neurons (Obata et al. 1967; Obata 1969). The presence of GABA was confirmed by microdissection techniques (Obata 1976). From these and subsequent findings it has been generally assumed that Purkinje cells are homogeneous, and that they all contain GABA and are inhibitory. Nevertheless, a critical analysis of the literature indicates that, in fact, whereas GABA could be localized to many cerebellar cortical structures reliably, the Purkinje cells remained the most inconsistent in this regard.

1. Autoradiography with ^{3}H-GABA uptake produced evidence for labeled cortical interneurons but did not reliably label Purkinje cells (Chan-Palay 1977; Hökfelt and Ljungdahl 1970; Sotelo et al. 1972).
2. Autoradiography with ^{3}H-muscimol binding consistently labeled intracortical interneurons but not Purkinje cells (Chan-Palay 1978).
3. Immunocytochemical studies with anti-GAD antibodies indicated that Purkinje cell somata are difficult to visualize unless colchicine is used in order to retard axoplasmic transport (Ribak et al. 1978). A recent study claims that all Purkinje cells can be labeled with the use of colchicine and a polyclonal antibody raised against a GAD preparation that is not the completely purified enzyme. However, closer examination of this material indicates that, in fact, unlabeled Purkinje cells are present (Oertel et al. 1981, 1982) and cells other than GAD immunoreactive ones may be labeled (Chan-Palay et al. 1982a, b).
4. Other studies with anti-GABA-T also indicated that not all Purkinje cells exhibit immunoreactivity to this serum even with the aid of colchicine (Barber and Saito 1976; Chan-Palay 1978). These reports were supported by other immunocytochemical studies in the literature indicating that labeling with anticyclic adenosine monophosphate (Bloom et al. 1972; Chan-Palay and Palay 1978) or with anticyclic guanosine monophosphate (Chan-Palay and Palay 1978) was also patchy.
5. In vivo injections of anti-GAD antibodies combined with anterograde and retrograde transport as a means of tracing chemically specific pathways indicated that although a large number of Purkinje cells contain GABA, some Purkinje cells contain only minor amounts and some contain none. The question was raised then whether or not this patchy labeling of Purkinje cells reflects cyclic changes in GABA content and fluctuating levels of GAD (Chan-Palay et al. 1979a).

The rising concern that other neuroactive substances are likely to be present in Purkinje cells has resulted in a search in our laboratory for candidates among the growing catalog of neuroactive peptides. One successful candidate for this role is motilin, a 22-amino acid polypeptide isolated from porcine gut, which stimulates enteric smooth muscle (Brown 1967; Brown et al. 1971), has endocrine effects when administered systematically (Christofides et al. 1979), has an excitatory effect on neurons of the cerebral cortex and spinal cord (Phillis and Kirkpatrick 1979), and is inhibitory in the cerebellum (Chan-Palay et al. 1982c). Another is taurine, which is detected by antibodies against cysteine sulfinic acid decarboxylase, the synthetic enzyme for this amino acid transmitter (Chan-Palay et al. 1982a, b).

Motilin-like immunoreactivity is detectable in the peripheral nervous system (Chey et al. 1980) and in the brain of a number of species, in the hypothalamus and pituitary (Jacobowitz et al. 1981), and in the cerebellum (Nilaver et al. 1981; Chan-Palay et al. 1981). In the cerebellum, motilin-like immunoreactivity can be found almost exclusively in Purkinje cells; approximately 60%–70% of the Purkinje cells are motilin positive and 30%–40% are not. These motilin cells occur mainly in the lateral cerebellum and the flocculonodular lobe with numerous terminals in deep cerebellar nuclei and Deiters nucleus. GAD-immunoreactive Purkinje cells also form approximately 60%–70% of the population, and 30%–40% of Purkinje cells are not GAD positive. GAD Purkinje cells are found in the vermal and paravermal regions, with a large representation in the flocculonodular lobe (Chan-Palay et al. 1981), and numerous terminals occur in the deep cerebellar nuclei and Deiters nucleus. None of these results is quantitatively altered by the administration of colchicine. A combined double-labeling study with anti-GAD and anti-motilin antibodies indicates that not only do some Purkinje cells contain either GAD or motilin immunoreactivity exclusively, but also certain cells, approximately 10%–20%, have both GAD and motilin (Chan-Palay et al. 1981) (Fig. 2a,b). These findings call for a reassessment of present assumptions that Purkinje cells are a population homogeneous in structure, connections, and function. The existence of chemical heterogeneity in Purkinje cells requires that this assessment take place at the single-neuron level and calls into question the present concept of how the cerebellum works.

Taurine (2-aminoethanesulfonic acid) has a simple structure, is chemically stable, and is the most abundant free amino acid in animals. Taurine has been suggested as a neurotransmitter in the mammalian CNS (Curtis 1979) and is inhibitory in the cerebellum. L-Cysteine sulfinic acid decarboxylase (CSADCase; EC 4.1.1.29) is the enzyme responsible for taurine synthesis in the brain through the decarboxylation of cysteine sulfinic acid to hypotaurine with subsequent oxidation to taurine or, alternatively, by the decarboxylation of cysteic acid to taurine (Jacobsen and Smith 1968). CSADCase has been successfully separated from GAD (EC 4.1.1.15) in bovine brain and purified to homogeneity. GAD is responsible for the synthesis of GABA. Antibodies raised against each substance show no cross-reactivity (Wu 1982). Thus taurine and GABA neurons are readily differentiated by means of immunochemical techniques, and we have recently reported that taurine neurons exist in the cerebellum (Chan-Palay et al. 1982a).

In the normal cerebellum CSADCase immunoreactivity was present in most but not all Purkinje neurons, particularly in primary dendritic segments and bifurcations and, to a lesser extent, in the cell somata. Purkinje axons and terminals

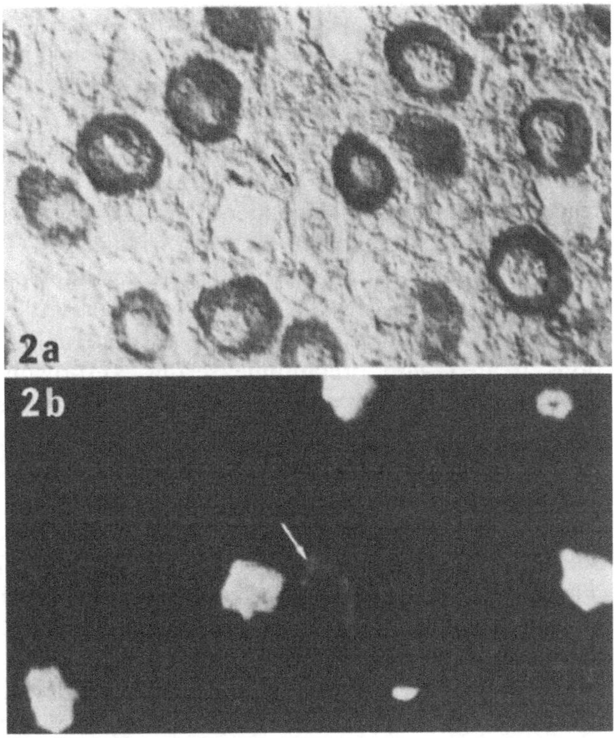

Fig. 2a,b. Purkinje cell field rich in motilin-immunoreactive cells. A pair of photomicrographs showing the same field of Purkinje cells in which 17 large neurons are labeled with motilin antibodies (*upper*) and five of the remaining Purkinje cells are labeled with GAD antibodies (*lower*). The *arrow* indicates a cell with low immunoreactivity. Mouse cerebellum lobule V, vermis. *a* immunoperoxidase; *b* immunofluorescence; ×200

were stained markedly less frequently both in the cortex and in the deep nuclei. A number of stellate, basket, and Golgi cells were CSADCase positive in somata, axons, and dendrites. Most remarkably, CSADCase reactivity appeared in many granule cell somata and dendrites in the granular layer and in their axons, the parallel fibers in the molecular layer. In some areas of the cortex, such as in lobules I and X (including the flocculus), most of these various cell types were heavily reactive. In the other lobules, fewer were labeled. In all of the deep cerebellar nuclei, a considerable number of larger neurons (diameters approx. 25–30 μm) were CSADCase-positive. Some of these neurons displayed a few CSADCase-positive terminals on their somata, presumably contributions from Purkinje cells in the overlying cortex; however, the stained axons or boutons in the cerebellar nuclei were not numerous. In colchicine-treated animals, the intensity of CSADCase immunoreactivity was increased, particularly over cell somata and primary dendrites of Purkinje cells, granule cells, and deep nuclear cells. No stained neuroglial cells were seen.

Electron microscopy confirmed these findings. Representatives of all the neuronal types described above showed CSADCase immunoreactivity. The nuclei of all labeled neurons were devoid of reaction product. In Purkinje cell somata, CSADCase reactivity was concentrated on the inner plasma membrane and in the cytoplasmic matrix around the Golgi apparatus, ribosomes, filaments, and smooth endoplasmic reticulum, generally avoiding organized Nissl substance and mitochondria. In the dendrites, CSADCase reactivity was found around microtubules, smooth endoplasmic reticulum, ribosomes, and filaments, and throughout the cytoplasmic matrix. In the somata and dendrites, CSADCase reactivity appeared to be condensed in 120-nm nonmembranous foci marked by electron-dense peroxidase reaction product. CSADCase immunoreactive postsynaptic elements and presynaptic and axonal elements were also observed. In axonal varicosities, CSADCase immunoreactivity was present in the presynaptic matrix and around the external surfaces of small synaptic vesicles. Control sections treated with preimmune serum, anti-ovalbumin antibodies, and anti-CSADCase antiserum preabsorbed with CSADCase showed no immunoreactivity.

The match between cerebellar neurons with CSADCase immunoreactivity and those with (^{3}H-)taurine uptake systems (Chan-Palay et al. 1982a) with respect to cell type and location supports the hypothesis that many Purkinje neurons, stellate, basket, and Golgi cells, granule cells, and some deep cerebellar nuclear neurons synthesize and utilize taurine. Taurine has been implicated as an inhibitory transmitter (Curtis and Watkins 1960; Chan-Palay et al. 1982c) and, therefore, its presence in neurons that are known to have inhibitory roles (Eccles et al. 1967) in the cerebellar circuits, such as Purkinje, stellate, basket, and Golgi neurons, might be expected. However, the presence of taurine in granule cells and in deep cerebellar projection neurons, which are generally considered to be excitatory elements (Wood et al. 1976), calls for a reexamination of their roles in the cerebellar circuits.

In the Purkinje cell, primary dendrites, their branch points, and the cell body represent the major synthetic and uptake sites for taurine. These dendritic branch points have a highly specialized structure and innervation (Palay and Chan-Palay 1974; Chan-Palay and Palay 1970). They are completely draped by climbing fibers, which synapse on thorns, and by numerous basket and stellate cell axons. It has been proposed that, at these locations, the integration between excitatory and inhibitory influences occurs, and the resultant effect may selectively tune in or out whole segments of dendritic tree distal to these branch points (Chan-Palay and Palay 1970; Llinás et al. 1968). It has been reported that calcium-dependent dendritic spikes develop at these branch points (Llinás and Sugimori 1979). We suggest that taurine might be associated with the calcium dendritic spike process. Finally, it should be noted that Purkinje cells are chemically heterogeneous. They have been known to contain gamma-aminobutyrate or motilin, and some even contain both. The present data demonstrate that many contain taurine.

The cerebellar cortex is organized on an orthogonal plan (Palay and Chan-Palay 1974) and the entire cerebellum has been divided into a number of narrow sagittal bands according to detailed study of the cerebellar white matter (Voogd 1964) (Fig. 3). Physiological experiments also indicate that the spinal and olivary afferents to the cortex are arranged in parasagittal zones (Oscarsson 1977). Neuroanatomic investigations show that olivocerebellar afferents follow a strict parasagittal pattern in the cortex (Armstrong et al. 1971; Chan-Palay et al. 1977;

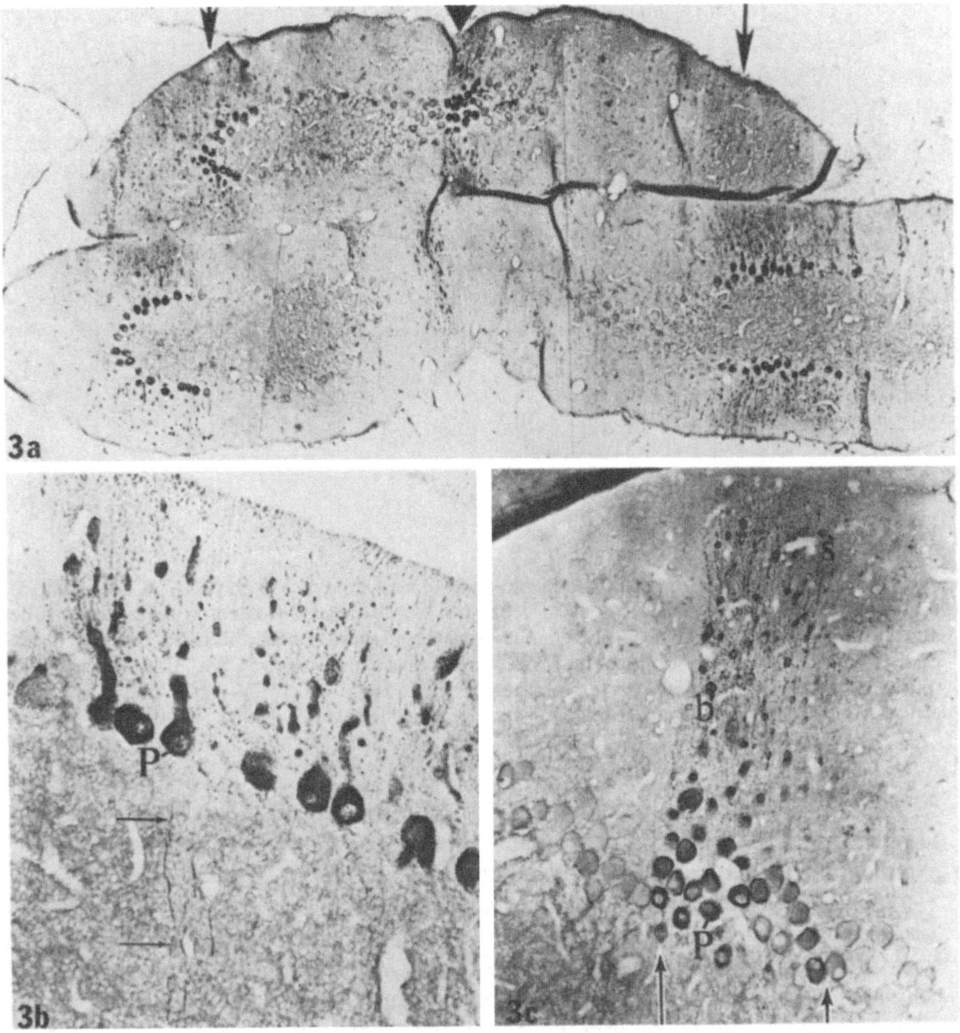

Fig. 3. a Coronal section through lobule IV in the rabbit vermis showing the median sagittal band (*thick arrow*) flanked on either side by a free interband and then by a second set of wider bands (*thin arrows*). *b* Higher magnification of two other bands in coronal section showing somata, axons (*arrows*), and dendrites of Purkinje cells (*P*) within a band. *c* Purkinje cells (*P*) in a band with stellate (*s*) and basket neurons (*b*). The edges of the band are sharp (*arrows*). (Anti-L-cysteine sulfinic acid decarboxylase; *a*, × 55; *b*, × 300; *c*, × 200)

Courville 1975), as do the afferents from the lateral reticular nucleus. Histochemical studies demonstrate that both acetylcholinesterase (Marani and Voogd 1973a) and 5'-nucleotidase (Marani and Voogd 1973b) are distributed in a banded pattern in the cerebellar cortex. Both enzymes appear to be associated with Purkinje cell dendrites. Our recent observations (Chan-Palay et al. 1981) disclosed a chemically specified sagittal microzonation which is defined by the distribution of Purkinje cells containing motilin alone or motilin together with GABA and another

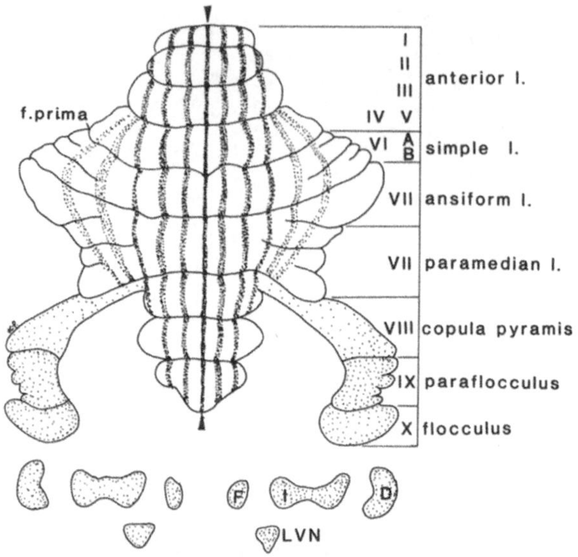

Fig. 4. Schematic summary of the L-cysteine sulfinic acid decarboxylase (CSADCase)-positive sagittal microzones or bands in mouse cerebellum. The bands are clearest in the anterior lobe and the vermis, less sharply defined in the hemispheres (*dense stipple*), and most difficult to discern in the paraflocculus and flocculus (*light stipple*) because of intense CSADCase reactivity in most Purkinje cells. The dentate (*D*), interpositus (*I*), fastigial (*F*), and lateral vestibular nuclei (*LVN*) contain numerous taurine cells. *f. prima*, primary fissure

set of sagittal zonations for taurine (Chan-Palay et al. 1982a, b). These observations lead to the inescapable conclusion that the cerebellum has an intrinsic chemical organization, which is based upon a sagittal banding scheme.

Vermal lobules I and X and the paraflocculus and flocculus contained the largest numbers of labeled neurons in all the cerebella. Sagittal microbands of CSADCase-immunoreactive neurons occurred in the cerebella of all species examined. The molecular layer, and to a lesser extent the granular layer, throughout all lobules of the anterior lobe contained symmetrically disposed CSADCase-positive sagittal bands alternating with negative zones. The white matter was not stained. In coronal section, a sharp, narrow band appeared precisely in the midline and contained intensely reactive Purkinje cell bodies, 12–16 across the band, with their dendrites extending into the molecular layer as well-defined vertical columns within the band. Intensely stained basket and stellate cells participated in the band. Their processes, particularly the axons of the former, formed a conspicuous network in the lower third of the molecular layer. The axons of Purkinje cells coursed through the granule cell layer, continuing the band into this layer. On either side of the median band there was a zone free from CSADCase-immunoreactive neurons, and beyond this laterally was another CSADCase-positive band symmetrical with one on the opposite side. This was followed laterally by another nonreactive band and, still farther laterally, by two more reactive bands separated by a clear zone. Thus, in the anterior lobe, there is a total of seven reactive bands, one in the midline and three pairs of bands on either side of it (Fig. 4).

The somata of Purkinje cells are intensely reactive in the bands and provide a useful measure of the band widths. In the adult animal the midline band is usually narrow, consisting of 12–16 cells across its width; the next lateral and subsequent pairs contain almost 50% more. (In the late stages of cerebellar development, when the Purkinje cell layer is not yet a monolayer, there are twice as many cells in each of these bands; presumably, subsequent growth of the cerebellum in the rostrocaudal direction decreases the number included in the width of a band in coronal sections.) The bands are narrowest and have the sharpest margins in lobule I and broaden progressively from lobule I to lobule V in the anterior lobe and from midline to lateral vermis. In sagittal sections through the anterior lobe, the differences in overall CSADCase immunoreactivity in different parts of a folium or in different folia are more distinct than in coronal sections. It is not difficult to decide whether the section passes through a band or not by the amount of staining in neurons in the successive folia.

In the posterior lobe of the vermis, the bands are present but are less distinct, and bands in the hemispheres are difficult to distinguish. In the hemispheres of lobules VI and VII, at least two bands can be delineated. They vary in width from 8 to 18 Purkinje cells in coronal section and conform to the complex lobulations of the folia. The paraflocculus and flocculus contain the highest proportion of intensely CSADCase-positive neurons, and reaction-free zones are difficult to discern (Fig. 4, light stipple). Estimates indicate that at least 60%–65% of Purkinje cells are CSADCase-positive. The deep cerebellar nuclei and lateral vestibular nucleus contain numerous medium-sized to large neurons with CSADCase immunoreactivity. It was not possible to discern a banding pattern in these nuclei.

That the taurine neurons of the cerebellar cortex are organized into precise sagittal microzones corroborates earlier observations that motilin- and GABA-containing Purkinje cells are similarly arranged (Chan-Palay et al. 1981). In fact, the motilin/GABA Purkinje cell band in the midline of the cerebellum overlaps with the midline CSADCase-positive band. Purkinje cells lying in the true midline of the vermal cerebellar cortex (approximately four to eight cells wide) contain coexistent taurine, motilin, and GABA. Thus these Purkinje cells have multiple chemical messengers. With such a variety of different neuroactive substances coexisting in Purkinje cells and with the sagittal organization precisely defined, it becomes imperative to determine how this intrinsic chemistry of the cerebellar cortex relates to the innervation patterns defined by afferent spinal, vestibular, reticular, olivary, or pontine fibers. The distribution of Purkinje cells with different chemical signatures must have a functional correlate because both afferent fiber systems and chemical characteristics are arranged in parallel, longitudinal, or parasagittal strips or compartments. The compartments for each feature, chemical or neuroanatomic, however, are not coincident in size, location, composition, or pattern but overlap and interdigitate in a complex and as yet unanalyzed manner.

The observations accumulated in the present and previous studies (Chan-Palay et al. 1981, 1982a) invalidate the claim that all Purkinje cells are uniformly GAD immunoreactive (Oertel et al. 1982) and GABA-ergic. The ordered distribution of Purkinje cells with different chemical signatures throughout the cerebellar cortex indicates that they, and probably other cerebellar cells, are arranged in a mosaic of interlocking specificities, reflecting their topographical and functional roles.

A considerable literature supports the identification of GABA as a major inhibitory transmitter in the cerebellum. However, recent cytochemical studies have indicated that Purkinje cells of the cerebellar cortex form a population that is heterogeneous chemically, morphologically, and probably also functionally. In addition to GABA, motilin and taurine are present in Purkinje cells of all mammalian species studied (mouse, rat, rabbit, monkey, and man). Immunocytochemical studies with polyclonal antibodies against motilin indicate that motilin and GABA exist separately in some Purkinje cells and coexist in others (Chan-Palay et al. 1981), because there is a significant population of motilin-immunoreactive Purkinje cells in the flocculus, paraflocculus, and anterior vermis and a major motilin innervation of the corresponding deep cerebellar and vestibular neurons. Furthermore, taurine (Chan-Palay et al. 1982a, b) may share this mediator action with GABA and motilin.

We compared the effects of motilin, taurine, and GABA upon neurons that receive projections from Purkinje cells (Chan-Palay et al. 1982c). The experimental paradigm chosen was iontophoretic application of a peptide or an amino acid alone or of two substances together, with extracellular recording of their effects on spike discharges from Deiters neurons of the rabbit's lateral vestibular nucleus (LVN). Deiters neurons are inhibited by activation of the Purkinje cells in the lateral vermis of the anterior lobe (Akaike et al. 1969) and were identified by antidromic activation through the vestibulospinal tracts and orthodromic stimulation through the ipsilateral vestibular nerve (Akaike et al. 1973).

Iontophoretic application of GABA, glycine, or taurine regularly induced a dose-dependent depession of spike frequencies in Deiters neurons. The effects were comparable in magnitude and time course with those reported for Deiters and other brainstem neurons (Obata et al. 1970; Haas and Hosli 1973). Iontophoretic application of motilin also induced a similar depression, as described below.

The vast majority of Deiters neurons tested were depressed by motilin in a dose-dependent fashion. Complete suppression of spike discharge was commonly observed with sufficient currents, usually 90–100 nA and on some occasions less than that. The time course of onset and recovery for motilin action was fast; no distinction could be made consistently between GABA and motilin on the spike-density histograms. Rebound excitation, which was commonly seen following the cessation of GABA application, however, was usually less with motilin. When motilin and GABA were applied concurrently, the inhibition observed was greater than when either GABA or motilin was applied alone (Fig. 5), and it developed in a simple additive manner.

The present observations establish that the peptide motilin exerts inhibitory effects on Deiters neurons. The time course of these inhibitory effects is as rapid as that induced by the depressant amino acids GABA, glycine, and taurine. Although relatively large iontophoretic currents are required to elicit an effect with the peptides, this does not necessarily indicate a low potency of the peptides tested. Because the peptides were tested at low concentrations (0.2–10 mM, because of their limited availability) in saline, it is likely that much of the current

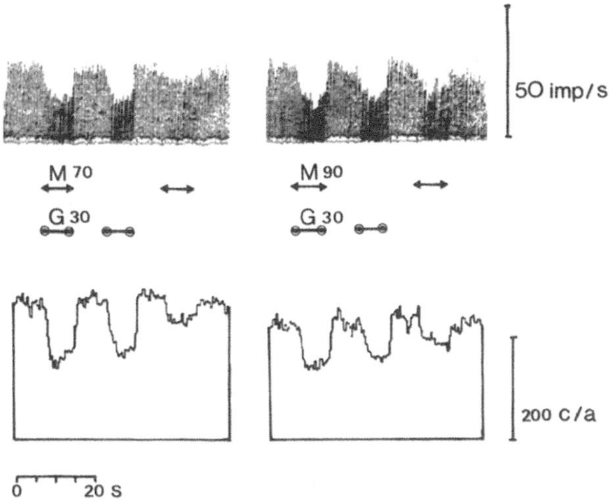

Fig. 5. Interaction of gamma-aminobutyric acid (*G*) and motilin (*M*). During simultaneous application of gamma-aminobutyric acid (30 nA) and motilin (70 and 90 nA), the observed depression of neuronal firing was slightly greater than that observed during application of gamma-aminobutyric acid or motilin alone.

was carried by Na^+ and Cl^- and that the final amounts of peptides ejected from the microelectrode by iontophoresis and electroosmosis were small. In addition, the peptides probably stick to the glass wall of the micropipette, thus reducing the effective concentration of peptide actually ejected. Motilin is likely to be an extremely potent inhibitor of Deiters neurons.

The present results showing the depressant action of somatostatin agree with a previous report on the action of this peptide in the cerebellar cortex and hypothalamus. However, the depressant action of motilin in the LVN is at variance with the reported excitatory action of motilin in the cerebral cortex and spinal cord. Previous iontophoretic studies with the enkephalins indicate that they have a predominantly depressant action on most brain areas, but excitatory effects on Renshaw cells and both excitatory and inhibitory effects on hippocampal neurons. It may be significant that the peptide motilin has only depressant effects on Deiters neurons.

Because recent immunocytochemical data suggest that Purkinje cells contain not only GABA but also motilin (Chan-Palay et al. 1981) and taurine (Chan-Palay et al. 1982a, b) and because rabbit Deiters neurons receive a massive innervation from vermal Purkinje cells, it can be postulated that these peptides and amino acids play significant roles in chemical mediation in the cerebellovestibular projections. The uniformly dose-dependent effects elicited by motilin and taurine, as well as by GABA, suggest their neurotransmitter roles. The consistent finding that these peptides and amino acids exert inhibitory effects supports the previous physiological data showing that Purkinje cells inhibit their target neurons in the deep cerebellar and vestibular nuclei (Ito and Yoshida 1964; Ito et al. 1964). The demonstration that the inhibitory actions of GABA and the peptides are sum-

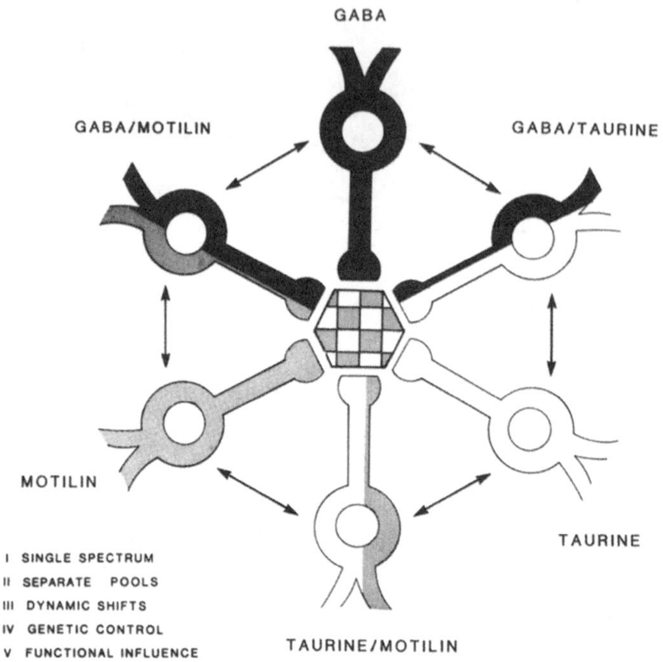

Fig. 6. Schematic diagram to illustrate the concept of dynamic interrelationships between taurine, motilin, and gamma-aminobutyric acid (*GABA*) in a single neuron. A neuron with both substances in coexistence may have fluctuating levels of one or both substances depending upon parameters of rhythm, time, and physiological demands for one or another mediator during specific types or phases of activity.

mated is consistent with the previous demonstration of the importance of GABA inhibition on Deiters neurons.

Finally, a comment is needed about the significance of the existence and coexistence of multiple neuroactive substances in the corticonuclear (Purkinje cell) pathway in the cerebellum. Although Purkinje cells contain a number of neuroactive substances in addition to GABA – e. g., motilin, taurine – the present study shows that all these substances produce only inhibitory effects with independent characteristics and intensities. Evidently, the Purkinje cell can use various substances to exercise its effect. It remains for future investigations to reveal the significance of these various agents and combinations of agents and to explore the nuances encoded in these multimodal messages.

In summary, we present the Purkinje cell, the familiar giant cell of the cerebellar cortex, as a model in which the phenomenon of coexistence of neuroactive substances has been readily studied. What are the cytological features of a pure GAD-immunoreactive, GABA-containing Purkinje cell? What are its connections? What are the cytological features of a pure motilin-immunoreactive Purkinje cell? What are its connections? What are the features of a pure taurine-containing Purkinje cell? What are the similarities and differences between these cells and what can be learned about their separate chemical and physiological functions? What are the ultrastructural correlates for dual neuroactive chemical content in the Purkinje cell? Does such a cell segregate the synthesis, transport, and release machinery for GABA, taurine, and motilin in different populations of such cellular organelles, or do they coexist? Can these compartments be distinguished immunocytochemically at the ultrastructural level, and are synaptic vesicles involved? Are the pure motilin-immunoreactive, pure GAD-immunoreactive, pure taurine and dual motilin/GAD-immunoreactive or GAD/taurine- and motilin/taurine-immunoreactive cells related in a cyclic manner so that they are not separate cell types but a single type of neuron in different phases fixed in time by an immunocytochemical experiment (see Fig. 6)? Motilin, taurine, and GAD cells occur in small foci. Can one trace the origins of these groups of Purkinje cells to specific precursor cells very early in development and test these hypotheses of cyclicity and genetic regulation? At least 30% of all cerebellar Purkinje cells have neither taurine, GAD, nor motilin immunoreactivity. What other neuroactive substances might they contain? What neuromediative role does motilin have? Is it inhibitory, noninhibitory, "modulatory," or excitatory? Could it be a substance with no part to play in neurotransmission and neuromodulation, but perhaps with a trophic function?

These are a few of the many questions that are raised by the fact that Purkinje neurons (and likely other cerebellar neurons) are chemically heterogeneous, and have a chemical repertory for communication that is at least trimodal. The finding calls for a profound reassessment of how the cerebellum works.

References

Akaike T, Fanardjian VV, Ito M, Kumada M, Nakajima H (1969) Electrophysiological analysis of the vestibulospinal reflex pathway of rabbit. I. Classification of tract cells. Exp Brain Res 17: 477–496

Akaike T, Fanardjian VV, Ito M, Kumada M, Nakajima H (1973) Cerebellar control of the vestibulospinal tract cells in rabbit. Exp Brain Res 18: 446–463

Armstrong DM, Harvey RJ, Schild RF (1971) Distribution in the anterior lobe of the cerebellum of branches from climbing fibers to the paramedian lobule. Brain Res 25: 203–206

Armstrong DM, Campbell NC, Edgley SA, Schild RF, Trott JR (1982) Investigations of the olivocerebellar and spinocerebellar pathways. In: Palay SL, Chan-Palay V (eds) The cerebellum – new vistas. Springer, Berlin Heidelberg New York, pp 196–232

Barber R, Saito K (1976) Light microscopic visualization of GAD and GABA-T in immunocytochemical preparations of rodent CNS. In: Roberts E, Chase TN, Towers DB (eds) GABA in nervous system function. Raven, New York, pp 113–131

Barber RP, Vaughn JE, Slemmon JR, Salvaterra PM, Roberts E, Leeman SE (1979) The origin, distribution and synaptic relations of substance P axons in rat spinal cord. J Comp Neurol 174: 467–493

Bloom FE, Hoffer BJ, Battenberg EF, Siggins GR, Steiner AL, Parker CW, Wedner HJ (1972) Adenosine 3',5'-monophosphate is localized in cerebellar neurons: immunofluorescence evidence. Science 177: 436–438

Brown JC (1967) Presence of a gastric motor-stimulating property in duodenal extracts. Gastroenterology 52: 225–229

Brown JC, Mutt V, Dryburgh JR (1971) The further purification of motilin, a gastric motor activity stimulating polypeptide from the mucosa of the small intestine of hogs. Can J Physiol Pharmacol 49: 399–405

Chan-Palay V (1977) Cerebellar dentate nucleus, organization, cytology, and transmitters. Springer, Berlin Heidelberg New York

Chan-Palay V (1978) Autoradiographic localization of gamma-aminobutyric acid receptors in the rat central nervous system using ^{3}H-muscimol. Proc Natl Acad Sci USA 75: 1024–1028

Chan-Palay V (1979) Combined immunocytochemical detection of substance P neurons, their processes and connections by in vivo injections of monoclonal antibodies: light and electron microscopy. Anat Embryol 156: 225–240

Chan-Palay V, Palay SL (1970) Interrelations of basket cell axons and climbing fibers in the cerebellar cortex of the rat. Z Anat Entwickl-Gesch 132: 191–227

Chan-Palay V, Palay SL (1978) Ultrastructural localization of gamma-aminobutyric acid receptors in the mammalian central nervous system by means of ^{3}H-muscimol binding. Proc Natl Acad Sci USA 75: 2977–2980

Chan-Palay V, Palay SL (1979) Light and electron microscope immunocytochemical localization of cyclic GMP: evidence for involvement of neuroglia. Proc Natl Acad Sci USA 76: 1485–1489

Chan-Palay V, Palay SL, Brown JT, Van Itallie C (1977) Sagittal organization of olivocerebellar and reticulo-cerebellar projections: autoradiographic studies with ^{35}S-methionine. Exp Brain Res 30: 561–576

Chan-Palay V, Jonsson G, Palay SL (1978) Serotonin and substance P coexist in neurons of the rat's central nervous system. Proc Natl Acad Sci USA 75: 1582–1586

Chan-Palay V, Palay SL, Wu J-Y (1979a) Gamma-aminobutyric acid pathways in the cerebellum studied by retrograde and anterograde transport of glutamic acid decarboxylase antibody after in vivo injections. Anat Embryol 157: 1–14

Chan-Palay V, Wu J-Y, Palay SL (1979b) Immunocytochemical localization of GABA-transaminase at cellular and ultrastructural levels. Proc Natl Acad Sci USA 76: 2067–2071

Chan-Palay V, Nilaver G, Palay SL, Beinfeld MG, Zimmermann EA, Wu J-Y, O'Donohue TL (1981) Chemical heterogeneity in cerebellar Purkinje cells: existence and coexistence of glutamic acid decaroboxylase-like and motilin-like immunoreactivities. Proc Natl Acad Sci USA 78: 7787–7791

Chan-Palay V, Lin CT, Palay SL, Yamamoto M, Wu J-Y (1982a) Taurine in the mammalian cerebellum: demonstration by autoradiography with tritiated taurine and immunocytochemistry with antibodies against the taurine synthesizing enzyme, cysteine sulfinic acid decarboxylase. Proc Natl Acad Sci USA 79: 2695–2699

Chan-Palay V, Palay SL, Wu J-Y (1982b) Sagittal cerebellar microbands of taurine neurons: immunocytochemical demonstration by using antibodies against the taurine synthesizing enzyme cysteine sulfinic acid decarboxylase. Proc Natl Acad Sci USA 79: 4221–4225

Chan-Palay V, Ito M, Tongroach P, Sakurai M, Palay SL (1982c) Inhibitory effects of motilin, somatostatin, leucine- and methionine enkephalin, and taurine on neurons of the lateral vestibular nucleus: their interaction with gamma-aminobutyric acid. Proc Natl Acad Sci USA 79: 3355–3359

Chey WY, Escoffery R, Roth F, Chang TM, You CH, Yajima H (1980) Motilin-like immunoreactivity (MLI) in the gut and neurons of peripheral and central nervous system. Regul Pept [Suppl] 1: 19

Christofides ND, Modlin IM, Fitzpatrick ML, Bloom SR (1979) Effect of motilin on the rate of gastric emptying and gut hormone release during breakfast. Gastroenterology 76: 903–907

Courville J (1975) Distribution of olivo-cerebellar fibers demonstrated by autoradiographic tracing method. Brain Res 95: 253–263

Curtis DR (1979) Gabargic transmission in the mammalian central nervous system. In: Krogsgaard-Larson P, Scheel-Kruger J, Kofod H (eds) GABA-neurotransmitters, pharmochemical, biochemical and pharmacological aspects. Academic, New York, pp 17–27

Curtis DR, Watkins JC (1960) The excitation and depression of spinal neurons by structurally related amino acids. J Neurochem 6: 117–141

Eccles J, Ito M, Szentágothai J (1967) The cerebellum as a neuronal machine. Springer, Berlin Heidelberg New York

Haas HL, Hosli L (1973) The depression of brainstem neurons by taurine and its interaction with strychnine and bicuculline. Brain Res 52: 399–402

Hökfelt T, Ljungdahl AA (1970) Cellular localization of labelled gamma-aminobutyric acid (^{3}H-GABA) in rat cerebellar cortex: an autoradiographic study. Brain Res 22: 391–396

Hökfelt T, Elde R, Johansson O, Luft R, Nilsson G, Arimura A (1976) Immunohistochemical evidence for separate populations of somatostatin-containing and substance P-containing primary afferent neurons in the rat. Neuroscience 3: 517–538

Hökfelt T, Ljungdahl AA, Steinbusch H, Verhofstad A, Nilsson G, Brodin E, Pernow B, Goldstein M (1978) Immunohistochemical evidence of substance P-like immunoreactivity in some 5-hydroxytryptamine-containing neurons in the rat central nervous system. Neuroscience 3: 517–538

Ito M, Yoshida M (1964) The origin of cerebellar induced inhibition of Deiters' neurons. I. Monosynaptic initiation of inhibitory postsynaptic potentials. Experientia 20: 515–516

Ito M, Yoshida M, Obata K (1964) Monosynaptic inhibition of the intracerebellar nuclei induced from the cerebellar cortex. Experientia 20: 575–576

Jacobowitz DM, O'Donohue TL, Chey WY, Chang TM (1981) Mapping of motilin-immunoreactive neurons of the rat brain. Peptides 2: 479–487

Jacobsen JG, Smith LH (1968) Biochemistry and physiology of taurine and taurine derivates. Physiol Rev 48: 424–511

Llinás R, Sugimori M (1979) Calcium conductances in Purkinje cell dendrites: their role in development and integration. In: Cuénod M, Kreutberg GW, Bloom FE (eds) Development and chemical specificity of neurons. Elsevier, Amsterdam, pp 324–334 (Progress in brain research, vol 51)

Llinás R, Freeman JA, Hillman DE (1968) Dendritic spikes and their inhibition in alligator Purkinje cells. Science 160: 1132–1135

Marani E, Voogd J (1973a) An acetyl cholinesterase band-pattern in the molecular layer of the cat cerebellum. J Anat 124: 335–345

Marani E, Voogd J (1973b) Some aspects of the localization of the enzyme 5'-nucleotidase in the molecular layer of the cerebellum of the mouse. Acta Morphol Neerl Scand 11: 365

McLaughlin B, Wood JG, Saito K, Barber R, Vaughn JE, Roberts E, Wu J-Y (1974) The fine structural localization of glutamate decarboxylase in synaptic terminals of rodent cerebellum. Brain Res 76: 377–391

McLaughlin B, Wood JG, Saito K, Roberts E, Wu J-Y (1975) The fine structural localization of glutamate decarboxylase in developing axonal processes and presynaptic terminals of rodent cerebellum. Brain Res 85: 355–371

Mugnaini E, Oertel W (1981) Distribution of GAD positive neurons in the rat cerebellar nuclei. Soc Neurosci Abstr 7: 122

Nilaver G, Zimmerman EA, Defendini R, Beinfeld MC, O'Donohue TL (1981) Motilin in brain: immunocytochemical studies. Soc Neurosci Abstr

Obata K (1969) Gamma-aminobutyric acid in Purkinje cells and motoneurones. Experientia 25: 1283

Obata K (1976) Association of GABA with cerebellar Purkinje cells: single cell analysis. In: Roberts E, Chase TN, Tower DB (eds) GABA in nervous system function. Raven, New York, pp 113–131

Obata K, Ito M, Ochi R, Sato N (1967) Pharmacological properties of the postsynaptic inhibition by Purkinje cell axons and the action of gamma-aminobutyric acid on Deiters neurones. Exp Brain Res 4: 43–57

Obata K, Takeda K, Shinozaki H (1970) Further study on pharmacological properties of the cerebellar-induced inhibition of Deiters' neurones. Exp Brain Res 11: 327–342

Oertel WH, Schmeckel DE, Mugnaini E, Tappaz ML, Kopin IJ (1981) Immunocytochemical localization of GAD in rat cerebellum with a new antiserum. Neuroscience 6: 2689–2700

Oertel WH, Mugnaini E, Schmeckel DE, Tappaz ML, Kopin IJ (1982) The immunocytochemical demonstration of GABA-ergic neurons – methods and application. In: Chan-Palay V, Palay SL (eds) Cytochemical methods in neuroanatomy. Liss, New York

Oscarsson O (1977) Spatial distribution of mossy and climbing fiber inputs into the cerebellar cortex. Exp Brain Res [Suppl] 1: 36–42

Palay SL, Chan-Palay V (1974) Cerebellar cortex, cytology, and organization. Springer, Berlin Heidelberg New York

Panula P, Emson P, Wu J-Y (1980) Demonstration of enkephalin, substance P and glutamate decarboxylase-like immunoreactivity in cultured cells derived from newborn rat neostriatum. Histochemistry 69: 169–179

Phillis JW, Kirkpatrick JR (1979) Motilin excites neurons in the cerebral cortex and spinal cord. Eur J Pharmacol 58: 469–472

Ribak CE, Vaughn JE, Saito K (1978) Immunocytochemical localization of glutamic acid decarboxylase in neuronal somata following colchicine inhibition of axonal transport. Brain Res 140: 315–332

Sotelo C, Privat A, Drian MJ (1972) Localization of (^{3}H)GABA in tissue culture of rat cerebellum using electron microscopy radioautography. Brain Res 45: 302–308

Voogd J (1964) The cerebellum of the cat. Structure and fibre connexions. Van Gorcum, Assen

Wood JG, McLaughlin BJ, Vaughn JE (1976) Immunocytochemical localization of GAD in electron microscopic preparations of rodent CNS. In: Roberts E, Chase N, Towers DB (eds) GABA in nervous system function. Raven, New York, pp 133–148

Wu J-Y (1982) Purification and characterization of cysteic acid and cysteine sulfinic acid decarboxylase and L-glutamate decarboxylase from bovine brain. Proc Natl Acad Sci USA 79: 4270–4274

Integration of Mossy Fiber and Climbing Fiber Inputs to Purkinje Cells

G. ANDERSSON, N.C. CAMPBELL, C.-F. EKEROT, G. HESSLOW and O. OSCARSSON, Lund

The publication in 1967 of *The Cerebellum as a Neuronal Machine* by Eccles, Ito and Szentágothai laid a firm basis for later research on the cerebellum and started a period of intense investigation of cerebellar anatomy and physiology. A major problem that has remained unsolved, despite great efforts, relates to the manner in which information from the two main kinds of afferents, mossy fibers and climbing fibers, is integrated in the efferent neurons of the cerebellar cortex, the Purkinje cells.

Figure 1A shows the basic circuit of the cerebellar cortex which has, in principle, remained unchanged in the vertebrate series since its first development in primitive fish (Eccles et al. 1967; Llinás and Hillman 1969; Palay and Chan-Palay 1974). The constancy of the circuit suggests that it has been a highly successful design. The basic circuit consists of the two afferent systems and a single efferent path (inhibitory interneurons have been omitted). The numbers in the figure give the degree of divergence (numbers within rectangles) and convergence (naked

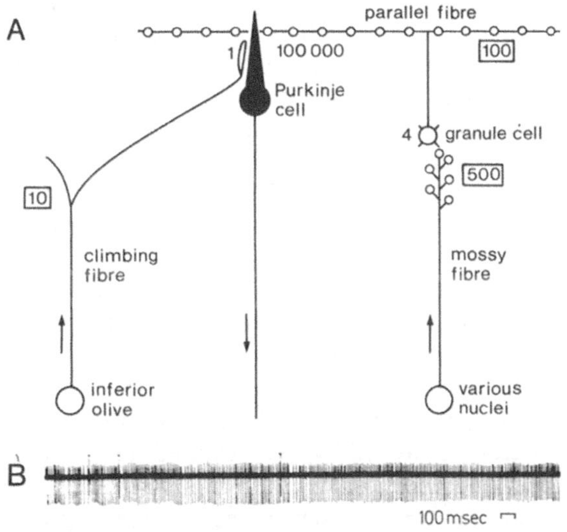

Fig. 1. A Basic circuit of cerebellar cortex showing excitatory connections between the two main afferent systems, the mossy fibers and climbing fibers, and the single efferent path, the Purkinje cells. *Numbers* give approximate degree of divergence (*numbers within rectangles*) and convergence (*naked numbers*) in the two afferent paths (Eccles 1977). *B* Extracellular recording from Purkinje cell showing typical background activity of "complex spikes" (*indicated by dots*) and "simple spikes". (Modified from Thach 1968)

Experimental Brain Research, Suppl. 9
© Springer-Verlag Berlin · Heidelberg 1984

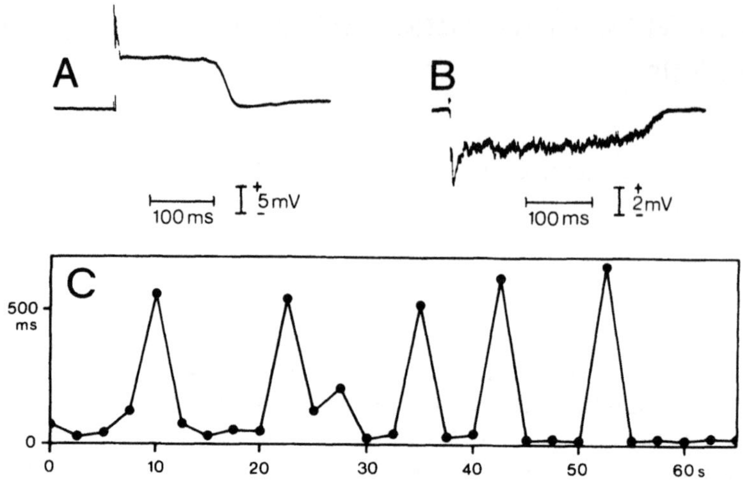

Fig. 2A–C. Plateau potentials evoked in Purkinje cell dendrites by impulses in climbing fibers. *A* Intracellular recording from proximal dendrite. *B* Extracellular recording from distal dendrite. *C* Variation in duration of consecutive plateau potentials evoked at intervals of 2.5 s and recorded extracellularly from distal dendrite. *Ordinate*, duration of response in milliseconds. *Abscissa*, time of consecutive responses in seconds. The observations were made on cats under pentobarbitone anaesthesia.

numbers) in the two afferent paths, as found in the commonly investigated mammals (Eccles 1977).

The mossy fibers originate from a variety of cell groups in the spinal cord and brain stem. These fibers activate the Purkinje cells disynaptically through the granule cells and their axons, the parallel fibers. The mossy fiber path is characterized by a high degree of divergence and an enormous convergence. About 100 000 parallel fibers pass through the dendritic tree to make synaptic contacts with the distal dendrites, the spiny branchlets, of each Purkinje cell. The pattern of synaptic input formed by the large number of parallel fibers contacting a single Purkinje cell must presumably be established, at least partly, by some kind of motor learning (Eccles 1977).

The climbing fibers have an entirely different organization. They originate, presumably, from one source only, the inferior olive, and make monosynaptic contacts with the smooth dendrites of the Purkinje cells. There is little divergence (each olivary neuron innervates about ten Purkinje cells) and there is no convergence (each Purkinje cell receives only one climbing fiber). The anatomical specificity is matched by a corresponding specificity in the information reaching each Purkinje cell from its climbing fiber. Each climbing fiber projecting to the anterior lobe has been shown to receive convergence of excitation from several spino-olivary and several cerebro-olivary paths, suggesting that an extremely complex integration occurs at the olivary level (Oscarsson 1980; Andersson and Nyquist 1983).

Activity in the mossy fibers is partly responsible for the "simple spike" activity, usually in the range of 20–100 Hz, which can be recorded from the Purkinje cell somata. The climbing fibers have a low discharge rate of about 1 Hz and are responsible for the so-called "complex spikes" in the Purkinje cells. The complex spikes, because of their low rate of discharge, cannot be expected to influence appreciably the frequency coding of the Purkinje cells. This is illustrated in Fig. 1B, showing extracellular recording from a Purkinje cell. The few complex spikes, indicated by dots, hardly add to the high-frequency simple spike activity, and the variability in latency of the climbing fiber responses evoked by stimulation of an afferent path to the inferior olive makes it unlikely that the climbing fibers signal exact timing of events (Ekerot and Oscarsson 1980). The climbing fibers can thus be of little importance for the instantaneous modulation of ongoing movement which is executed by the output from the Purkinje cells. This control is presumably entirely dependent on the mossy fiber paths which carry precise information about peripheral events and activity in motor centers (Lundberg 1971; Oscarsson 1973; Ekerot et al. 1979).

What is, then, the function of the climbing fibers? The unique one-to-one anatomical relationship between the climbing fibers and the Purkinje cells, together with the powerful synaptic responses generated in the Purkinje cells by impulses in climbing fibers, suggests a very specific function for this input. An explanation is offered by the recent finding of Ekerot and Oscarsson (1980, 1981) that impulses in climbing fibers evoke not only complex spikes in the Purkinje cell somata, but also prolonged plateau-like depolarizations in the Purkinje cell dendrites. Such plateau potentials are, under physiological conditions, evoked exclusively by the large EPSPs generated by impulses in climbing fibers (Campbell et al. 1983b).

The plateau potentials have different characteristics in the coarse, proximal dendrites and in the thin, distal dendrites, as illustrated in Fig. 2. Intracellular recordings are readily obtained from the proximal dendrites and reveal responses consisting of an initial spike-like component followed by a plateau-like depolarization with a relatively constant duration of about 100 ms (Fig. 2A). Electrode penetrations are seldom successful in the distal dendrites. However, extracellular recordings demonstrate responses consisting of unitary negative potentials of long duration which presumably correspond to intradendritic depolarizations (Fig. 2B). The duration of the negative plateaus vary from about 20 ms to several hundreds of milliseconds (Fig. 2C).

The plateau potentials in the Purkinje cells are probably generated by a calcium conductance increase (Llinás and Sugimori 1980) and are reminiscent of the prolonged plateau potentials generated by voltage-dependent calcium currents in embryonic neurons (cf. Spitzer 1979). In the adult stage of higher vertebrates, voltage-dependent calcium conductances seem to be well developed in dendrites (cf. Llinás 1979) and have been demonstrated in many regions of the nervous system, e.g. the neocortex, thalamus, hippocampus, inferior olive and cerebellum (for references see Heinemann and Pumain 1980; Llinás and Jahnsen 1982). However, plateau potentials of such long duration and constant amplitude as in embryonic neurons have only been observed in Purkinje cells. The typical plateau potentials in Purkinje cells can be found in recordings from cerebella not only of mammals, but also of reptiles and fish (for references see Ekerot and Oscarsson 1981). The plateau potentials thus represent a phylogenetically old mechanism presumably fundamental to the climbing fiber synapses.

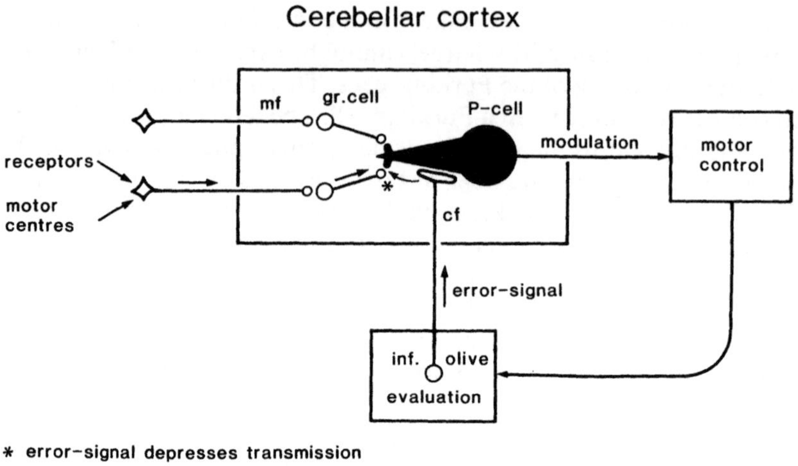

Fig. 3. The "learning" theory of cerebellar cortex. Impulses in mossy fibers (*mf*) largely determine the output from the Purkinje cells, which in turn modifies the execution of motor acts. The correctness of the motor acts is evaluated by the inferior olive and errors in motor performance are signalled through the climbing fibers (*cf*) to the Purkinje cells and result in heterosynaptic suppression of those parallel fiber synapses which have just been active and therefore partly responsible for the misperformance. *gr. cell*, granule cell; *P-cell*, Purkinje cell.

The plateau potentials may provide a means of integration of the mossy and climbing fiber inputs. In the "learning" theories advanced by Marr (1969), Albus (1971) and Ito (1972, 1982), the mossy and climbing fibers have entirely different tasks (see Fig. 3). Impulses in mossy fibers are largely responsible for the output from the Purkinje cells, which in turn influences motor centers and modulates the execution of motor acts. The inferior olive evaluates the motor performance and errors are signalled through the climbing fiber to the Purkinje cell and result, according to Albus (1971), in a heterosynaptic suppression of the parallel fibre synapses which have just been active and therefore partly responsible for the misperformance. Repetition of these events increases the suppression of the transmission with, finally, elimination of the erroneous movement.

This hypothesis poses a number of problems, e. g. how are errors in motor performance detected and what is the cellular mechanism responsible for the heterosynaptic suppression?

With respect to error detection, observations on the floccular control of the vestibulo-ocular reflex and on the anterior lobe control of limb and body movements suggest possible answers. In the first case, the error signal conveyed by the inferior olive is simply the slip of the retinal image when, for example, the eye movements do not compensate adequately for head rotation (Ito 1982). In the second case, the situation is more complex, since the repertoire of movements is much greater. What would then be considered as an error must be based on a comparison between the intended and performed movement. This is considered in the comparator hypothesis (Miller and Oscarsson 1970; Oscarsson 1980), which suggests that the inferior olive acts as a comparator of convergent descending and ascending inputs. The descending pathways would carry information about motor command signals and the ascending pathways about activity induced by these signals in lower motor centers and the resulting movement. When there is a mismatch between the command signals and the ascending information, the inferior olive would send an error signal to the cerebellum. These assumptions are supported by the findings that the cerebral and spinal inputs to the olivary regions projecting to the anterior lobe match closely with respect to somatotopy and uni- or bilaterality (Andersson and Nyquist 1983).

The theories about the cellular mechanisms are supported by the recent experimental findings of Ito et al. (1982), who showed that simultaneous activation of climbing fibers and mossy fibers results in a long-lasting reduction in synaptic efficacy of the activated parallel fiber synapses. The mechanisms by which information about climbing fiber activity is transferred to the parallel fiber synapses might be the climbing-fiber-evoked plateau potentials. The plateau potentials in the dendrites probably cause an increase in the intradendritic calcium concentration. The calcium ions might represent a second messenger which is responsible for the long-lasting reduction in synaptic efficacy.

Recent experimental findings (Campbell et al. 1983a) have revealed an additional facet of the climbing fiber-mossy fiber interaction. In this study, it was shown that the duration of the climbing fiber plateau potentials in Purkinje cell dendrites can be modulated by parallel fiber input. The results indicate that the duration of the plateau potentials probably depends on the membrane potential, and it is suggested that changes in potential are mediated by the local parallel fiber activity. The marked variation in the duration of the plateau potentials in distal dendrites (Fig. 2C) might be a consequence of this interaction. If there exist local differences in parallel fiber input, then individual dendritic branches may act as independent integrators of mossy fiber and climbing fiber information.

Acknowledgments. This work was supported by grants from the Swedish Medical Research Council (Project 1013) and the Medical Faculty, University of Lund, Sweden. N. Campbell was supported by a NATO Science Fellowship.

References

Albus JS (1971) A theory of cerebellar function. Math Biosci 10: 25–61

Andersson G, Nyquist J (1983) Origin and sagittal termination areas of cerebro-cerbellar climbing fibre paths. J Physiol (Lond) 337: 257–285

Campbell NC, Ekerot C-F, Hesslow G (1983a) Interaction between responses in Purkinje cells evoked by climbing fibre impulses and parallel fibre volleys. J Physiol (Lond) 340: 225–238

Campbell NC, Ekerot C-F, Hesslow G, Oscarsson O (1983b) Dendritic plateau potentials evoked in Purkinje cells by parallel fibre volleys. J Physiol (Lond) 340: 209–223

Eccles JC (1977) An instruction-selection theory of learning in the cerebellar cortex. Brain Res 127: 327–352

Eccles JC, Ito M, Szentágothai J (1967) The cerebellum as a neuronal machine. Springer, Berlin Heidelberg New York

Ekerot C-F, Oscarsson O (1980) Prolonged dendritic depolarizations evoked in Purkinje cells by climbing fibre impulses. Brain Res 192: 272–275

Ekerot C-F, Oscarsson O (1981) Prolonged depolarization elicited in Purkinje cell dendrites by climbing fibre impulses in the cat. J Physiol (Lond) 318: 207–221

Ekerot C-F, Larson B, Oscarsson O (1979) Information carried by the spinocerebellar paths. In: Pompeiano O (ed) Reflex control of posture and movement. Elsevier, Amsterdam, pp 79–90 (Progress in brain research, vol 50)
Heinemann U, Pumain R (1980) Extracellular calcium activity changes in cat sensorimotor cortex induced by iontophoretic application of aminoacids. Exp Brain Res 40: 247–250
Ito M (1972) Neural design of the cerebellar motor control system. Brain Res 40: 81–84
Ito M (1982) Cerebellar control of the vestibulo-ocular reflex – around the flocculus hypothesis. Annu Rev Neurosci 5: 275–296
Ito M, Sakurai M, Tongroach P (1982) Climbing fibre induced depression of both mossy fibre responsiveness and glutamate sensitivity of cerebellar Purkinje cells. J Physiol (Lond) 324: 113–134
Llinás R (1979) The role of calcium in neuronal function. In: Schmitt FG, Worden FG (eds) The neurosciences fourth study program. MIT Press, Cambridge, pp 555–571
Llinás R, Hillman DE (1969) Physiological and morphological organization of the cerebellar circuits in various vertebrates. In: Llinás R (ed) Neurobiology of cerebellar evolution and development. American Medical Association, Chicago, pp 43–73
Llinás R, Jahnsen H (1982) Electrophysiology of mammalian thalamic neurones in vitro. Nature 297: 406–408
Llinás R, Sugimori M (1980) Electrophysiological properties of in vitro Purkinje cell dendrites in mammalian cerebellar slices. J Physiol (Lond) 305: 197–213
Lundberg A (1971) Function of the ventral spinocerebellar tract. A new hypothesis. Exp Brain Res 12: 317–330
Marr D (1969) A theory of cerebellar cortex. J Physiol (Lond) 202: 437–470
Miller S, Oscarsson O (1970) Termination and functional organization of spino-olivocerebellar paths. In: Fields WS, Willis WD Jr (eds) The cerebellum in health and disease. Green, St Louis, pp 172–200
Oscarsson O (1973) Functional organization of spinocerebellar paths. In: Iggo A (ed) Somatosensory system. Springer, Berlin Heidelberg New York, pp 339–380 (Handbook of sensory physiology, vol II)
Oscarsson O (1980) Functional organization of olivary projection to the cerebellar anterior lobe. In: Courville J, de Montigny C, Lamarre Y (eds) The inferior olivary nucleus: anatomy and physiology. Raven, New York, pp 279–289
Palay SL, Chan-Palay V (1974) Cerebellar cortex. Cytology and organization. Springer, Berlin Heidelberg New York
Spitzer NC (1979) Ion channels in development. Annu Rev Neurosci 2: 363–397
Thach WT (1968) Discharge of Purkinje and cerebellar nuclear neurons during rapidly alternating arm movements in the monkey. J Neurophysiol 31: 785–797

Cerebrocerebellar Interactions in Premovement Organization of Conditioned Hand Movements in the Monkey

KAZUO SASAKI, Kyoto

Summary. The analysis of laminar field potentials in the cerebral cortex on stimulation of thalamic and cerebellar nuclei, together with the knowledge of cerebrocerebellar neuronal circuits, enables field potential studies on the activity of the cerebral cortex which organizes "voluntary movements." The investigation of the cortical activities preceding operantly conditioned hand movements of monkeys is presented and discussed.

Field potentials on the surface and at 2.5–3.0 mm depth in the cerebral cortex were recorded in various cortical areas with chronically implanted electrodes, and the potentials which preceded hand movements in response to a light stimulus were observed during the process of learning the conditioned movement. A monkey was trained to lift a lever by wrist extension within the duration of the 510–900 ms light stimulus, which was delivered in front of the monkey at a random interval of 2.5–6.0 s.

Significant short-latency responses to the light stimulus appeared bilaterally in certain areas of the prefrontal and prestriate cortices at an early stage of learning in which the monkey still lifted the lever randomly, and they gradually became larger on further training. Short-latency responses were also noted in the bilateral premotor cortices from an early stage. The increasing activities in the association cortices appear to lead the cognitive learning process.

When the monkey started to respond to the stimulus with the appropriate movement, early surface-positive (s-P), depth-negative (d-N) premovement potentials were recorded in the forelimb motor cortex. As the movement was performed faster and more stably, late s-N, d-P premovement potentials, which are mediated by the neocerebellum and superficial thalamocortical (T-C) projections, appeared after the early s-P, d-N potentials and gradually became larger and steeper in the forelimb motor cortex contralateral to the moving hand. The increase in the s-N, d-P potentials and the decrease in reaction times (faster and more stable movement) were closely correlated. This indicates that the recruitment of cerebrocerebellar interactions is important for the attainment of fast and skilled movement. Resection of the cerebellar hemisphere after the accomplished state of the skilled movement degraded the movement to the unskilled state. Resection before training prevented the monkey from attaining the skill, but the early cognitive learning was able to proceed, although slowly.

It is concluded that the cognitive learning process of conditioned movements is led mainly by the association and premotor cortices, and that the activity of cerebrocerebellar neuronal circuits is recruited by repeated training and works to

Experimental Brain Research, Suppl. 9
© Springer-Verlag Berlin · Heidelberg 1984

make the movement fast and skillful. The process of motor learning is discussed and an important role of the thalamus in the process is suggested.

Key words: *Cerebrocerebellar interaction – Cortical field potential – Visually initiated hand movement – Motor learning – Monkey*

Cerebrocerebellar Neuronal Circuits and Laminar Field Potential Analysis of Thalamocortical and Corticocortical Responses in the Cerebral Cortex

Microelectrophysiological studies on reciprocal connections between the cerebral cortex and the cerebellum were carried out in cats and monkeys (see Sasaki 1979). In cats, the studies added two major points to the previous morphological and electrophysiological investigations: (a) the neocerebellum (posterior lobe including the lateral nucleus) is interconnected in both directions with the association cortex (rostral parts of the middle suprasylvian and lateral gyri) (Sasaki et al. 1972a, b, 1975); (b) there are pontine nuclei neurons which almost exclusively relay the volley from the association cortex to the neocerebellum (Sasaki et al. 1975; Oka et al. 1975). Thus the neuronal pathway, association cortex-pontine nuclei (inferior olive)-neocerebellum-thalamus-motor cortex, can be emphasized in conveying the impulses for initiation of "voluntary movement" (see Hassler 1956; Evarts and Thach 1969; Eccles 1973, 1979; Allen and Tsukahara 1974; Brooks and Thach 1981).

It would be necessary to have equivalent knowledge about the cerebrocerebellar neuronal circuits in monkeys in order to test the functional significance of the assumed pathway in actual performance of the movement. Studies on the circuits in monkeys revealed in principle much the same results as in cats. A few specific points are the following:

1. The association cortices that send efferents to the neocerebellum are mainly in the frontal lobe in monkeys instead of the parietal lobe as in cats.
2. The forelimb area of the motor cortex is the most powerful activator of the neocerebellum and receives in turn the strongest input from the neocerebellum.
3. The premotor cortex and the prefrontal association cortex also send significant efferents to the neocerebellum and receive inputs from it (Sasaki et al. 1976, 1977, 1979; see Sasaki 1979).

These results are schematically summarized in the upper part of Fig. 1.

Meanwhile, laminar field potential analysis in the cerebral cortex was also performed in cats and monkeys, and the mode of termination of thalamocortical (T-C) and corticocortical (C-C) projections was investigated electrophysiologically. Two different kinds of T-C responses were distinguished in thalamic and cerebellar induced cortical potentials, and they are called deep and superficial T-C responses (Sasaki et al. 1970). Electrophysiological and morphological substantiation of the two T-C projection neurons were made in cats (Sasaki et al. 1972b; Mizuno et al. 1973; Kawaguchi et al. 1981). The idea was extended to T-C projections of monkeys (Sasaki et al. 1976; see Sasaki 1976/77, 1979), as illustrated in Fig. 1 (major part of the lower half). The superficial T-C neuron originates in

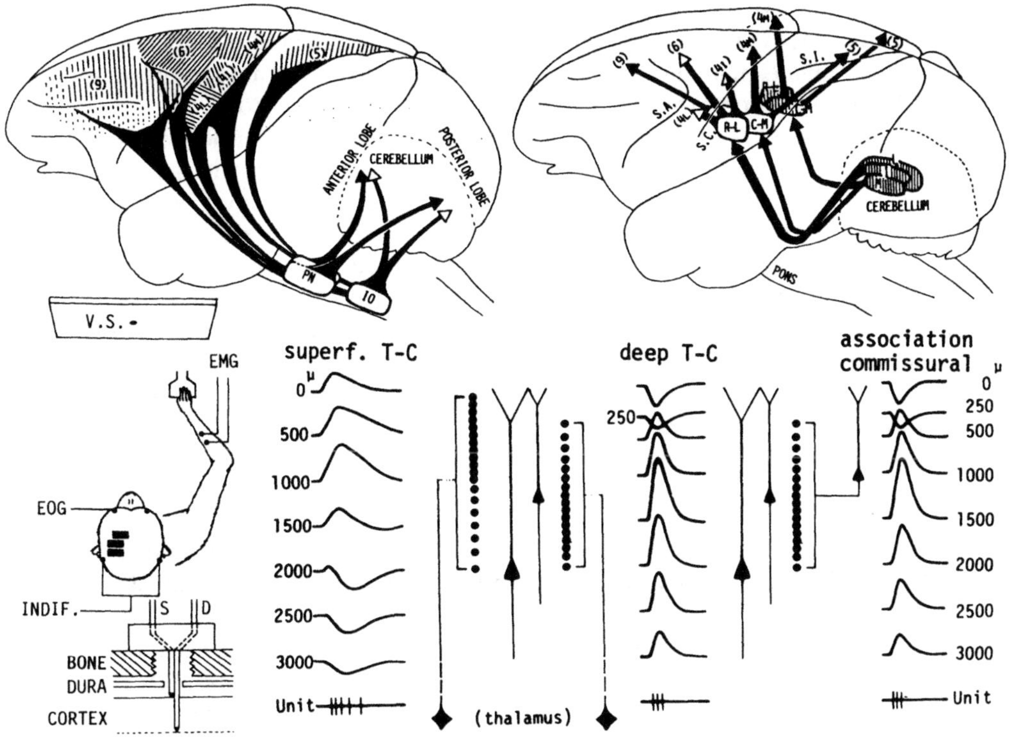

Fig. 1. Schemata of cerebrocerebellar interconnections in monkeys (*upper diagrams*) and those of laminar field potentials of two thalamocortical (superficial and deep T-C) and corticocortical (association and commissural) inputs (*lower major part*), and diagrams for methods of chronic experiments (*lower left*). *Upper diagrams*: *numerals* indicate area numbers of the cortex; *PN*, pontine nuclei; *IO*, inferior olive; *S.A.*, arcuate sulcus; *S.C.*, central sulcus; *S.I.*, intraparietal sulcus; *M,I,L*, medial, interpositus, and lateral nucleus; *R-L*, *C-M*, two nuclear complexes of the thalamus. *Lower diagram*, *numerals* indicate depths from the cortical surface in micrometers; *EMG*, electromyogram; *EOG*, electrooculogram; *INDIF.*, indifferent electrode; *S*, surface; *D*, depth; *T-C*, thalamocortical. *Unit*, schematic pattern of firing of cortical pyramidal neuron. Presumed excitatory synaptic terminals on apical dendrites of cortical pyramidal neurons in layer III and V are diagrammatically shown by *dots* for three afferent inputs. Laminar field potentials are due to electrical dipoles generated in pyramidal neurons by the EPSPs. (Modified from Sasaki 1979)

the thalamus and terminates in superficial layers of the cortex, producing EPSPs in superficial parts of apical dendrites of cortical pyramidal neurons. Thus synchronized activation of the superficial T-C neurons evokes surface-negative, depth-positive (s-N, d-P) field potentials in the cortex. On the contrary, the deep T-C neuron terminates in deep layers of the cortex and produces EPSPs in deep parts of apical dendrites and somata of cortical pyramidal neurons, eliciting s-P, d-N field potentials. Besides the two kinds of T-C projections, corticocortical (C-C) projections from the same hemisphere (association) and the contralateral one (commissural or transcallosal) constitute one of the main inputs to the cortex. They provoke s-P, d-N field potentials in the cortex, resembling the deep T-C responses (Sasaki et al. 1981b).

These results suggested that we should utilize the laminar field potential analysis to study activities of the cerebrocerebellar neuronal circuits associated with actual movements in unanesthetized animals, especially hand movements of monkeys, because we knew that the neocerebellar (dentate) projection onto the forelimb motor cortex is mediated exclusively by superficial T-C projections (Sasaki et al. 1976). Chronically implanted electrodes on the surface and in the deep part (2.5–3.0 mm) of the cortex (sometimes also in an intermediate layer) would be capable of differentiating at least superficial T-C responses from deep T-C and C-C responses (Fig. 1, lower left diagram). It was hoped that such chronically implanted electrodes in various parts of the bilateral hemispheres would give us the three sorts of information which would be otherwise difficult to obtain: (a) details of long-lasting changes in electrical activities in the cortex; (b) simultaneous recording of the activities of several cortical loci; (c) qualitative differentiation of cortical inputs by the polarity of field potentials. These investigations have been performed, as will be shown below.

Cortical Field Potentials Preceding Visually Initiated Hand Movement as Recorded with Chronically Implanted Electrodes in the Cortex of Monkeys

Silver needle electrodes (outer diameter of 0.25 mm and insulated except at their pointed tips) were placed on the surface (S) and at 2.5–3.0 mm depth (D) at every locus of prefrontal, premotor, forelimb motor and somatosensory, parietal association, occipital association (prestriate), striate cortex, etc., bilaterally (Figs. 1 and 4). They were fixed to the bone with dental cement and used to record cortical field potentials for several months (Fig. 1, lower left). Indifferent electrodes were buried in the bone just behind the ears on both sides. The electrooculogram (EOG) was monitored through electrodes placed on the rostrolateral edge of the frontal bone above the orbits on both sides. Cortical potentials and EOGs were amplified with ac amplifiers of 2.0-s time constant and recorded on magnetic tapes. The electromyogram (EMG) from wrist extensor muscles of the moving hand was recorded and rectified.

A monkey was trained to perform wrist extension as a reaction time movement in response to the light stimulus given by a green-light-emitting diode in front of the monkey. Stimulus duration was at first 900 ms, and was then shortened to 700 and 510 ms as the monkey became faster in the movement. The stimulus was repeated at a random interval of 2.5–6.0 s, and the movement within the stimulus duration resulted in delivery of a small amount of fruit juice about 600 ms after the movement. Initially, the monkey performed self-paced movements without regard to the light stimulus (see reaction time histograms of stages I and II in Fig. 5). With repeated training, the monkey began to respond to the light stimulus with reaction time movements, and further training led it to the steady state with fast and stable movements (see stages III and IV in Fig. 5). Cortical potentials, EOGs, and EMGs were averaged with respect to either the onset of light stimulus or the movement (lever elevation). Averaging was usually based on 100 trials for every experimental session. The histogram of reaction times was plotted in 16-ms bins. Sites of all cortical recording electrodes were checked histologically. In sev-

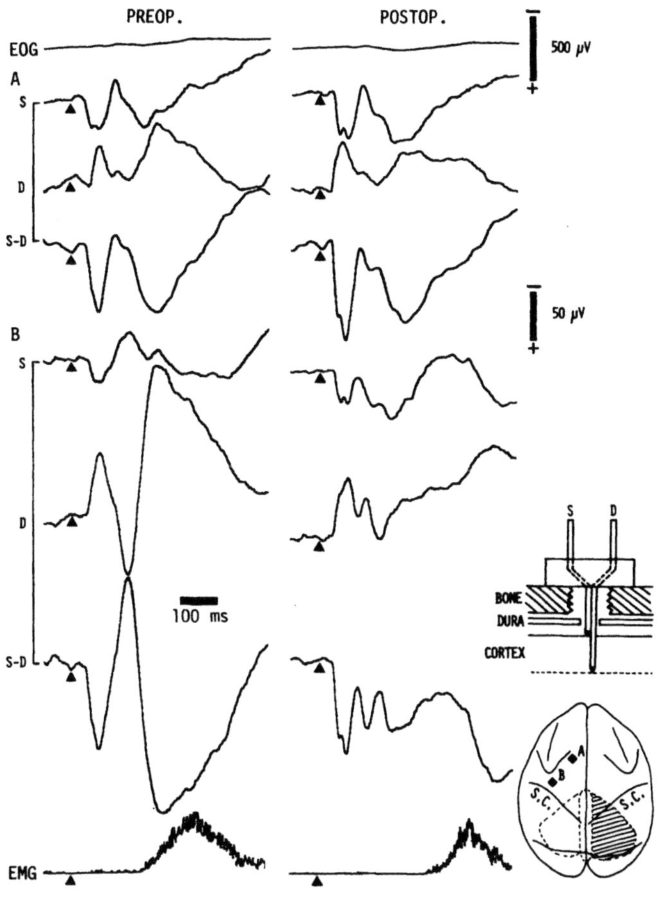

Fig. 2. Cortical field potentials preceding visually initiated hand movements in the premotor (*A*) and forelimb motor (*B*) cortices contralateral to the moving hand (see *inset*). The potentials before (*PREOP.*, *left column*) and 4 days after (*POSTOP.*, *right column*) unilateral cerebellar hemispherectomy (*shaded* in the *lower inset diagram*) are recorded from the same monkey. *S* and *D* rows present field potentials on the surface and at depth (2.5–3.0 mm from the surface) respectively as in the *upper inset diagram*. *S-D* gives surface minus depth potentials. 100 ms for all records. 500 μV for electrooculogram (EOG) and 50 μV for cortical potentials. *S.C.*, central sulcus. EOGs, all cortical potentials, and electromyograms (EMGs) were averaged 100 times with respect to the onset of the stimulus pulse as indicated by *triangle*. (Sasaki et al. 1981a)

eral cases, the hemispherical part of the cerebellum on the side ipsilateral to the moving hand was resected during or before the experiment and the extent of the resection was examined histologically.

Premovement cortical field potentials are exemplified from a monkey in Fig. 2. The records in the left column (PREOP.) were obtained at the well-trained stage and those in the right column (POSTOP.) after the cerebellar hemispherectomy, all being averaged 100 times starting at the stimulus pulse (triangle). In the premotor cortex contralateral to the moving hand, s-P, d-N potentials appeared at about 40 ms latency (S and D rows of A). Similar potentials were led

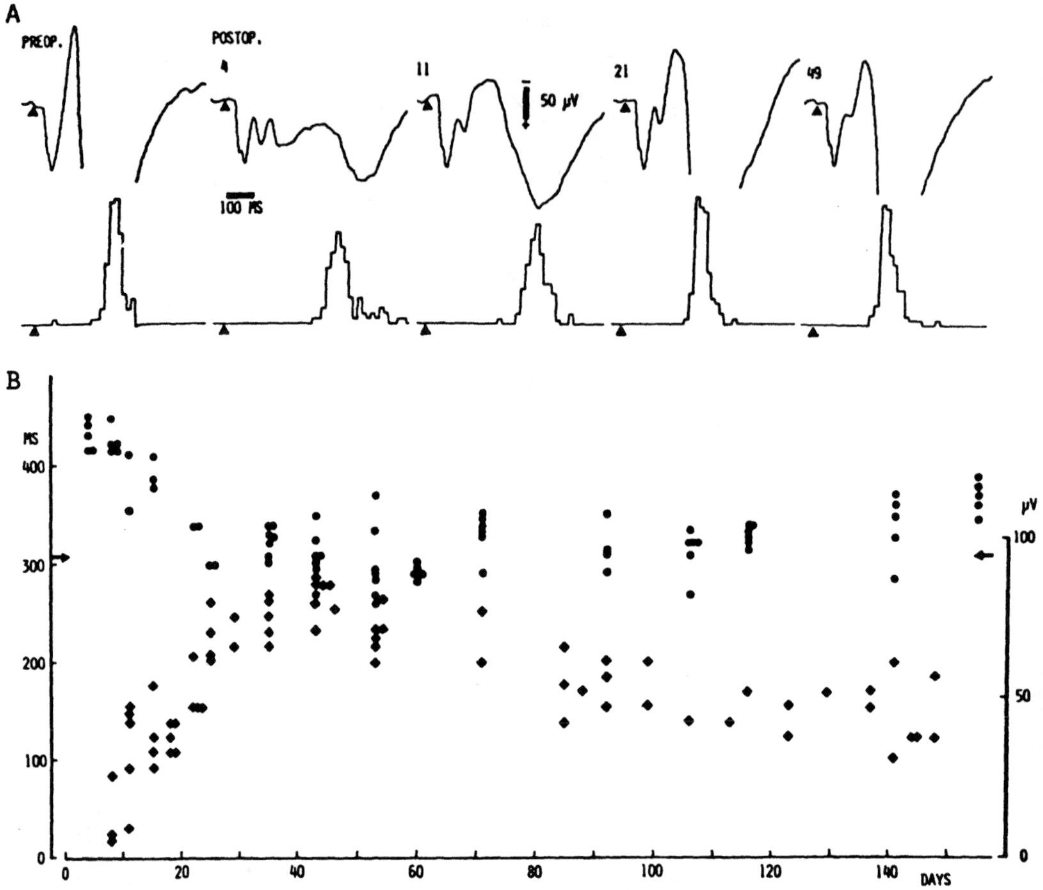

Fig. 3. A Visually initiated premovement potentials (surface minus depth) in the forelimb motor cortex (*upper row*) and histograms of reaction times (*lower row*) as influenced by cerebellar hemispherectomy when the interpositus was spared. 50 μV for potentials and 100 ms for potentials and histograms. One hundred samples for every potential and histogram. The *leftmost column* is control (*PREOP.*) and the others are 4, 11, 21, and 49 days after the operation (*POSTOP.*). *Triangle* indicates the onset of light stimulus. *B* Reaction times (*filled circles*) and peak heights of the late s-N, d-P potentials (*filled squares*) plotted against days after cerebellar hemispherectomy (*abscissa*). Data from the same monkey as in *A*. Mean preoperative reaction time is indicated by *horizontal arrows*. (Sasaki et al. 1982)

with the surface (S) and depth (D) electrodes placed in the forelimb motor cortex contralateral to the hand (S and D rows of B), and they were followed by s-N, d-P potentials at a latency of about 120 ms. Subtraction of depth potential from surface potential gives the S-D trace, which reveals the transcortical potentials without appreciable contamination of EOG, some other potentials generated in the central nervous structures and muscles, etc. Rectified and averaged EMG shows that the early and late potentials in the motor cortex preceded the wrist extension. The early potentials are interpreted as being due to either deep T-C or C-C responses, or to both admixed, whereas the late potentials are due at least mainly to superficial T-C responses. The superficial T-C responses in the forelimb motor cortex can be assumed to be evoked by the impulses mediated by the dentate nucleus (neocerebellum) (Sasaki et al. 1976). In fact, this interpretation was supported by the results of cerebellar hemispherectomy contralateral to the motor cortex, i.e., ipsilateral to the moving hand, as shown in the POSTOP. column. The late s-N, d-P premovement potentials were entirely eliminated by the cerebellar hemispherectomy. After the operation, the monkey could lift the lever but the reaction times were longer, as seen in EMG records. Averaged reaction times in five monkeys with the cerebellar hemispherectomy were prolonged by 90–250 ms. This corresponds to the delay in contraction in the human cases with acute cerebellar injuries reported by Holmes (1917).

When the cerebellar hemispherectomy included both dentate and interpositus nuclei, the prolonged reaction time scarcely recovered for several months after the operation. However, the control reaction time was reattained gradually over a few weeks when the operation involved only the dentate nucleus but spared the interpositus nucleus, as shown in Fig. 3 (Sasaki et al. 1982). This recovery of reaction time is accompanied by reappearance of the s-N, d-P potentials in the forelimb motor cortex, as seen in Fig. 3. These were accounted for by new participation of the interpositus nucleus in activation of the motor cortex for initiation of the fast hand movement (Sasaki et al. 1982), which is consistent with the process of learning to perform fast and stable movements, as will be presented below (cf. Fig. 7).

Development and Change of Premovement Cortical Field Potential During Learning Process Involving Visually Initiated Hand Movement

Visually initiated hand movements, when a monkey is sufficiently trained, are preceded by characteristic field potentials in various cortical areas on both sides, as shown in Figs. 4 and 5. As seen in the rightmost column of Fig. 5 (stage IV, sufficiently trained stage), different profiles of the potentials can be recorded in dif-

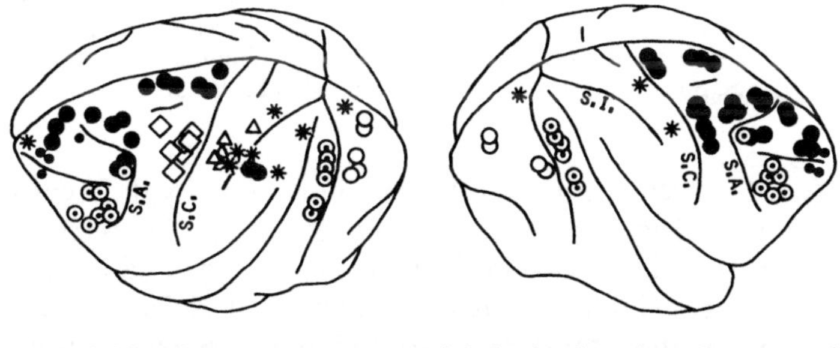

Fig. 4. Distribution of visually initiated premovement potentials in the dorsolateral aspects of hemispheres contralateral and ipsilateral to the moving hand. They were summarized from 14 monkeys at the well-trained stage. Different *symbols* represent different kinds of potentials. Loci without marked potentials are indicated by *asterisks*. Some of the electrode sites could not be plotted in the figures because these were too crowded in a few areas. Sizes of potentials were not considered but only very small potentials were given by small symbols. Characteristic features of respective potentials with different symbols may be referred to the records in Fig. 5, column IV. (Sasaki and Gemba 1982)

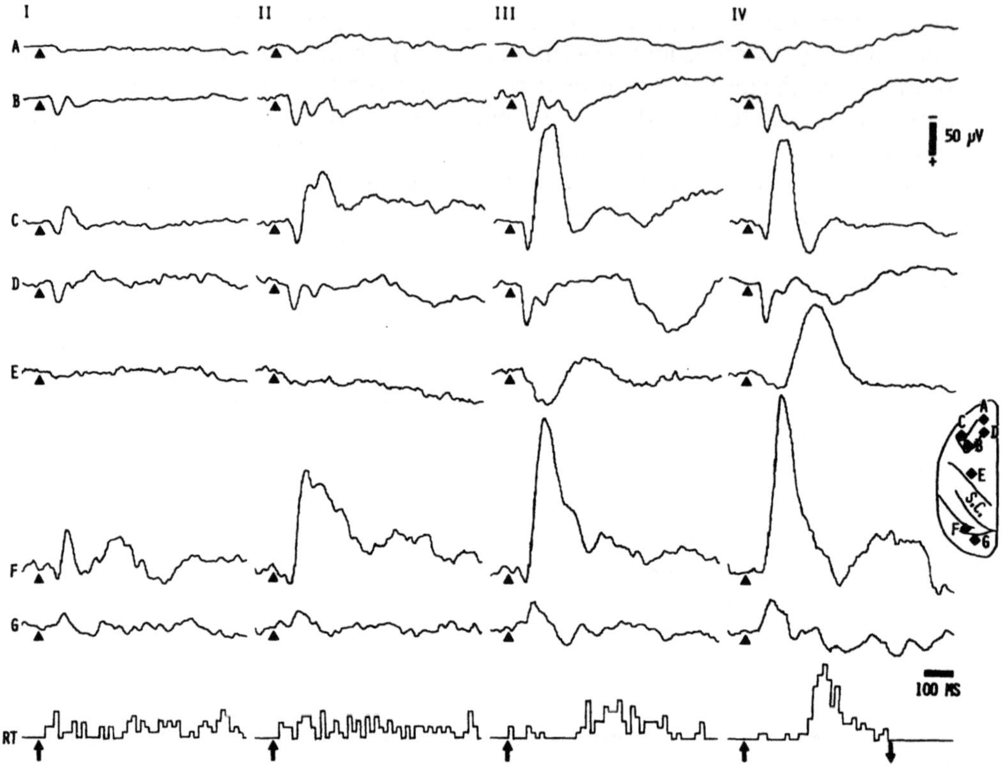

Fig. 5. Premovement surface minus depth potentials in seven cortical loci on the left hemisphere of a monkey performing the right hand movement. Columns I–IV present four sessions at different stages in the learning process; I was taken at the second session, II was 21 days after I, III 3 days after II, and IV 35 days after III. Two sessions a week were regularly held with this monkey, which lifted the lever several hundred times in one session. Potentials were averaged 100 times from every session for the respective cortical loci indicated by *alphabetical symbols in the inset diagram* and the potential row. Light stimulus was given for 900 ms at stages I–III but for 510 ms at stage IV. Only the movements which occurred during the light stimulus were counted in reaction time (*RT*) histograms and averaged potentials (the later part of 900 ms is curtailed in columns I–III). *Triangles and upward arrows* indicate the onset of the light stimulus, whereas *downward arrows* mark the end of the stimulus. 50 μV for all potentials and 100 ms for all traces. *S.C.*, central sulcus. (Sasaki and Gemba 1982)

ferent cortical areas, and they are illustrated by different symbols in Fig. 4. S-D records in Fig. 5 IV reveal the following: s-P, d-N potentials at a latency of about 40 ms in loci A and B in the prefrontal cortex and in the premotor cortex of its wider areas (D) (filled circles, Fig. 4); s-P, d-N followed by s-N, d-P potentials in locus C of the prefrontal cortex and in the prestriate cortex (F) (open circles with dot in center); s-P, d-N followed by long-latency s-N, d-P ones in the forelimb motor cortex (E) (open square), etc. These potentials in the well-trained accomplished state developed and changed during training processes, and the motor learning process is illustrated in four stages as exemplified in Fig. 5.

A naive monkey at the beginning of training usually shows no appreciable potentials in the cortical areas except the striate cortex when the light stimula-

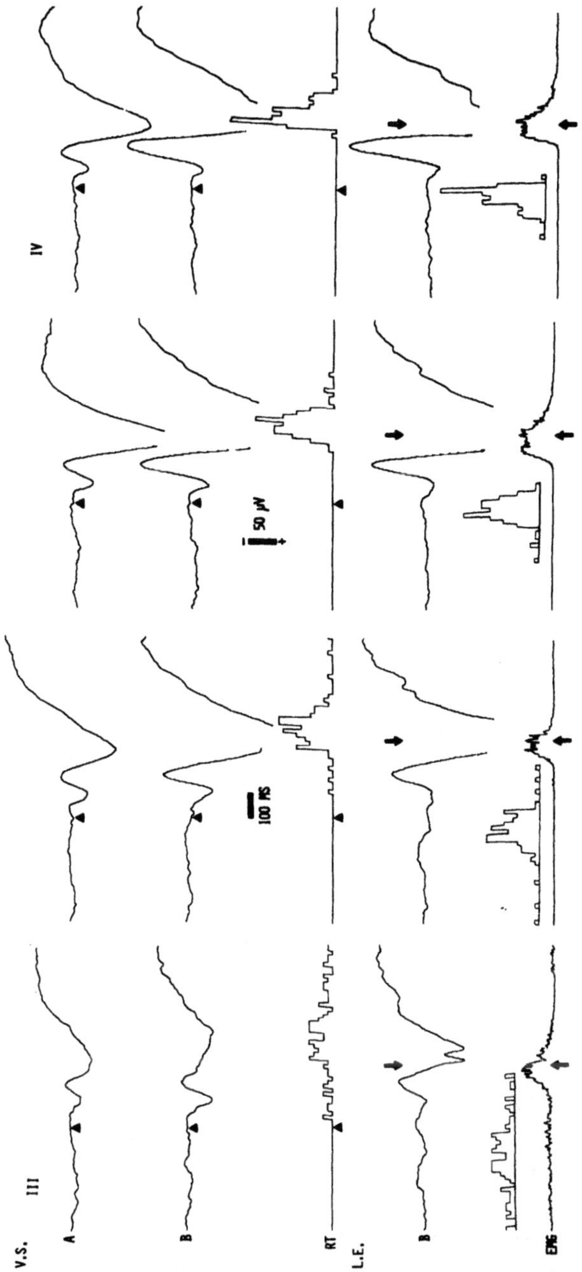

Fig. 6. Premovement surface minus depth potentials from the premotor (*A*) and forelimb motor (*B*) cortices are exemplified on 4 different days from stage III (*leftmost column*) to stage IV (*rightmost column*). *V.S.*, two rows of the potentials (*A* and *B*) were averaged by the pulse of stimulus onset (*triangle*). *L.E.*, row B gives the same data as the row *B* in *V.S.* but is averaged by the pulse of movement (*arrow*). Histograms of onset times of the light stimulus as preceding the movement time are shown above the electromyograms (*EMGs*) averaged by the movement pulse. Duration of the light stimulus was 900 ms in the leftmost and second columns and 510 ms in the third and rightmost columns. 100 ms for all traces and 50 μV for all cortical potentials. (Sasaki and Gemba 1983)

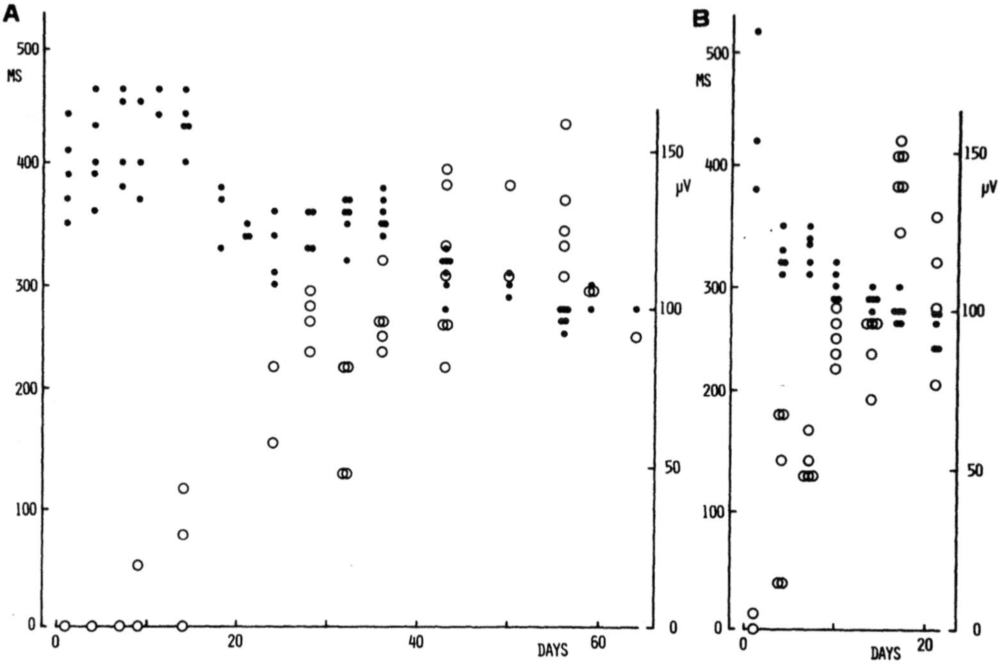

Fig. 7A,B. Changes in reaction times in milliseconds (*filled circles*) and heights of s-N, d-P field potentials in the forelimb motor cortex in microvolts (*open circles*) are plotted against days from stage III to IV (*abscissae*). Examples of two different monkeys are presented in *A* and *B* respectively. The reaction time represents that of the peak in a histogram of 100 samples in every session. For the reaction time and the potential height, several points are presented in respective sessions on each day. (Sasaki and Gemba 1983)

tion is delivered randomly. In the monkey presented in Fig. 5, the second training session (second day) revealed some evoked potentials not only in the striate cortex (column I, row G) but also in the prefrontal (A,B,C), premotor (D), and prestriate (F) cortices. The C and F loci were especially remarkable and the potentials gradually increased in size with repeated training, as shown in column II (stage II, 3 weeks after I). At both stages I and II, the monkey lifted the lever at its own pace without regard to the light stimulus, as noted in the reaction time histograms (RT). However, at stage II, the monkey may have become more interested in the light stimulus and probably recognized it as "meaningful," at least gazing at the light more eagerly.

In the next session to II (2 days later), the monkey suddenly began to respond to the light stimulus with the movement (column III, stage III), i.e., it had obviously learned the reaction time movement, although the movement occurred with longer and more variable latencies than at stage IV. Up to stage III, the stimulus duration was 900 ms; it was 510 ms at stage IV. At stage III, the s-P, d-N followed by s-N, d-P potentials in the association cortex (C and F rows) reached the maximal size. The reaction time movement in stage III was preceded by s-P, d-N potentials (ca. 40 ms latency) in the motor cortex but few or no s-N, d-P potentials were seen at this stage. The s-P, d-N potentials in the premotor cortex in-

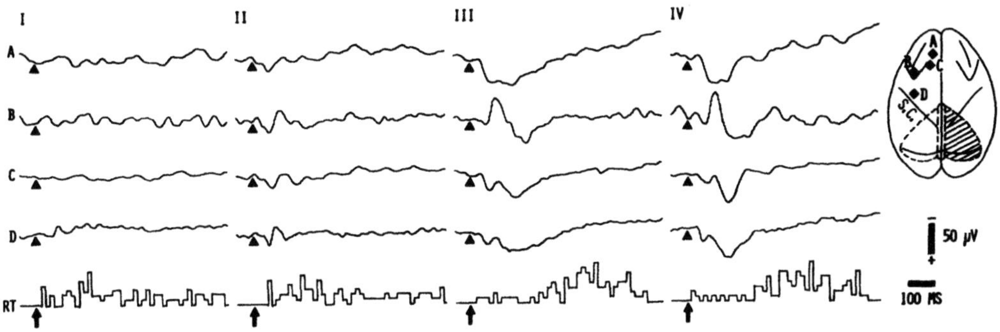

Fig. 8. Premovement surface minus depth potentials in four loci of the left cerebral hemisphere of the monkey whose cerebellar hemisphere on the right side (*shaded by oblique lines*) had been extirpated before it was trained to perform the right hand movement. *S.C.*, central sulcus. Averaged 100 times by the pulse of onset of the light stimulus (*triangle and arrow*) lasting for 900 ms. The later part of columns I–IV is curtailed. For all potentials: 50 μV; for all traces: 100 ms. (Sasaki and Gemba 1982)

creased a little in size at stage III (D row). After stage III, repeated training for several weeks (3–8 weeks in different monkeys) made reaction times shorter and less variable, and concomitantly the s-N, d-P potentials in the contralateral forelimb motor cortex were gradually enhanced (III–IV in row E).

As shown in Fig. 6, another monkey reached stage III as noted in the leftmost column (III) and revealed early s-P, d-N potentials in the premotor (A) and forelimb motor (B) cortices with variable long reaction times. The potentials in the contralateral motor cortex are presented by averaging the same data with the pulses of stimulus onset (V.S.) and movement (L.E.). The latter averaging discloses a small s-N, d-P premovement potential in the motor cortex which is hardly seen in the averaging with the stimulus pulse. In any case, it is evident that the late s-N, d-P premovement potentials gradually increase in size during repeated training from stage III to IV, and that the enhancement of the potentials is closely correlated with faster and less variable reaction times. The wrist movement itself did not change so much, as seen in averaged EMG records. Different monkeys took 3–8 weeks from stage III to IV and reached the steady state, which continued for many months without much change in the premovement potentials and reaction times. The longest (A) and shortest (B) time courses of the changes in the s-N, d-P potentials and the reaction times are plotted during stages III–IV for two monkeys in Fig. 7. These are considered to be the time courses of the recruitment of cerebrocerebellar interactions in attaining fast and stable hand movement ("skillfulness"). It may be noted that the time courses appear similar to those of the recovery of reaction times and of the s-N, d-P potentials in the hemispherectomized monkey in which the dentate nucleus was severed but the interpositus spared (Fig. 3). That would reveal learning processes of the interpositus system on the occasion in which the dentate nucleus, which works in the normal state as a link of the cerebrocerebellar interactions, is eliminated (Sasaki et al. 1982).

After the cerebellar hemisphere had been extirpated on one side, training was started in two monkeys for the hand movement on the same side as the hemispherectomy. It took a longer time to train such monkeys than normal monkeys.

However, the successive appearance of premovement cortical potentials in various areas occurred in exactly the same order and manner as in normal monkeys except for the late s-N, d-P potentials in the contralateral forelimb motor cortex. Even with repeated training for several months, the hemispherectomized monkeys never revealed these potentials, and accordingly the reaction time movement was hardly improved, as shown in Fig. 8 (column IV). Later histological examination proved that the dentate nucleus and the major part of the interpositus nucleus were destroyed by the operation in the two monkeys.

Processes of Motor Learning

It is interesting to note, as in Fig. 5, that the latencies of premovement potentials in various cortical areas have largely equal values of about 40 ms (35–47 ms) in the different loci except in the somatosensory cortex, which usually shows a longer latency. The same ranges of latencies among the striate (primary visual), association (prefrontal and prestriate), premotor, and motor cortices suggest that the impulses evoked by the visual stimulus arrive at these cortical areas almost simultaneously and in parallel during the late stage in which the learning has been accomplished (Fig. 5 IV). The latencies of the characteristic potentials in respective cortical areas changed little during the learning process. In fact, it was consistently observed in every monkey that responsive cortical areas were successively added during the learning process, and that respective premovement potentials increased in size gradually up to relatively certain levels at different stages without marked changes in the latencies (see Fig. 5). These findings indicate that visually evoked impulses arrive at the primary visual cortex at the beginning of learning, and simultaneously or several days later at association and premotor cortices, then several weeks later at the motor cortex, and thus let the motor cortex execute the conditioned movement. At the later stage of motor learning, the cerebrocerebellar neuronal circuits are recruited to activate the motor cortex more efficiently in order to perform the movement quickly and skillfully.

The sequential participation of these different cortical areas with much the same latencies indicates that visually evoked impulses are conveyed in a stepwise and additional manner to the different areas by a kind of "switching" mechanism, i.e., volleys evoked by the visual stimulus are first conveyed only to the primary visual cortex, and then delivered in succession to the association and premotor cortices and later to the motor cortex, during stages I–III of the learning process by the switching mechanism working somewhere in the subcortical structures. The early components of premovement potentials recorded at latencies of about 40 ms in the different cortical areas were mostly s-P, d-N. According to depth profiles in the cortex, they must be either deep T-C responses or C-C ones, or admixtures of the two (Sasaki 1979; Sasaki et al. 1981b). The fact of the latencies mentioned above suggests that the early components are more likely to be constituted mainly by deep T-C responses than by C-C ones. Therefore, the switching mechanism should occur in the thalamus and its closely related structures. Thalamic nuclei, especially associational and intralaminar nuclei of the thalamus might be candidates for such a function, since various paths of information flow, including those from the cerebral cortices, can converge into these nuclei, which in turn project divergently onto various cortical areas both directly and indirectly (Hassler 1972; Kievit and Kuypers 1977; Akert and Hartmann-Von Monakow 1980).

Acknowledgments. The author would like to express cordial thanks to the organizers of this memorable symposium. Thanks are also due to Prof. N. Mizuno for help and advice on morphological examinations and Dr. H. Gemba for collaboration throughout chronic experiments.

References

Akert K, Hartmann-Von Monakow K (1980) Relationships of precentral, premotor and prefrontal cortex to the mediodorsal and intralaminar nuclei of the monkey thalamus. Acta Neurobiol Exp (Warsz) 40: 7–25

Allen GI, Tsukahara N (1974) Cerebrocerebellar communication systems. Physiol Rev 54: 957–1006

Brooks VB, Thach WT (1981) Cerebellar control of posture and movement. In: Brooks VB (ed) The nervous system II. Am Physiol Soc, Bethesda, pp 877–956 (Handbook of physiology)

Eccles JC (1973) The cerebellum as a computer: patterns in space and time. J Physiol 228: 1–32

Eccles JC (1979) Introductory remarks. In: Massion J, Sasaki K (eds) Cerebro-cerebellar interactions. Elsevier, Amsterdam, pp 1–18

Evarts EV, Thach WT (1969) Motor mechanisms of the CNS: cerebrocerebellar inter-relations. Annu Rev Physiol 31: 451–498

Hassler R (1956) Die extrapyramidalen Rindensysteme und die zentrale Regelung der Motorik. Dtsch Z Nervenheilk 175: 233–258

Hassler R (1972) Hexapartition of input as a primary role of the thalamus. In: Frigyesi T, Rinvik E, Yahr MD (eds) Corticothalamic projections and sensorimotor activities. Raven, New York, pp 551–579

Holmes G (1917) The symptoms of acute cerebellar injuries due to gunshot injuries. Brain 40: 461–535

Kawaguchi S, Miyata H, Kawamura M, Harada Y, Noda T (1981) Electrophysiological and morphological demonstration of the two different types of thalamocortical projections in the cat. Neurosci Lett [Suppl] 6: 19

Kievit J, Kuypers HGJM (1977) Organization of the thalamocortical connexions to the frontal lobe in the rhesus monkey. Exp Brain Res 29: 299–322

Mizuno N, Mochizuki K, Akimoto C, Matsushima R, Sasaki K (1973) Projections from the parietal cortex to the brain stem nuclei in the cat, with special reference to the parietal cerebro-cerebellar system. J Comp Neurol 147: 511–522

Oka H, Sasaki K, Matsuda Y, Yasuda T, Mizuno N (1975) Responses of pontocerebellar neurones to stimulation of the parietal association and the frontal motor cortices. Brain Res 93: 399–407

Sasaki K (1976/77) Electrophysiological studies on the cerebellothalamocortical projections. Appl Neurophysiol 39: 239–250

Sasaki K (1979) Cerebro-cerebellar interconnections in cats and monkeys. In: Massion J, Sasaki K (eds) Cerebro-cerebellar interactions. Elsevier, Amsterdam, pp 105–124

Sasaki K, Gemba H (1982) Development and change of cortical field potentials during learning processes of visually initiated hand movements in the monkey. Exp Brain Res 48: 429–437

Sasaki K, Gemba H (1983) Learning of fast and stable hand movement and cerebro-cerebellar interactions in the monkey. Brain Res 277: 41–46

Sasaki K, Staunton HP, Dieckmann G (1970) Characteristic features of augmenting and recruiting responses in the cerebral cortex. Exp Neurol 26: 369–392

Sasaki K, Kawaguchi S, Matsuda Y, Mizuno N (1972a) Electrophysiological studies on cerebello-cerebral projections in the cat. Exp Brain Res 16: 75–88

Sasaki K, Matsuda Y, Kawaguchi S, Mizuno N (1972b) On the cerebello-thalamo-cerebral pathway for the parietal cortex. Exp Brain Res 16: 89–103

Sasaki K, Oka H, Matsuda Y, Shimono T, Mizuno N (1975) Electrophysiological studies of the projections from the parietal association area to the cerebellar cortex. Exp Brain Res 23: 91–102

Sasaki K, Kawaguchi S, Oka H, Sakai M, Mizuno N (1976) Electrophysiological studies on the cerebello-cerebral projections in monkeys. Exp Brain Res 24: 495–507

Sasaki K, Oka H, Kawaguchi S, Jinnai K, Yasuda T (1977) Mossy fibre and climbing fibre responses produced in the cerebellar cortex by stimulation of the cerebral cortex in monkeys. Exp Brain Res 29: 419–428

Sasaki K, Jinnai K, Gemba H, Hashimoto S, Mizuno N (1979) Projection of the cerebellar dentate nucleus onto the frontal association cortex in monkeys. Exp Brain Res 37: 193–198

Sasaki K, Gemba H, Hashimoto S (1981a) Influences of cerebellar hemispherectomy upon cortical potentials preceding visually initiated hand movements in the monkey. Brain Res 205: 425–430

Sasaki K, Gemba H, Hashimoto S (1981b) Premovement slow cortical potentials on self-paced hand movements and thalamocortical and corticocortical responses in the monkey. Exp Neurol 72: 41–50

Sasaki K, Gemba H, Mizuno N (1982) Cortical field potentials preceding visually initiated hand movements and cerebellar actions in the monkey. Exp Brain Res 46: 29–36

Cerebellar Plasticity and Motor Learning

MASAO ITO, Tokyo

Introduction

The challenge of examining complex structures and functions of the central nervous system at the level of neurons and synapses has been the major theme of the Eccles school since its Canberra era. The monograph *The Cerebellum as a Neuronal Machine* (Eccles et al. 1967) summarizes the remarkable progress of cerebellar studies made in the 1960s on this theme. The detailed information about cerebellar neuronal networks introduced by this monograph gives the definite impression that the cerebellum has an extremely ordered internal structure as intricate as a modern computer, and has resulted in formulation of Marr's and others' theoretical models of the cerebellar neuronal network.

During the past decade, studies of the cerebellum have been expanded to solve various fundamental problems, such as development of neuronal elements, ionic mechanisms of neuronal signal generation, identification of neurotransmitters, quantitative estimation of divergence and convergence of neuronal connections, and specification of synaptic plasticity. At the same time, a number of new techniques have been introduced for analyzing neuronal connections in and around the cerebellum, and for testing cerebellar functions in alert behaving animals. Studies of the cerebellum have thus developed in multiple directions toward the goal of understanding not only how the cerebellum is constructed but also how the cerebellar circuitry operates and eventually serves for diverse bodily functions.

In our laboratory, we have tried to afford experimental evidence on synaptic plasticity in the cerebellar cortex on one hand, and to establish a control system model of the cerebellum for a particular case of eye movement, i.e., adaptive modification of the vestibuloocular reflex, on the other. These two lines of investigation have converged into a hypothesis that the cerebellum acts as a modifiable element of a motor or autonomic control system affording the system with the ability to adapt and learn. This article briefly introduces the outcome of these experimental studies and speculations.

Synaptic Plasticity in the Cerebellar Cortex

Marr's (1969) original assumption that the parallel fiber-Purkinje cell synapses increase the transmission efficacy under conjoint influences of parallel fiber and climbing fiber impulses on the same Purkinje cell dendrites has been modified by Albus (1971) to the effect that the transmission efficacy decreases instead of in-

Experimental Brain Research, Suppl. 9
© Springer-Verlag Berlin · Heidelberg 1984

creasing. A long-lasting depression of the parallel fiber-Purkinje cell transmission has been found accordingly to follow conjunctive stimulation of parallel fibers and climbing fibers (Ito et al. 1982; Ito and Kano 1982; Ekerot et al. 1983).

The reason why the long-lasting depresion has been overlooked in earlier experiments appears to be twofold. First, climbing fibers were stimulated at such high frequencies that a general depression of excitability occurred in Purkinje cells, obscuring changes in parallel fiber-Purkinje cell transmission. Secondly, field potentials in the cerebellar cortex were referred to, but these might provide a rather insensitive index of parallel fiber-Purkinje cell transmission. It is now clear that the long-lasting depression can be induced effectively with relatively low frequency stimulation within a physiological range (1–4 Hz) and can sensitively be represented by reduction of firing index values in individual Purkinje cells. Relatively small changes have been detected in the negative field potentials of the molecular layer which represents postsynaptic excitatory currents driven by parallel fiber impulses (Ito and Kano 1982), but these changes are rather small and often difficult to detect.

Following the full-time course of the long-lasting depression has not been an easy task because of technical difficulties in keeping the stimulating and recording conditions sufficiently stable. In limited cases, though, the depression has been followed over 3–4 h, giving the impression that it lasts for a longer period of time. It seems evident that the long-lasting depression underlies the postulated plasticity of the parallel fiber-Purkinje cell transmission, at least a substantial part of it.

A pharmacologic study further revealed that a long-lasting depression takes place in the sensitivity of Purkinje cells to glutamate, the putative neurotransmitter of parallel fibers, when glutamate is applied in conjunction with climbing fiber stimulation (Ito et al. 1982). It is apparent that subsynaptic chemosensitivity is involved in the long-lasting depression observed after conjunctive stimulation of parallel fibers and climbing fibers. Molecular mechanisms of the depression in subsynaptic chemosensitivity of Purkinje cells have yet to be investigated, but two possibilities have been raised (Ito et al. 1982). First, calcium ions flowing into Purkinje cell dendrites during climbing fiber activation would affect the subsynaptic receptor molecules in Purkinje cell dendrites in the manner suggested for desensitization in neuromuscular junctions (Miledi 1980; Ekerot and Oscarsson 1981). Secondly, climbing fibers may liberate a chemical substance which reacts with both subsynaptic receptor molecules and parallel fiber neurotransmitter, thereby rendering the receptors insensitive to the parallel fiber neurotransmitter (Ito et al. 1982). The examination of these two possibilities will be an interesting future task.

Modeling of the Cerebellar Neuronal Network

Marr's (1969) and Albus' (1971) models of the cerebellar neuronal network have recently been modified by Fujita (1982), who formulated an adaptive filter model using learning principles similar to those of Marr-Albus models. Fujita's model, unlike Marr-Albus models, can deal with frequency-modulated time analog signals, and therefore enables us to make a realistic simulation of the performance of cerebellar tissues. In brief, the mossy fiber-granule cell-Golgi cell network is assumed to constitute a transducer stage which converts a mossy fiber input signal into a set of signals of parallel fibers with varied phase shifts. A Purkinje cell then selects signals from parallel fibers through synaptic contact with high transmission efficacy. The transmission efficacy is modified through conjunctive activation of parallel fibers and climbing fibers. It is notable that either an increase as assumed by Marr (1969) or a decrease as postulated by Albus (1971) leads similarly to convergence of responses of a Purkinje cell to a desired response, when a climbing fiber for a Purkinje cell keeps signalling deviation of its responses from the desired one. Fujita's model (1982) visualizes that the relation between mossy fiber inputs and Purkinje cell outputs is modified according to climbing fiber signals.

Floccular Control of the Vestibuloocular Reflex

A small area of the cerebellar cortex, called "microzone" (Oscarsson 1976), projects to a small subgroup of cerebellar or vestibular nuclear cells. These cortical microzones and cortical cell subgroups receive common excitatory inputs through both climbing fiber and mossy fiber afferents. It would be convenient to define an assembly of a cortical microzone and a nuclear cell subgroup as a functional unit of the cerebellum; it will be termed a "corticonuclear microcomplex."

The major input to corticonuclear microcomplex is provided by mossy fibers and the output by nuclear axons. Through these input and output connections, a corticonuclear microcomplex is inserted into an autonomic or motor system formulated in extracerebellar tissues. If climbing fiber afferents inform about errors in the eventual performance of the system, and thereby modify signal transfer characteristics of the corticonuclear microcomplex as assumed in the Marr-Albus-Fujita model, the corticonuclear microcomplex will function as a modifiable element of the system, making adaption and learning possible. A good example of this scheme has been found in the vestibular system (Fig. 1). A corticonuclear

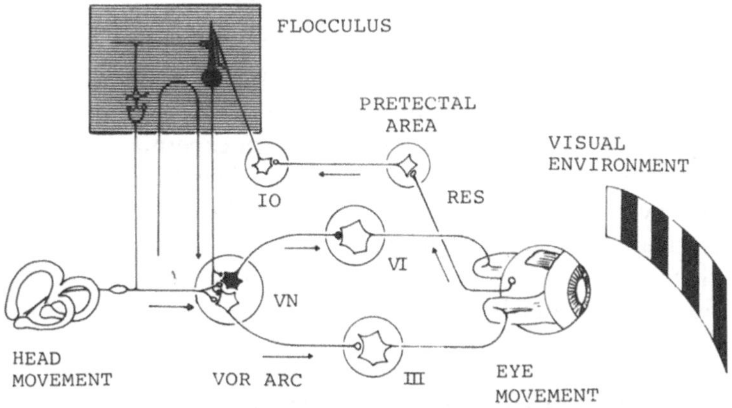

Fig. 1. Neuronal diagram of the floccular control system for the horizontal vestibuloocular reflex (*VOR*). *RES*, retinal slip signal; *IO*, inferior olive; *VN*, vestibular nucleus; *III, VI*, oculomotor and abducens nuclei.

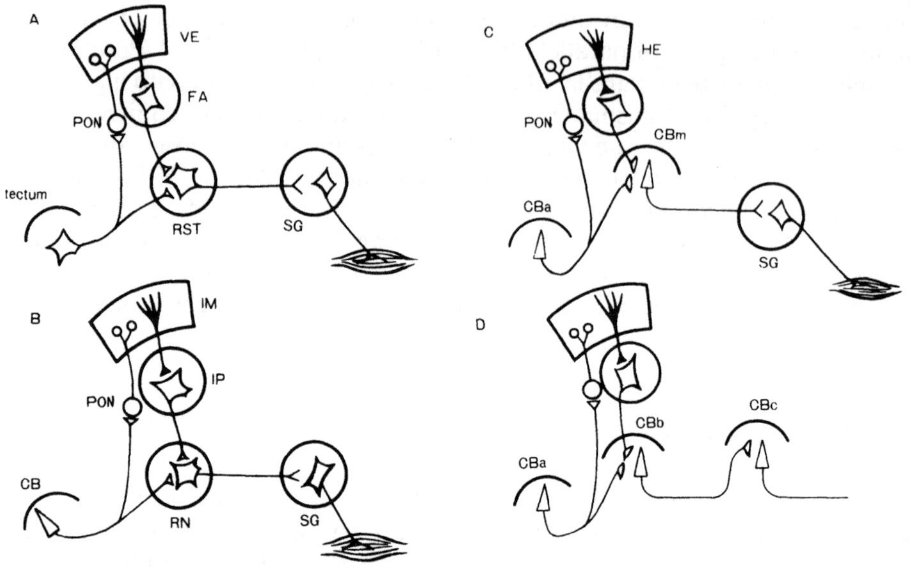

Fig. 2A–D. Sidepath insertion of corticonuclear complexes into central pathways. *VE*, vermis; *FA*, fastigial nucleus; *PON*, pontine nuclei; *HE*, cerebellar hemisphere; *IM*, intermediate cortex of the cerebellum; *IP*, nucleus interpositus; *RST*, reticulospinal tract cell; *SG*, segmental neuron pool; *RN*, red nucleus; *CBa, CBb, CBc*, cerebral cortical areas; *CBm*, cerebral motor area.

microcomplex consisting of a microzone in the flocculus and a group of vestibulo-ocular relay cells, both receiving vestibular inputs, forms a part of the major vestibuloocular reflex pathway. Retinal error signals are forwarded to the flocculus via climbing fibers. Thus it is hypothesized that the flocculus acts as a modifiable element of the vestibuloocular reflex pathway, and accordingly is the site of adaptive modification of the vestibuloocular reflex which occurs under visual vestibular interaction conditions (Ito 1970, 1972, 1982).

The flocculus hypothesis has been supported by two lines of experimental evidence. First, ablation of the flocculus or interruption of the visual climbing fiber pathway abolished the adaptive modification of the vestibuloocular reflex (Ito 1982). Secondly, responses of flocculus Purkinje cells to vestibular mossy fiber inputs altered in parallel with an adaptive change of the vestibuloocular reflex gain. The second line of evidence derived from experiments in rabbits (Dufossé et al. 1978) was not confirmed in a later experiment on monkeys (Miles et al. 1980). However, a more recent work closely referring to the microzonal structure of monkey's flocculus (Balaban et al. 1981) could successfully reveal adaptive changes in Purkinje cell behavior in monkey's flocculus similar to those in rabbit flocculus (Watanabe 1984).

Modifiable Sidepath Model of the Cerebellum

The scheme that the flocculus acts as a modifiable sidepath of the vestibuloocular reflex arc can be generalized to relations of a cerebellar corticonuclear microcomplex to any central pathway which constitutes a control system. The central pathway can be a reflex arc or a pathway which conveys command signals from a cerebral cortex to a brain stem area or from a cortical area to another cortical area (Fig. 2). Errors in performance of such a system would be signalled to the cortico-nuclear microcomplex via a climbing fiber pathway, and accordingly modify the signal transfer characteristics of the microcomplex. A corticonuclear microcomplex would thus make adaption and learning possible, enabling the system to maintain an adequate performance under changing environmental conditions. It is conceivable that a cerebellar corticonuclear microcomplex is inserted as a side-path to central pathways such as reticulotectospinal, corticorubrospinal, corticocorticospinal, and corticocorticocortical pathways through pontine and cerebellar nuclei (Fig. 2). It seems to be meaningful to examine highly complex cerebellar connections from this paricular point of view.

References

Albus JS (1971) A theory of cerebellar function. Math Biosci 10: 25–61

Balaban CD, Ito M, Watanabe E (1981) Demonstration of zonal projections from the cerebellar flocculus to vestibular nuclei in monkeys (*Macca fuscata*). Neurosci Lett 27: 101–105

Dufossé M, Ito M, Jastreboff PJ, Miyashita Y (1978) A neuronal correlate in rabbit's cerebellum to adaptive modification of the vestibulo-ocular reflex. Brain Res 150: 611–616

Eccles JC, Ito M, Szentágothai J (1967) the cerebellum as a neuronal machine. Springer, Berlin Heidelberg New York

Ekerot K-F, Oscarsson O (1981) Prolonged depolarization elicited in Purkinje cell dendrites by climbing fibre impulses in the cat. J Physiol (Lond) 318: 207–221

Ekerot K-F, Ito M, Kano M (1983) Long-lasting depressions of parallel fiber-Purkinje cell transmission caused by conjunctive stimulation of parallel fibers and climbing fibers. Neurosci Lett [Suppl] 9

Fujita M (1982) Adaptive filter model of the cerebellum. Biol Cybern 45: 195–206

Ito M (1970) Neurophysiological aspects of the cerebellar motor control system. Int J Neurol 7: 162–176

Ito M (1972) Neural design of the cerebellar motor control system. Brain Res 40: 81–84

Ito M (1982) Cerebellar control of the vestibulo-ocular reflex – Around the flocculus hypothesis. Ann Rev Neurosci 5: 275–296

Ito M, Kano M (1982) Long-lasting depression of parallel fiber-Purkinje cell transmission induced by conjunctive stimulation of parallel fibers and climbing fibers in the cerebellar cortex. Neurosci Lett 33: 253–258

Ito M, Sakurai M, Tongroach P (1982) Climbing fibre induced depression of both mossy fibre responsiveness and glutamate sensitivity of cerebellar Purkinje cells. J Physiol (Lond) 324: 113–134

Marr D (1969) A theory of cerebellar cortex. J Physiol (Lond) 305: 197–213

Miledi R (1980) Intracellular calcium and desensitization of acetylcholine receptors. Proc Soc Lond [Biol] 209: 447–452

Miles F, Braitman DJ, Dow BM (1980) Long-term adaptive changes in primate vestibuloocular reflex. IV. Electrophysiological observations in flocculus of adapted monkeys. J Neurophysiol 43: 1477–1493

Oscarsson O (1976) Spatial distribution of climbing and mossy fibre inputs into the cerebellar cortex. In: Creutzfeldt O (ed) Afferent and intrinsic organization of laminated structures in the brain. Exp Brain Res [Suppl 1], pp 36–42

Watanabe E (1984) Neuronal events correlated with long-term adaptation of the horizontal vestibulo-ocular reflex in the prima primate flocculus. Brain Res 297: 169–174

The Cerebellum and Adaptive Tuning of Movements

VERNON B. BROOKS, London (Ont.)

Today we are celebrating Sir John's 80th birthday, an occasion that makes many of us reflect on the directions of our scientific lives after their course was altered by the impact of the Eccles comet. This volume bears tribute to the diverse inspirations that have propelled so many in new directions.

Eccles' writings on the integrative functions of the cerebellum stress its role in the interplay between programmed movements and their ongoing adjustments, their adaptations based on comparison of "all the learnt skills" with the actual performance (cf. Eccles 1977). In the inaugural Stevenson Lecture (given at the University of Western Ontario on 5 October 1972, unpublished), Eccles spoke on "The Role of the Brain in Movement and Skill." He explained how learned movements are executed by motor programs that involve the cerebellum (among other brain structures), and he emphasized how moving "with gentleness and style" has to be especially learned. Style that pleases the beholder often coincides with a manner of execution that is most suited to the body, that is most efficient. We intuitively tend to attribute beauty and pleasure to a performance that bespeaks a well-tuned relation of the brain and the body that it lives in. Melvill Jones and Watt (1971) have drawn attention to a possible relation between popular dance rhythms and the capabilities of our motor system.

How is one to define skill? Earlier books on the subject offer no definitions (e.g., Welford 1968; Connolly 1970; Poulton 1974). Sometimes it is simply equated with use of learned movements (Schmidt 1982). Following Eccles, we will take "skill" to mean the use of learned movements in the most efficient way for the user in the performance of goal-directed behavior. "Most efficient," skillful performance requires adjustments at many levels of the planning hierarchy. At a complex plane for example, the basal ganglia are needed for integration of motor acts into previously learned behavioral performance (perhaps as gates for some hippocampal outflow?). Greeting a person with the successive motor acts of standing up and proffering a hand, for instance, may fail in a patient with Parkinson's disease because the sequence may break between the two motor acts (cf. Marsden 1982). Cerebellar contributions appear to lie in the less complex "packaging" of planned motor acts by sequencing and composing intended, compound movements and their constituent simple movements by adjusting cortical and subcortical targets of cerebellar influence (also cf. Wiesendanger 1981). In this paper, some new physiological findings are offered that may have a bearing on motor skill, and some avenues of exploration are considered that may add to the meaning of skill, of doing things "with style."

Experimental Brain Research, Suppl. 9
© Springer-Verlag Berlin · Heidelberg 1984

The cerebellum helps us gain motor skill in at least two distinct ways: by assisting with (a) the learning of motor tasks and (b) their adaptation for optimal execution under various conditions. The first cerebellar contribution involves establishment of motor programs, including subprograms for proper timing of starts and stops of movements that we have learned to make in the right direction and with optimal forces and speed, to reach the desired endpoint. Compound, multijoint movements appear skillful and elegant when their constituent simple movements are timed to blend the whole into one smooth performance. Cerebellar dysfunction decomposes previously learned, skilled compound movements into a clumsy, unskilled sequence of individual simple movements (Holmes 1939; Eccles 1977). Evidence that anticipatory subprograms, including those for predictive timing of starts and stops, emanate from the neocerebellum has been published by my colleagues and myself (Brooks 1969, 1972; Brooks et al. 1973; Meyer-Lohmann et al. 1977; Vilis and Hore 1980) and has been reviewed recently (Brooks and Thach 1981).

The second cerebellar contribution toward gaining motor skill is its role in the adaptation of learning programs to suit changing conditions. For example, patients with cerebellar disorders lose the adaptibility of functional stretch reflexes in their legs so that they can no longer adjust sway of their bodies brought on by tilt of the surface upon which they stand (Nashner and Grimm 1978). Generally, dysfunction of the neocerebellum and its olivary input causes degradation of programs and failure to adapt to changing conditions, so that they finally fall into disuse for lack of success. I would like to address the question of adaptation here. Since movements are learned and adapted in similar ways by human subjects and by monkeys, the sequence can be analyzed in the animal model.

To study programming, we must be able to recognize programmed movements. Fortunately, this is easy because programmed simple movements have a distinctive velocity profile with only one peak, giving them their "continuous" appearance. These movements of moderate speed last about 0.5 s, and their single-joint action resembles that of a damped spring, for monkeys and for humans (Kozlovskaya et al. 1970; Brooks et al. 1973; reviewed by Brooks 1979). An example is shown at the top of Fig. 1 next to a diagram of a monkey performing a step-tracking task with arm movements pivoted about the elbow joint. The graphs in Fig. 1 show that take-offs and landings of accurately programmed movements were learned and programmed together right from the beginning when nonprogrammed movements still predominated (Brooks et al. 1983). The task situation posed no narrow behavioral requirements, in fact the animals essentially chose their own modes of behavioral performance and of movement execution. (The handle was turned in self-paced cadence in a horizontal arc into alternate target zones within 1 s after presentation of the target, where it had to be held for at least a few seconds before the return move).

Figure 2 (left) shows that the animals learned the behavioral task requirements (direction, move and hold times) at the same time as they learned to program the movements with which to carry out the task. Yet accurately programmed movements (landing in target after a flight with only one velocity peak) came into common use only when the subjects began to evince signs of understanding the be-

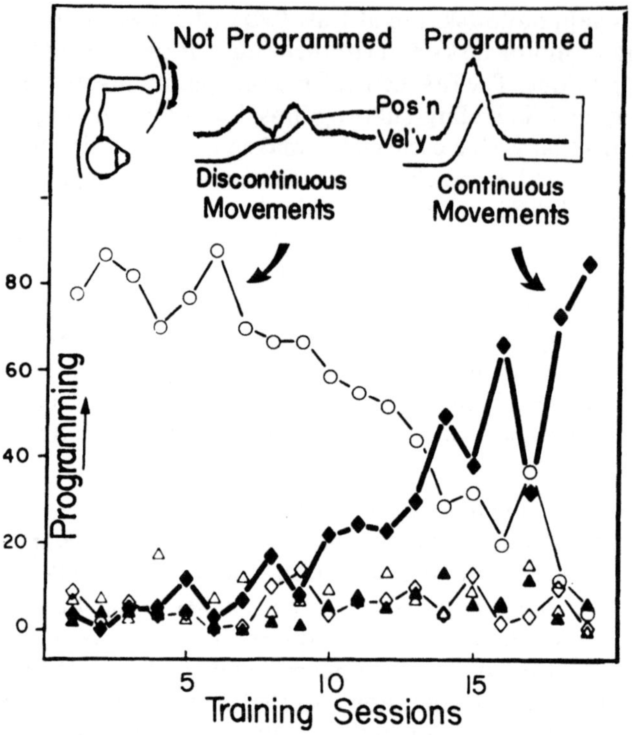

Fig. 1. Programmed arm movements replace nonprogrammed movements made by a monkey with practice of a step-tracking task. Continuous and discontinuous movements: *filled diamonds* and *open circles*. Partly programmed movements: *triangles* and *open diamonds*. All symbols represent means of about 200 values per training session. (Brooks et al. 1983)

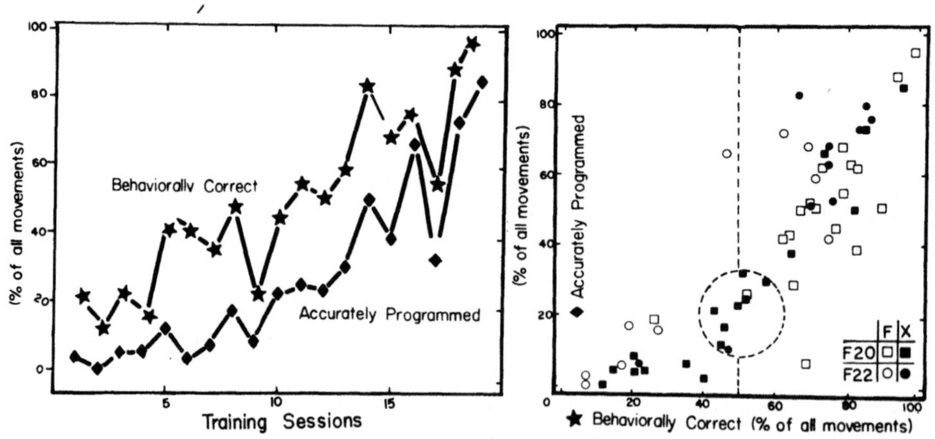

Fig. 2. Common use of programmed movements depends on understanding the behavioral requirements of the task. *Left*: correct behavioral sequence of the task is learned by a monkey (F22) at the same time as accurate programming of movements with which to carry it out. *Right*: relation of behaviorally correct and of accurately programmed flexions (*F*) and extensions (*X*) of two monkeys (F20, 22). Same conventions and sessions as in Fig. 1. (Brooks et al. 1983)

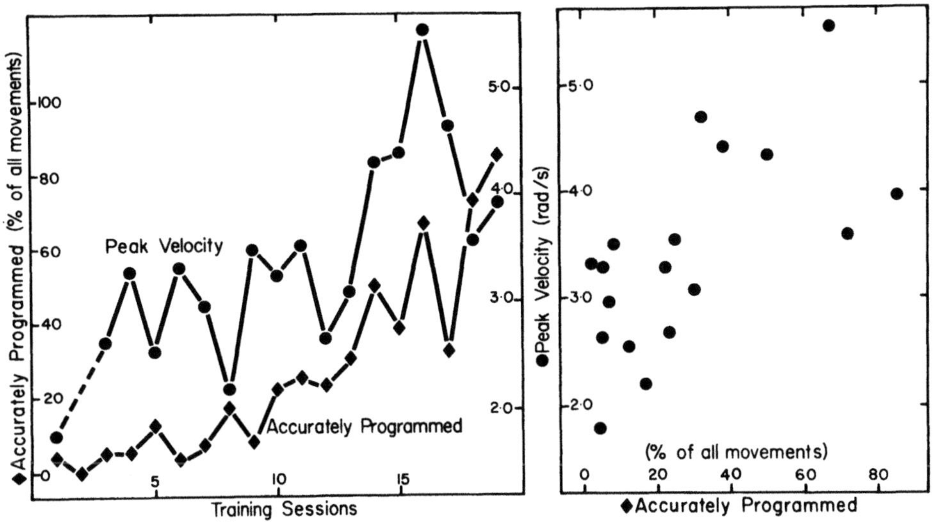

Fig. 3. Peak velocities af accurately programmed arm movements of a monkey increase right from the start of training. *Left*: data from same session as Fig. 1. *Right*: relation of velocities and incidence of accurately programmed movements. Same conventions and sessions as in Figs. 1 and 2. (Brooks and Watts 1983)

havioral sequence required by the task, i. e., when they began to behave correctly more often than by chance (50%, in the circled area of Fig. 2, right).

During the process of learning to program, another process began that changed the speeds of behaviorally correct, accurately programmed movements, which we shall presently identify as a form of adaptation (Brooks and Watts 1983). Figure 3 (left) demonstrates the trend for increased speeds from the beginning of training. The relation to behaviorally correct, accurate, programmed movements is plotted in Fig. 3 (right), revealing average increases of about 1 rad/s or more.

We were struck by the regular occurrence of average changes of this magnitude in the training records of three monkeys, although they were not rewarded for speeding up (Fig. 4). Such increases are limited by the target positions of course, because excursions grow in proportion to speed and yet movements must stop in the prescribed targets. Speeding up of individual movements can be observed within single training sessions. In the example of Fig. 5 (left), individual, behaviorally correct, accurate movements were faster on average in the second half of the session than in the first half by about the same amount as those in Fig. 4.

The adaptive nature of the speed changes becomes apparent when a single parameter of the task is changed without altering the task otherwise. In the example represented in Fig. 5 (left), speed had increased while the animal moved the handle against a steady load. When that load was removed, however, the previously acquired speed was excessive for the new condition and now decreased to the original starting level (Fig. 5 right). The data presented thus far show that programs and adaptations of speed appear together at the beginning of task-learning, and

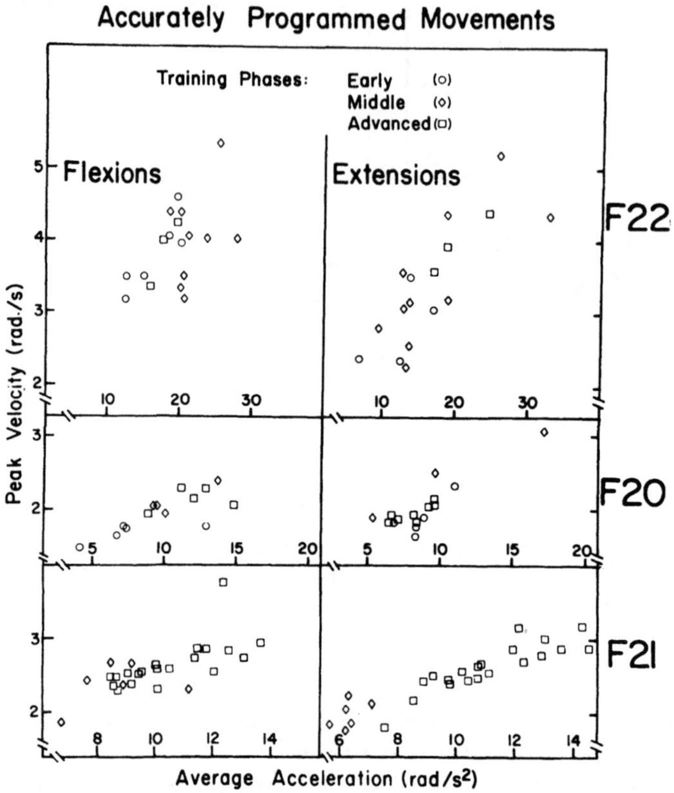

Fig. 4. Peak velocities of accurately programmed arm movements rise in similar ways during training of three naive monkeys (F20, 21, 22). *Symbols* represent means of about 200 values per training session. (Brooks and Watts 1983)

that adapted speeds carry over on average from one session to the next. These task-oriented adaptations seem to be a form of motor learning that may add skill to a previously learned performance. Speeding up may be a consequence of becoming less "tense," i.e., of diminishing cocontraction of opposing muscles, which usually first prevails when performing an unfamiliar task.

According to the expectation expressed above, we found that adaptive speeding up is indeed caused by an increase of premovement inhibition of the antagonist muscle. (The inhibition was first described for human subjects by Hufschmidt and Hufschmidt 1954). This relaxes the joint, making it less stiff, i.e., more compliant. These changes of spinal excitability before intended movements are well-known for normal human subjects from tests with H-reflexes, and are called "tuning," a term first used by Bernstein in 1961 according to Kots (1977). Many further studies have appeared (e.g., Kots and Zhukov 1971; Simoyama and Tanaka 1974; Tanji and Taniguchi 1976), see also recent discussions by Smith (1981) and Hogan (1980, 1982). Figure 6 illustrates electromyographic (EMG) evidence for adapting of tuning, or adaptive tuning, during a training session of a monkey. Movement speeds increased (panel A), because the initial agonist burst became

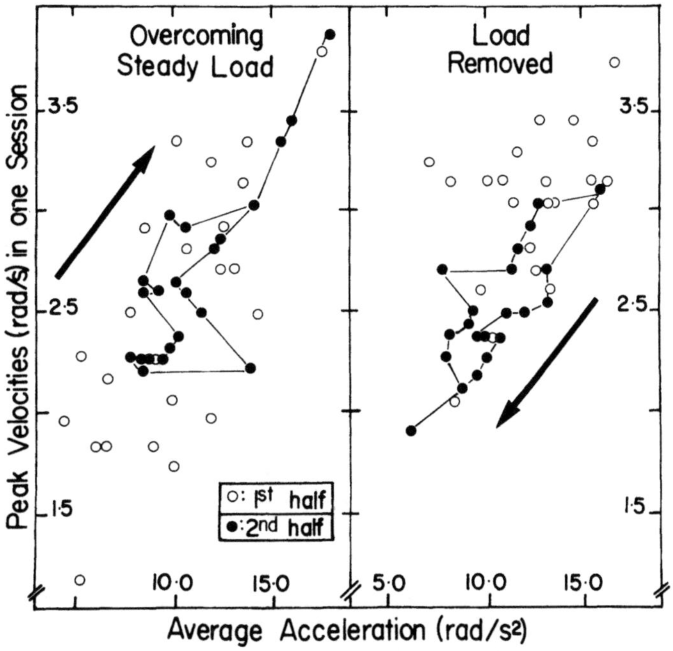

Fig. 5. Adaptive decrease in peak velocities of accurately programmed arm movements of a monkey after unexpected removal of a load of about 20 g pulling against the handle. *Open and filled circles* represent individual movements made during first and second half of one work session. (Brooks and Watts 1983)

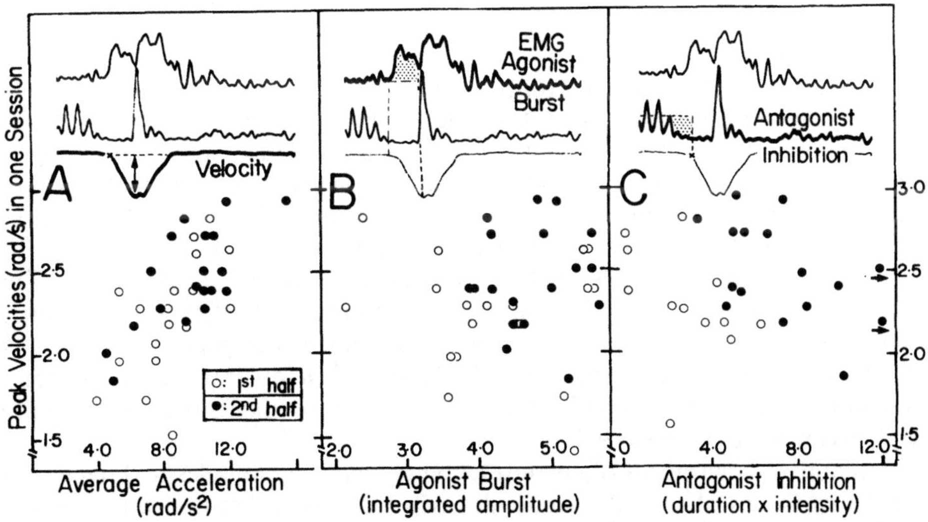

Fig. 6. Adaptive tuning of accurately programmed arm movements of a monkey during one training session. Increased peak velocities (*Panel A*) are caused by increased initial agonist burst (*B: dotted area*), and by prolonged premovement antagonist inhibition (*C. dotted area*). Same conventions as in Fig. 5. *EMG*, electromyogram. (Brooks and Watts 1983)

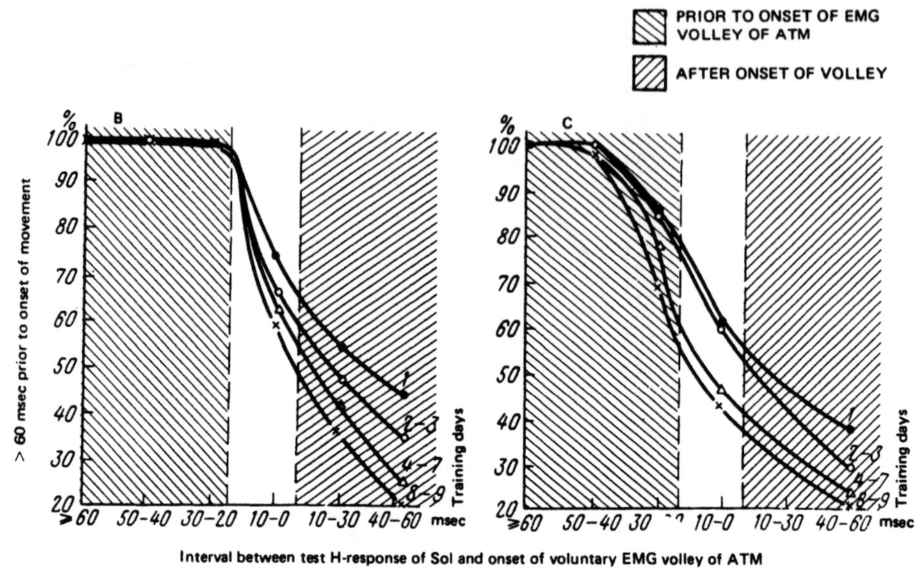

Fig. 7. Adaptive tuning of human ankle movements. Effect of training on reflex excitability of the antagonist (soleus) motoneuron pool (*B*), and the state of the 1a interneuron [called the "spinal reciprocal apparatus"] dealing with the agonist: anterior tibial muscle (*C*) during the organization and at the onset of voluntary, rapid flexion of the foot against a 2-kg weight. *Symbols* represent means of about 250 flexions per training session averaged from data of six normal subjects. *EMG*, electromyogram; *ATM*, anterior tibial muscle; *Sol*, soleus. (Kots 1977)

more intense (B), and because of weakened opposition from the antagonist (C). In this case, practice during the session prolonged and deepened the antagonist premovement inhibition, while in other cases it sometimes deepened and/or became prolonged.

Our analysis of monkey performance shows that continued practice of a voluntary task improves the ease of its execution, making it smoother, more skilled. The similarity to the changes in human performance brought about by analogous enhancement of antagonist premovement inhibition provides an animal model for such adaptive changes in human subjects. For instance, Person (1958) published EMG evidence for reduced cocontraction of opposing muscles and for deepened premovement antagonist silence during continued practice of a manual task. This was confirmed by Kots (1977) with extensive studies of H-reflexes during persistent practice of foot flexions. His illustration of intensified premovement antagonist inhibition is reproduced in Fig. 7. Children develop anticipatory silence of antagonists between 2½ and 3 years of age, when movements become more precise, according to Gatev (1972).

Adaptive replacement of cocontraction of opposing muscles with reciprocal action by means of central commands invites the guess that the 1a-interneuron is being facilitated through the corticospinal (and in monkeys also the rubrospinal) tracts (Jankowska et al. 1976). Another possibility, advanced by Fel'dman (1980), is a shift of the balance between separate descending systems for cocontraction and for reciprocal action. The existence of such systems (see Fig. 12) has

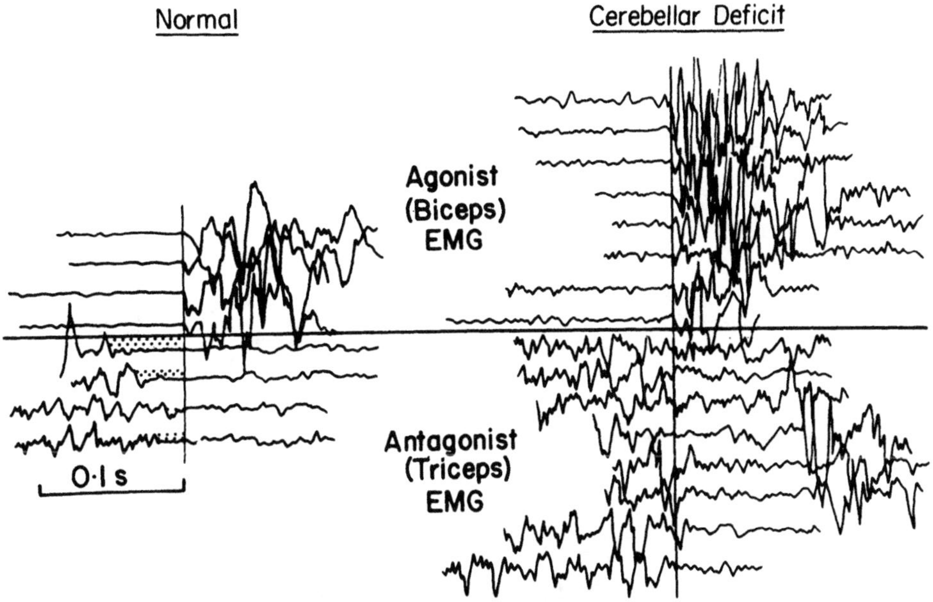

Fig. 8. Adaptive premovement antagonist inhibition is lost after cerebellar injury in human subjects. *Left*: electromyograms (*EMGs*) of individual first biceps brachii bursts and of triceps brachii activity (*upper and lower traces*) for the same flexions against a 2-kg weight. *Right*: patient with mild ataxia has no triceps inhibition in equivalent successive trials. (Hallett et al. 1975a, b)

indeed been put forward recently for the cerebral cortex by Humphrey and Reed (1983). These thoughts about cortical control in turn suggest its cerebellar modulation (cf. Taborikova 1973), which we know to be essential for premovement antagonist inhibition in human subjects, as illustrated in Fig. 8 from the work of Hallett et al. (1975a, b). Smith (1981) has suggested that intermediate Purkinje cells are inhibited when cocontraction of antagonists is called for.

We therefore reviewed our studies of cerebellar dysfunction of monkeys, produced by blocking the neocerebellar dentate output through local cooling (reviewed by Brooks 1983). This procedure degrades programmed movements, making them miss the intended endpoint. Nevertheless, the animals continue with the strategy of using programmed movements as long as they can aim a sufficient number of them into target to gain their rewards. When cerebellar dysfunction degrades the tactical details of programmed movements beyond control, however, monkeys abandon the program strategy in favor of nonprogrammed movements. Near that level of dysfunction, patients as well as animals make both some programmed as well as some nonprogrammed movements. We noticed that at that level, produced by cerebellar cooling, peak velocities of accurately aimed programmed movements were approximately 1 rad/s lower than normal, which just about equals the speeding-up gained by previous adaptive tuning (Fig. 9). Diminished speed is a well-known part of the clinical picture of cerebellar dysfunction described by Gordon Holmes (1939).

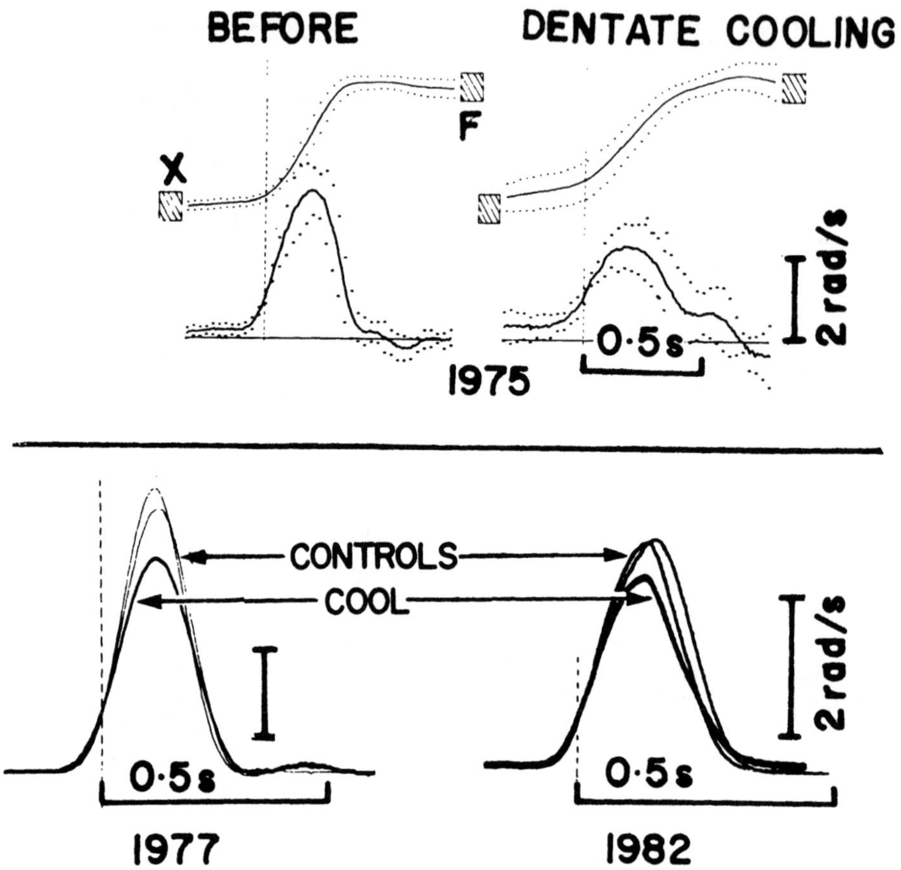

Fig. 9. Dentate cooling reduces peak velocity of accurately programmed arm movements of monkeys by about 1 rad/s. *Upper row*: averages (±SD) of about 20 position and velocity traces; *lower row*: averages of about 100 velocity traces of controls before and after, and traces during cooling. *F*, flexions; *X*, extensions. (Meyer-Lohmann et al. 1975, 1977; Miller and Brooks 1982)

The slow-down of about 1 rad/s can be seen progressively during dentate cooling in Fig. 10 (indicated by the arrow pointing down to the filled triangles). Rewarming returns the values to normal, in the direction of the upward arrow. Adaptive tuning of the last controls before and after cooling (open squares and triangles) was slight, and is not illustrated here. In such cases, the slow-down may represent loss of accumulated adaptive tuning which is of this magnitude (cf. Fig. 4). The main neocerebellar olivary input comes from the principal olivary nucleus (PO_A), whose dysfunction mimics that of the dentate (Kennedy et al. 1982). Figure 11 shows that during cooling of PO_A some accurate programmed movements were still made, but with velocities reduced by about 1 rad/s just like during dentate cooling. Rewarming the nucleus restored the original speeds.

In summary, we conclude that adaptive tuning may be a means of adding skill to movements that have already been learned and programmed accurately. Task-oriented, adaptive changes of movement speed are brought about by tuning of

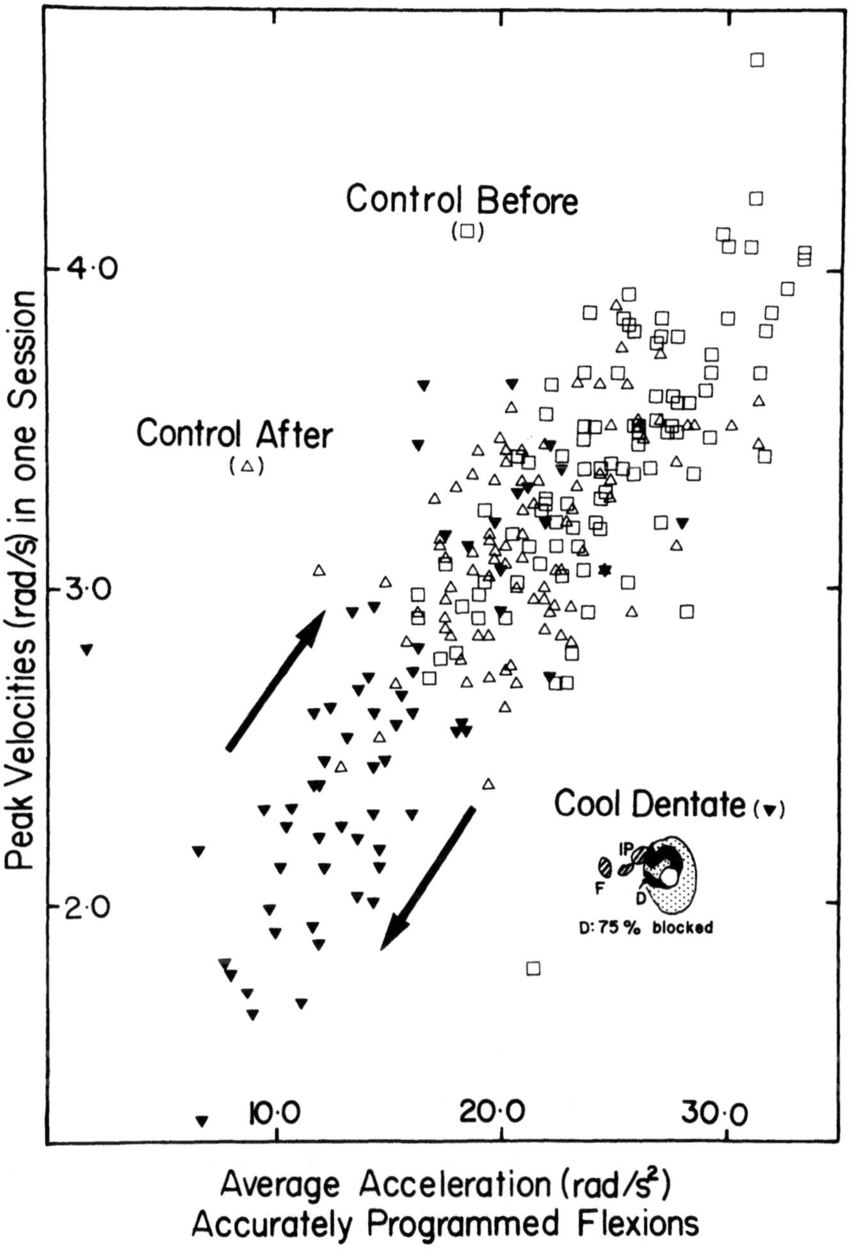

Fig. 10. Loss of adaptive velocity tuning of accurately programmed arm movements of a monkey during dentate cooling (see *inset diagram*). *Arrows* indicate reduction by about 1 rad/s. *Symbols* represent single movements. *IP*, interpositus; *F*, fastigial; *D*, dentate. (Computed from a trial in Miller and Brooks 1982)

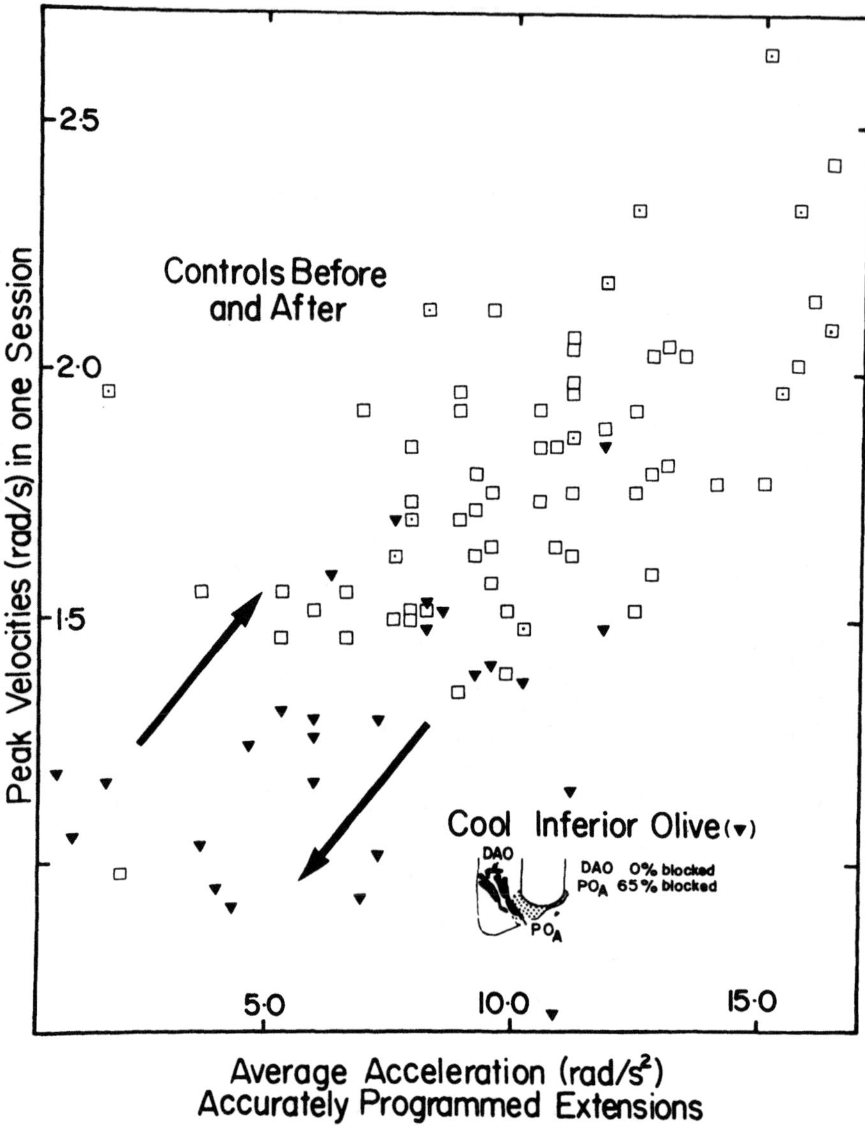

Fig. 11. Loss of adaptive velocity tuning of accurately programmed arm movements of a monkey during cooling of the principal nucleus of the inferior olive (see *inset diagramm*). Conventions as in Fig. 10. *DAO*, dorsal accessory olive; PO_A, principal olivary nucleus. (Computed from a trail in Kennedy et al. 1982)

the balance of opposing muscles (starting with premovement inhibition of the antagonist), from which it may be concluded that the mechanical stiffness of the joint is reduced. Adaptive tuning is carried forward on average from one work session to the next, and it depends on normal function of the neocerebellum, including its olivary input and dentate output. One possible mechanism whereby this might come about could be selective cerebellar triggering of cortical precen-

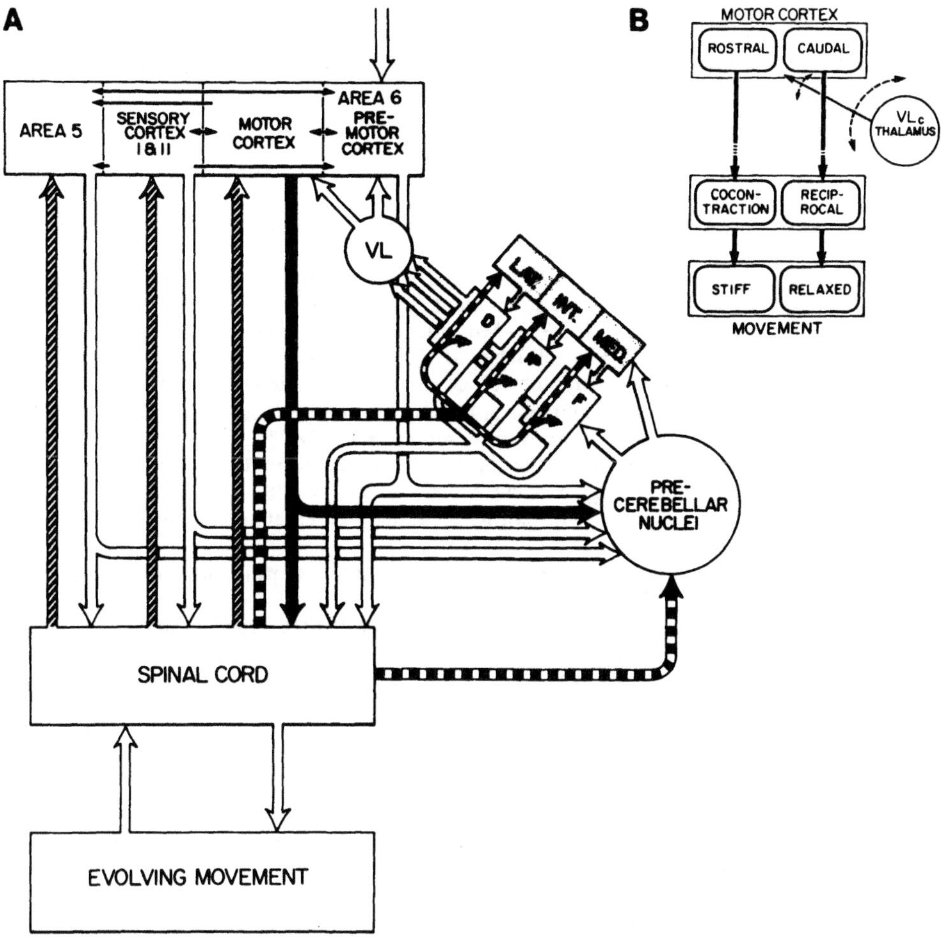

Fig. 12A,B. The "cerebellar circuit" drawn as an oversimplified cartoon to highlight routes through which adaptive tuning could be conveyed to the spinal cord. The diagram in *A* is partial and oversimplified: nuclear groups have been combined and connections omitted, and laterality is ignored. *VL*, ventral lateral nucleus. The diagram in *B* indicates how cerebellar connections through the caudal part of the ventral lateral nucleus of the thalamus (VL_c) might select outputs from the more rostral or caudal parts of the motor cortex which would favor cocontraction or reciprocal action of opposing muscles. (Brooks 1984)

tral neurons that cause either cocontraction or reciprocal action of opposing muscles. Such a selection could, for instance, involve the ventrolateral (VL) nucleus of the thalamus, as sketched diagrammatically in Fig. 12B.

References

Brooks VB (1969) Anticipatory movement control. Discussion. In: Leibovic K (ed) Information processing in the nervous system. Springer, Berlin Heidelberg New York, pp 268–269

Brooks VB (1972) Some new experiments on cerebellar motor control. Electroencephalogr Clin Neurophysiol [Suppl] 32: 109–116

Brooks VB (1979) Motor programs revisited. In: Talbott RE, Humphrey DR (eds) Posture and movement. Raven, New York, pp 13–49

Brooks VB (1983) Study of brain function by local, reversible cooling. Rev Physiol Biochem Pharmacol 95: 1–109

Brooks VB (1984) Cerebellar functions in motor control. Human Neurobiology 2: 251–260

Brooks VB, Thach WT (1981) Cerebellar control of posture and movement. In: Brooks VB (ed) Motor control. Am Physiol Soc, Bethesda, pp 877–946 (Handbook of physiology, sect 1, vol II)

Brooks VB, Watts S (1983) Task-oriented adaptations of cocontraction of opposing muscles (tuning) depend on the neocerebellum. Soc Neurosci Abstr 9: 606

Brooks VB, Kozlovskaya IB, Atkin A, Horvath FE, Uno M (1973) Effects of cooling the dentate nucleus on tracking-task performance in monkeys. J Neurophysiol 36: 974–995

Brooks VB, Ross HG, Kennedy PR (1983) Movement programming depends on understanding of behavioral requirements. Physiol Behav 31: 561–563

Connolly K (1970) Mechanisms of motor skill development. Academic, New York

Conrad B, Matsunami K, Meyer-Lohmann J, Wiesendanger M, Brooks VB (1974) Cortical load compensation during voluntary elbow movements. Brain Res 71: 507–514

Eccles JC (1977) Cerebellar function in the control of movement. In: Rose F (ed) The physiological aspect of clinical neurology. Blackwell, Oxford, pp 157–178

Fel'dman AG (1980) Superposition of motor programs – I. Rhythmic forearm movements in man. Neuroscience 5: 81–90

Gatev V (1972) Role of inhibition in the development of motor co-ordination in early childhood. Dev Med Child Neurol 14: 336–341

Hallett M, Shahani BT, Young RR (1975a) EMG analysis of stereotyped voluntary movements in man. J Neurol Neurosurg Psychiatry 38: 1154–1162

Hallett M, Shahani BT, Young RR (1975b) EMG analysis of patients with cerebellar deficits. J Neurol Neurosurg Psychiatry 38: 1163–1169

Hogan N (1980) Tuning muscle stiffness can simplify control of natural movement. Adv Bioeng (Am Soc Mech Eng, Winter Meeting) 279–282

Hogan N (1982) Moving with control: using control theory to understand motor behavior. Behav Brain Sci 5: 550–551

Holmes G (1939) The cerebellum of man. The Hughlings Jackson memorial lecture. Brain 62: 1–30

Hufschmidt HI, Hufschmidt T (1954) Antagonist inhibition as the earliest sign of a sensory-motor reaction. Nature 174: 607

Humphrey DR, Reed DJ (1983) Separate cortical systems for the control of joint movement and joint stiffness: reciprocal activation and co-activation of antagonistic muscles. In: Desmedt J (ed) Motor control in health and disease. Raven, New York, Adv Neurol 39: pp 347–372

Jankowska E, Padel Y, Tanaka R (1976) Disynaptic inhibition of spinal motoneurones from the motor cortex in the monkey. J Physiol (Lond) 258: 467–487

Kennedy PR, Ross HG, Brooks VB (1982) Participation of the principal olivary nucleus in neocerebellar motor control. Exp Brain Res 47: 95–104

Kots YM (1977) The organization of voluntary movement. Plenum, New York

Kots YM, Zhukov VI (1971) Supraspinal control of the segmental centres of the muscle-antagonists in man – III. "Tuning" of the spinal apparatus of reciprocal inhibition in the period of organization of voluntary movement Biofizika 16: 1085–1091

Kozlovskaya IB, Uno M, Atkin A, Brooks VB (1970) Performance of a step-tracking task by monkeys. Comm Behav Biol 5: 153–156

Marsden CD (1982) The mysterious motor function of the basal ganglia: the Robert Wartenberg Lecture. Neurology 32: 514–539

Melvill Jones G, Watt DGD (1971) Observations on the control of stepping and hopping movements in man. J Physiol (Lond) 219: 709–727

Meyer-Lohmann J, Conrad B, Matsunami K, Brooks VB (1975) Effects of dentate cooling on precentral unit activity following torque pulse injections into elbow movements. Brain Res 94: 237–251

Meyer-Lohmann J, Hore J, Brooks VB (1977) Cerebellar participation in generation of prompt arm movements. J Neurophysiol 40: 1038–1050
Miller AD, Brooks VB (1982) Parallel pathways for movement initiation in monkeys. Exp Brain Res 45: 328–332
Nashner LM, Grimm RJ (1978) Analysis of multiloop dyscontrols in standing cerebellar patients. In: Desmedt J (ed) Cerebral motor control in man. Karger, Basel, pp 300–319 (Progr clin neurophysiol, vol 4)
Person RS (1958) An electromyographic investigation of coordination of the activity of antagonist muscles in man during the development of a motor habit. Pavlov J Higher Nerv Act 8: 13–23
Poulton EC (1974) Tracking skill and manual control. Academic, New York
Schmidt RA (1982) Motor control and learning. Human Kinetics, Champaign
Simoyama M, Tanaka R (1974) Reciprocal 1a inhibition at the onset of voluntary movements in man. Brain Res 82: 334–337
Smith AM (1981) The coactivation of antagonist muscles. Can J Physiol Pharmacol 59: 733–747
Taborikova H (1973) Supraspinal influences on H-reflexes. In: Desmedt J (ed) Human reflexes, pathophysiology of motor systems, methodology of human reflexes. New Developm Electromyogr Clin Neurophysiol 3: 328–335
Tanji J, Taniguchi K (1976) Presetting of the spinal 1a inhibitory pathway in relation to the direction of an intended movement. Neurosci Lett 3: 321–327
Vilis T, Hore J (1980) Central neural mechanisms contributing to cerebellar tremor produced by limb perturbations. J Neurophysiol 43: 279–291
Welford AT (1968) Fundamentals of skill. Methuen, London
Wiesendanger M (1981) The pyramidal tract. Its structure and function. In: Towe A, Luschei ES (eds) Motor coordination. Plenum, New York, pp 401–491 (Handbook of behavioral neurobiology, vol 5)

Recent Aspects of the Function of the Inferior Olive

PIERGIORGIO STRATA, Torino

Introduction

As soon as Eccles et al. (1966a) made the fundamental discovery that the "inactivation responses" described by Granit and Phillips (1956) were due to the powerful excitatory action of the synapse between the climbing fiber and the Purkinje cell, it became clear that we were faced with a peculiar cerebellar input, not only from the morphological but also from the functional point of view. Since then, an enormous number of papers has been devoted to investigating the functional properties of this cerebellar input, but on the other hand its specific role in cerebellar operation is still a matter of debate. I shall review here some recent experiments performed in our laboratory in order to add perhaps another drop to the understanding of the functional role of the olivocerebellar input.

The Climbing Fiber Input

The climbing fibers form one of the two most studied inputs to the cerebellar cortex. The cells of origin are located in the inferior olivary nucleus (Szentágothai and Rajkovits 1959; Batini et al. 1976; Courville and Faraco-Cantin 1978; Montarolo et al. 1980). The other input is made by the mossy fibers. These two inputs show very different characteristics from both a morphological and a functional point of view (Eccles et al. 1967a; Bloedel 1973; Strata 1976a, b; Gilman et al. 1981; Llinás 1981).

One of the most peculiar characteristics of the Purkinje cell activity due to the climbing fibers, the so-called complex spike, is that it fires in a rather narrow frequency range. Under resting conditions its frequency is around 1/s, and during intense activation of peripheral receptors it can reach a frequency of only 8/s (Ferin et al. 1971). Similar, though slightly lower, maximal values have been found under physiological conditions, during sleep (Marchesi and Strata 1971) or during the performance of novel motor tasks (Gilbert and Thach 1977).

The fact that the complex spikes have such a low frequency of discharge suggests that the important function of the olivocerebellar neurons beyond that of transmitting information in the form of spikes, is that of exerting a conditioning effect on those neurons of the cerebellar cortex to which they project. At the level of the Purkinje cell, such a conditioning action may consist of: (a) a short-term effect exerted through short-lasting changes in membrane polarization; (b) a change in the properties of the synapse between the parallel fibers and the Purkinje cell, which might account for possible longer-term changes.

Experimental Brain Research, Suppl. 9
© Springer-Verlag Berlin · Heidelberg 1984

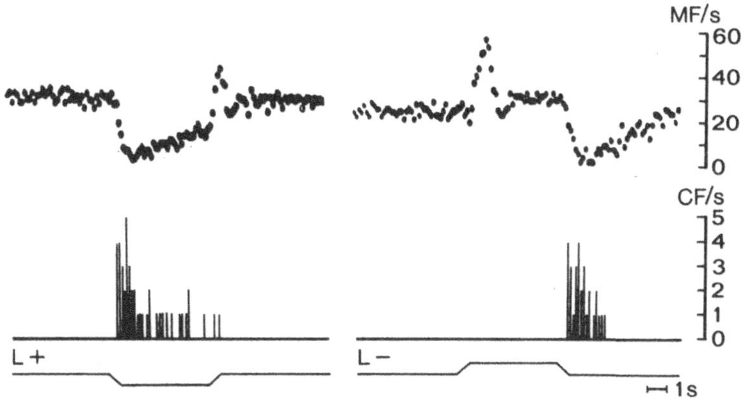

Fig. 1. Effect of galvanic stimulation of the labyrinth on Purkinje cell activity. The pattern of response of the mossy-fiber- (*MF*) and climbing-fiber- (*CF*) driven activities is shown in relation to the stimulation of the left labyrinth with anodal (*L*+) and cathodal (*L*-) currents. The MF-driven activity, shown in the *top records*, was elaborated by the computer. Each bin corresponds to an interval of 125 ms. CF-driven activity was counted directly from the film in intervals of 125 ms. Each interval is represented by a *vertical line*. Both types of activity are the average responses to ten stimuli. Histological control showed that this unit was localized in the dorsal part of the nodulus. (Ferin et al. 1970)

Short-Term Changes

Figure 1 shows a typical response of a Purkinje cell of the nodulus in the cat in response to galvanic stimulation of the labyrinth (Ferin et al. 1970, 1971). It is clear how the frequency of firing of the complex spikes is reciprocal to the other type of activity, the so-called simple spikes. Such behavior may in part be due to opposite effects exerted on the two afferent channels by the same stimulus. However, it may also be due, at least in part, to an inhibitory effect exerted by the climbing fibers on the Purkinje cells. This inhibitory effect can be observed by examining the spontaneous discharge of the Purkinje cells, where a pause is usually present following each complex spike.

When the frequency of firing of the complex spikes increases, there is a decrease in the probability of firing of the simple spikes. Following administration of harmaline, the complex spike frequency may reach a value of 8–10/s and the simple spike activity may be totally suppressed (Lamarre and Mercier 1971; Lamarre and Weiss 1973; Llinás and Volkind 1973).

The pause which follows each complex spike is generated by two different mechanisms. In part, it is due to the direct effect of the climbing fiber synapse, which generates an increased calcium-dependent potassium permeability (Llinás and Sugimori 1980a, b). A second mechanism is due to the activation of corticocerebellar interneurons by climbing fiber collaterals (Eccles et al. 1967a; Chan-Palay and Palay 1971; Murphy and Sabah 1971; Bloedel and Roberts 1971; Latham and Paul 1971; Burg and Rubia 1972; Rubia et al. 1974).

It is therefore clear that the climbing fibers, despite their low frequency of discharge, exert a strong conditioning effect on the excitability of the Purkinje cells.

Long-Term Changes

The idea that long-term changes may occur in the Purkinje cells as a consequence of the climbing fiber activity has been proposed, on theoretical grounds, as a basis of motor learning (Marr 1969; Albus 1971; Gilbert 1974). In order to account for changes lasting hours or days or even longer, it has been proposed that they consist not in the change of the level of polarization of the membrane, but rather in the mechanism of synaptic transmission. In other words, the climbing fiber synapse would be able to change the intrinsic properties of the synapse between the parallel fibers and the Purkinje cell.

There are two recent experiments in support of this hypothesis:

1. A few hours after a lesion of the inferior olive, the properties of the Purkinje cells are so much altered that their inhibitory action on the target neurons is strongly reduced (Dufossé et al. 1978; Ito et al. 1978, 1979; Ito 1979, 1982).
2. Combined stimulation of the inferior olive and of the vestibular nerve induces a depression of the synaptic transmission in the parallel fiber-Purkinje cell synapse, due to a reduced chemosensitivity of the Purkinje cell dendrites. It is interesting that such a depression lasts for 1 h (Ito et al. 1982; see also the chapter by Ito in this volume).

It has been demonstrated at a behavioral level that a lesion of the inferior olive induces severe motor disabilities (Wilson and Magoun 1945; Carrea et al. 1947; Murphy and O'Leary 1971; Armstrong 1974, 1978; Llinás et al. 1975; Soechting et al. 1976; Barmack and Simpson 1980; Kennedy et al. 1982). In addition, an inferior olive lesion impairs the phenomena of compensation which follow hemilabyrinthectomy (Llinás et al. 1975) and the plasticity of the vestibuloocular reflex (Ito and Miyashita 1975; Haddad et al. 1980; Demer and Robinson 1982). In order to understand better why an inferior olive lesion is so powerful in disrupting motor behavior and the phenomena of plasticity, it would seem to be important to investigate the effect of such a lesion on the activity of those corticocerebellar neurons impinged upon by the olivocerebellar pathway and of the neurons of the intracerebellar and subcerebellar nuclei.

Possible Trophic Effects of the Olivocerebellar Fibers on Their Target Neurons

As a first step we have tried to confirm the results of Ito et al. (1978), who showed that following a chemical lesion of the inferior olive by means of 3-acetylpyridine (3-AP) (Desclin and Escubi 1974), there is a reduction of the inhibitory properties of the Purkinje cells on Deiters neurons. However, we failed to confirm such a decrease in inhibition (Montarolo et al. 1981b). In Wistar rats, the inferior olive was injured by means of 3-AP 3 days to 6 months before the acute experiment. Stimulation of the cerebellar cortex was effective in decreasing the spontaneous firing for at least 20 ms in 50.9% (27 out of 53 tested units) of the Deiters neurons antidromically activated from the spinal cord. In intact rats, the corresponding value was 52.3% (22 out of 42 tested units). In the rats with lesions, cerebellar stimulation was also effective in depressing the antidromic activity evoked in

Deiters neurons from spinal cord stimulation and the orthodromic activity evoked by vestibular nerve stimulation.

Therefore, our experiments show that the Purkinje cells maintain their inhibitory properties also when deprived of their climbing fiber input. The discrepancy with the conclusion reached by Ito et al. (1978) is difficult to explain, since our experiments were performed under similar conditions. However, it is possible that those differences between intact rats and rats with lesions not due to the Purkinje cell inhibitory properties are detected when particular parameters of stimulation are used or when the time after the lesion is not considered. In fact:

1. Stimulation of the cerebellar cortex in intact animals activates not only the Purkinje cells, as in the animals with lesions, but also the climbing fiber terminals. Since the latter have several branches to different Purkinje cells, the same stimulus may result in being more powerful in the intact rats. This difference may become appreciable when the intensity of cerebellar stimulation is rather weak.
2. As we shall see later, when the inferior olive activity is abolished, there is an increase in the simple spike activity of the Purkinje cell (Colin et al. 1980; Montarolo et al. 1982). Such an increase vanishes gradually to disappear at the end of 1 month (Benedetti et al. 1983a). It is likely that during this period the Deiters neurons are hyperpolarized and therefore have a decreased spontaneous firing. Under these conditions, cerebellar stimulation may add little to the already present background inhibition. Therefore, one may obtain different results depending on how long after the lesion the experiment is performed.

In further experiments (P.G. Montarolo and P. Strata, unpublished observations) we have studied possible trophic effects on other neurons innervated by the olivocerebellar fibers. We have investigated the effect of an inferior olive lesion on the inhibitory properties of basket and Golgi cells, which are known to receive collaterals of the olivocerebellar fibers (Hàmori and Szentágothai 1966; Eccles et al. 1967a; Chan-Palay and Palay 1971; Palay and Chan-Palay 1974; Brodal 1981).

The inhibitory action exerted by the basket cells on the Purkinje cells has been measured by the reduction in amplitude of the antidromic field (N_1 wave) elicited in the cerebellar cortex by stimulating the underlying white matter following parallel fiber stimulation (Eccles et al. 1966b, 1967b). Figure 2 shows the time course of the basket cell inhibition in a rat with a lesion of the inferior olive. No difference from the intact rat was detected, either in the intensity or in the time course.

Similar conclusions were reached for the Golgi cell inhibition. This was measured by the depression of the positive field (P_2 wave) evoked in the granular layer by mossy fiber stimulation following an activation of the Golgi cells by stimulation of the parallel fibers, either directly or through transfolial stimulation (Eccles et al. 1967b).

From these experiments of ours, it is possible to conclude that in rats with a permanent lesion of the inferior olive there are no signs of reduced inhibitory activity in any of the three types of corticocerebellar neuron innervated by the olivocerebellar fibers: the Purkinje, the basket and the Golgi cells. These results are supported by the fact that in tissue cultures the Purkinje cells can form func-

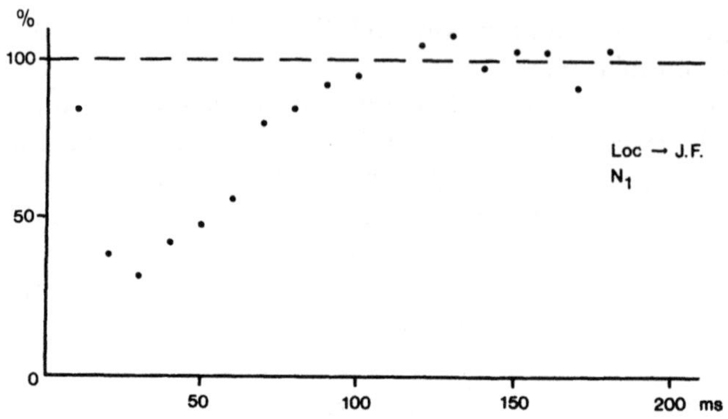

Fig. 2. Measurement of the basket cell inhibition in the cerebellar cortex following a bilateral lesion of the inferior olive by means of 3-acetylpyridine. The *filled circles* represent the amplitude of the antidromic field (N_1 wave) evoked in the cerebellar cortex by white matter stimulation in the juxtafastigial region (*J.F.*) following a conditioning stimulus to the parallel fibers (*Loc*) at the intervals indicated on the *abscissa*. The amplitude of the N_1 wave is expressed as percentage relative to the control (*ordinate*). (Courtesy of P.G. Montalolo and P. Strata)

tional inhibitory synapses on their target neurons, although in the absence of climbing fibers (Marshall et al. 1980). Ito (1982) suggests that the reduction of Purkinje cell inhibition after deprivation of climbing fiber afferents may not be all or nothing, but graded. If so, we suggest that the reduction is very small, because in our experiments, where we studied the Purkinje cell inhibition of Deiters neurons, the percentage of cells inhibited for at least 20 ms was exactly the same in intact rats and in those with lesions of the inferior olive (Montarolo et al. 1981b).

Effect of Inactivation of the Inferior Olive on the Electrical Activity of the Cerebellar Cortex and of the Intracerebellar and Subcerebellar Nuclei

Purkinje cells

We have seen above that often an increase in the complex spike frequency is accompanied by a decrease in the simple spike activity (Ferin et al. 1970, 1971; Ghelarducci et al. 1975; Gilbert and Thach 1977; Leonard and Simpson 1982). However, it is difficult to establish how much of this decrease is due to the depressant effect of the climbing fibers. This action may be studied by abolishing the spontaneous complex spike activity or by selective stimulation of the climbing fibers.

Colin et al. (1980) have shown that following inferior olive lesion by means of 3-AP, the Purkinje cells increased their simple spike firing, whereas repetitive inferior olive stimulation had a long-lasting depressant effect. The latter result has been confirmed by Rawson and Tilokskulchai (1981, 1982). Such a depressant effect has been attributed to the direct effect of the climbing fiber on the Purkinje cell, although effects mediated through climbing fiber collaterals to corticocerebellar interneurons or to brain stem nuclei cannot be ruled out.

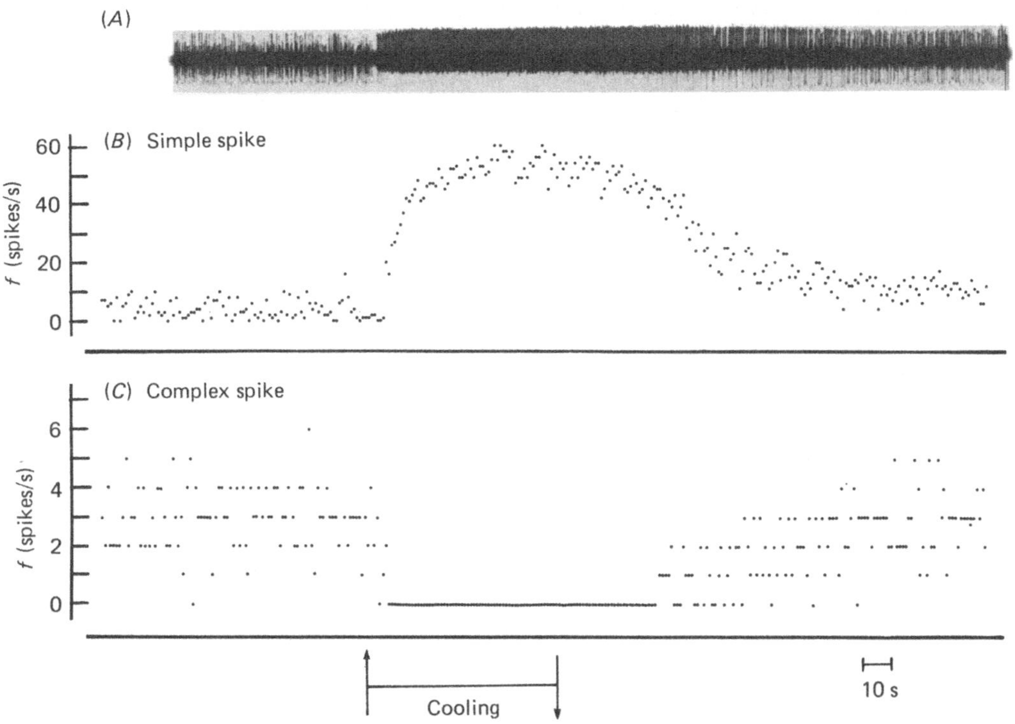

Fig. 3A–C. Effect of cooling the left inferior olive on the Purkinje cell activity of the cerebellar cortex of the right side. *A* Specimen of Purkinje cell activity. Simple spikes have a mainly upwards (negative) direction and complex spikes a mainly downwards direction. *B* The frequency (*f*) of the simple spikes has been counted every second on a specimen recorded at a more expanded time base. *C* Similar counts for the complex spikes. Period of cooling is indicated by the *horizontal bar*. (Montarolo et al. 1982)

This problem has also been investigated in our laboratory, with a different experimental approach, whereby the inferior olive was reversibly inactivated by applying a cooling probe to the surface of the medulla (Montarolo et al. 1981a, 1982). The experiments were performed under barbiturate anaesthesia. Figure 3 shows the effect of a reversible cooling of the left inferior olive on the Purkinje cell activity of the cerebellar cortex of the contralateral side. The frequency of the complex spikes decreases to disappear in a few seconds. At the same time, the frequency of the simple spikes increases to reach a plateau level in about 10 s. At the end of the cooling there is first a gradual decrease of the simple spike firing and then the reappearance of the complex spikes. Both types of activity reach the control values again. The increase in the simple spike frequency was a consistent phenomenon in all the Purkinje cells, which showed a disappearance of the complex spikes. In a sample of 22 Purkinje cells, the average simple spike frequency was 28.02/s (±13.7 SD) before cooling, to reach a plateau level of 57.62/s (±11.9 SD) following cooling. There are two experiments which show that the increase of the simple spike activity is due to a specific suppression of the inferior olive activity.

1. When the activity of two Purkinje cells is recorded simultaneously from each hemicerebellum, an increase in the simple spike activity is present only in the Purkinje cell of the side contralateral to the cooled inferior olive, where the complex spikes disappear.
2. When cooling is applied to the same bulbar area in rats submitted to a chemical lesion of the inferior olive some weeks or months before, no significant variation is observed in the simple spike frequency of the Purkinje cells of the contralateral side.

These experiments show that the spontaneous activity of the inferior olive has a predominant tonic depressant effect on the Purkinje cell activity. We have seen that such a depression may be exerted through at least two main mechanisms: (a) the activation of a Ca-K system (Llinás and Sugimori 1980a, b); and (b) the inhibition mediated by the corticocerebellar interneurons (Eccles et al. 1967a; Murphy and Sabah 1971; Bloedel and Roberts 1971; Latham and Paul 1971; Burg and Rubia 1972; Rubia et al. 1974). The former mechanism depends on the synapse between the climbing fiber and the Purkinje cell, whereas the latter one is mediated through collaterals of the olivocerebellar fibers to the interneurons. The effect due to the Ca-K system depends on the suppression of the activity of only that olivary neuron which innervates the tested Purkinje cell. On the other hand, the effect mediated by the interneurons depends on the number of inactivated inferior olive cells, since each interneuron is impinged upon by climbing fiber collaterals of many inferior olive cells.

The importance of each of these two mechanisms has been evaluated in the following experiment. In rats with an inferior olive lesion, it is possible to record occasionally the activity of Purkinje cells with complex spikes, due to the firing of few surviving inferior olive cells (Benedetti et al. 1983c). Following cooling of the inferior olive region, the complex spikes disappear, but the simple spike frequency shows a modest increase. This means that the mere suppression of the complex spike makes a small contribution to the simple spike frequency increase. The much larger effect which is present when a higher number of inferior olive cells is inactivated is taken as evidence that the corticocerebellar interneurons are under a strong tonic influence from the olivocerebellar fibers.

Another explanation for this larger increase may be related to the calcium ion concentration. These ions enter the Purkinje cell during climbing fiber activation (Llinás and Sugimori 1980a, b). Furthermore, climbing fiber activity is associated with a decreased calcium concentration in the Purkinje cell environment (Stöckle and ten Bruggencate 1980). A change in extracellular and intracellular calcium concentration in the cerebellar cortex might be present also during complex spike suppression, and such a change may have an influence on the synaptic transmission between the parallel fibers and the Purkinje cells.

The inferior olive may thus be regarded not only as a system which generates patterns of phasic excitation-inhibition sequences, but also as a system which exerts a remarkable tonic inhibitory action on the Purkinje cells (Strata and Montarolo 1982). It is interesting that as far as the simple spikes are concerned, there is a good correlation between the amount of their frequency increase and their background frequency: the higher the background activity, the lower the amount of increase. In fact, following cooling most units tend to fire at a frequency around 50/s, a value which does not depend on the initial frequency.

Since barbiturates enhance the inhibition mediated by gamma-aminobutyric acid (GABA), it is possible that the simple spike frequency increase in these experiments, performed under barbiturate anaesthesia, depends on an enhanced inhibitory tone exerted on the Purkinje cells by the corticocerebellar interneurons. In fact, both basket and Golgi inhibitions are mediated by GABA (Bisti et al. 1971). However, the results reported above were confirmed also in rats under urethane anaesthesia, which is not known to affect GABAergic transmission, and in unanaesthetized rats (Savio and Tempia 1983a, b). An increase in the simple spike frequency has also been observed in lightly anaesthetized rabbits, following inferior olive inactivation by local anaesthetic (Leonard and Simpson 1982), and it has been confirmed more recently in rats with a lesion of the inferior olive due to 3-AP (Bardin et al. 1982; Bendetti et al. 1983a).

The duration of the simple spike frequency increase has been studied in rats submitted to inferior olive lesion performed by means of cryocoagulation or electrolytic or chemical (3-AP) methods (Benedetti et al. 1983a). It has been observed that the simple spike frequency increase which follows the lesion vanishes gradually, to disappear at the end of the 1st month.

Intracerebellar and Lateral Vestibular Nuclei

Once it was established that inferior olive inactivation induces an increase in simple spike frequency in the Purkinje cells, the next step was to study the consequences on the Purkinje cell target neurons in the intracerebellar and lateral vestibular nuclei (Benedetti et al. 1983b). These neurons also receive excitatory collaterals of the olivocerebellar fibers.

Cooling the inferior olive of one side resulted in a complete suppression of the spontaneous activity of all the 14 units recorded from the contralateral Deiters nucleus, whose average frequency before cooling was 11.7 ($\pm$8.7 SD). The spontaneous activity recovered completely after the end of cooling.

In some of these units, inactivation of the inferior olive suppressed the antidromic unitary activity elicited by spinal cord stimulation. The suppression paralleled the decrease in the spontaneous activity and it has been attributed to the strong hyperpolarization of the recorded neuron, mainly because of the increased inhibitory drive from the Purkinje cells, and also to the withdrawal of the excitatory action of the collaterals of the olivocerebellar fibers.

Similar results were obtained when recording from the interpositus and the fastigial neurons. The average resting discharge of 20 cells was 20.7/s ($\pm$4.0 SD). Cooling of the left inferior olive resulted in a complete suppression of the spontaneous activity in 17 out of the 20 tested units.

Figure 4 shows a summary of the behavior of all the 34 units recorded in the intracerebellar and lateral vestibular nuclei in relation to cooling (filled circles). The empty circles show the behavior of another 34 units recorded from the same nuclei in rats, whose inferior olive was previously destroyed by 3-AP. The difference in the pattern of response of the two groups of units is attributable to a specific effect of cooling of the inferior olive.

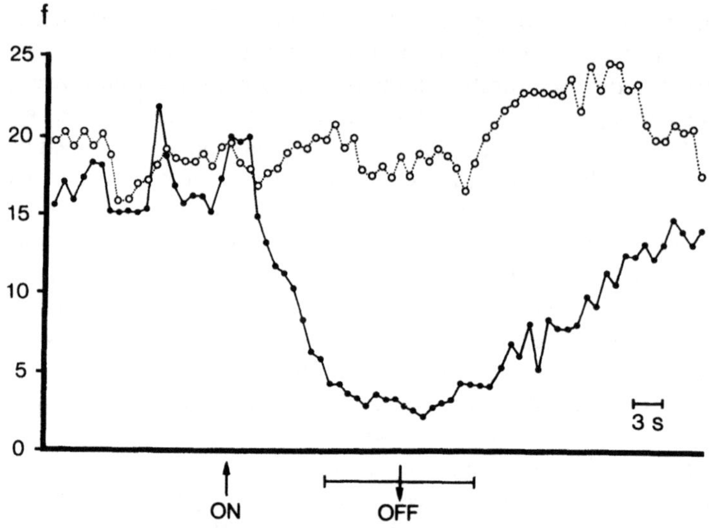

Fig. 4. A comparison between the activity of neurons recorded from the intracerebellar and Deiters nuclei in intact (*filled circles*) rats and (*open circles*) in rats whose inferior olive has been previously destroyed by 3-acetylpyridine. *Each circle* indicates the average frequency (*f*) of firing counted every second for all the units of the intact rats (34 units) and of the rats with lesions (34 units) which had a spontaneous discharge. In each unit the frequency computation started 20 s before the cooling. Cooling time varied from 7 to 40 s in different units. Its mean duration is indicated by the *downwards arrow* and the standard deviation by the *horizontal line*. (Benedetti et al. 1983b)

It was remarkable to find that inactivation of the inferior olive was powerful enough to suppress the spontaneous activity in a great majority of the intracerebellar and lateral vestibular neurons. This fact suggests that the tonic spontaneous activity of the inferior olive is very powerful in controlling the level of excitability not only of the Purkinje cells, but also of their target neurons. The fact that the activity of the intracerebellar neurons is nearly silenced following inferior olive inactivation means that inferior olive lesion is very similar to cerebellectomy. These results are in line with those observations performed at a behavioral level where it has been shown that an inferior olive lesion results in motor deficiencies which are very similar to those obtained by cerebellectomy (Wilson and Magoun 1945; Carrea et al. 1947; Murphy and O'Leary 1971; Llinás et al. 1975; Soechting et al. 1976; Kennedy et al. 1982). The fact that the tonic activity of the inferior olive cells is necessary to keep a sufficient level of activity in these nuclei suggests that also in physiological conditions small variations of the activity of the climbing fibers may result in a significant change of excitability of the nuclei of the intracerebellar and other subcerebellar structures. As a consequence it is likely that also the postural activity and any information processed by subcerebellar structures may be controlled by the inferior olive.

Postural Activity

In intercollicular decerebrate cats, cooling of the caudal part of the left inferior olive resulted in a decrease in the electromyographic activity of the contralateral triceps muscle of the forelimb. Usually, no changes in activity took place in the ipsilateral triceps muscle. This effect was much more evident when cooling was performed following the administration of harmaline. In this case, a clear-cut tremor developed bilaterally in the triceps muscles. Inactivation of the inferior olive of one side was very effective in eliciting a complete suppression of the tremor, with associated background activity only in the contralateral forelimb (Batini et al. 1983).

Effect of Inactivation of the Inferior Olive on the Metabolic Activity of the Cerebellar Nuclei

Following lesions of the inferior olive, an increased deoxyglucose uptake has been demonstrated at the level of the intracerebellar and lateral vestibular nuclei (Batini et al. 1981). The following interpretations may be given to explain such a finding.

1. The increased metabolism is related to an increased activity of the nuclear cells. This interpretation would fit in with the hypothesis of a loss of inhibitory properties by the Purkinje cell, following climbing fiber deafferentation (Ito 1982).
2. The increased metabolism is attributable to an increased activity of the Purkinje cell axon terminals. This would be in line with the finding of an increased Purkinje cell activity following lesion of the inferior olive (Colin et al. 1980).
3. Both mechanisms are responsible for the increased deoxyglucose uptake.

In order to support one of these three possible interpretations, we have determined the deoxyglucose uptake following inferior olive inactivation by cooling. In this experimental condition there is an increased activity of the Purkinje cell axons (Montarolo et al. 1982) and there is a remarkable depression of the cell activity in the cerebellar nuclei (Benedetti et al. 1983b). In addition, a possible loss of inhibitory function of the Purkinje cells is not likely, since the inactivation of the olive lasted at most 45 min. This time is too short for the assumed loss of inhibition due to the lack of a trophic action by the climbing fibers (Ito et al. 1979). The result of the inferior olive inactivation by cooling was an increased deoxyglucose uptake in the intracerebellar nuclei (Batini et al. 1982). Therefore, such an increased metabolism is attributable to the presynaptic component of the nuclei made by the Purkinje axon terminals and by the likely increase in metabolic activity required to maintain the intracellular chloride concentration below resting potential.

Discussion

Despite their slow and irregular firing, the neurons of the inferior olive nucleus are very important in conditioning the activity of the Purkinje cells. On one hand, they induce *phasic* patterns of excitation-inhibition sequences. On the other hand, they exert a powerful *tonic* control on the Purkinje cells and consequently on the intracerebellar and subcerebellar structures.

The *phasic* pattern may be very important in shaping the discharge of the Purkinje cells. In the case, for instance, of the vestibuloocular reflex, elicited by horizontal sinusoidal rotation, the complex spikes may change the phase of modulation of the Purkinje cells and therefore they may alter the gain of the reflex (Ito 1982). The climbing fibers may thus contribute to the phase shift and therefore to gain regulation.

The *tonic* effect exerted by the inferior olive neurons on the excitability of the Purkinje cells and on the neurons of the intracerebellar and subcerebellar nuclei may also play an important role in motor control. We have seen that inactivation of the inferior olive has the consequence of reducing or abolishing the firing of the neurons of the intracerebellar and Deiters nuclei. Therefore, such an inactivation should produce behavioral effects similar to those due to the lesion of the intracerebellar nuclei. In fact, both Soechting et al. (1976) and Kennedy et al. (1982) have shown that similar motor disabilities are obtained when a lesion is made in the inferior olive and in the intracerebellar nuclei. The effects of inferior olive inactivation should also be similar to those of cerebellectomy (Wilson and Magoun 1945; Carrea et al. 1947; Murphy and O'Leary 1971) except for the fact that in the latter condition also the long-axoned Purkinje cells are eliminated, whereas during inferior olive lesion they should be hyperactive.

A tonic inhibitory action of the climbing fibers on the Purkinje cells is suggested also by the experiments of Barmack and Simpson (1980). Following a lesion of the dorsal cap of the inferior olive, there is a conjugate drift of the eyes towards the side contralateral to the lesion. On the basis of the results obtained by recording from and stimulating the dorsal cap of the inferior olive (Barmack and Hess 1980a, b), the following interpretation has been given of this behavioral effect: damage to the dorsal cap of the right inferior olive has the effect of increasing the net discharge of the Purkinje cells of the left flocculus, causing a greater depression of the activity of the neurons of the medial vestibular nuclei. This has the consequence of inducing a leftwards drift of the eyes (Barmack and Simpson 1980).

Our results may also be taken to account for the finding of Ito and Miyashita (1975), who showed that following an inferior olive lesion in the rabbit there is a reduction in the gain of the horizontal vestibuloocular reflex. This result has been considered a complicating issue in the cerebellar function (Ito 1982). In fact, it has been assumed that the climbing fiber projection to zone II of the flocculus, where Purkinje cells related to horizontal movements are located, is influenced by the visual and not by the vestibular input (Ghelarducci et al. 1975). Therefore it has been predicted that a lesion of the inferior olive should have no effect on the vestibuloocular reflex in the dark. Attention has thus been drawn on the fact that a lesion of the inferior olive induces a permanent depression of the Purkinje cell inhibitory properties (Ito et al. 1978). It has been shown that this effect is equivalent to that obtained by removing the flocculus: the Purkinje cells, during the reflex, would have an enhancing effect on the medial vestibular neurons, since although they have an inhibitory action they respond out of phase with head rotation.

A different interpretation is offered by our experimental findings showing that the lesion of the inferior olive does not significantly reduce the Purkinje cell inhibitory action. On the other hand, the increased Purkinje cell firing due to the inferior olive inactivation is likely to exert an enhanced tonic inhibition on the medial vestibular neurons, with the consequence of reducing the vestibuloocular gain.

The inferior olive seems to be very important for the adaptive modifications which occur in motor learning (Ito 1979, 1982; Harvey 1980; Andersen 1982) and in the phenomena of plasticity which occur during the process of recovering from lesions (Llinás et al. 1975). However, it is a matter of debate whether the cerebellar cortex is a site of learning or whether the learning occurs at precerebellar level and the appropriate information is carried through the climbing fibers. A current argument in favor of the latter hypothesis is the following: if learning is stored in the cortex as a modification of the synapses between the parallel fibers and the Purkinje cells, the removal or the inactivation of the inferior olive should abolish new learning, but it should not alter the learning already stored in the cerebellum. Following, for instance, inactivation of the inferior olive by a local anaesthetic, the adaptive modifications of the vestibuloocular reflex are no longer present and this fact has been taken as evidence that the site of the plastic phenomena is not in the cerebellum (Demer and Robinson 1982). On the basis of our experiments, this conclusion seems to be untenable. In fact, inactivation of the inferior olive induces an increased firing of the Purkinje cells, and it is therefore expected that any processing of information which is under cerebellar control would be modified. Therefore, the site of the learning process cannot be determined on the basis of an inferior olive lesion or inactivation.

Our experiments, however, are not in contradiction of the hypothesis that learning occurs at precerebellar level and is carried to the cerebellum through the climbing fibers. It is currently assumed that the participation of the cerebellum in learning consists in a modification of the processing of information at the level of the synapses between the parallel fibers and the Purkinje cells. Such a change may be induced by the climbing fiber input in two ways: (a) by changing the intrinsic properties of the synapse (see chapter by Ito in this volume); and (b) by a functionally mediated change in the excitability of the Purkinje cells. The climbing fiber input seems to be well suited to perform this latter function simply by changing its tonic firing rate. In other words, learning may take place at a precerebellar level and may be transmitted to the Purkinje cells by a change in frequency of the complex spikes and consequently as a change in performance of the cerebellar cortex.

Acknowledgements: The experiments have been supported by grants of the Ministry of Education and of the National Research Council.

References

Albus JS (1971) A theory of cerebellar function. Math Biosci 10: 25–61

Andersen P (1982) Cerebellar synaptic plasticity – putting theories to the test. Trends Neurosci 5: 324–325

Armstrong DM (1974) Functional significance of connections of the inferior olive. Physiol Rev 54: 358–417

Armstrong DM (1978) The mammalian cerebellum and its contribution to movement control. Int Rev Physiol 17: 239–294

Bardin JM, Batini C, Billard JM (1982) Disfacilitation cérébelleuse après destruction pharmacologique de l'olive inférieure. C R Acad Sci (Paris) 295: 391–393
Barmack NH, Hess DT (1980a) Multiple-unit activity evoked in dorsal cap of inferior olive of the rabbit by visual stimulation. J Neurophysiol 43: 151–164
Barmack NH, Hess DT (1980b) Eye movements evoked by microstimulation of dorsal cap of inferior olive in the rabbit. J Neurophysiol 43: 165–181
Barmack NH, Simpson JI (1980) Effects of microlesions of dorsal cap of inferior olive of rabbits on optokinetic and vestibuloocular reflexes. J Neurophysiol 43: 182–206
Batini C, Corvisier J, Destombes J, Gioanni H, Everett J (1976) The climbing fibers of the cerebellar cortex, their origin and pathways in cat. Exp Brain Res 26: 407–422
Batini C, Buisseret-Delmas C, Conrath-Verrier M, Corvaja N (1981) The effect of harmaline and 3-acetylpyridine on the olivo-cerebellar-nuclear system in rats studied with ^{14}C-2-deoxyglucose. In: Szentágothai J, Hàmori J, Palkovits M (eds) Regulatory functions of CNS subsystem. Akademiai Kiadò, Pergamon Press, Budapest, pp 145–149 (Adv Physiol Sci, vol 2)
Batini C, Benedetti F, Buisseret-Delmas C, Montarolo PG, Strata P (1982) Effect of local cooling of the inferior olive on the metabolic activity at the intracerebellar nuclei. Neurosci Lett [Suppl] 10: p 61
Batini C, Bernard JF, Montarolo PG, Strata P (1983) The olivocerebellar pathway exerts a tonic control on the postural activity. Neurosci Lett [Suppl] 14: p 20
Benedetti F, Montarolo PG, Rabacchi S, Savio T (1983a) Long term functional changes in the Purkinje cell to climbing fibre deprivation. Neurosci Lett [Suppl] 14: p 24
Benedetti F, Montarolo PG, Strata P, Tempia F (1983b) Inferior olive inactivation decreases the excitability of the intracerebellar and lateral vestibular nuclei in the rat. J Physiol (Lond) 340: 195–208
Benedetti F, Montarolo PG, Strata P, Tosi L (1983c) Collateral reinnervation in the olivocerebellar pathway in the rat. In: Giuffrida Stella AM, Haber B, Hashim G, Perez-Polo JR (eds) Nervous system regeneration. Liss, New York, pp 461–464
Bisti S, Iosif G, Marchesi GF, Strata P (1971) Pharmacological properties of the inhibitions in the cerebellar cortex. Exp Brain Res 14: 24–37
Bloedel JR (1973) Cerebellar afferent systems: a review. Prog Neurobiol 2: 1–68
Bloedel JR, Roberts WJ (1971) Action of climbing fibers in cerebellar cortex of the cat. J Neurophysiol 34: 17–31
Brodal A (1981) Neurological anatomy in relation to clinical medicine. Oxford University Press, New York
Burg D, Rubia FJ (1972) Inhibition of cerebellar Purkinje cells by climbing fiber input. Pflügers Arch 337: 367–372
Carrea RME, Reissig M, Mettler FA (1947) The climbing fibres of the simian and feline cerebellum. Experimental inquiry into their origin by lesions of the inferior olive and deep cerebellar nuclei. J Comp Neurol 87: 321–365
Chan-Palay V, Palay SL (1971) Tendril and glomerular collaterals of climbing fibers in a granular layer of the rat's cerebellar cortex. Z Anat EntwGesch 133: 247–273
Colin F, Manil J, Desclin JC (1980) The olivocerebellar system. Delayed and slow inhibitory effects: an overlooked salient feature of the cerebellar climbing fibers. Brain Res 187: 3–27
Courville J, Faraco-Cantin F (1978) On the origin of the climbing fibers of the cerebellum. An experimental study in the cat with an autoradiographic tracing method. Neuroscience 3: 797–809
Demer JL, Robinson DA (1982) Effects of reversible lesions and stimulation of olivocerebellar system on vestibuloocular reflex plasticity. J Neurophysiol 47: 1084–1107
Desclin J, Escubi J (1974) Effects of 3-acetylpyridine on the central nervous system of the rat, as demonstrated by silver methods. Brain Res 77: 349–364
Dufossé M, Ito M, Miyashita Y (1978) Diminution and reversal of eye movements induced by local stimulation of rabbit cerebellar flocculus after partial destruction of the inferior olive. Exp Brain Res 33: 139–141
Eccles JC, Llinás R, Sasaki K (1966a) The excitatory synaptic action of climbing fibres on the Purkinje cells of the cerebellum. J Physiol (Lond) 182: 268–296
Eccles JC, Llinás R, Sasaki K (1966b) The action of antidromic impulses on the cerebellar Purkinje cells. J Physiol (Lond) 182: 316–345

Eccles JC, Ito M, Szentágothai J (1967a) The cerebellum as a neuronal machine. Springer, Berlin Heidelberg New York

Eccles JC, Sasaki K, Strata P (1967b) A comparison of the inhibitory actions of Golgi cells and of basket cells. Exp Brain Res 3: 81–94

Ferin M, Grigorian RA, Strata P (1970) Purkinje cell activation by stimulation of the labyrinth. Pflügers Arch 321: 253–258

Ferin M, Grigorian RA, Strata P (1971) Mossy and climbing fiber activation in the cat cerebellum by stimulation of the labyrinth. Exp Brain Res 12: 1–17

Ghelarducci B, Ito M, Yagi N (1975) Impulse discharges from flocculus Purkinje cells of alert rabbit during visual stimulation combined with horizontal head rotation. Brain Res 87: 66–72

Gilbert PFC (1974) A theory of memory that explains the function and structure of the cerebellum. Brain Res 70: 1–18

Gilbert PFC, Thach WT (1977) Purkinje cell activity during motor learning. Brain Res 128: 309–328

Gilman S, Bloedel JR, Lechtenberg R (1981) Disorders of the cerebellum. Davis, Philadelphia

Granit R, Phillips CG (1956) Excitatory and inhibitory processes acting upon individual Purkinje cells of the cerebellum in cats. J Physiol (Lond) 135: 520–547

Haddad GM, Demer JL, Robinson DA (1980) The effect of lesion of the dorsal cap of the inferior olive on the vestibuloocular and optokinetic systems of the cat. Brain Res 185: 265–275

Hàmori J, Szentágothai J (1966) Identification under the electron microscope of climbing fibers and their synaptic contacts. Exp Brain Res 1: 65–81

Harvey R (1980) Cerebellar regulation in movement control. Trends Neurosci 3: 281–284

Ito M (1979) Is the cerebellum really a computer? Trends Neurosci 2: 122–126

Ito M (1982) Cerebellar control of the vestibulo-ocular reflex – around the flocculus hypothesis. Annu Rev Neurosci 5: 275–296

Ito M, Miyashita Y (1975) The effects of chronic destruction of the inferior olive upon visual modification of the horizontal vestibulo-ocular reflex of rabbits. Proc Jap Acad 51: 716–720

Ito M, Orlov I, Shimoyama I (1978) Reduction of the cerebellar stimulus effect on rat Deiters neurones after chemical destruction of the inferior olive. Exp Brain Res 33: 143–145

Ito M, Nisimaru N, Shibuki K (1979) Destruction of inferior olive induces rapid depression in synaptic action of cerebellar Purkinje cells. Nature 277: 568–569

Ito M, Sakurai M, Tongroach P (1982) Climbing fibre induced depression of both mossy fibre responsiveness and glutamate sensitivity of cerebellar Purkinje cells. J Physiol (Lond) 324: 113–134

Kennedy PR, Ross H-G, Brooks VB (1982) Participation of the principal olivary nucleus in neocerebellar motor control. Exp Brain Res 47: 95–104

Lamarre Y, Mercier LA (1971) Neurophysiological studies of harmaline induced tremor in the cat. Can J Physiol Pharmacol 49: 1049–1058

Lamarre Y, Weiss M (1973) Harmaline-induced rhythmic activity of alpha and gamma motoneurones in the cat. Brain Res 63: 430–434

Latham A, Paul DH (1971) Spontaneous activity of Purkinje cells and responses to impulses in climbing fibres. J Physiol (Lond) 213: 135–156

Leonard CS, Simpson JI (1982) Effects of suspending climbing fiber activity on the discharge patterns of floccular Purkinje cells. Soc Neurosci Abstr 8: p 830

Llinás R (1981) Electrophysiology of the cerebellar networks. In: Brookhart JM, Mountcastle VB (eds) The nervous system. Am Physiol Soc, Bethesda, pp 831–876 (Handbook of physiology, sect 1, vol II)

Llinás R, Sugimori M (1980a) Electrophysiological properties of *in vitro* Purkinje cell somata in mammalian cerebellar slices. J Physiol (Lond) 305: 171–195

Llinás R, Sugimori M (1980b) Electrophysiological properties of *in vitro* Purkinje cell dendrites in mammalian cerebellar slices. J Physiol (Lond) 305: 197–213

Llinás R, Volkind RA (1973) The olivo-cerebellar system: functional properties as revealed by harmaline-induced tremor. Exp Brain Res 18: 69–87

Llinás R, Walton K, Hillman DE, Sotelo C (1975) Inferior olive: its role in motor learning. Science 190: 1230–1231

Marchesi GF, Strata P (1971) Mossy and climbing fiber activity during phasic and tonic phenomena of sleep. Pfügers Arch 323: 219–240
Marr DA (1969) A theory of cerebellar cortex. J Physiol (Lond) 202: 437–470
Marshall KC, Wojtowicz JM, Hendelman WJ (1980) Patterns of functional synaptic connections in organized cultures of cerebellum. Neuroscience 5: 1847–1857
Montarolo PG, Raschi F, Strata P (1980) On the origin of the climbing fibres of the cerebellar cortex. Pflügers Arch 383: 136–142
Montarolo PG, Palestini M, Strata P (1981a) Effects of inferior olive cooling on the Purkinje cell activity. Neurosci Lett [Suppl] 7: p 120
Montarolo PG, Raschi F, Strata P (1981b) Are the climbing fibres essential for the Purkinje cell inhibitory action? Exp Brain Res 42: 215–218
Montarolo PG, Palestini M, Strata P (1982) The inhibitory effect of the olivocerebellar input on the cerebellar Purkinje cells in the rat. J Physiol (Lond) 332: 187–202
Murphy MG, O'Leary JL (1971) Neurological deficit in cats with lesions of the olivocerebellar system. Arch Neurol 24: 145–157
Murphy JT, Sabah NH (1971) Cerebellar Purkinje cell responses to afferent inputs. I. Climbing fiber activation. Brain Res 25: 449–467
Palay SL, Chan-Palay V (1974) Cerebellar cortex: cytology and organization. Springer, Berlin Heidelberg New York
Rawson JA, Tilokskulchai K (1981) Suppression of simple spike discharges of cerebellar Purkinje cells by impulses in climbing fibre afferents. Neurosci Lett 25: 125–130
Rawson JA, Tilokskulchai K (1982) Climbing fibre modification of cerebellar Purkinje cell responses to parallel fibre inputs. Brain Res 237: 492–497
Rubia FJ, Höppener U, Langhof H (1974) Lateral inhibition of Purkinje cells through climbing fiber afferents. Brain Res 70: 153–156
Savio T, Tempia F (1983a) The inhibitory effect of the olivocerebellar fibres on the cerebellar Purkinje cells in the rat under urethane anaesthesia. Neurosci Lett [Suppl] 14: p 327
Savio T, Tempia F (1983b) Inibizione tonica della via olivocerebellare sulle cellule di Purkinje nel ratto sveglio o sotto anestesia da uretano. Boll Soc Ital Biol Sper (to be published)
Soechting J, Ranish N, Palminteri R, Terzuolo C (1976) Changes in a motor pattern following cerebellar and olivary lesions in the squirrel monkey. Brain Res 105: 21–44
Stöckle H, ten Bruggencate G (1980) Fluctuation of extracellular potassium and calcium in the cerebellar cortex related to climbing fiber activity. Neuroscience 5: 893–901
Strata P (1976a) A general review of the physiological function of the neuronal machine in the cerebellar cortex. Exp Brain Res [Suppl] 1: 103–112
Strata P (1976b) Das Kleinhirn. In: Gauer OH, Kramer K, Jung R (eds) Sensomotorik. Urban und Schwarzenberg, München, pp 421–463 (Physiologie des Menschen, vol 14)
Strata P, Montarolo PG (1982) Functional aspects of the inferior olive. Arch Ital Biol 120: 321–329
Szentágothai J, Rajkovits U (1959) !ber den Ursprung der Kletterfasern des Kleinhirns. Z Anat EntwGesch 121: 130–141
Wilson WC, Magoun HW (1945) The functional significance of the inferior olive in the cat. J Comp Neurol 83: 69–77

IV. The Physiology of Nerve Cells

Lord E. D. Adrian with John Eccles
at a Ciba Foundation Symposium 1967

What Central Inhibitory Pathways Tell Us About Mechanisms of Transmitter Release

H. KORN, Paris

Summary. Investigations undertaken to clarify the mechanisms of transmitter release at CNS synapses and to determine the physical correlates of quantal release parameters are presented in this paper.

Experimental factors that usually compromise attempts to characterize central synaptic transmission are minimal in the Mauthner (M) cell preparation, which is, in fact, uniquely suited for such studies. In particular, simultaneous intracellular recordings can be obtained from this neuron and from adjacent identifiable inhibitory interneurons. Amplitude fluctuations of unitary inhibitory postsynaptic potentials (IPSPs) produced by directly evoked presynaptic impulses were subjected to quantal analysis. A computational procedure was developed to minimize effects of background noise through a deconvolution process and to allow optimal fits of the data according to the predictions of either Poisson or binomial equations. These two models were compared on the basis of their likelihood criteria and of statistical tests which showed that: (a) the binomial relation gives a more adequate description of the fluctuations than does the Poisson; and (b) transmitter release is quantal in the CNS.

The release parameters were correlated with histological features of the presynaptic cells, which were iontophoretically injected with horseradish peroxidase (HRP) and reconstructed: a striking equivalence was found between the binomial term n and the number of stained presynaptic boutons. This identity was verified for 30 cells which met the mathematical requirement for the analysis, with binomial n's ranging from 3 to 28. It was thus concluded that each synaptic knob functions as an all-or-none independent releasing unit.

On the basis of a simple equivalent circuit model incorporating the postsynaptic conductances and the IPSP driving force, E, equations were derived which provided a method for calculating q_{ipsp}, the quantal inhibitory conductance (previously, the resting input conductance, G_m, was determined and it was demonstrated that the full-sized collateral IPSP is a reliable indicator of E). Use of these equations provided support for the postulate that the action of a single quantum (and therefore of one releasing terminal) is similar to that of the contents of a single synaptic vesicle.

Transverse sections of HRP-filled terminals revealed that they exhibit typical aspects of Gray type II synapses with distinct pre- and postsynaptic differentiations and that they contain pleiomorphic vesicles. The size of the synaptic vesicles and of the presynaptic dense projections, as well as their geometrical organization, were the same as reported for other CNS synapses with different staining techniques. On the other hand, transverse sections stained with ethanolic

Experimental Brain Research, Suppl. 9
© Springer-Verlag Berlin · Heidelberg 1984

phosphotungstic acid showed that almost all endings contained only one presynaptic complex. Therefore, the equivalence between the binomial parameter n and the sum of stained presynaptic boutons indicates a correspondence between this parameter and the number of active sites present in the terminals of the releasing interneurons.

The notion that, following a presynaptic impulse, the contents of only one are released by each presynaptic grid was reconciled with the paradoxical fact that many vesicles are present in each terminal by the proposition that a mechanism exists by which exocytosis of one vesicle inhibits the discharge of its neighbors throughout the whole presynaptic grid. This postulate is supported by morphometric data suggesting that the opening of one vesicle produces a transient enlargement of the space where it occurs and a compression of neighboring presynaptic densities, the presynaptic grid thus being far from a rigid structure.

A comparison of the physiological properties characterizing the presynaptic grids of central synapses and the dense bars of neuromuscular junctions favors the concept that these elements are "morphological synaptic units" functioning under the same constraints, especially since the binomial n also reflects the sum of the releasing active zones in a number of investigated neuromuscular preparations.

Key words: *Central synapses – Binomial – Quantal release – Presynaptic grid – Exocytosis*

Introduction

Quantal analysis of central nervous system (CNS) synapses is more difficult than that of the neuromuscular junction because technical limitations usually preclude stimulation of single identified inputs. In addition, fluctuations in postsynaptic potential amplitudes are often complicated by noise, due in part to spontaneous activity generated by various afferent sources. Consequently, data about central transmitter release in rat (Katz and Miledi 1963), cat (Kuno 1964a, b; Kuno and Miyahara 1969; Mendell and Weiner 1976), or amphibian (Shapovalov and Shiriaev 1980) motoneurons have essentially been interpreted on the basis of concepts established for the periphery.

A major advance in understanding the physiology of the CNS came from intracellular descriptions of inhibitory postsynaptic potentials (IPSPs) by Eccles and his collaborators; their pioneering evidence that chemically mediated inhibition (a) is exerted through an anion-sensitive postsynaptic conductance (ref. in Eccles 1964) and (b) is due to the specific action of specialized interneurons (ref. in Eccles 1969) still constitutes the main body of knowledge on this subject. Our work, which was aimed at clarifying the mechanisms of transmitter release by central cells was, in part, based on these grounds. In a series of experiments performed at inhibitory synapses (Korn et al. 1981, 1982; Faber and Korn 1982a, b; Triller and Korn 1982a, b), we have attempted to determine whether secretion of neurotransmitter (in this case probably glycine, ref. in Faber and Korn 1978) is quantal in the CNS, and what further information is given by such systems, which would either be specific to central neurons, or which would be applicable to peripheral junctions. The first question was alone of importance in view of the surprising claim that transmission at Ia synapses is nonquantal, and cannot be described by Poisson or binomial statistics (Edwards et al. 1976a, b).

Amplitude fluctuations of chemically mediated postsynaptic potentials evoked by single presynaptic impulses can be described by mathematical relations that share one common assumption, namely that the response is made up of an integral number of equal basic units, called quanta. This concept was derived from observations at the neuromuscular junction, where variations of postsynaptic potentials (PSPs) occur in discrete steps, each of which corresponds to a quantum of the same size, q. A mathematical analysis of such fluctuating responses was facilitated by the fact that spontaneously occurring miniature endplate potentials turn out to be equivalent to single quanta. Poisson statistics adequately described the composition of evoked potentials: that is, the probability, p_x, that a given response is made up of 1,2,3...x miniature components is given by $p_x = e^{-m} \times m^x/x$ (with m being the average number of packets released or quantal content). This theorem implies a low probablity of release, p, and a large number, n, of packets, hence the hypothesis that each unit of transmitter represents a synaptic vesicle in the nerve terminal (del Castillo and Katz 1954). Despite its extreme beauty, this mode of analysis has serious limitations, one of which is that it is impossible to give independent meaning to n and p, since only m appears in Poisson equations. Also, in a number of preparations (ref. in McLachlan 1978) binomial predictions of the form $p_x = \binom{n}{x}\ p_x\ (1-p)^{n-k}$ fit the data more adequately. This model may be more representative of normal physiological events because agreement with the Poisson is generally produced by artificially lowering p pharmacologically. Thus, for our purpose, it became critical to develop new approaches that would allow us to use the binomial model, particularly in order to determine possible correlations between this model's parameters and structural components of presynaptic neurons.

The accessibility of the teleost Mauthner cell (M cell) soma, its main dendrites (Furshpan and Furukawa 1962), and its inhibitory afferents (Faber and Korn 1980; Korn and Faber 1975, 1976) for simultaneous pre- and postsynaptic intracellular recordings at a central inhibitory synapse make it favorable material for the solution of some of these basic questions. Once it was established that such afferents, the so-called passive hyperpolarizing potential (PHP) cells, mediate chemical as well as electrical inhibition of the M cell (Korn and Faber 1976), it became possible to study the properties of the unitary IPSPs mediated by them, and of their fluctuations (it should be noted here that in the M cell the voltage recorded during the unitary IPSP is a direct measure of the resistive postsynaptic current – see Faber and Korn 1978). A computer program was used which allowed us, once the contribution of noise was removed by deconvolution, to determine whether the Poisson or the binomial representation gave the best fit with the histogram of the evoked responses. This part of our work was possible because the M cell system satisfies the requirements for a reliable extraction of binomial parameters, including steady state conditions (McLachlan 1978), which can be maintained during the sampling period.

In order to unravel the physical bases of the quantal release parameters, we then combined our physiological studies with systematic histological reconstruction of the investigated interneurons stained with horseradish peroxidase (HRP). To our surprise, we found that in each experiment the value of the binomial parameter n and the number of presynaptic knobs established on the M cell are equal. An important concept follows, being that each individual terminal bouton functions as an independent all-or-none unit. These results led to other investigations which indicated that the action of a single releasing terminal is functionally similar to that of a single synaptic vesicle's contents. Finally, data suggesting that there exists a so far unsuspected mechanism by which exocytosis of one vesicle inhibits the discharge of its neighbors throughout the whole presynaptic grid were also obtained.

Methods

The techniques and theoretical considerations necessary for this work have been extensively described in previous publications (see reference below) and will briefly be summarized here.

Electrophysiological Techniques and Data Analysis

Experiments were performed on goldfish (*Carassius auratus*) 12–18 cm in length. The preparation and basic recording procedure were similar to those employed before (Furshpan and Furukawa 1962; Korn and Faber 1975). Simultaneous intracellular recordings from or stimulation of the M cell and PHP inhibitory neurons were obtained as illustrated by Fig. 1A, using micropipettes filled respectively with 3 M KCl and HRP. The M cell was antidromically activated by bipolar stimulation of the spinal cord, and PHPs evoked in neighboring neurons (Fig. 1B) by antidromic spikes were used as a criterion to establish that the investigated presynaptic cells mediate chemical inhibition of the M cell (Korn and Faber 1976). Because the IPSP in the M cell is hard to detect as a potential change (Furukawa and Furshpan 1962), Cl^- was iontophoretically injected through the recording microelectrode until large and stable depolarizing collateral IPSPs could be recorded; thus unitary IPSPs appeared as depolarizing potentials (Fig. 1C1, 1C2). Unitary IPSPs were evoked at the rate of one or two per second (unless otherwise stated) and stored on tape for measurements and statistical analysis.

Values of peak amplitudes (sample size generally = 100–300) and of estimated noise variances were fed to a CII Mitra 15 or a MINC computer, which was programmed (a) to calculate with a free search the probability density function of each series of counts and the optimal corresponding parameters that fit the histogram, assuming either a Poisson or binomial distribution; and (b) to estimate whether either model satisfied an independent test, the Kolmogorov. If necessary, the measured responses were corrected for any nonlinearity between postsynaptic voltage and conductance change (Martin 1955).

More specifically, the statistical analyses were based on a model where fluctuations in both evoked IPSPs and background noise summated algebraically ac-

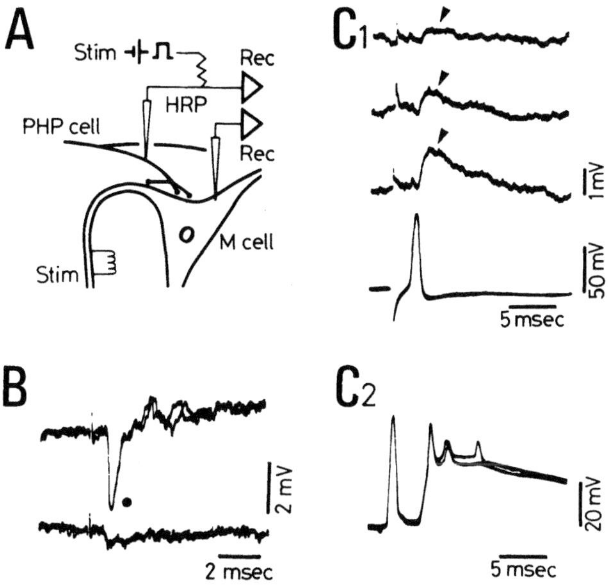

Fig. 1A–C. Fluctuating properties of unitary depolarizing postsynaptic potentials evoked in the Mauthner (M) cell by stimulation of a single interneuron. *A* Schematic representation of the experimental arrangement used for simultaneous intracellular recordings (*Rec*) from the M cell and a presynaptic inhibitory element (*PHP cell*), both cells being identified by their characteristic responses to antidromic stimulation (*Stim*). The presynaptic electrode was used for intracellular stimulation using transmembrane current pulses and also for iontophoretic injection of horseradish peroxidase (*HRP*). *B Upper and lower traces*: typical passive hyperpolarizing potential (*filled circle*) recorded when the spinal stimulus was above threshold for M cell antidromic invasion and corresponding extracellular field (several superimposed traces in each). *C1* Amplitude variations of monosynaptic unitary inhibitory postsynaptic potentials (IPSPs) (*arrowheads*), following directly evoked presynaptic spikes, one of which is shown in the *lower trace*. *C2* M cell antidromic spike followed by a full-sized collateral IPSP which was large enough to fire the cell (*two superimposed traces*)

cording to the equation $X = qK + B$, where X is the observed random variable, q the elementary or quantum size, K either the Poisson or binomial variable, and B the noise process; the latter could be approximated as a white (uncorrelated) Gaussian process (Jazwinski 1970) with a constant variance, σ^2, which was assessed from direct measurements of spontaneous activity.

On the other hand, estimation of parameters defining K (Korn et al. 1981, 1982) on the basis of a sample of N responses is a problem of "constraint deconvolution," requiring definition of a criterion of fit, for which the likelihood criterion, L, was chosen (Kendall and Stuart 1977). Its definition depended upon the statistical characteristics of the background noise; if the observed individual responses had values of x_i with $i = 1 \ldots, N$, L could be expressed by:

$$L(\{x_i\}, n, p, q) = \prod_{i=1}^{N} \sum_{k=0}^{n} \begin{bmatrix} \text{binomial} \\ \text{or} \quad \text{equations} \\ \text{Poisson} \end{bmatrix} \cdot \frac{1}{\sqrt{2\pi}\sigma} \exp - \frac{(x_i - qk)^2}{2\sigma^2} .$$

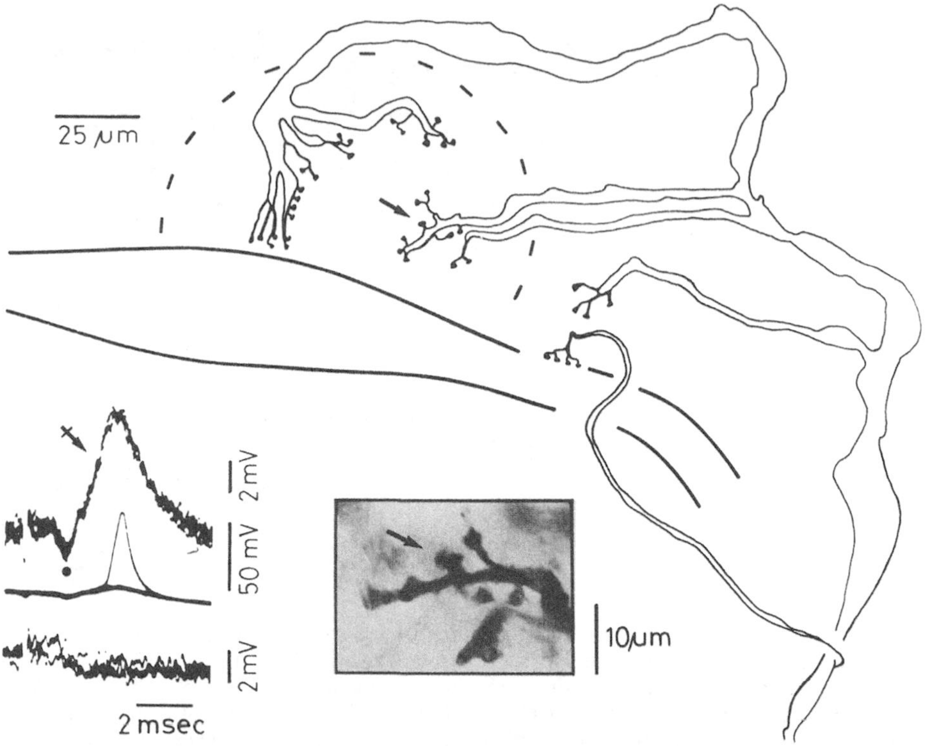

Fig. 2. Synaptic investment of a Mauthner (M) cell by a stained inhibitory interneuron. *Upper diagram*: camera lucida reconstruction of a passive hyperpolarization potential exhibiting neuron belonging to the collateral network of the M cell intracellularly stained with horseradish peroxidase (HRP) (the figure is focused on this neuron's terminals). The M cell soma dendritic membranes are schematized by *bold lines* and its axon cap by *dashed lines*. For simplification, terminal branches and boutons are *solid*, swellings representing the latter (39 of them were counted). A large fraction of the boutons located within the axon cap were remote from the M cell soma, their only possible target in this region being thus represented by M cell cap dendrites. *Lower left hand side*: recordings obtained with the HRP-filled microelectrode prior to iontophoretic injection of the dye. A passive hyperpolarizing potential (*filled circle*) and a subsequent excitatory postsynaptic potential, which was large enough to generate a spike (*crossed arrow*), were evoked in the collateral neuron by spinal stimulations (several superimposed sweeps). *Upper and middle traces*: intracellular recording at high and low gain respectively; *lower trace*: field potential recorded outside that cell. *Inset*: light micrograph of a section used for the camera lucida reconstruction; corresponding processes of the neuron are indicated by *arrows*

These criteria were maximized using a nonlinear optimization technique (Nelder and Mead 1965), which yielded the optimal values of the binomial and Poisson parameters and indicated which model provided a better fit.

In conclusion, it is important to point out that the analytical approach adopted here differs from the method used by Edwards et al. (1976a, b) when they studied unitary excitatory postsynaptic potentials in spinal motoneurons. They derived histograms of variations in estimated charge transfer due to the PSPs alone by deconvolving the total signal and noise. Instead, we have fitted the observed fluctuations in response amplitude with models incorporating the additive contributions of the noise and different quantal processes.

Histological Techniques

After each experiment, HRP was iontophoretically injected into the selected PHP-exhibiting cells (one per fish) using cathodal current pulses (Korn et al. 1978; Triller and Korn 1981), and the preparation was left undisturbed for about 1 h. For optical observations, fish were perfused and HRP was revealed according to standard procedure (Triller and Korn 1982a). Two representations were commonly adopted to focus on the terminals impinging on the M cell, that is, the reconstruction itself (Figs. 2 and 3C) and a stick diagram which provides a simplified replica of the latter (Figs. 3D and 4A2–B2). Terminal boutons were clearly distinguished in most cases, as shown in the inset in Fig. 2. Some of them were in contact with the M cell soma and main dendrites and others were located within the axon cap. Terminals taken as indicating synaptic sites were recognized as either rounded classical boutons or, more rarely, as abrupt endings without swelling. The validity of this procedure was demonstrated by electron microscopic evidence that (a) all such terminals exhibit a well-differentiated synaptic apparatus characteristic of the chemical type (Korn et al. 1981; Triller and Korn 1981, 1982a; see also Fig. 7); and (b) boutons en passant were never apparent in this material. There was little opportunity to overestimate the number of terminals because those within the axon cap necessarily innervated the M cell soma or thin cap dendrites, which are their only possible targets in this area (Nakajima 1974).

For electron microscopy the HRP was revealed in 100-μm-thick sections, which were then placed in 1% OsO_4 in phosphate buffer, dehydrated in ethanol, and embedded in Araldite. Each of them was flattened in a drop of resin between two plastic slides and polymerized, and the labeled processes were located with a light microscope, drawn in a camera lucida, and photographed. Slices of interest were then mounted on an Araldite block and serial sectioned for electron microscopic analysis.

Visualization of the Presynaptic Grid

A method which selectively stains the presynaptic junction and allows a systematic visualization *en face* of presynaptic grids stained with ethanolic phosphotungstic acid (E-PTA) (Vrensen et al. 1980) was used. After perfusion the brains were kept overnight at 4 °C in fixative and sectioned with a Vibratome. Slices 100 μm thick were dehydrated in graded ethanol and incubated in E-PTA for 2 h at 60 °C and finally flat embedded in Araldite. The M cell and its axon cap were localized and a small pyramid was then trimmed around them. Sections (0.5 μm thick) without additional staining were observed with the electron microscope operated at 100 kV. Due to the thickness of the sections, many isolated grids in *en face* views could be observed simultaneously to their full extent.

Results

Passive hyperpolarization potential exhibiting neurons, which mediate Cl^--dependent inhibition of the M cell, belong to two classes: one group contains cells that are part of the M cell's collateral network (Korn and Faber 1975; Korn et

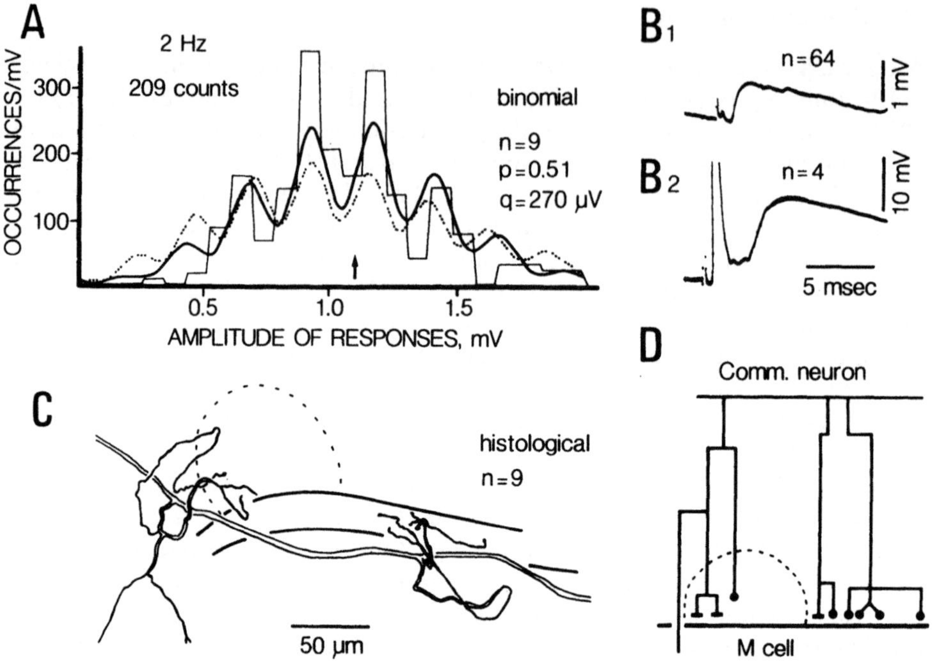

Fig. 3A–D. Correlation of the statistical properties of postsynaptic responses with morphological features of a presynaptic neuron. *A* Probability histogram for 209 counts and superimposed computer-modeled curves using Poisson (*dotted line*) and binomial equations (*solid line*). The likelihood criterion was better for the binomial, which yielded a value of 9 for parameter n, while p and q were 0.51 and 270 μV respectively. Both sets of predictions satisfied the Kolmogorov statistics (with $P_B > 0.11$ and $P_P > 0.30$), but tests for possible equivalence between models indicated that the binomial was more satisfactory. *B1* On-line computer-averaged (n = 64) unitary inhibitory postsynaptic potentials (IPSPs) evoked by stimulation of a passive hyperpolarization potential exhibiting cell at the rate of 2 Hz (presynaptic spikes not illustrated); the overall mean amplitude for 209 trials (*arrow in A*) was 1.11 mV. *B2* Computer-averaged histogram of the maximum amplitude IPSP (n = 4) following activation of the recurrent collateral network. *C,D* Camera lucida reconstruction of terminals established by the horseradish-peroxidase-filled interneuron on the recorded Mauthner cell (*C*) and corresponding schematic drawing (*D*). Nine terminal boutons (*closed circles and bars*), three of which were within the axon cap (*dashed line*), were visualized. The binomial parameter n and the number of synaptic knobs were thus the same

al. 1978); those of the second category are second-order commissural interneurons (Triller and Korn 1978; Zottoli and Faber 1980). Results obtained from these two groups were essentially the same.

Data Selected for Morphofunctional Correlations

The most important characteristics of the potentials recorded during this work are exemplified in Fig. 1, which illustrates, in addition to the PHP intracellularly recorded from a medullary cell (Fig. 1B), the unitary (Fig. 1C1) and the maximum amplitude collateral (Fig. 2C2) IPSPs in the adjacent Cl^--loaded M cell. The full time course of the unitary IPSPs, which remained relatively unchanged in each series, and some of their fluctuations are also evident. Two difficulties could complicate evaluation of individual response amplitudes. Spontaneous synaptic potentials could appear so close to the peak of the evoked IPSPs that one had to determine whether they represented artifacts or delayed evoked components; thus any second peak occurring 0.5 ms or more after the summit of the response was discarded, unless it could be attributed to slow impulse conduction in the finer branches of the stimulated interneuron. Another problem was that responses lost in background activity could lead to incorrect estimates of the number of failures. However, as our model took noise into consideration, even potentials of low amplitudes could be treated as responses during our measurements.

When the HRP-filled inhibitory interneurons were reconstructed histologically from several sections which also included their target M cell, presynaptic geometrical factors could be compared with the physiological results. Figure 2 shows the camera lucida reconstruction of a typical collateral neuron with individual terminal synaptic knobs which could be clearly discerned; all impinged on the M cell itself or upon its dendrites (not represented), 31 of them within the axon cap region and eight outside it. Their total number, which defines the histological n, ranged from three to 93 (Triller and Korn 1981, 1982a). However, if n exceeds about 30, the binomial analysis of fluctuating responses can lead to undistinguishable solutions with several equivalent parametric values, due to the mathematical phenomenon called "degenerescence" (Kendall and Stuart 1977). As a consequence, only cases with a binomial product $\leqq 6.6$ were sampled for this study.

Quantal Nature of Evoked Transmission and Structural Correlate of Binomial n

The data illustrated in Figs. 3 and 4 summarize two of our most significant results: (a) the binomial prediction, which was always satisfactory, provided a statistically more adequate definition of IPSP amplitude fluctuations than did the Poisson; and (b) the values of the optimal binomial parameter n was the same as the number of stained presynaptic boutons.

The first conclusion is derived from experiments similar to that of Fig. 3; here, the standard deviation of the noise was $\cong 80\,\mu V$ and the unitary IPSPs evoked by presynaptic impulses varied between extremes of 0.46 and 2.03 mV, with a mean of 1.11 mV for 209 trials (arrow in A). The average records of Fig. 3B1, 3B2 reflect the mean amplitude and the time course of the unitary and of the full collateral IPSPs respectively. A visual evaluation of the computer-determined optimal curves, calculated to fit the probability density function and which are shown in A, already suggests that the Poisson predictions (dotted line) were less acceptable than the binomial ones (solid curve), although the two models passed the Kolmogorov test; this impression was confirmed by statistical tests based on the comparison of the optimal likelihood criteria which were not equivalent, the latter being (as in six other experiments – for details, see Korn et al. 1982) more satisfactory than the former. The second conclusion was obtained from comparisons between the binomial parameters and the geometrical features of the presynaptic neurons: the camera lucida reconstruction of the cell investigated in Fig. 3, which was a commissural neuron (Fig. 3C), and the corresponding semidiagrammatic drawing (Fig. 3D) show that the number of terminal synaptic boutons was nine and equalled the binomial parameter n extracted by computer analysis.

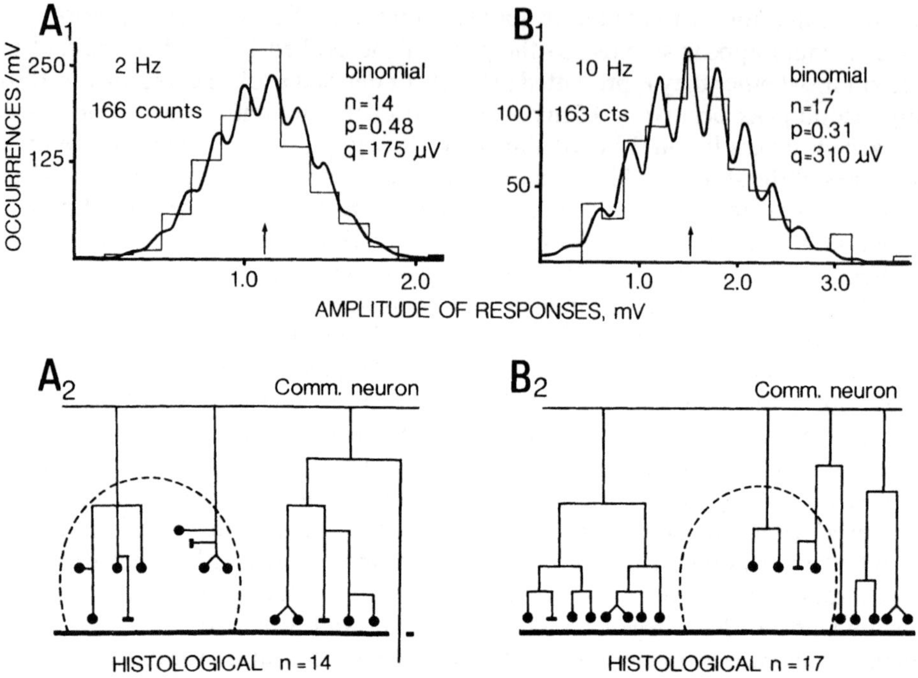

Fig. 4A,B. Further evidence that the binomial parameter n is equivalent to the number of presynaptic endings. *A1* Probability density histogram for 166 trials and computer-determined optimal curve and parameters assuming a binomial curve (Kolmogorov test satisfactory with $p>0.65$). *A2* Schematic representation of the reconstructed terminals issued by the horseradish-peroxidase-filled interneuron and in contact with the Mauthner (M) cell. This commissural (*Comm.*) neuron established synaptic boutons (*closed circles*) in relation to the M cell soma (*thick line*); nine of them were located within the axon cap (*dashed circle*), and five were situated outside this region. *B1,B2* Comparison of quantal parameters provided by the probability density functions (176 counts) and best fits (*B1*) using a binomial model (Kolmogorov test satisfied with $p>0.75$) and of the histological features of another stimulated neuron which was also a commissural cell (*B2*). Note that in each of these two experiments the number of presynaptic terminals (14 and 17 respectively) strictly equalled that of the binomial term n

Other interneurons had a larger number of terminals than that above, often clustered within tightly packed boutons which were sometimes difficult to distinguish at the light microscope level. Yet, as demonstrated by the two examples of Fig. 4, the total number of releasable quanta n ascertained by the computer treatment of unitary response fluctuations was still in agreement with the histological n. In these commissural cells, the binomial parameter n, which was 14 in Fig. 4A1 and 4A2 and 17 in Fig. 4B1 and 4B2, corresponded to the number of histologically determined synaptic boutons, thus indicating that despite some close grouping all of them functioned as distinct releasing units.

The data from 30 experiments involving 24 commissural and six collateral interneurons are summarized by the plot of Fig. 5, which demonstrates that the equivalence between the n's derived by the statistical and the histological approaches was valid for values ranging from 3 to 28. In fact, there was an absolute

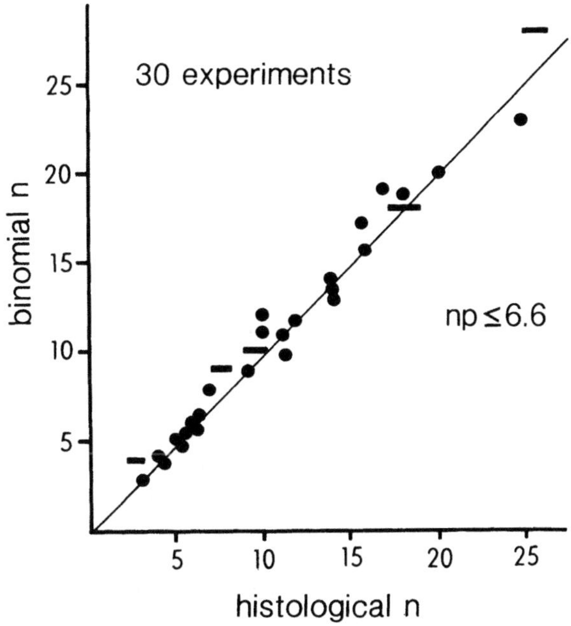

Fig. 5. One-to-one correspondence between the binomial parameter n and the number of synaptic boutons. Plot of correlation between the binomial (*ordinate*) and histological (*abscissa*) n's as determined in 30 experiments for which the product np was less than or equal to 6.6. The *horizontal bars* encompass the possible range of histological values in several cases of uncertainties and the *solid line* is the identity relationship. The correlation suggests that for each of the investigated neurons the quantal unit corresponds to the transmitter released by one synaptic bouton

identity of these parameters for 18 cells. A slight discrepancy was noticeable for the remaining ones, the binomial n being more often one or two units larger than the well-defined histological n (and the corresponding points thus appear above the identity line). These differences were always slight and they suggest that either (a) some terminals were lost in the process of neuronal reconstruction, or (b) some of them infrequently bear more than one presynaptic grid (see Discussion). Of the two approaches, the most reliable indicator of n would thus be the binomial one. In any case, with this model, each synaptic knob can be equated to a quantal unit which releases in an all-or-none manner, even when in a tight cluster arising from the same terminal, and the quantum, defined by binomial terms, corresponds to the basic unit processed by the postsynaptic neuron.

Estimates of binomial parameters were resolved from those experiments. The probability of release, p, was consistently high, ranging from 0.17 to 0.75 with a mean of 0.42 (SD = 0.16; n = 30), which favors a binomial description of the fluctuations rather than a Poisson (McLachlan 1978). On the other hand, quantum sizes q were expressed by normalizing these parameters with respect to the full-sized IPSP evoked by spinal stimulation (this notation allows calculations of the underlying quantal conductances; see next paragraph.). With this procedure, q appeared to be quite constant for the most experiments: except for two extreme cases, it ranged from 0.6% to 1.9% that of the collateral IPSP, with a mean of

1.24 (SD=0.40; n=30). This finding implies that transmitter action is relatively stable at each subsynaptic site and favors the concept described further of a vesicular correlate of q.

Evidence That the Contents of a Single Vesicle Account for a Binomial Quantum

Experimental fits with the binomial model indicate that the quantal size is relatively constant or, otherwise stated, that each fundamental unit represents either the effect of a fixed (and therefore relatively high) number of released vesicles, or that of only one. These alternative possibilities were tested by relating the binomial terms to the basic properties of unitary IPSPs: the population of channels opened by the transmitter released at each synaptic bouton was estimated in order to determine whether the latter corresponds to the presumed action of one (or more) synaptic vesicle(s). The problem was approached with a simple analog circuit of the M cell, which is illustrated in Fig. 6A; a detailed description of this model can be found in an earlier paper (Faber and Korn 1982a). Suffice it to indicate here that: (a) the M cell's membrane capacitance and distributed dendritic cable properties are ignored, since the neuron's time constant is extremely short relative to the IPSP kinetics (Faber and Korn 1980), and because the synapses are electrotonically close to the recording site (Korn et al. 1981); and (b) for facility, all potentials are measured with respect to the resting level and resistances are expressed as their reciprocals, i. e., as conductances (in siemens), since they are directly proportional to the number of open channels in the membrane.

With this circuit, some equations were derived (Faber and Korn 1982a, b). The amplitude of a PSP would be $V = E \times g_{psp}/(g_{psp} + G_m)$. Solving for the synaptic conductance yields $g_{psp} = V \times G_m/(E - V)$, which for small responses, such as quantal ones, becomes

$$g_q = (q/E)G_m \qquad (1)$$

where g_q and q are the quantal conductance and voltage respectively. Since q was known from the binomial and G has been determined with standard techniques, the critical parameter to estimate was E, which for these Cl^--dependent IPSPs varied greatly from one experiment to the next. However, use of double intracellular impalements led to the important finding that the peak amplitude of the full-sized collateral IPSP, which is easily obtained by antidromic activation of the M cell (Fig. 1A–C2), is about one-half the IPSP driving force. Thus Eq. (1) can be rewritten as

$$g_q = (0.5q/V_{coll})\, G_m \qquad (2)$$

where V_{coll} is the amplitude of the collateral IPSP. G_m was shown to be 6.08×10^{-6} S and, as indicated above, the mean value of the normalized quantum size is 1.15%. Then, the conductance increase associated with the action of a single binomial quantum would be in the range of 3.5×10^{-8} S. Assuming a single-channel conductance of 25 pS (see Neher and Stevens 1977; Dudel et al. 1977, 1980) implies that approximately 1400 channels are opened by each releasing synaptic bouton, with extremes of 400 and 2800. This finding leads to the tentative conclusion (Korn et al. 1982; Faber and Korn 1982a) illustrated in Fig. 6B which states

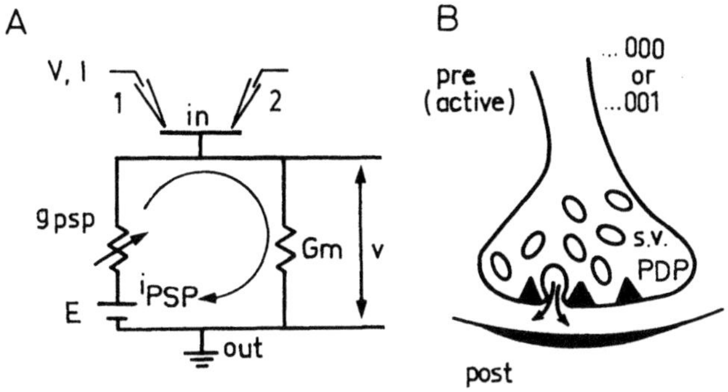

Fig. 6A,B. Electrical analog of the postsynaptic cell and graphical illustration of the proposition that a quantal unit equals the action of a single vesicle's transmitter content. *A* Circuit diagram of the synaptic membrane under conditions for generating a depolarizing inhibitory postsynaptic potential (IPSP). E and g stand, respectively, for the driving force and the increment of conductance contributed by a single unitary IPSP, while G_m is the resting conductance of the unactivated part of the postsynaptic membrane. V is the transmembrane potential change recorded during the response by an intracellular (*in*) microelectrode, with reference to the external potential (*out*). It follows that, if G_m and E are known, g_{psp} and even the quantal conductance can be assessed in order to determine the number of channels opened across the membrane. These two variables were measured with two microelectrodes in the Mauthner (M) cell, one for current and one for voltage. *B* At high magnification, a terminal bouton (*pre*) appears to contain many synaptic vesicles (*s.v.*) and regularly spaced presynaptic dense projections (*PDP*). When the terminal bouton is depolarized sufficiently (*active*), exocytosis of one vesicle may occur, and the transmitter diffuses toward the postsynaptic (*post*) side. The binary notation (...000 or ...001) emphasizes that although there is a large number of apparently available vesicles, a maximum of one can be singled out after an impulse. (Modified from Faber and Korn 1982a)

that at each individual synapse (or bouton) the physical counterpart of the binary-releasing property of each bouton is that apparently no more than one vesicle can normally undergo exocytosis after one presynaptic impulse.

Active Zones at M Cell's Inhibitory Synapses

Horseradish-peroxidase-stained terminals of PHP-exhibiting cells are segregated within and outside the axon cap of the M cell. According to a recent classification (Nakajima 1974), they exhibit the distinct characteristics of the unmyelinated club endings and small vesicle boutons, which both display the typical aspects of Gray type II synapses. We have considered in detail whether or not these synaptic complexes exhibit any fine structural features which would make them different from those arising from other chemically transmitting central axons; this was not the case, nor was it the case in commissural and collateral neurons (Triller and Korn 1981, 1982a). The geometrical organization of the presynaptic organelles and that of the specialized regions which are the site of interaction between the synaptic vesicles and the presynaptic membrane were examined for further correlations between histological events and their morphological counterparts.

All the boutons examined were completely filled by numerous pleiomorphic vesicles; the mean diameter of the latter ranged from 40 to 46 nm, their elongation coefficient from 1.4 to 1.5, and their surface area from $13–15 \times 10^2$ nm^2; vesicles

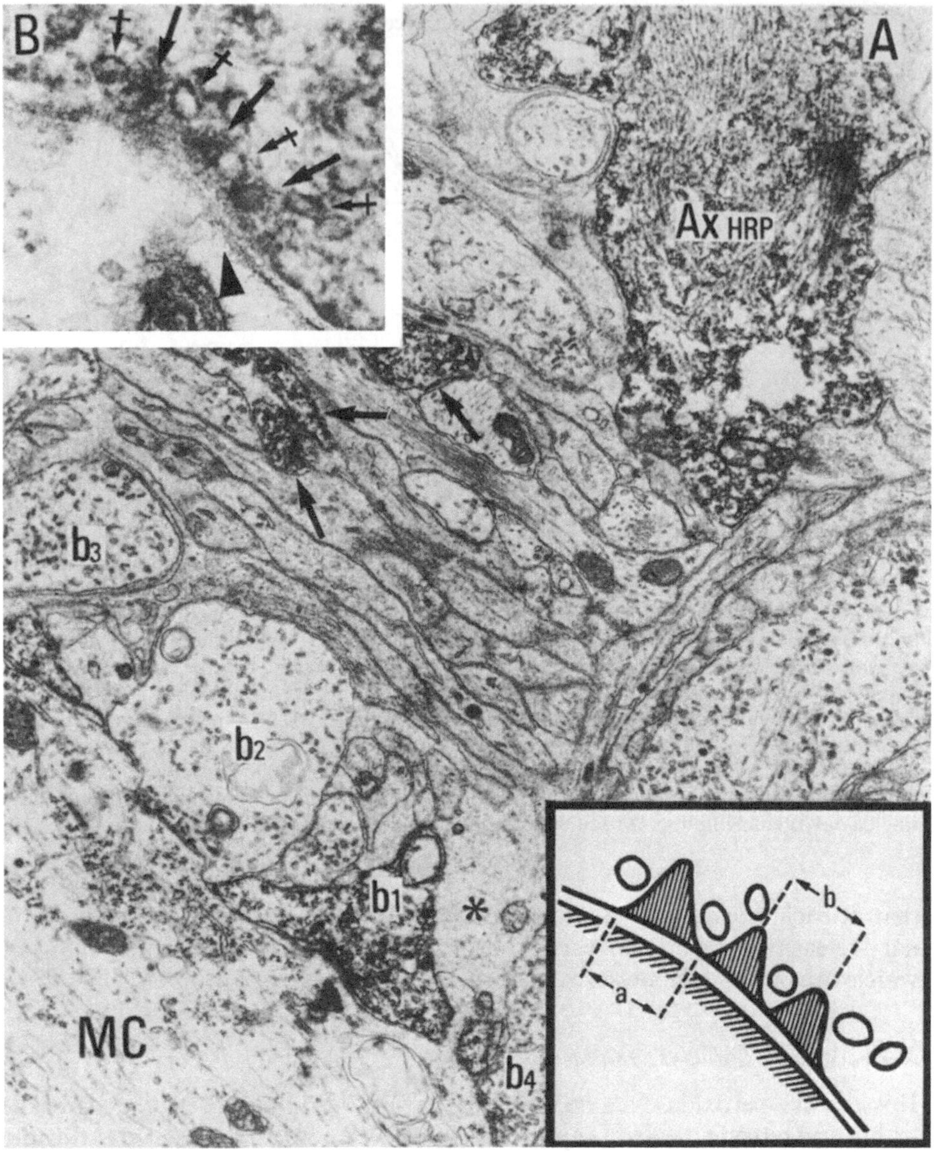

Fig. 7A,B. Ultrastructural characteristics of a stained unmyelinated club ending synapsing on the Mauthner (M) cell's soma, and evidence that synaptic vesicles situated between the dense projections are in a favorable position for exocytosis. *A* A synaptic bouton (*b1*) filled with horseradish peroxidase (HRP) and terminating on the M cell's soma (*MC*) at the peripheral part of the axon cap; as expected from a terminal from this class, this portion of its afferent processes (*AxHRP*) is devoid of myelin. Note that this element is surrounded by other unlabeled synaptic contacts (*b2, b3, b4*) and that three distinct small axonal branches (*arrows*) belonging to the same stained neuron are also visible. The extracellular space (*asterisk*) is filled with a granular material (×19000). *B* Higher magnification of the HRP-filled synaptic region illustrated in *A*. Three presynaptic dense projections (*arrows*) protrude toward the cytoplasmic side and synaptic vesicles (*crossed arrows*) nestle between and beside them. A postsynaptic differentiation (*arrow head*) faces the presynaptic grid (×74700). *Inset*: semischematic representation of the synaptic complex shown in *B*. Symbols *a* and *b* stand for the base and for the distance between presynaptic dense potentials respectively. The postsynaptic differentiation is represented by *hatched lines*. (Triller and Korn 1982a)

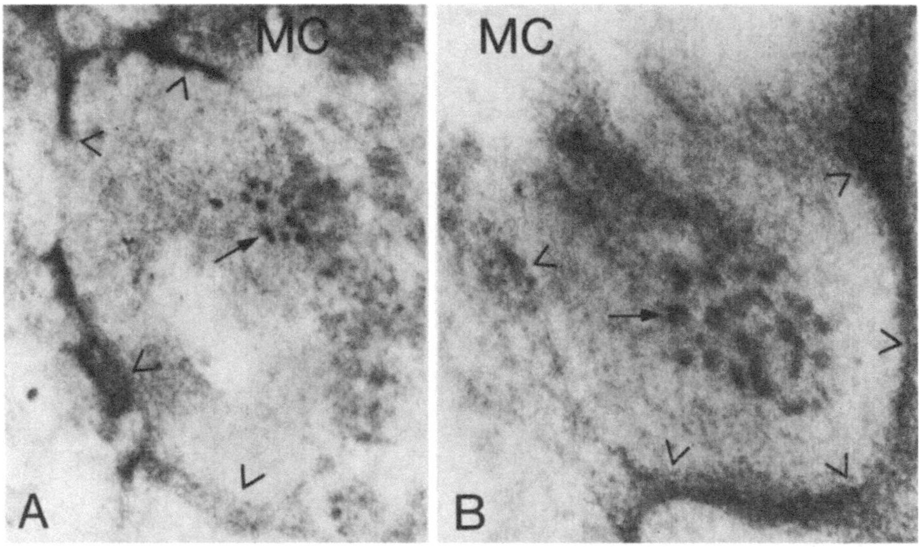

Fig. 8A,B. En face views of ethanolic phosphotungstic acid (E-PTA)-stained terminals showing that each synaptic bouton contains only one presynaptic vesicular grid. Electron micrographs of 0.5-μm-thick E-PTA-treated tissue in the peripheral part of the axon cap. Two profiles are clearly delineated by their dark borders (*V-shaped arrows*), since in this material the extracellular space is markedly stained; each of them contains a single macular active zone in an en face view composed of ten and of about 21 presynaptic dense potentials (*arrows*) respectively. Note that there is no relation between the size of a given grid and the diameter of its corresponding terminal bouton. This figure suggests that the equivalence found during quantal analysis between binomial term n and the number of stained presynaptic boutons can also be interpreted as expressing a one-to-one correspondence with the number of presynaptic grids. *MC*, Mauthner cell. *A* ×32800, *B* ×69300

of adjacent unstained terminals of the same classes had a mean diameter ranging from 39 to 46 nm, an elongation coefficient of 1.5–1.6, and a surface area of 12–16.5 × 10^2 nm^2. Thus HRP allows a reliable analysis of synaptic vesicles, despite a slight tendency of some of them to be more spherical in the presence of the enzyme.

Presynaptic specializations were compared with those of other investigated central synapses: in the latter, they are formed (Gray 1959, 1963) by electron-opaque patches called "presynaptic dense projections" (PDPs), which are disposed in triangular arrays and which are linked at their base (Bloom et al. 1966, 1968; Pfenninger et al. 1969). Thus they are referred to as part of a "presynaptic grid" (Akert et al. 1972), which guides the vesicles toward the plasma membrane, as suggested by freeze fracture studies (Pfenninger et al. 1972). Transverse sections, such as those of Fig. 7, showed that these structures were roughly triangular in shape, with the base facing the membrane, and as expected, synaptic vesicles were often wedged between PDPs (Fig. 7B and inset).

Following E-PTA-staining, there was a dark electron-dense reaction in the extracellular space of the axon cap and of the synaptic bed of the M cell which permitted us to determine the characteristics of single terminals and their synaptic grids. A total of 86 synaptic grids were analyzed in *en face* views similar to those

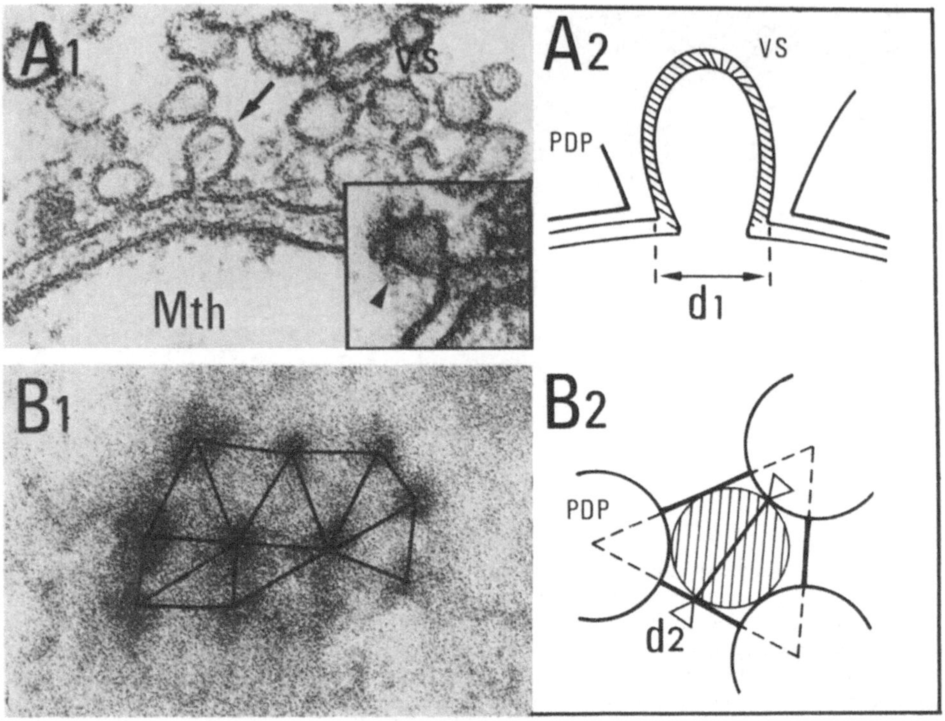

Fig. 9A,B. Different parameters computed in order to determine whether the presynaptic grid expands as a synaptic vesicle fuses with the plasma membrane. *A1, A2* Transverse section of an active zone passing through the opening of a synaptic vesicle. *A1*, electron micrograph of a terminal bouton containing pleiomorphic vesicles, one of which (indicated by an *arrow*) is opened toward the extracellular space. *Inset*: coated vesicle (*arrowhead*) visualized in the same preparation (×180000). *A2*, schematic representation of a synaptic vesicle with its membrane (*hatched area*) in direct continuity with that of the presynaptic terminal (the boundaries between them are indicated by *vertical broken lines*). *B1,B2* En face view of the presynaptic grid. *B1*, semithin section stained with ethanolic phospotungstic acid showing the presynaptic dense projections; their triangular disposition is emphasized by *solid lines* (×187400). *B2*, methods used in order to determine the size of the area which is left for exocytosis within the zone outlined by three dense projections and the filamentous crossbridges linking them (*hatched circle*), with its diameter indicated by a *thick line between two triangles*. *d1*, diameter of the vesicular neck; *d2*, diameter of the largest available free space; *Mth*, Mauthner cell; *PDP*, presynaptic dense projections; *VS*, synaptic vesicle. (Triller and Korn 1982b)

of Fig. 8. They only occupied a fraction of the whole interneuronal interface; they were more often macular in shape and a few of them were annular, with the PDPs involved in their formation leaving a free space in their middle part. The more important conclusion reached was that at least 95% of the terminals, if not more, contain one presynaptic grid: only four of them appeared as possibly exhibiting two distinct and separate complexes. Thus the striking equivalence found during the quantal analysis of fluctuating unitary IPSPs, between the binomial term n and the number of stained presynaptic boutons, can also be expressed as indicating a one-to-one correspondence with the number of presynaptic grids present in terminals of the investigated inhibitory interneurons.

In *en face* views, the grids were formed by discrete polygonal patches, each of them representing a single PDP seen from above. Their mean number of PDPs was 14.2 (SEM = 0.65; n = 86) per synaptic grid. The photomicrographs of Fig. 9B1 indicate that the PDPs were organized following a regular triangular array; this pattern is similar to that reported at all chemically mediated central synapses.

Transverse sections of E-PTA-stained presynaptic complexes also provided important fine details. The PDPs were well separated and there were holes between them. They appeared as roughly triangular-shaped structures: their bases ranged from 44 to 70 nm (mean = 56 nm; SEM = 1.70; n = 46), their center-to-center distances varied between 70 and 90 nm (mean = 80 nm; SEM = 2.33; n = 36), and the free space between them ranged from 10 to 30 nm (mean = 20 nm; SEM = 1.5; n = 30). These values are comparable to those obtained in other material, whether untreated or stained with HRP (Triller and Korn 1982a).

Can Exocytosis of One Vesicle Inhibit the Discharge of Its Neighbors?

The conclusion that at synapses of inhibitory interneurons no more than one synaptic vesicle can be released by each impulse must be reconciled with the presence of so many vesicles in the proximity of the releasing site. An attractive and so far little considered explanation would be that the release of second "quanta" is somehow blocked when a neighboring synaptic vesicle undergoes exocytosis. Therefore, we have considered the possibility that the "paracrystalline structure" of the presynaptic grid is not a rigid formation but rather that the space between PDPs expands when the vesicle fuses with the adjacent synaptic membrane (Triller and Korn 1982a). The reliability of earlier measurements performed at those synaptic complexes allowed us to undertake a morphometric analysis of putative dynamic changes that would be produced at synaptic complexes during or after exocytosis.

As shown by Fig. 9, spontaneous exocytosis of synaptic vesicles (Fig. 9A1) is relatively common and the diameter d1 of their neck was assessed in transverse sections (Fig. 9A2) of synaptic terminals. The free space between PDPs was evaluated on *en face* views of E-PTA-stained boutons (Fig. 9B1) and the diameter d2 of the larger circular area available for exocytosis was determined using a computer program which took into consideration the size of the PDPs and the shape of each triangle determined by their association (Fig. 9B2). Preliminary results (Triller and Korn 1982b) strongly suggest that the mean diameter of the necks of exocytotic vesicles is larger than that of the spaces which are left free within the frame of the presynaptic grid. The parameter d1 ranged from 7.9 to 58.6 nm with a mean of 20.5 ± 10.5 nm (n = 380, from 40 synaptic grids). By contrast, d2 ranged from 17.0 to 67.2 nm with a mean of 28.0 ± 9.6 nm (n = 40). The ratio d1/d2 was thus 0.73 (with a statistical significance of $p < 1‰$, given by tests rejecting equality of the means). It follows that vesicular discharges only occur at the level of the larger free spaces between PDPs. On the other hand, incompatibility of the means suggests that the free spaces are widened by the collars of exocytosing vesicles: this notion is strengthened by the fact that about half of the free spaces had a diameter $d2 < 15$ nm. These small areas could be produced by a compression due to exocytosis occurring in remote regions of the grids: as shown in Fig. 10, the

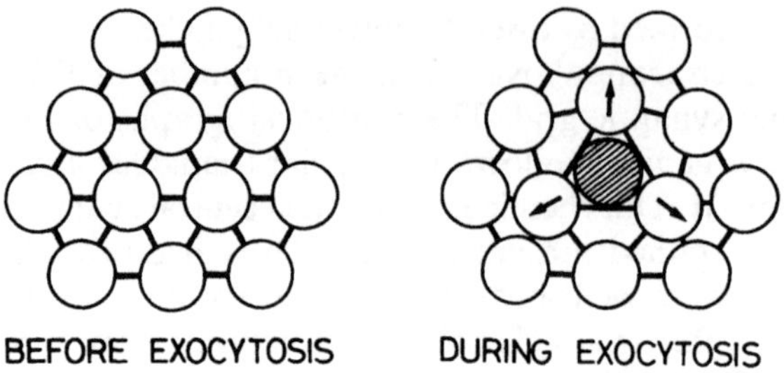

Fig. 10. Diagrammatic representation of postulated presynaptic changes when a vesicle undergoes exocytosis. *Above*: at rest, presynaptic dense projections (PDPs) (*open circles*) are regularly organized along triangular arrays, with equal center-to-center distances between them. *Below*: as a vesicle bursts out (*hatched circle*), movements of adjacent PDPs are produced in each direction indicated by *arrows*. As a consequence, the area between PDPs surrounding the releasing unit is widened, while the extent of the remaining free space is reduced. (Modified from Triller and Korn 1982b)

opening of one vesicle could produce a transient enlargement of the space where it occurs and a distortion of the neighboring elements which would then be propagated to the whole presynaptic grid. This deformation could be a regulatory mechanism by which the number of exocytotic vesicles is limited after each impulse.

Discussion

The data reported here constitute a firm basis upon which the notion of a quantal nature of transmitter release, which was derived from studies of neuromuscular and invertebrate junctions, can be extended to central synapses. This conclusion alone is of interest: it contributes to reinforce the growing concept of a "remarkable uniformity" (Eccles 1964) of the junctional basic operations by which excitable cells interact with each other throughout the whole nervous system. The discussion of this paper is focused on this important precept, since we have already considered in detail elsewhere (Faber and Korn 1982a; Korn et al. 1982; Triller and Korn 1982a) problems related to the validity, precision, and limitations of the mathematical models used during this work, as well as those of their associated histological techniques.

Analogies Between Neuromuscular Release Sites and Central Presynaptic Grids

Chemical synapses are morphologically characterized by active zones (Couteaux and Pécot-Dechavassine 1970a, b) with a specific arrangement of membrane densities around which synaptic vesicles are clustered and undergo exocytosis; these densities exhibit various geometrical forms, such as dense bands at neuromuscular junctions (or bars, also first named release sites, in crustacea, see Atwood 1976) or hexagonal lattices in central synapses (ref. in Llinas and Heuser 1977). Electron-dense specializations with vesicles clustered around them were first noticed in motor nerve to muscle terminals by Birks et al. (1960), and opening of vesicles was demonstrated to occur at their borders (Couteaux and Pécot-Dechavassine 1970a, b). These repeated bars are parallel, 100 nm wide, and disposed, with their tight row of associated vesicles, perpendicularly to the long axis of the terminal axon (Dreyer et al. 1973; Heuser and Reese 1973; Heuser et al. 1974).

That "release sites" could represent a correlate for n at the neuromuscular junction was often suggested on the basis of indirect arguments. For instance, binomial n's calculated at crayfish motor nerve terminals (Johnson and Wernig 1971; Wernig 1972, 1975; Zucker 1973) most likely correspond to the number of release sites presumably detected by the postsynaptic microelectrode (Atwood and Johnston 1968; Jahromi and Atwood 1974). At the neuromuscular junction of the frog, there also seems to be a positive correlation between calculated n's (Katz and Miledi 1965) or m (Kuno et al. 1971) and the number of sites present along the length of a presynaptic fiber that contribute to the recordings (Birks et al. 1960; Wernig 1975). More recently, the mean quantal m was correlated with the number and size of presynaptic dense bars in single lobster muscle fibers (Meiss and Govind 1980). The same correlation was consistently noted during the formation of neuromuscular junctions: a progressive increase in binomial n (Bennett and Florin 1974) observed during phrenic reinnervation of mammalian diaphragm was associated with the growth of the regenerating severed axon (Bennett et al. 1973). A comparable correlation was seen during the normal development of this junction (Bennett and Pettigrew 1974); similarly, n was directly proportional to the size of nerve terminals during both the formation and regression of synapses at reinnervated axolotl striated muscles (Bennett and Raftos 1977).

Our data provide a direct proof that active zones are the physical correlate of n; they strongly substantiate the above-mentioned reports and by extending this notion to central synapses they allow peripheral and central junctions to be equated in terms of the binomial model of transmitter release. Figure 11 exemplifies this functional convergence: each integer of the binomial n pertains to either a single dense band or to one synaptic bouton, provided the latter contains one synaptic grid and no more. This morphological condition is satisfied not only at the M cell's inhibitory terminals but, for instance, at synapses between the parallel fibers and the dendritic spines of cerebellar Purkinje cells (Palay and Chan-Palay 1974) or at dendritic spines of small pyramidal cells (Peters and Kaiserman-Abramof 1969). However, this relationship would be modified at more complex junctions: although data are poorly substantiated in the literature regarding the number of presynaptic complexes present in the terminals of central neurons, the possibility of multiple active zones has been mentioned, for example at freeze-fractured single neuroneuronal synapses of cat's subfornical organ (Akert 1973). The binomial n would then be higher than the number of synaptic boutons *per se*, as shown in Fig. 11, for a junction bearing two presynaptic grids. Thus, although the different geometry of active zones at neuromuscular junctions and at central synapses does not reflect fundamental differences in the statistics of transmitter release, caution is required in interpreting the results of binomial analysis of fluctuating potentials in the absence of additional ultrastructural studies of the investigated system.

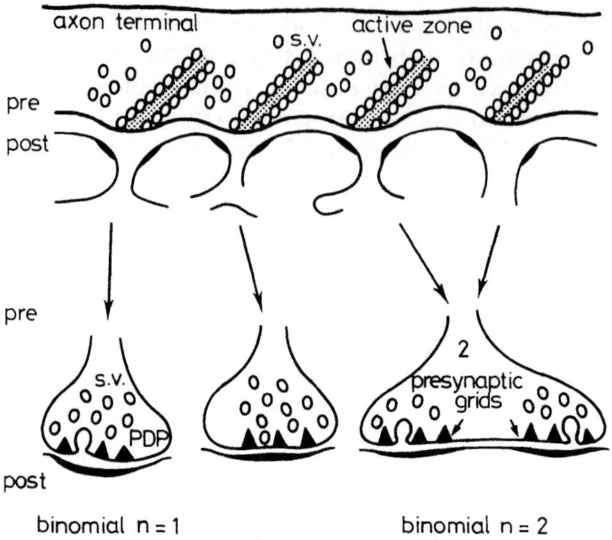

Fig. 11. Diagram illustrating that in terms of the binomial model, release sites of the neuromuscular junction and presynaptic vesicular grids of the central synapses are functionally equivalent. *Above*: longitudinal section through a terminal branch of a frog neuromuscular junction. At the presynaptic side of the junction (*pre*) synaptic vesicles (*s.v.*) cluster around a dense cytoplasmic formation (*hatched area*), which forms regularly spaced parallel bands facing muscular folds, each density and its adjacent vesicles thus defining an active zone where chemical transmitter is most likely released. A dense material (*dark areas*), which may represent the receptive complex of the postsynaptic element (*post*), is also shown. *Below*: drawings of terminal boutons and their synaptic complexes; the latter, which represent the active zones of the synapse, are composed of arrays of presynaptic dense projections (*dark triangles*) which delineate the presynaptic vesicular grid, with synaptic vesicles nestling between them. The figure shows that the binomial term n is equivalent to 1 at synaptic boutons bearing only one synaptic grid (*left diagrams*) but that the value of this parameter, which reflects the number of active zones present in the presynaptic terminals, is expected to increase in the case of more complicated synapses (*lower right hand side of the diagram*).

Can the "One Vesicle" Hypothesis Be Generalized?

Our calculations suggesting that the quantal conductance increase at each activated synaptic knob is brought about by the contents of a single vesicle are consistent with those obtained at other inhibitory junctions (Barker and McBurney 1979; Dudel et al. 1977). In fact, the assumption of such a one-to-one relationship has already been considered in the past, but without direct evidence, and furthermore, the exclusive use of toxins or drugs in order to capture exocytosing vesicles (see Heuser et al. 1979) has focused attention on massive but often artificial transmitter output, rather than on more natural events. Assuming that Ia excitatory terminals impinging upon spinal motoneurons have the same density of quanta as at a neuromuscular junction, Eccles (1961a) stated for the first time that "excitation of one terminal knob may be considered to release no more than one quantum," and with further computations he still put the upper limit of mobilized quanta at two or three (Eccles 1961b). This conclusion, which was also reached using the Poisson statistics to analyze fluctuations of evoked Ia-EPSPs (Kuno

1964a), is not restricted to central synapses: with a simple binomial model, no more than one quantum is released per active zone and per impulse at the crayfish neuromuscular junction (Wernig 1975, 1976).

The above proposition seems to be in contradiction to the presence of many synaptic vesicles in axonal endings, but this apparent paradox can be approached by speculations taking into account (a) physiological data obtained at the neuromuscular junction, and (b) the morphology of the presynaptic complex. First, implicit in the possible equivalence between the binomial n and the number of release sites is the postulate that active zones release in an all-or-none manner (see also Wernig and Stirner 1977), as do central synapses. Similarly, a statistical analysis at the frog motor endplate, with quick-freezing technique, did not favor the hypothesis that several vesicles composed miniature endplate potentials (Heuser et al. 1979). Moreover, the number of vesicles that can line up along both sides of all synaptic densities at one frog endplate is roughly about 20000 (Heuser and Reese 1977), but during the first few impulses of a tetanus 300 vesicles discharge at each impulse (Katz 1962, 1971), or otherwise stated, only one vesicle out of $\cong 66$ is released.

Secondly, in PHP-exhibiting cells, the mean number of PDPs was 14. Given their triangular arrangement and assuming that vesicles occupy the free space between and around them, it was possible to evaluate that 44 vesicles are associated with a typical ellipsoid grid and 62 with an annulated one. These values vary within a range of 29–83, which is compatible with the release of only one vesicle per grid if the probability of discharge calculated for the neuromuscular junction holds for the CNS presynaptic grids.

At the present time, one can only advance two general schemes to account for the apparent unity of active zones. One is that a terminal is constrained to release the contents of no more than one vesicle for reasons not yet elucidated; although our model of a mechanism by which exocytosis of one vesicle can block the discharge of adjacent ones is quite simplistic at present, it fits a number of reported data, including morphometric analysis of the presynaptic grid. The other is that this content alone is sufficient to saturate all receptors subsynaptic to a single bouton, but computations (Kuffler and Yoshikami 1975) as well as experimental results have shown (Hartzell et al. 1975) that postsynaptic receptors are not saturated by molecules of transmitter contained in a quantum. In any case, both alternatives are compatible with our results and with the postulate that the functional binomial unit is equivalent to the histological one.

References

Akert K (1973) Dynamic aspects of synaptic ultrastructure. Brain Res 49: 511–518

Akert K, Pfenninger K, Sandri C, Moor H (1972) Freeze etching and cytochemistry of vesicles and membrane complexes in synapses of the central nervous system. In: Pappas GD and Purpura DP (eds) Structure and function of synapses. Raven, New York, pp 67–86

Atwood HL (1976) Organization and synaptic physiology of crustacean neuromuscular systems. Prog Neurobiol 7: 291–391

Atwood HL, Johnston HS (1968) Neuromuscular synapses of a crab motor axon. J Exp Zool 167: 457–470

Barker JL, McBurney RN (1979) Phenobarbitone modulation of postsynaptic GABA receptor function on cultured mammalian neurons. Proc R Soc Lond [Biol] 206: 319–327
Bennett MR, Florin T (1974) A statistical analysis of the release of acetylcholine at newly formed synapses in striated muscle. J Physiol (Lond) 238: 93–107
Bennett MR, Pettigrew AG (1974) The formation of synapses in striated muscle during development. J Physiol (Lond) 241: 515–545
Bennett MR, Raftos J (1977) The formation and regression of synapses during the re-innervation of axolotl striated muscle. J Physiol (Lond) 265: 261–295
Bennett MR, McLachlan EM, Taylor RS (1973) The formation of synapses in reinnervated mammalian muscle. J Physiol (Lond) 233: 481–500
Birks R, Huxley HE, Katz B (1960) Differentiation of nerve terminals in the crayfish opener muscle and its functional significance. J Physiol (Lond) 150: 134–144
Bloom FE, Aghajanian GK (1966) Cytochemistry of synapses: selective staining for electron microscopy. Science 154: 1575–1577
Bloom FE, Aghajanian GK (1968) Fine structural and cytochemical analysis of the staining of synaptic junctions with phosphotungstic acid. J Ultrastruct Res 22: 361–375
Couteaux R, Pécot-Dechavassine M (1970a) Vésicules synaptiques et poches au niveau des zones actives de la jonction neuro-musculaire. CR Acad Sci (Paris) 271: 2346–2349
Couteaux R, Pécot-Dechavassine M (1970b) L'ouverture des vésicules synaptiques au niveau des zones actives. In: Favard P (ed) Microscopie electronique, vol III. Seventh International Electronic Microscopy, Grenoble, pp 709–710
del Castillo J, Katz B (1954) Quantal components of the end-plate potential. J Physiol (Lond) 124: 560–573
Dreyer FK, Peper K, Akert K, Sandri C, Moor H (1973) Ultrastructure of the active zone in the frog neuromuscular junction. Brain Res 62: 373–380
Dudel J, Finger W, Stettmeier H (1977) GABA induced membrane current noise and the time course of the inhibitory synaptic current in crayfish muscle. Neurosci Lett 6: 203–208
Dudel J, Finger W, Stettmeier H (1980) Inhibitory synaptic channels activated by γ-aminobutyric acid (GABA) in crayfish muscle. Pflügers Arch 387: 143–151
Eccles JC (1961a) The effect of frequency of activation on transmission across synapses. In: Chagas C, de Carvalho AP (eds) Proc symp bioelectrogenesis. Elsevier, Amsterdam, pp 297–309
Eccles JC (1961b) The mechanism of synaptic transmission. Ergebn. Physiol 51: 299–430
Eccles JC (1964) The physiology of synapses. Springer, Berlin Göttingen Heidelberg, pp 173–200
Eccles JC (1969) The inhibitory pathways of the central nervous system. Thomas, Springfield, p 135
Edwards FR, Redman SJ, Walmsley B (1976a) Statistical fluctuations in charge transfer at Ia synapses on spinal motoneurones. J Physiol (Lond) 259: 665–688
Edwards FR, Redman SJ, Walmsley B (1976b) Non quantal fluctuations and transmission failures in charge transfer at Ia synapses on spinal motoneurones. J Physiol (Lond) 259: 689–704
Faber DS, Korn H (1978) Electrophysiology of the Mauthner cell: basic properties, synaptic mechanisms and associated networks. In: Faber DS, Korn H (eds) Neurobiology of the Mauthner cell. Raven, New York, pp 47–131
Faber DS, Korn H (1980) Single-shot channel activation accounts for duration of inhibitory postsynaptic potentials in a central neuron. Science 208: 612–615
Faber DS, Korn H (1982a) Transmission at a central inhibitory synapse. I – Magnitude of the unitary postsynaptic conductance change and kinetics of channel activation. J Neurophysiol 48: 654–678
Faber DS, Korn H (1982b) Binary mode of transmitter release at central synapses. Trends Neurosci 5: 157–159
Furshpan EJ, Furukawa T (1962) Intracellular and extracellular responses of the several regions of the Mauthner cell of the goldfish. J Neurophysiol 25: 732–771
Gray EG (1959) Axo-somatic and axo-dendritic synapses on the cerebral cortex: an electron microscope study. J Anat 93: 420–433
Gray EG (1963) Electron microscopy of presynaptic organelles of the spinal cord. J Anat 97: 101–106

Hartzell HC, Kuffler SW, Yoshikami D (1975) Postsynaptic potentiation: interaction between quanta of acetylcholine at the skeletal neuromuscular synapse. J Physiol (Lond) 251: 427–463

Heuser JE, Reese TS (1973) Evidence for recycling of synaptic vesicle membrane during transmitter release at the frog neuro-muscular junction. J Cell Biol 57: 315–344

Heuser JE, Reese TS (1977) Structure of the synapse. In: Brookhart JM, Mountcastle VB (eds) The Nervous System, vol. I, part 1. Md: American Physiological Society, Bethesda, pp 261–294 (Handbook of physiology, section I)

Heuser JE, Reese TS, Landis, DMD (1974) Functional changes in frog neuromuscular junctions studies with freeze fracture. J Neurocytol 3: 109–131

Heuser JE, Reese TS, Dennis MJ, Jan Y, Jan L, Evans L (1979) Synaptic vesicle exocytosis captured by quick freezing and correlated with quantal transmitter release. J Cell Biol 81: 275–300

Jahromi SS, Atwood HL (1974) Three dimensional ultrastructure of the crayfish neuromuscular apparatus. J Cell Biol 63: 599–613

Jazwinski AH (1970) Stochastic processes and filtering theory. Academic, New York, pp 81–85

Johnson EW, Wernig A (1971) The binomial nature of transmitter release at the crayfish neuromuscular junction. J Physiol (Lond) 218: 757–767

Katz B (1962) The transmission of impulses from nerve to muscle and the subcellular unit of synaptic action. Proc Soc Lond [Biol] 155: 455–477

Katz B (1971) Quantal mechanism of neural transmitter release. Science 173: 123–126

Katz B, Miledi R (1963) A study of spontaneous miniature potentials in spinal motoneurons. J Physiol (Lond) 168: 389–422

Katz B, Miledi R (1965) The effect of temperature on the synaptic delay at the neuromuscular junction. J Physiol (Lond) 181: 656–670

Kendall M, Stuart A (1977) The advanced theory of statistics, vol 1. Charles Griffin, London, pp 109–110

Korn H, Faber DS (1975) An electrically mediated inhibition in goldfish medulla. J Neurophysiol 38: 452–471

Korn H, Faber DS (1976) Vertebrate central nervous system: same neurons mediate both electrical and chemical inhibitions. Science 194: 1166–1169

Korn H, Triller A, Faber DS (1978) Structural correlates of recurrent collateral interneurons producing both electrical and chemical inhibitions of the Mauthner cell. Proc Soc Lond [Biol] 202: 533–538

Korn H, Triller, A, Mallet A, Faber DS (1981) Fluctuating responses at a central synapse: n of binomial fit predicts number of stained presynaptic boutons. Science 213: 898–901

Korn H, Mallet A, Triller A, Faber DS (1982) Transmission at a central inhibitory synapse. II – Quantal description of release, with a physical correlate for the binomial n. J Neurophysiol 48: 679–707

Kuffler SW, Yoshikami D (1975) The number of transmitter molecules in a quantum: an estimate from iontophoretic application of acetylcholine at the neuromuscular synapse. J Physiol (Lond) 251: 465–482

Kuno M (1964a) Quantal components of excitatory synaptic potentials in spinal motoneurons. J Physiol (Lond) 175: 81–99

Kuno M (1964b) Mechanism of facilitation and depression of the excitatory synaptic potential in spinal motoneurons. J Physiol (Lond) 175: 100–112

Kuno M, Miyahara, JT (1969) Non-linear summation of unit synaptic potentials in spinal motoneurons of the cat. J Physiol (Lond) 201: 479–493

Kuno M, Turkanis SA, Weakly JN (1971) Correlation between nerve terminal size and transmitter release at the neuromuscular junction of the frog. J Physiol (Lond) 213: 545–556

Llinas R, Heuser JE (1977) Depolarization release coupling systems in neurons. Neuro Res Prog Bull, MIT Press 15: 557–687

McLachlan EM (1978) The statistics of transmitter release at chemical synapses. In: Parker R (ed) Intern rev physiol, neurophysiol III, vol 17. University Park Press, Baltimore, pp 49–117

Martin AR (1955) A further study of the statistical composition of the endplate potential. J Physiol (Lond) 130: 114–122

Meiss DE, Govind CK (1980) Heterogeneity of excitatory synapses at the ends of single muscle fibers in lobster *Homarus Americanus*. J Neurobiol 11: 381–395
Mendell LM, Weiner R (1976) Analysis of pairs of individual Ia EPSPs in single motoneurons. J Physiol (Lond) 225: 81–104
Nakajima Y (1974) Fine structure of the synaptic endings on the Mauthner cell of the goldfish. J Comp Neurol 156: 375–402
Neher E, Stevens CF (1977) Conductance fluctuations and ionic pores in membranes. Annu Rev Biophys Bioeng 6: 345–381
Nelder JA, Mead R (1965) A simplex method for function minimization. Computer Journal 7: 308–313
Palay SL, Chan-Palay J (1974) Cerebellar cortex. Cytology and organization. Springer, Berlin Heidelberg New York, pp 348
Peters A, Kaiserman-Abramof IR (1969) The small pyramidal neuron of the cerebellar cortex. The synapses upon dendritic spines. Z Zellforsch Mikrosk Anat 100: 487–506
Pfenninger K, Sandri C, Akert K, Eugster CH (1969) Contribution to the problem of structural organization of the presynaptic area. Brain Res 12: 10–18
Pfenninger K, Akert K, Moor H, Sandri C (1972) The fine structure of freeze-fractured presynaptic membranes. J Neurocytol 1: 129–149
Shapovalov AI, Shiriaev BI (1980) Dual mode of junctional transmission at synapses between single primary afferent fibres and motoneurones in the amphibian. J Physiol (Lond) 306: 1–15
Triller A, Korn H (1978) Mise en évidence électrophysiologique et anatomique de neurones vestibulaires inhibiteurs commissuraux chez la tanche (*Tinca tinca*). CR Acad Sci (Paris) 286: 89–92
Triller A, Korn H (1981) Morphologically distinct classes of inhibitory synapses arise from the same neurons: ultrastructural identification from crossed vestibular interneurons intracellularly stained with HRP. J Comp Neurol 203: 131–155
Triller A, Korn H (1982a) Transmission at a central inhibitory synapse. III – Ultrastructure of physiologically identified terminals. J Neurophysiol 48: 708–736
Triller A, Korn H (1982b) Déformations de la grille présynaptique liées à l'exocytose dans les synapses centrales. CR Acad Sci (Paris) 259: 655–660
Vrensen G, Nunes J, Cardozo J, Müller L, Van der Want J (1980) The presynaptic grid: a new approach. Brain Res 184: 23–40
Wernig A (1972) Changes in the statistical parameters during facilitation at the crayfish neuromuscular junction. J Physiol (Lond) 226: 751–759
Wernig A (1975) Estimates of statistical release parameters from crayfish and frog neuromuscular junctions. J Physiol (Lond) 244: 207–221
Wernig A (1976) Localization of active sites in the neuromuscular junction. Brain Res 118: 63–72
Wernig A, Stirner H (1977) Quantum amplitude distributions point to functional unity of the synaptic "active zones". Nature 269: 820–822
Zottoli SJ, Faber DS (1980) An identifiable class of statoacoustic interneurons with bilateral projections in the goldfish medulla. Neuroscience 5: 1287–1302
Zucker RS (1973) Changes in the statistics of transmitter release during facilitation. J Physiol (Lond) 229: 787–810

Reorganization of Neuronal Membrane Properties Following Axotomy*

D.S. FABER, Buffalo

Summary. Morphological and physiological changes in the goldfish Mauthner cell following axotomy by spinal transection include chromatolysis, reactive deafferentation of excitatory inputs to the soma, the appearance of soma-dendritic impulses, and a decrease in electrical excitability of the initial segment. These changes persist for at least 200 days following axotomy and indicate a shift in the control of regional membrane specializations in this neuron. The different components of the axon reaction have been studied in the same individual neurons, and it appears that the increased soma-dendritic excitability is always associated with chromatolysis, and that "somatic stripping" is the one consistent feature of the response to axotomy.

Key words: *Mauthner cell – axotomy – chromatolysis – neuronal excitability – deafferentation*

Introduction

One of the measures of Eccles' remarkable ongoing contributions to neurobiology lies in the continuing relevance of his early studies, which provide both a stimulus and a standard of reference for present research endeavors. Certainly, the classic works of Coombs et al. (1957a, b) on the generation of impulses in spinal motoneurons serve such functions. For example, in those papers Eccles and his colleagues demonstrated, with direct and orthodromic stimulations of normal motoneurons, that an action potential was generated in the initial segment (IS) of the motor axon, as the excitable soma-dendritic (SD) membrane had a much higher threshold for spike initiation. Thus the concept emerged that there is a regional specialization of the neuronal membranes, with the IS (defined as the axon hillock plus the nonmyelinated axon) being selectively or particularly excitable (see also Araki and Otani 1955; Brock et al. 1952, 1953; Fatt 1957; Fuortes et al. 1957). Presumably, there is a much higher density of voltage-sensitive Na^+ channels in this region than in the adjacent membrane. However, the mechanisms underlying this and other examples of regional membrane specialization remain poorly understood.

One experimental approach to questions concerned with the regulation of specific membrane properties of nerve cells is based upon findings that when the axon

* Supported by NIH Grant NS 15335

© Springer-Verlag Berlin · Heidelberg 1984

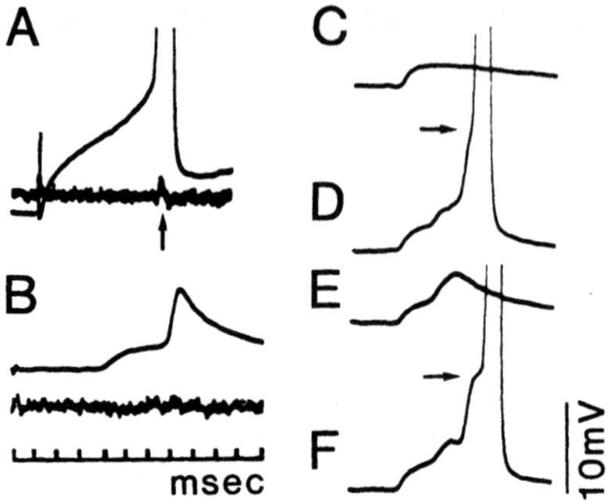

Fig. 1 A–F. Enhanced soma-dendritic excitability of a chromatolyzed spinal motoneuron. *A,B* (*upper traces*) and *C–F* are intracellular records from a flexor digitorum longus motoneuron whose motor axon in ventral root L7 was severed 16 days previously. *Lower traces* in *A* and *B* are extracellular recordings from a filament in the ventral root which contained the motor axon. *A*, A depolarizing current pulse evoked a full-sized action potential (*peak not shown*) which was actively propagated along the axon (*arrow*). *B–F*, Responses to stimulation of the afferent nerve flexor digitorum longus, at a strength which produced a maximum group I volley. They illustrate variable partial dendritic spikes (*B,E*) which both failed to be elicited (*C*) or at other times evoked full-sized orthodromic impulses (*D,F*). Only the latter were propagated actively along the axon. (Modified from Eccles et al. 1958)

of a neuron is severed the cell body exhibits a profound series of morphological and chemical changes, often referred to as the "axon reaction" and generally including chromatolysis (see reviews by Grafstein 1975; Lieberman 1971). Concomitant with these changes, neuronal membrane properties are also altered; indeed, the first comprehensive intracellular study of chromatolyzed motoneurons by Eccles et al. (1958) clearly demonstrated that the selective excitability of the IS is diminished or even lost. Their conclusion was based upon findings, illustrated in part in Fig. 1, that in chromatolyzed motoneurons (a) there are exceptionally excitable patches of SD membrane, and (b) the IS spike is smaller in amplitude than in normal cells. The added excitability was manifest as partial spikes (Fig. 1B, E) which were apparently generated at a number of different loci and exhibited a continually varying degree of excitability. In contrast to full-sized action potentials, these partial spikes were not actively propagated along the motor axon (compare Fig. 1A and B). They were faster than the excitatory postsynaptic potentials (EPSPs) which triggered them (compare Fig. 1B and C), and they were sometimes sufficient to evoke an orthodromic IS-SD spike (Fig. 1D–F). The contrast between this effect and the decreased IS excitability, which is manifested by a failure of antidromic invasion, suggests that the chromatolyzed neuron is in a dedifferentiated state where regions of voltage-sensitive channels are not sharply delineated and might provide a useful model for studying mechanisms underlying membrane organization and remodeling.

In recent years I have been analyzing the axon reaction of the teleost Mauthner (M-) cell in collaboration with Titmus, Wood, and Zottoli (Faber and Zottoli

1981; Titmus and Faber 1982, 1983; Wood and Faber 1982, 1983; Zottoli and Faber 1977, 1978), with a number of specific goals in mind. First, we had the opportunity of correlating the physiological and morphological descriptions of the axon reaction of an identified vertebrate neuron which is contained entirely within the central nervous system and undergoes, at best, limited axonal regeneration (Zottoli 1981), as is true for most central neurons. These factors can be contrasted with those relevant to motoneurons, which maintain trophic interactions with a nonneuronal target cell type, skeletal muscle fibers, and whose axons in the periphery do regenerate. Indeed, we have found that while chromatolysis is considered to be a transient state for motoneurons and presumably disappears when functional contacts are reestablished at muscle endplates, the axon reaction of the M cell is instead irreversible. Secondly, we have been able to ask whether physiological changes in membrane properties associated with the axon reaction are linked to either chromatolysis or axonal sprouting, by following individual cells with both morphological and physiological techniques. Thirdly, the M-cell is an extreme example of a neuron with electrical excitability restricted to the IS-axon hillock region, as numerous investigators have been unable to demonstrate that the SD membranes are electrogenic (Diamond 1968; Fukami et al. 1965; Furshpan and Furukawa 1962; D.S. Faber, unpublished observations). Thus it seemed that if axotomy could lead to a significant change in the cellular distribution of excitable channels in the M cell, this neuron, which is ultimately accessible for biochemical analysis (Koenig 1978), would provide a superior model for investigating mechanisms underlying membrane specialization and plasticity. The results reviewed here indicate that this is indeed the case: starting at 20–40 days after axotomy by spinal cord transection, the M-cell can produce SD spikes comparable to those seen in normal motoneurons, and the spread of voltage-sensitive channels to the SD region is associated with a decreased excitability of the IS (Faber and Zottoli 1981; M.J. Titmus, D.S. Faber, and S.J. Zottoli, unpublished observations). The morphological and physiological components of this reaction persist for at least a few hundred days and do not appear to be reversible, in contrast to the transient responses of motoneurons and certain identified invertebrate neurons (Kuwada and Wine 1981; Mendell et al. 1976; Kuno and Llinás 1970a, b; Kuno et al. 1974a, b).

Methods

Common goldfish, *Carassius auratus*, of 12.5–17.5 cm body length were used. The M cell is a paired medullary neuron whose axon crosses the midline and courses the length of the spinal cord. The results reviewed here are based upon spinal transections which were performed, during anesthesia, at the high cervical level and left a proximal stump of the M-axon approximately 7–9 mm long. More recent studies done in collaboration with Zottoli indicate that a localized lesion of the M-axon alone, using a technique he has recently perfected (Zottoli 1981), has the same effects. During the initial weeks following surgery, aquarium water was treated with potassium dichromate in order to minimize the likelihood of infection. The fish were paralyzed caudal to the cut; feeding and locomotion involved both head and pectoral fin movements. Some fish eventually regained swimming ability, presumably due to partial spinal cord regeneration, but all tests indicate that there is no functional regeneration of the M-axon caudal to the transection. The fish were maintained at 60°–62 °F throughout these experiments. For the morphological studies the fish were perfused with 2.5% glutaraldehyde and 1% paraformaldehyde in 0.10 M sodium cacodylate at a pH of 7.2, and the brains were embedded in Spurr. Serial cross sections 3–5 μm thick were cut up to the level of the M cell. Thin sections were also prepared for electron microscopy.

The electrophysiological experiments involved intra- and extracellular recordings with KCl-filled microelectrodes, from the M-cell axon cap, soma, and lateral dendrite. The axon hillock of the M-cell was typically identified by the large all-or-none negative field potential recorded extracellularly when that region was activated antidromically by stimulation of the M-axon in the spinal cord or orthodromically by stimulation of the ipsilateral eighth nerve (Faber and Korn 1978; Furshpan and Furukawa 1962). All subsequent recording sites were referenced to that position. In this series of experiments, both stimulations were routinely employed. Since in our other experiments antidromic stimulation of the M-axon is typically produced with an electrode placed on the caudal spinal cord, it was first necessary to perform controls in which the stimulating electrode was placed on the cord surface at a high cervical level. As expected, the characteristics of antidromic invasion of the M-cell IS-axon hillock region were the same as determined with the more caudal stimulating sites. In addition, a preliminary series of acute experiments indicated that the safety factor for antidromic invasion was not altered simply by truncation of the axon, at least not for the first few hours after spinal transection. Finally, M-cell input impedance was measured with two independent intracellular microelectrodes located approximately 50 μm lateral to the axon hillock; one electrode was for current injection and the other was for voltage recording.

Results

Morphological Observations

The structural changes in the M-cell following axotomy at the high spinal level are quite similar to those described for other vertebrate neurons, with the exception that the axon reaction of the M-cell has a greatly expanded time course. As shown in Fig. 2A, the normal M-cell soma generally has clumps of Nissl substance (corresponding to "free" ribosomes and rough endoplasmic reticulum, RER, at the electron microscopic level) distributed fairly evenly throughout its cytoplasm, and the nuclear envelope has a relatively smooth oval outline. Chromatolysis and the other aspects of the axon reaction cannot be detected until about 16–20 days after axotomy, and are clearly manifest only by 40–60 days. At the light microscopic level, the chromatolysis is seen as a dispersion of the Nissl substance, which becomes concentrated in a perinuclear ring (Fig. 2B) of ribosomes and RER; the soma is often swollen, and the nuclear envelope is no longer smooth, but has marked infoldings (Faber and Zottoli 1981; M.R. Wood, S.J. Zottoli, and D.S. Faber, unpublished observations). In addition, recent studies with Wood have

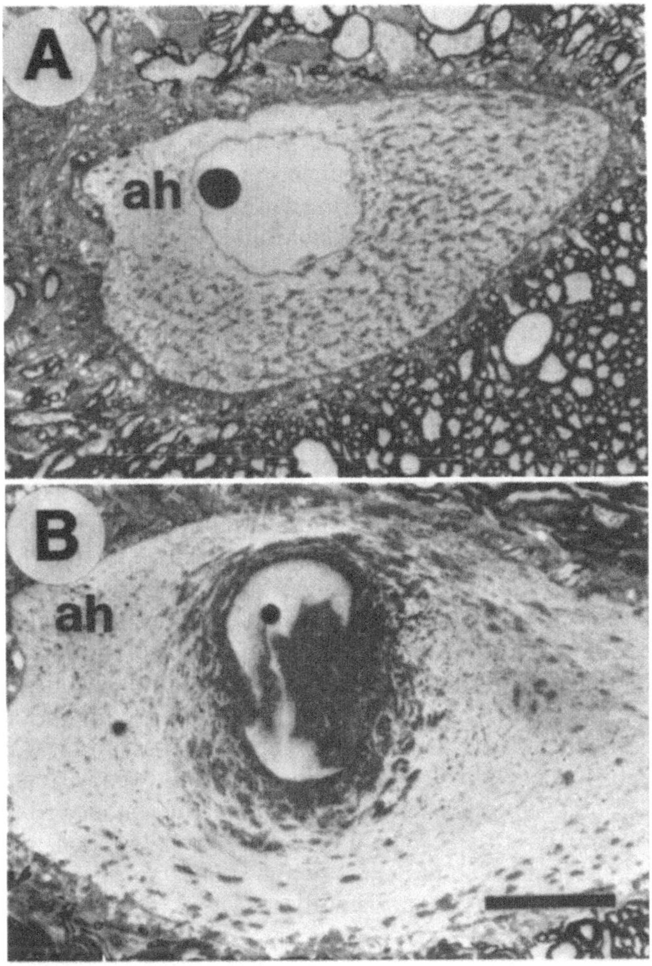

Fig. 2A,B. Morphological aspects of the Mauthner (M) cell's axon reaction. Photomicrographs of 3-μm sections cut at the level of the M-cell nucleolus and stained with toluidine blue. *A* In this right M-cell of a control preparation, the nuclear envelope is relatively smooth and situated in the dorsal half of the soma. Small clumps of Nissl substance are distributed evenly throughout most of the soma, except near the axon hillock (*ah*). *B* An axotomized right M-cell, obtained 70 days after spinal transection. The nucleus exhibits a marked infolding of its membranes and there is a perinuclear buildup of basophilic material. The soma appears swollen and chromatolyzed. Dorsal is up and lateral to the right in both *A* and *B*; the *scale* is 25 μm. (Modified from Faber and Zottoli 1981)

shown that there is a reactive deafferentation of the M-cell soma, with a selective loss of terminals bearing features associated with excitatory synapses (Wood and Faber 1982). This deafferentation is a slow ongoing process, and may be indirectly related to a patchy clustering of rough endoplasmic reticulum near the somatic membrane and to an increased frequency of granular-coated vesicles in the M-cell in close association with normal-looking terminal contacts (Wood and Faber 1983). Indeed, we suggest that following axotomy the M-cell is placed into a state characterized by continuous membrane modifications which maintain the integrity of some input synapses while allowing disjunction of others and, as discussed in the following section, result in the redistribution of voltage-sensitive channels. It remains to be seen whether these changes are due to axotomy itself or are rather consequences of changes in metabolic organization and increased insertion of specific proteins into the membrane elsewhere, such as at the IS, as the neuron shifts to a condition favoring axonal outgrowth. Our preliminary results do, however, indicate that although chromatolysis, altered excitability, and sprouting are components of the M-cell's axon reaction which are not consistently detected, reactive deafferentation is always observed.

One of the advantages of the M-cell model for studies of the axon reaction is that it is possible to cause lesions in the axon at defined distances from the cell's soma. We initially performed spinal transections at the midbody level and found no signs of an axon reaction (Zottoli and Faber 1978). In view of our subsequent observations of a marked and persistent, but delayed, response following lesions at 7–9 mm from the soma (opercular level), S.J. Zottoli (unpublished observations) chose to look more carefully at the effects of varying the lesion site, using quantitative morphological measures for the threshold and peak intensity of the response. He confirmed our earlier findings with midbody level cuts and found that the times of occurrence of both parameters increased as the level of axotomy was shifted from the brain stem (4.7 mm from the cell soma) to the posterior edge of the opercula (7–9 mm), and to a distance equal to one-quarter body length (~18.5 mm). Specifically, a response could be detected at about 10, 15–20, and 40 days, respectively, for the three sites, and it was typically maximal about 30–40 days later. These results are certainly consistent with notion that the axon reaction depends upon the length of the intact proximal stump (Kuwada and Wine 1981; Lieberman 1971) and may be related to the fact that the different physiological components of the spinal motoneuron response to axotomy have not been observed consistently in various laboratories.

Alterations in M-cell Excitability

Two clear results emerged from an analysis of M-cell action potentials following axotomy. The first is that there is a significant increase in spike height, which in an initial experimental series went from an average of about 43 mV to 60 mV, for cells studied over a time period of 16–209 days after spinal transection (Faber and Zottoli 1981). Representative examples of control and experimental antidromic impulses recorded from the M-cell soma are shown in Fig. 3. The control in Fig. 3A is similar to that recorded routinely in our laboratory; numerous lines of evidence have demonstrated that this is an IS spike, generated by the IS-axon hillock region and conducted electrotonically to the soma and dendrites (Diamond 1968; Faber and Korn 1978; Furshpan and Furukawa 1962). The responses in Fig. 3B1 and B2 encompass the range of times after axotomy during which we have followed changes in M-cell excitability. Both action potentials are appreciably greater than 60 mV in amplitude and are larger than those ever recorded in controls. We initially thought there might be a tendency for spike height to increase gradually throughout that time period (Faber and Zottoli 1981), but Tit-

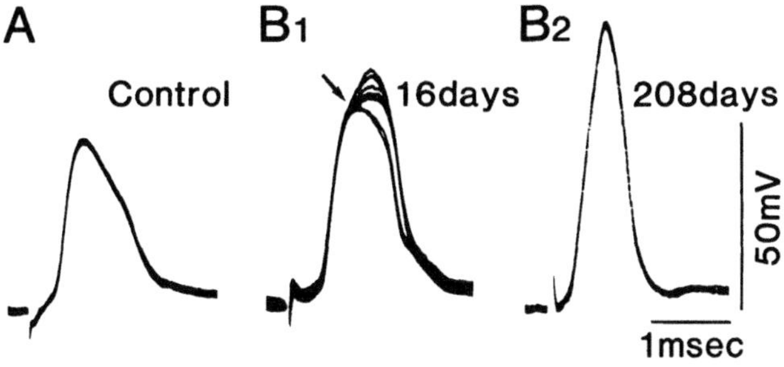

Fig. 3A,B. Intracellular recordings from the Mauthner (M) cell soma illustrating enhanced soma-dendritic excitability following axotomy. *A* Control antidromic spike, 45 mV in amplitude. B_1, B_2 Antidromic impulses recorded from two different preparations 16 and 208 days respectively after the M-axon had been severed by spinal cord transection approximately 8 mm posterior to the level of the cell's soma. The spike heights of 64 (B_1) and 76 (B_2) mV are significantly greater than those recorded in controls, and the *superimposed traces* in B_1 demonstrate that the added component fluctuated in amplitude, with a clear inflection point (*arrow*) appearing at a spike amplitude of about 40 mV. Two or more responses are superimposed in all traces. (Modified from Faber and Zottoli 1981)

mus has now looked at more than 50 additional axotomized cells and there is no such correlation. Rather, it seems that the increase in spike height occurs in a more or less all-or-none manner at about 20–30 days. In addition, this shift is linked to the presence of a chromatolytic response: those cells which do not exhibit an augmentation in spike height are also those which do not undergo chromatolysis (M.J. Titmus, D.S. Faber, and S.J. Zottoli, unpublished observations).

The waveform of the antidromic or directly evoked action potential recorded from the M-cell soma following axotomy was quite variable, and a number of indications demonstrated that the added component could be distinguished from the IS spike which triggered it. For example, as shown in Fig. 3B1, there were often spontaneous fluctuations in spike height, with the underlying IS spike remaining stable and roughly equal in amplitude to that of controls. The added labile component was also quite sensitive to repetitive stimulation at frequencies of 20 Hz or greater, and it contributed only a small depolarization to the falling phase of impulses recorded simultaneously from the M-axon. Thus, as discussed in more detail in subsequent sections, it seems appropriate to classify this component as an SD spike.

The second clear effect of axotomy on M-cell action potential generation was that the excitability of the IS and axon hillock membranes was reduced. In a series of experiments where a clear IS-SD break could be detected after axotomy, M.J. Titmus (unpublished observations) found that the amplitude of the IS spike averaged about 82% of that in controls, while its maximum rate of rise decreased from about 150 V/s to around 94 V/s. In the same cells, the SD spike added 30–50 mV to the amplitude of the antidromic action potential, and its maximum rate of rise was approximately 200 V/s (M.J. Titmus, D.S. Faber, and S.J. Zottoli, unpublished observations). Another indication of the reduced excitability of the IS and axon hillock was that the antidromic axon impulse frequently failed to invade

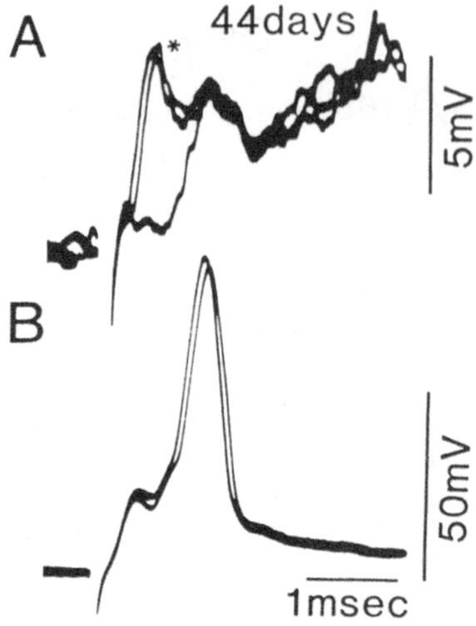

Fig. 4A,B. Reduced excitability of the initial segment following axotomy. Intracellular recordings of Mauthner (M) cell somatic responses to surface stimulation of the proximal spinal cord, 44 days after axotomy by spinal transection. The *superimposed traces* in *A* were obtained with stimulus strength at threshold for antidromic activation of the M-axon. An all-or-none axon spike (*) is apparently incapable of triggering an action potential in the axon hillock-initial segment region. This axon spike is followed by a complex excitatory postsynaptic potential (EPSP). In *B* stimulus strength was increased and the EPSP evoked an orthodromic spike which was greater than 60 mV in amplitude, i.e., significantly larger than normal. Superimposed records of two or more responses are shown in each case.

this region in axotomized M-cells. That is, the safety factor for impulse propagation was reduced, as is shown in Fig. 4A for a cell 44 days after axotomy. Spinal cord stimulation at threshold for antidromic activation of the M-axon produced a small axon spike which was conducted passively to the soma. This representation of the axon spike was similar in amplitude to that recorded from the soma in normal cells, but it failed to initiate an IS spike. There may be a tendency to consider such a failure of antidromic invasion as being indicative of a generalized decrease in excitability (Baker et al. 1981; Farel 1980), but the records in Fig. 4B, which are from the same experiment, clearly indicate that this need not be the case. The rostral spinal stimulus also activated an excitatory synaptic input to the M-cell, and when the stimulus strength was increased an orthodromic action potential was evoked. The amplitude of the spike was significantly greater than 50 mV, i.e., much larger than normal. That is, this neuron exhibited a decreased IS excitability and enhanced, or novel, SD excitability.

Constancy of Passive M-cell Membrane Properties

Before we could conclusively attribute the added excitability to an SD spike, additional information was needed, including an assessment of the effects of axotomy on passive M-cell membrane properties. It was clear that resting potential was unaltered: this parameter averaged 78.62 ± 4.85 mV and 78.24 ± 2.76 mV in control and experimental cells, respectively (Faber and Zottoli 1981). In order to measure M-cell input resistance, simultaneous intracellular recordings with two

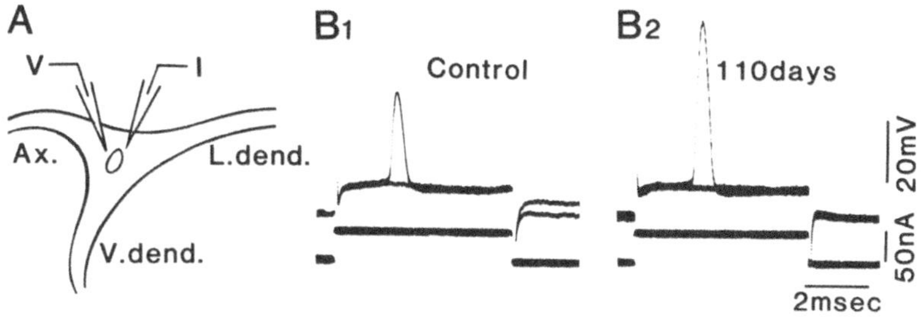

Fig. 5A,B. Mauthner (M) cell passive membrane properties and threshold are unchanged following axotomy. *A* Schematic design illustrating that two intracellular microelectrodes were used in these experiments, one for recording M cell voltage (*V*) and one for injection of constant current pulses (*I*). *L.dend.*, lateral dendrite; *V.dend.*, ventral dendrite. B_1, B_2 Sample recordings from control and axotomized (110 days after spinal transection) M-cells of electrotonic and spike potentials elicited by current pulses at threshold; in each case action potentials were elicited in one of the two superimposed records. Note that the magnitudes of the current pulses and the response amplitudes were the same for both cells; in contrast spike amplitude in the axotomized cell was almost twice that of the control. (Modified from Faber and Zottoli 1981)

microelectrodes were required, as indicated in Fig. 5A. This requirement stems from the fact that control M-cell input resistance is typically in the range of 120–200 KΩ (reviewed in Faber and Korn 1978, 1982; Faber and Zottoli 1981), with our most recent studies yielding values around 170–200 KΩ. The records in Fig. 5B1 and B2 have been chosen to illustrate the finding that axotomy did not alter input resistance. The same stimulus current was used for both and the passive electrotonic potentials recorded from the control (B1) and axotomized (B2) cells are essentially equivalent. In general, input resistance was measured for a voltage range within ±10–20 mV of resting potential; it averaged about 175 KΩ for both normal and axotomized M-cells (M.J. Titmus, D.S. Faber, and S.J. Zottoli, unpublished observations). Thus the increased spike height is not a consequence of changes in passive membrane properties.

The records in Fig. 5B1 and B2 also show a slightly puzzling observation. The thresholds for IS spike initiation with direct stimulation were the same in both cases illustrated and for a large population of normal and axotomized neurons. This observation would seem to contradict the finding that axotomy results in a lowered safety factor for antidromic invasion of the IS and axon hillock regions. Two possible explanations of this apparent discrepancy are that there is a change in axon spike parameters, which we have not detected yet, or that the electrical couplings between axon and IS (antidromic stimulus) and/or soma and IS (direct stimulus) have been altered. Recent observations with M. Wood (unpublished) indicate that following axotomy the size and geometry of the IS change in a manner which might support these observations, particularly if the focal point of the spike initiation site (Funch and Faber 1982) is shifted in the direction of the axon hillock.

Extracellular Fields Associated with M-cell Action Potentials

The spatial profile of antidromic spike height recorded intracellularly from the M-cell soma and lateral dendrite suggested that the SD spike was generated at the

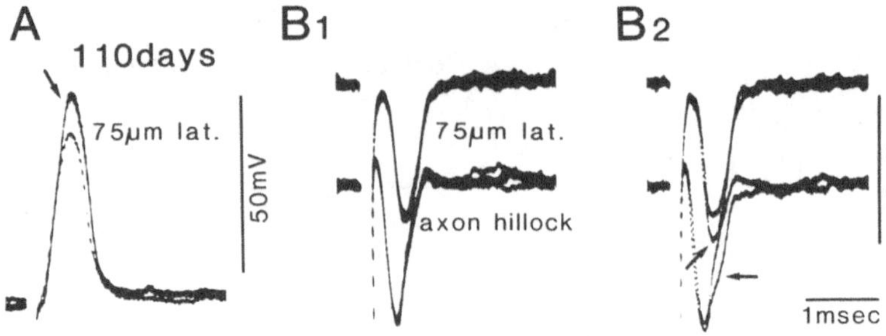

Fig. 6A,B. Evidence that the enhanced electrical excitability following Mauthner (M) cell axotomy is localized to the normally inexcitable soma-dendritic membranes. *A* Intracellular recordings from the proximal lateral dendrite of a M-cell 110 days after axotomy by spinal transection. The recording site was 75 μm lateral to the center of the axon cap, as judged by the amplitudes of extracellular field potentials. *Superimposed traces* illustrate spontaneous fluctuations in antidromic spike height. The spike component added to the underlying impulse generated by the initial segment-axon hillock region is indicated by the *arrow*; it is presumably generated at or near the recording site. B_1, B_2 Simultaneous extracellular recordings of M-cell antidromic spikes at the indicated sites. The responses in B_1 correspond to the smaller antidromic impulse illustrated in *A*, and demonstrate a delay in impulse propagation laterally from the axon hillock. In B_2 the response amplitudes fluctuate, with the larger extracellular negativity (*first arrow*) corresponding to the larger intracellular record of *A*. The field recorded simultaneously in the axon hillock exhibits a shoulder on its falling phase (*second arrow*). This delayed additional regativity is smaller than the extra component seen laterally, suggesting that the spike generator is closer to the lateral recording site. (Modified from Faber and Zottoli 1981)

soma and proximal lateral dendrite, and more distal sources were not detected. It therefore remained to demonstrate that the active membrane was indeed in the indicated regions and was spatially distinct from the IS spike generator. Comparison of extracellular field potentials recorded during spontaneous fluctuations in spike height (e.g., Fig. 6A) confirmed our hypothesis, as shown in Fig. 6B1 and B2. Specifically, the extracellular field recorded lateral to the IS-axon hillock region exhibits a delayed negativity, suggesting active propagation laterally (Fig. 6B1). In addition, the added labile spike component produces a negative extracellular field potential which is maximal at the lateral recording site, which is presumably closer to the spike generator (Faber and Zottoli 1981).

Ionic Basis of the SD Spike

There is evidence that the somatic spikes which appear in certain invertebrate neurons following axotomy involve voltage-sensitive Na^+ currents (Goodman and Heitler 1979; Kuwada 1981), but comparable studies have not been conducted with vertebrate cells. Although it is not possible completely to control the extracellular medium bathing the M cell in situ, since the neuron is about 1.5 mm below the medullary surface, we have used both superfusion and iontophoresis of specific blocking agents to characterize the ionic basis of the IS spike (Kaars and Faber 1981; Faber and Korn 1978; M.J. Titmus, D.S. Faber, and S.J. Zottoli, unpublished observations). Our results can be summarized as follows: (a) the inward current of the action potential is carried by Na^+ ions, as the spike is totally abolished by tetrodotoxin (TTX); (b) spike repolarization involves a voltage-sensitive K^+ current which is blocked by tetraethyl ammonium (TEA) and the aminopyridines; and (c) evidence that topical Co^{2+} applications reduce the duration of the spike recorded in the presence of TEA indicate that there is also an inward Ca^{2+} current, which cannot produce a regenerative spike potential alone (i.e., in the presence of TTX and TEA). More recently, we have applied the same techniques for the analysis of the SD spike produced by axotomy (Titmus and Faber 1982) and have focused our efforts on the effects of TTX; the results indicate that the SD spike is Na^+ dependent, as it is also abolished by TTX. In addition, we have confirmed the localization of this spike component to the soma and proximal lateral dendrite by studying the time course of the TTX blocking action when it is applied iontophoretically within the axon cap and external to it at defined distances lateral to the axon hillock. Therefore, it appears that axotomy results in a decrease in the maximum Na^+ conductance (g_{Na}) in the IS and axon hillock and at the same time leads to the appearance of a significant voltage-sensitive g_{Na} in the SD membrane.

Discussion

M-cell Membrane Reorganization Underlying SD Spiking

Given our evidence suggesting that following axotomy voltage-sensitive Na^+ channels appear in the SD membrane and at the same time become less concentrated in the IS and axon hillock region, it is appropriate to consider whether this effect is due to a redistribution of a normal complement of channels or rather to the synthesis and/or insertion of a greatly increased number of channels. That is, is a shift in channel distribution alone sufficient to explain the observed changes? Certainly, spike height would increase if voltage-sensitive Na^+ channels were located closer to the recording site. Rough estimates of Na^+ channel densities may provide some insight into this problem. As mentioned before, the somatic input resistance, which provides a resistive load on the Na^+ channels, is no more than 150–200 KΩ. If we take 175 KΩ as a representative value and assume that spike amplitude reaches approximately one-half of the Na^+ driving force, the peak sodium conductance is about $5–6 \times 10^{-6}$ S and, assuming that the single-channel conductance is 14 pS (Stampfli 1981), the number of voltage-sensitive Na^+ channels in the SD membrane must be about 0.4×10^6. If all of these channels had migrated from the IS and axon hillock (2000 μm^2 surface area), their density alone in that region would have been about $200/\mu m^2$. And, since the maximum rate of rise of the IS spike is reduced by about 33% after axotomy, the original Na^+ channel density would have been more in the range of $600/\mu m^2$, which is similar to present estimates for nodes of Ranvier (Stampfli 1981), where the channels are clearly densely packed. Thus after axotomy the total number of functional Na^+ channels in the region encompassing the IS to the soma and proximal lateral dendrite may not be appreciably higher than under normal conditions. The increased excitability of the SD membrane may also reflect the loss of axonal sinks for Na^+ channels, as normally about 10^6 channels are concentrated at distinct nodal sites distal to the lesion (Funch and Faber 1983). The interesting molecular problem, therefore, is whether Na^+ channels are still transported along the axon or whether vectoral transport shifts to favor channel insertion at the level of the cell body.

It should be stressed that the ability of the M-cell soma to produce a sizable action potential is remarkable, given the low input resistance of this neuron and its low specific membrane resistance (at most 200–300 $\Omega \, cm^2$ – reviewed in Faber and Korn 1978; see Crank and Faber 1983). The estimated number of Na^+ channels (0.4×10^6) corresponds to a density at the soma of about 20–40 channels per square micrometer, which does not seem high until one takes into consideration that much of the somatic membrane is specialized for postsynaptic functions and unavailable for impulse generation. These considerations suggest that although the induced SD spike is generally restricted to the soma and proximal lateral dendrite, voltage-sensitve Na^+ channels may have a patchy distribution over that region, similar to that proposed previously by Eccles et al. (1958) for the axotomized spinal motoneuron. This patchy distribution, with localized clusters of high concentrations of Na^+ channels, might underly our observations that the SD spike amplitude is labile and varies spontaneously. It would appear that these clusters are not functionally stable; regulation of their voltage sensitivity might well depend upon local chemical signals, since all evidence suggests that the soma is isopotential. In this context it is interesting to note that somatic action potentials in axotomized invertebrate neurons (Goodman and Heitler 1979; Kuwada and Wine 1981), which do not typically have synaptic inputs, and postsynaptic specializations, on the soma, are not labile.

It also seems that there is an as yet ill-defined connection between the distributions of voltage sensitivity and excitatory synapses in axotomized neurons. Reactive deafferentation typically involves only excitatory synapses, as is the case with the M-cell; there is only one report of a loss of inhibition onto axotomized neurons (Baker et al. 1981), and in that instance there is no evidence for enhanced excitability. Therefore, future studies might be focused on the differential properties of these membranes, particularly from the perspective of their potential for plasticity.

The Persistent Nature of the M Cell's Axon Reaction

Kuwada and Wine (1981) reported that axotomy-induced changes in the membrane properties of crayfish central neurons are transient, regardless of whether the cells regenerate and reinnervate their target cells. In contrast, Kuno et al. (1974a, b) found that the original electrical properties they studied in axotomized cat spinal motoneurons (conduction velocity, spike afterpotential duration, and overshoot) were only restored if regeneration and reinnervation of muscle by the motor nerve did occur. However, this reversal of the effects of axotomy on a given neuron was not dependent upon whether that neuron had successfully reinnervated its target cell, as long as the nerve had established functional contacts. Finally, synaptic transmission onto axotomized sympathetic ganglion cells is also only restored when the damaged nerve reestablishes functional output synapses (Purves 1975). There therefore seems to be a distinction between the requirements for reversing the reactions of vertebrate and invertebrate cells. The persistence of the M-cell's axon reaction would then be a consequence of its failure to reestablish functional output connections distal to the level of spinal transection. This apparent correlation is strengthened by recent observations suggesting that the M-axon's output synapses rostral to the lesion site are partially dysfunctional (Titmus and Faber 1983). Thus trophic influences from spared synapses may be minimized when lesions are caused in the axon 7–9 mm distal to the M-cell soma, and there would be no signal competing with that which triggers and maintains the axon reaction.

Chromatolysis and Physiological Consequences of Axotomy

Electrophysiologists have had a tendency to treat axotomized neurons as being chromatolyzed (Eccles et al. 1958; Kuno and Llinás 1970a, b). However, it is clear that chromatolysis is not always a part of the axon reaction; in our experiments it is not uncommon for only one of the paired M-cells to exhibit chromatolysis and changes in excitability following spinal transection. On the other hand, we find that reactive deafferentation is the only consistent component of the axon reaction. These distinctions between the different aspects of the axon reaction may explain why some authors have observed partial dendritic spikes in axotomized spinal motoneurons (Eccles et al. 1958; Farel 1978; Kuno and Llinás 1970a) and others have not (Farel 1980; Mendell et al. 1976; Kuno et al. 1974a). In general, the ability to obtain such correlations directly with the M-cell model illustrates again the advantages it offers for cellular studies of this nature.

Perspectives

This brief review of some of our recent work should serve to emphasize the fact that the original contribution by Eccles et al. (1958) to the literature on chromatolyzed neurons still serves as the reference point for work on excitability changes following axotomy. In our case, we have been able to provide a detailed description of the axon reaction of an identifiable central vertebrate neuron, extending some of the work by Eccles and his collaborators and obtaining correlations between structural and physiological changes. As many of Eccles' major early contributions were in the area of cellular physiology, it is encouraging to think that as we move into a new era of cellular neurobiology, the problems he and his colleagues posed can be addressed anew. Although he taught us to always look ahead and to think about the next experiment, I remain grateful for the foundation he provided many of us and look back upon it with respect and affection.

Acknowledgments. I am greatful to M. Titmus, M. Wood, and S. Zottoli for allowing me to share many of their still unpublished data.

References

Araki T, Otani T (1955) Response of single motoneurones to direct stimulation in toads spinal cord. J Neurophysiol 18: 472–485

Baker R, Delgado-Carcia J, McCrea R (1981) Morphological and physiological effects of axotomy on cat abducent motoneurons. In: Precht W, Flohr H (eds) Lesion-induced neuronal plasticity in sensorimotor systems. Springer, Berlin Heidelberg New York, pp 51–63

Brock LG, Coombs JS, Eccles JC (1952) The recording of potentials from motoneurones with an intracellular electrode. J Physiol 117: 431–460

Brock LG, Coombs JS, Eccles JC (1953) Intracellular recording from antidromically activated motoneurones. J Physiol 122: 429–461

Coombs JS, Curtis DR, Eccles JC (1957a) The interpretation of spike potentials of motoneurones. J Physiol 139: 198–231

Coombs JS, Curtis DR, Eccles JC (1957b) The generation of impulses in motoneurones. J Physiol 139: 232–250

Crank WD, Faber DS (1983) Passive membrane properties of the Mauthner cell. Soc Neurosci Abst 9: 234

Diamond J (1968) The activation and distribution of GABA and L-glutamate receptors on goldfish Mauthner neurones: an analysis of dendritic remote inhibition. J Physiol 195: 669–723

Eccles JC, Libet B, Young RR (1958) The behaviour of chromatolysed motoneurones studied by intracellular recording. J Physiol 143: 11–40

Faber DS, Korn H (1978) Electrophysiology of the Mauthner cell: basic properties, synaptic mechanisms and associated networks. In: Faber DS, Korn H (eds) Neurobiology of the Mauthner cell. Raven, New York, pp 47–131

Faber DS, Korn H (1982) Transmission at a central inhibitory synapse. I. Magnitude of unitary postsynaptic conductance change and kinetics of channel activation. J Neurophysiol 48: 654–678

Faber DS, Zottoli SF (1981) Axotomy-induced changes in cell structure and membrane excitability are sustained in a vertebrate central neuron. Brain Res 223: 436–443

Farel PB (1978) Reflex activity of regenerating frog spinal motoneurons. Brain Res 158: 331–341

Farel PB (1980) Selective synaptic changes following spinal motoneuron axotomy. Brain Res 189: 67–77

Fatt P (1957) Sequents of events in synaptic activation of a motoneurone. J Neurophysiol 20: 61–80

Fukami Y, Kurukawa T, Asada Y (1965) Excitability changes of the Mauthner cell during collateral inhibition. J Gen Physiol 48: 581–600

Funch PG, Faber DS (1982) Action-potential propagation and orthodromic impulse initiation in Mauthner axon. J Neurophysiol 47: 1214–1231

Funch PG, Faber DS (1983) Pharmacological evidence for saltatory conduction along the goldfish Mauthner axon. Soc Neurosci Abst 9: 21

Fuortes MGF, Frank K, Becker MC (1957) Steps in the production of motoneuron spikes. J Gen Physiol 40: 735–752

Furshpan EJ, Furukawa T (1962) Intracellular and extracellular responses of the several regions of the Mauthner cell of the goldfish. J Neurophysiol 25: 732–771

Goodman CS, Heitler WJ (1979) Electrical properties of insect neurones with spiking and nonspiking somata: normal, axotomized, and colchicine-treated neurones. J Exp Biol 83: 95–121

Grafstein B (1975) The nerve cell body response to axotomy. Exp Neurol 48: 32–51

Kaars C, Faber DS (1981) Myelinated central vertebrate axon lacks voltage-sensitive potassium conductance. Science 212: 1063–1065

Koenig E (1978) Molecular biology of the Mauthner axon. In: Faber DS, Korn H (eds) Neurobiology of the Mauthner cell. Raven, New York, pp 167–182

Kuno M, Llinás R (1970a) Enhancement of synaptic transmission by dendritic potentials in chromatolysed motoneurones of the cat. J Physiol 210: 807–821

Kuno M, Llinás R (1970b) Alterations of synaptic action in chromatolysed motoneurones of the cat. J Physiol 210: 823–838

Kuno M, Miyata Y, Munoz-Martinez J (1974a) Differential reaction of fast and slow α-motoneurones to axotomy. J Physiol 240: 725–739

Kuno M, Miyata Y, Munoz-Martinez J (1974b) Properties of fast and slow alpha-motoneurones following motor reinnervation. J Physiol 242: 273–288

Kuwada JY (1981) Ionic and metabolic dependence of axotomy-induced somatic membrane changes. J Physiol 317: 463–473

Kuwada JY, Wine JJ (1981) Transient, axotomy-induced changes in the membrane properties of crayfish central neurones. J Physiol 317: 435–461

Lieberman AR (1971) The axon reaction: a review of the principal features of perikaryal responses to axon injury. Int Rev Neurobiol 14: 50–125
Mendell LM, Munson JB, Scott JG (1976) Alterations of synapses on axotomized motoneurones. J Physiol 255: 67–79
Purves D (1975) Functional and structural changes of mammalian sympathetic neurones following interruption of their axons. J Physiol 252: 429–463
Stampfli R (1981) Overview of studies on the physiology of conduction in myelinated nerve fibers. In: Waxman SG, Ritchie JM (eds) Demyelinating diseases: basic and clinical electrophysiology. Raven, New York, pp 11–23 (Advances of neurology, vol 31)
Titmus MJ, Faber DS (1982) Enhanced excitability following axotomy of a central vertebrate neuron. Soc Neurosci Abst 8: 414
Titmus MJ, Faber DS (1983) Altered transmission at spared synapses after Mauthner-cell axotomy. Soc Neurosci Abst 9: 603
Wood MR, Faber DS (1982) Reactive deafferentation in the teleost Mauthner cell following axotomy. Soc Neurosci Abst 8: 749
Wood MR, Faber DS (1983) Granular coated vesicles in the goldfish Mauthner cell following axotomy. Brain Res 273: 341–346
Zottoli SF (1981) Regeneration following selective axotomy of an identified vertebrate central neuron. Soc Neurosci Abst 7: 259
Zottoli SF, Faber DS (1977) Lack of functional Mauthner cell recovery following spinal cord regeneration in the goldfish. 27th IUPS congress, Paris, 2498 (abstract)
Zottoli SF, Faber DS (1978) Morphological and electrophysiological stability of the Mauthner cell following axotomy by spinal cord transection. Soc Neurosci Abst 4: 481

Posttetanic Potentiation, Presynaptic Inhibition, and the Modulation of the Free Ca^{2+} Level in the Presynaptic Terminals

R. KRETZ, E. SHAPIRO, J. CONNOR, and E.R. KANDEL, New York

Introduction

The importance of specific neuronal connections for the behavior of vertebrates has been repeatedly demonstrated, perhaps most impressively by Roger Sperry (1963, 1965). His work and that of Hubel, Wiesel (Hubel 1982; Wiesel 1982), and others suggests that at least rudimentary forms of these connections are determined by genetic and developmental processes and do not require experience for their formation. Similar conclusions have been reached from studies of the connections between neurons in invertebrates (Palka 1979; Horvitz 1982; Goodman and Pearson 1982). These findings pose an interesting question. If the formation of connections between most neurons in the nervous system is largely determined by genetic and developmental processes, how is behavior modified? Even brief behavioral modifications last minutes, and some learning may endure for years. How can patterns of neuronal activity produce a long-term modification in the function of a set of specified connections? A number of solutions to this question have been proposed. The two that have been explored most extensively are based on the idea that modifications of behavior stem from either the dynamic or the plastic properties of neurons.

The dynamic hypothesis derives from the physiological studies of Forbes (1922) and Lorente de No (1934, 1938), whose work suggested that neural activity could be sustained by the circulation or reverberation of impulses in a closed chain of interconnected self-reexciting neurons. This mechanism would not require anatomic change: it would not even require a functional change in the properties of neurons. However, as studies of the physiological properties of neural networks have advanced, it has become clear that what had appeared anatomically to be self-reexciting loops usually contain one or more inhibitory connections that would prevent reexcitation. In addition, memory was found to survive epileptic convulsions and other drastic experimental manipulations likely to interrupt the circulation of impulses in closed chains of neurons. As a result, the possibility that dynamic activity might provide a common neural basis of even short-term memory seems unlikely, although it has not yet been excluded (for an exceptional case see Tsukahara 1981).

The plasticity hypothesis derives from the writings of Cajal (1911) and two of his contemporaries, Lugaro (1899) and Tanzi (1893). An influential extension of this line of thought was proposed in 1948 by Jerzy Konorski. Thus Konorski wrote in 1948:

Experimental Brain Research, Suppl. 9
© Springer-Verlag Berlin · Heidelberg 1984

The application of a stimulus...leads to changes of a two-fold kind in the nervous system...the first property, by virtue of which the nerve cells *react* to the incoming impulses with a certain cycle of changes, we call *excitability*, and their changes arising in the centres because of this property we shall call *changes due to excitability*. The second property, by virtue of which certain permanent functional transformations arise in particular systems of neurons as a result of appropriate stimuli or their combination, we shall call *plasticity* and the corresponding changes *plastic changes*.

A current version of this hypothesis (for review see Kandel 1976) states that even though the anatomic connections between neurons develop according to a definite plan, the strength or effectiveness of the connections is not entirely predetermined. These and other properties of neurons may be altered by experience. The plasticity hypothesis predicts that neurons, and in particular their synapses, should prove capable of changing their functional properties as a result of altered activity.

In contrast to work based on the dynamic hypothesis, studies of the plastic capabilities of neurons have turned out to be surprisingly rewarding. A variety of experiments have now shown that many chemical synapses have a remarkable capacity for functional modification. These changes in the strength of synaptic connections may persist for periods ranging from minutes to hours (short-term plasticity) or from days to years (long-term plasticity). The short-term forms of synaptic plasticity have been studied most actively and have been found to fall into two major classes – homosynaptic and heterosynaptic – depending on how the change in synaptic strength is produced. Homosynaptic plasticity, which includes posttetanic potentiation and homosynaptic depression, results from activity within the neuron. Heterosynaptic plasticity, which includes presynaptic inhibition and presynaptic facilitation, results from activity in other neurons. Despite the fact that the mechanisms underlying the two forms of synaptic plasticity are widespread and found in both invertebrates and vertebrates, the details of their mechanisms are only now beginning to be understood.

The first question to be addressed was: Where in each of these two forms does the critical plastic change occur? What part of the synapse is altered when the synapse undergoes a change in effectiveness? Chemical synapses consist of three elements: a presynaptic element that releases transmitter, a postsynaptic element containing receptor molecules that respond to the transmitter, and a zone of apposition. Does plasticity involve a presynaptic change – an alteration in the amount of chemical transmitter released – or does it involve a postsynaptic change resulting from an alteration in the sensitivity of the receptor molecules? The first approach to this question was provided in 1954 when del Castillo and Katz (1954a, b) demonstrated quantal transmission at the nerve-muscle synapse of the frog, a finding subsequently generalized to all chemical synapses examined (Eccles 1964). As an action potential sweeps into the presynaptic terminal, it leads to the release not of individual transmitter molecules but of multimolecular packets each containing about 5000 molecules. The discovery of quantal transmission was important for two reasons. First, it gave a key insight into the nature of chemical transmission. Secondly, it established a set of techniques for analyzing the relative contribution to synaptic transmission of changes in transmitter release on the one hand and receptor sensitivity on the other. For example, del Castillo and Katz (1954a, b), Liley (1956a, b), and others performed quantal analysis of two instances of homosynaptic plasticity (depression and posttetanic potentiation) and found them both to be presynaptic. In 1961 Dudel and Kuffler (1961a, b) and subsequently Castellucci and Kandel (1976) applied a similar analysis to presynaptic inhibition and facilitation and found that these two forms of heterosynaptic plasticity also involved alterations in transmitter release. These several studies indicated that four commonly studied forms of short-term synaptic plasticity, involving homo- and heterosynaptic modes of activation, involved alteration in the release of transmitter by the presynaptic terminals. The sensitivity of the receptors was in all these cases unchanged.

What modifies the rate of transmitter release for periods of time ranging from minutes to hours? Is it a persistent alteration in membrane potential, a change in the configuration of the action potential, or a change in Na^+ influx? The specific conceptual advance that allowed an approach to this question came with the recognition by Katz and Miledi (1965) that the Ca^{2+} influx is the critical agent for triggering transmitter release. This insight in turn suggested the possibility that the various forms of plasticity could result from the modulation of the level of free Ca^{2+} in the terminals. This modulation could be achieved in at least two ways: by alterations in Ca^{2+} influx or by alterations in Ca^{2+} buffering within the terminals. Factors that affect Ca^{2+} influx include changes in K^+ current that alter the size and duration of presynaptic spikes, as well as changes in the activation or inactivation of the Ca^{2+} channels themselves (Katz and Miledi 1965; Klein and Kandel 1978, 1980; Takeuchi and Takeuchi 1962; Shapiro et al. 1980a, b; Klein et al. 1980; Llinás et al. 1981a, b). Changes in Ca^{2+} buffering in the presynaptic terminal could result from an alteration in the Ca^{2+}-buffering organelles in the presynaptic terminal (the endoplasmic reticulum and the mitochondria), in the proteins that bind Ca^{2+}, or in Ca^{2+} exchange across the membrane (Katz and Miledi 1968; Rahamimoff 1968; Alnaes and Rahamimoff 1975; Kretz et al. 1982; Charlton et al. 1982; Rahamimoff et al. 1980; Atwood et al. 1983; Misler and Hurlbut 1983).

Which, if any, of the possible mechanisms for altering the level of free Ca^{2+} apply for the major types of short-term plasticity? To explore these relationships optimally, it is necessary to combine analysis of Ca^{2+} influx as examined under voltage clamp with analysis of free Ca^{2+} levels as assayed with indicators of intracellular Ca^{2+} such as the Ca^{2+}-activated K^+ current or a calcium indicator dye such as Arsenazo III. In previous studies of synaptic plasticity in the vertebrate nervous system – including studies of presynaptic inhibition in the vertebrate spinal cord and posttetanic potentiation at the neuromuscular junction – it has not been possible to explore Ca^{2+} influx or Ca^{2+} buffering direcly because the presynaptic terminals are inaccessible to intracellular recording.

For example, a reduction in spike amplitude has been suggested as a mechanism for presynaptic inhibition (Eccles 1964; Takeuchi and Takeuchi 1966b; Baxter and Bittner 1981). Yet little is known in vertebrates of the relationship between the size of the action potential and Ca^{2+} influx (for discussion see Nicoll and Alger 1979). Similarly, it is important to know the normal resting potential of the presynaptic cell, and the activation-inactivation properties of presynaptic channels such as the early K^+ channel. If the resting potential is normally hyperpolarized, and early K^+ channels are present at the terminal, a presynaptic depolarization would inactivate K^+ channels to a greater extent than it would the Na^+ and Ca^{2+} channels. As a result, transmitter release might be expected to increase (Shapiro et al. 1980; Shimahara 1981). But, if the presynaptic membrane potential is normally depolarized, so that the early K^+ channels are inactivated (or are simply not present), the same modulatory depolarization might primarily inactivate the Na^+ and Ca^{2+} channels and thereby reduce transmitter output (Takeuchi and Takeuchi 1962; Charlton et al. 1982).

Details of the ionic mechanisms responsible for plasticity are also important, since presynaptic depolarization mediated by increased Cl^- conductance might decrease transmitter output by shunting the action potential (Eccles 1964; Kennedy et al. 1980), while a depolarization mediated by a decreased K^+ conductance might increase transmitter output by enhancing invasion of the terminal (Klein and Kandel 1978, 1980). Finally, as Ryall (1978) has pointed out in his review of presynaptic inhibition, it is also conceivable that in certain instances changes in presynaptic membrane potential may be epiphenomena, unrelated to changes in transmitter release; or, as was shown by Kretz et al. (1982), these changes may actually oppose plastic changes in transmitter release.

To examine the mechanisms of transmitter modulation, we have therefore used the technically advantageous nervous system of the invertebrate *Aplysia* to explore the role of changes in Ca^{2+} influx and Ca^{2+} buffering in five instances of plasticity:

1. Control of transmitter release by the resting membrane potential of the presynaptic neuron
2. Homosynaptic depression
3. Posttetanic potentiation
4. Presynaptic facilitation
5. Presynaptic inhibition.

In this paper we summarize results on three of these: control of transmitter release by resting membrane potential, posttetanic potentiation, and presynaptic inhibition. Using an identified cholinergic neuron and its identified follower cells, we found that we could voltage clamp the cell body of the presynaptic neuron and record transmembrane currents while controlling transmitter release from the terminals of the neuron with graded depolarizing commands. This fortunate circumstance permitted us to relate alterations in transmitter release to changes in the Ca^{2+} and K^+ currents in the presynaptic neuron. In addition to monitoring changes in Ca^{2+} influx, we have inferred changes in the buffering capability of the presynaptic neuron by using two indicators: (a) the Ca^{2+}-activated K^+ current; and (b) the Ca^{2+}-sensitive metallochromic dye Arsenazo III.

By these means, we found that control of transmitter release by membrane potential, posttetanic potentiation, and presynaptic inhibition all involve alterations of transmitter release due to modulation of the level of free Ca^{2+} in the terminal. However, the details of the mechanisms differ. Posttetanic potentiation primarily involves alterations in Ca^{2+} buffering; presynaptic inhibition and membrane potential primarily act through alterations in Ca^{2+} influx. We have also applied this approach to the other two forms of plasticity (homosynaptic depression and presynaptic facilitation) by studying other cells in the abdominal ganglion, and in our discussion we refer to these results as well in an attempt to provide an overall view into the mechanisms underlying these five interrelated types of synaptic plasticity in *Aplysia* neurons.

Results

The Cell body of Aplysia *Neurons as a Vantage Point for Studying the Presynaptic Terminal*

The membrane of the cell body of *Aplysia* neurons contains Ca^{2+} channels whose properties seem to resemble those of the presynaptic terminal (Geduldig and Gruener 1970; Stinnakre and Tauc 1973; Llinás et al. 1976, 1981a). Moreover, in certain cells changes in the Ca^{2+} current of the cell body have been shown to parallel the changes in transmitter release at the terminals (Klein and Kandel 1978; Shapiro et al. 1980a). In addition, in some of these neurons the presynaptic terminals controlling transmitter release appear to be so close to the cell body electrically that it is possible to modify transmitter release from the terminals by injecting current into the cell body (Shimahara and Tauc 1975; Shimahara and Peretz 1978; Shapiro et al. 1980a). These observations suggested to us that we might be able to examine the relationships between transmitter release and specific ionic currents of the presynaptic membrane by combining an analysis of ionic currents of the cell body of the presynaptic neuron, using voltage clamp analysis and pharmacologic blockade of specific ion channels, with an assay of transmitter release from the presynaptic cell, using intracellular recordings of the synaptic potential in the postsynaptic cell (Fig. 1). These combined techniques have allowed us to examine changes in specific ionic conductances associated with membrane potential control of transmitter release, posttetanic potentiation, and presynaptic inhibition. In addition, since the neurons of *Aplysia* are large and can be readily injected with Arsenazo III, we have also been able to monitor the level of free Ca^{2+} in the presynaptic neuron.

The presynaptic cell we examined, cell L10, is a multiaction cholinergic interneuron. Intracellular stimulation of this neuron produces excitation in some follower cells (R15 and RB cells), inhibition in others (L1–L6, LB_{VC}, LD_{HI}, etc.), and conjoint excitation-inhibition in yet a third group of cells (see Kandel et al. 1967; Wachtel and Kandel 1967; Kehoe 1972; Koester et al. 1974; Koester and Kandel 1977). Under conditions that optimize space clamp control, with the Na^{+} channels of L10 blocked with tetrodotoxin (TTX) and the axon cut off, graded depolarizing commands applied to the membrane of the presynaptic cell body under voltage clamp cause graded transmitter release from the terminals, as determined by the size of the synaptic potential recorded intracellularly from the postsynaptic cell (Fig. 2). By using other specific pharmacologic blocking agents [tetraethylammonium (TEA)] to block delayed rectification, 4-aminopyridine (4-AP) to block the early K^{+} channel, and Ba^{2+} substitution for Ca^{2+} to block the Ca^{2+}-activated K^{+} channel [Thompson 1977; Adams et al. 1980], we were able to examine the Ca^{2+} current of L10 directly during transmitter release and its plastic modulation by various types of homo- and heterosynaptic activation.

We found that the threshold for transmitter release was near the threshold for the Ca^{2+} current and, in the range of depolarizing voltages in which Ca^{2+} current

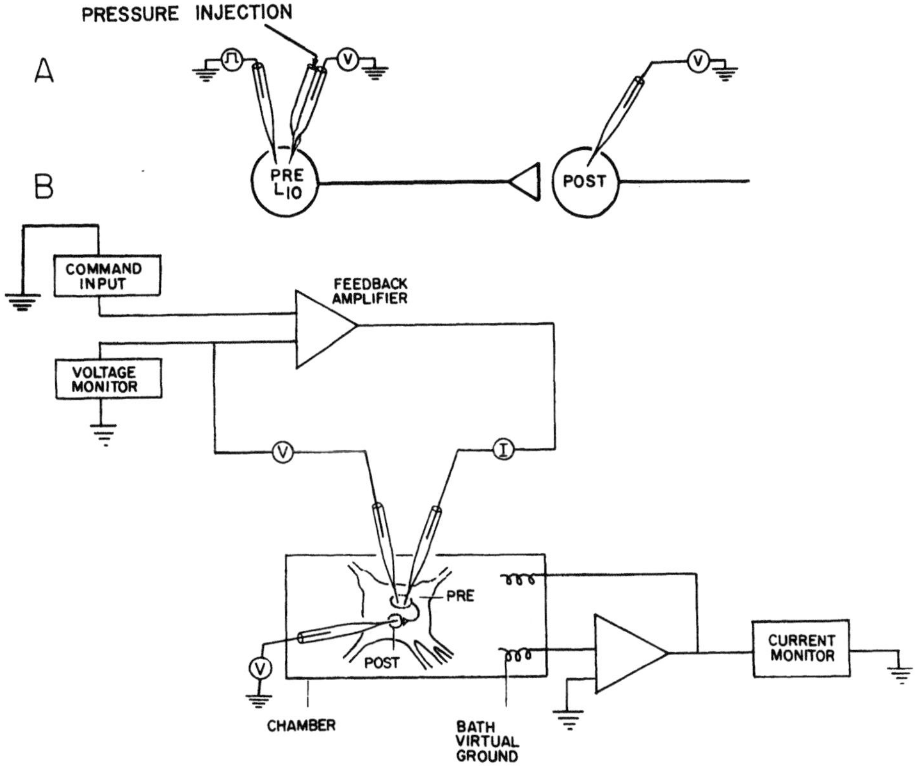

Fig. 1. A Experimental system. In the *Aplysia* abdominal ganglion presynaptic cell L10 can be impaled with two electrodes for current passing and recording. Double-barreled electrodes can be used, one barrel for recording and the second barrel for pressure injection of various pharmacologic agents. Postsynaptic cells in the ganglion can be recorded from to provide an assay of transmitter release from L10. *B* L10 can be voltage clamped with a two-electrode voltage clamp circuit, while recording from postsynaptic cells. The voltage clamp circuit we used is described in Byrne et al. (1979). *V*, voltage sensing electrode.

was increasing, transmitter release also increased (Fig. 2). Indeed, the sigmoid function relating transmitter release to presynaptic depolarization is similar to that reported at the squid giant synapse (Katz and Miledi 1969; Llinás et al. 1976, 1981a, b). Moreover, as is the case with the squid giant synapse, the function relating transmitter release to presynaptic depolarization overlaps the voltage sensitivity of the presynaptic Ca^{2+} current (Fig. 2; Klein et al. 1980). These several findings support the idea that at least some of the release sites of L10 are electrically close to the soma.

To examine the specific mechanisms of synaptic modulation and release, we combined voltage clamp experiments with injection of various agents such as Ethylene-bis(oxyethylenenitrile)tetraacetic acid (EGTA), Arsenazo III, and other specific pharmacologic probes. Using this approach, we have examined three types of synaptic plasticity: (a) the control of transmitter release by the membrane potential of the presynaptic neuron; (b) posttetanic potentiation; and (c) presynaptic inhibition.

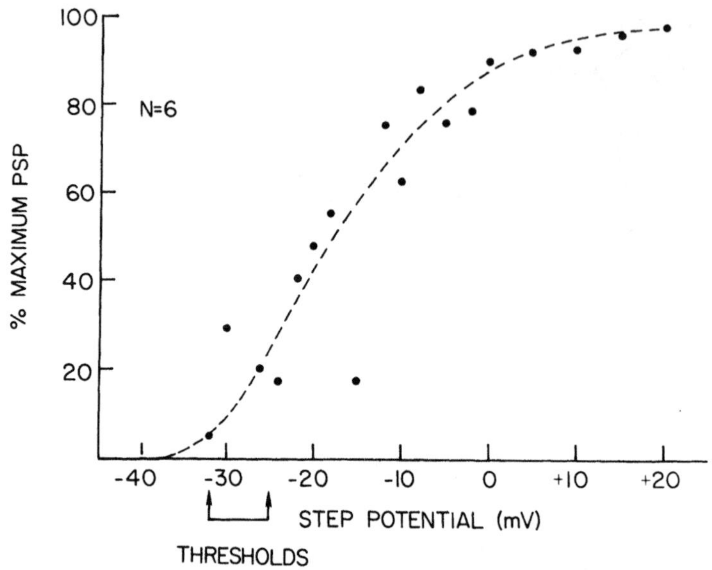

Fig. 2. Control of transmitter release by cell body membrane potential in L10. L10 was voltage clamped at holding potential -40 mV in seawater solutions containing tetrodotoxin. Step depolarizations above threshold elicit transmitter release which increases in a graded fashion as the size of the depolarization increases. Thresholds for release are in the range of the threshold for activation of the transient Ca^{2+} current. The curve is pooled data from six preparations whose thresholds for transmitter release varied in the indicated range. In each experiment postsynaptic potentials (PSP) were normalized as percentage of maximum release for each preparation. *Each point* is the average of one to four experiments

The Control of Transmitter Relase by the Membrane Potential of the Presynaptic Cell

In 1975 Shimahara and Tauc described a simple form of synaptic plasticity whereby the membrane potential of the presynaptic neuron exerts a powerful influence over the effectiveness of the connections made by the neuron. They found that hyperpolarizing the presynaptic cell decreases the synaptic potential elicited by the presynaptic action potential, whereas depolarizing the presynaptic cell enhances the synaptic action. Although this effect was opposite to that described in the giant synapse of the squid (Hagiwara and Tasaki 1958; Takeuchi and Takeuchi 1962), similar potential-dependent control of transmitter release by membrane potential has been described in the leech (Nicholls and Wallace 1978a) and may occur at the neuromuscular junction (del Castillo and Katz 1954c; Liley 1956b; Katz 1962). Membrane potential dependence of release occurs between cell L10 and its follower cells (Waziri 1977; Shapiro et al. 1980a). In L10 this type of plasticity is not only interesting in its own right, but it also plays a role in the expression of both posttetanic potentiation and presynaptic inhibition, since both of them involve changes in the membrane potential of the presynaptic neuron.

Exploring the modulation of transmitter release by membrane potential in cell L10, we found that this form of plasticity involves two distinct mechanisms, each of which affects the level of free Ca^{2+} in the terminals: (a) control of Ca^{2+} influx;

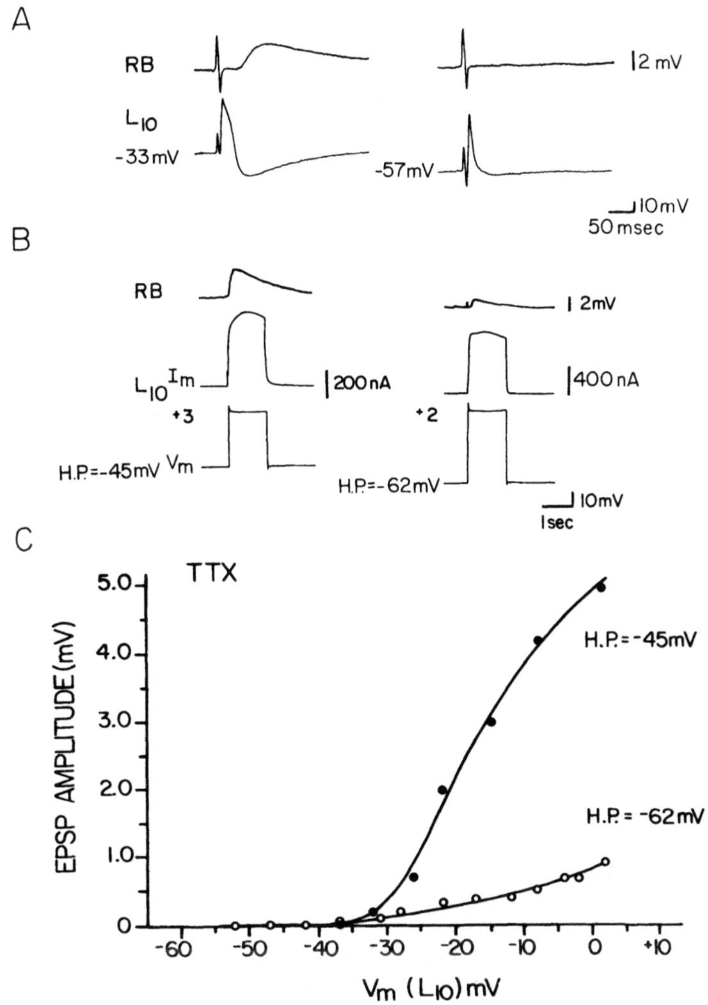

Fig. 3A–C. L10 membrane potential modulates release. *A* Spikes elicited from depolarized L10 membrane potentials elicit large postsynaptic potentials (PSPs). As L10 is hyperpolarized, spikes become narrower and smaller and transmitter release is reduced. *B* Transmitter release evoked by voltage clamp pulses of constant amplitude and duration also can be modulated by membrane potential. Voltage clamp pulses (*lower traces, L_{10}* V_m) in artificial seawater solution with TTX from depolarized holding potentials (*H.P.*) elicit large PSPs (*top traces*). When L10 is hyperpolarized pulses to the same level elicit larger outward clamp currents (*middle traces, L_{10}* I_m; note different calibrations) and reduced PSPs. V_m, membrane voltage; I_m, membrane current. *C* Input-output curve from two holding potentials. Same experiment as in part *B*. (Shapiro et al. 1980a)

and (b) control of Ca^{2+} buffering. When the membrane potential of L10 is progressively increased, transmitter output at the cell's various branches is reduced (Fig. 3). Even a slight increase of the 5 mV in membrane potential sometimes produces significant depression of transmitter release. With larger hyperpolarizing changes of 20–40 mV, transmitter release can be completely blocked (Fig. 3A).

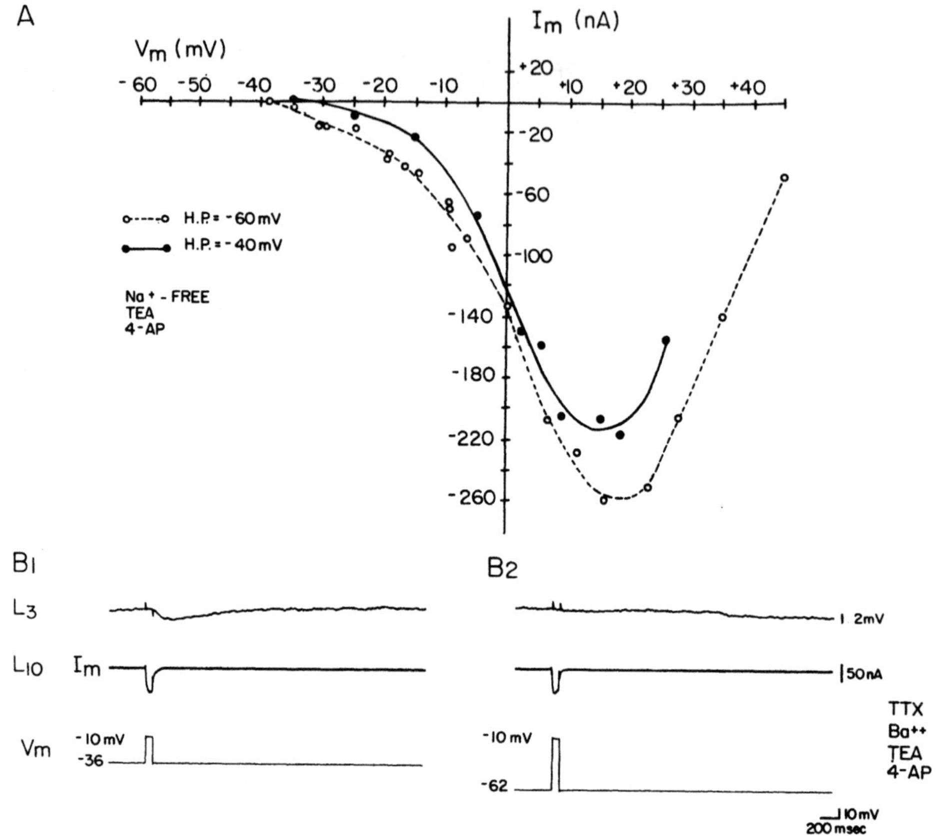

Fig. 4A, B. Transient Ca^{2+} current is not modulated by membrane potential. *A* In Na^+-free solutions containing TEA and 4-AP, transient inward currents are carried by Ca^{2+} ions. The peak transient inward currents elicited by step depolarizations from holding potentials (*H.P.*) of -40 (*filled circles*) or -60 (*open circles*) are plotted against the size of the step depolarization. Hyperpolarization does not reduce the size of the transient inward Ca^{2+} current. V_m, membrane voltage; I_m, membrane current. *B* With all K^+ channels blocked in solutions containing TTX, Ba^{2+}, TEA, and 4-AP membrane potential can still modulate transmitter release. Voltage clamp steps to -10 mV (*lower traces*) from holding potentials of -36 or -62 mV elicit inward currents (*middle traces*) carried by Ba^{2+} ions. The size of this inward current is the same from the two holding potentials. Postsynaptic potentials (*top traces*) elicited by the depolarizations are larger from the depolarized holding potential, and are reduced by presynaptic hyperpolarization. (Shapiro et al. 1980a)

Conversely, depolarizing the cell's membrane potential increases transmitter release. The changes in membrane potential affect the duration and height of the presynaptic action potential: spikes become narrower and shorter with hyperpolarization, and broader and taller with depolarization (Fig. 3A). Both of these potential-induced alterations are likely to alter transmitter release because they alter the Ca^{2+} current (Horn and Miller 1977; Klein and Kandel 1978; Llinás et al. 1981b). However, as we will discuss below, these alterations account for only part of the effects of membrane voltage on release. Part of the effect of membrane voltage is independent of changes in spike height and duration.

This partial independence can be seen in voltage clamp experiments, in which the size and duration of the depolarizations that elicit transmitter release can be held constant while the membrane potential is varied. Even under these conditions, alterations in membrane potential are effective in modulating transmitter release (Fig. 3B,C). At depolarized holding potentials, currents elicited by step depolarization under voltage clamp are less outward than at hyperpolarized holding potentials (Fig. 3B; Connor and Stevens 1971a, b; Adams et al. 1980; Shimahara 1981). This difference is net current presumably accounts for the difference in the configuration of the action potential at the two holding potentials in unclamped cells (Fig. 3A). However, the ability of the membrane potential to modulate transmitter release under voltage clamp conditions, in which the duration and amplitude of the presynaptic command pulse is held constant, suggested to us that mechanisms other than modulation of the K^+ current contribute to this form of plasticity.

To assess the contribution of each of the two voltage-dependent K^+ channels to membrane potential modulation of release, we examined transmitter release in solutions containing TEA or 4-AP. When TEA alone was used, the delayed outward current was substantially reduced, but hyperpolarization still removed inactivation of the early outward potassium current. As a result, the outward currents were larger from hyperpolarized holding potentials than from depolarizing potentials, and transmitter release was still reduced by hyperpolarization. When 4-AP alone was used, the early K^+ current was blocked, but since the delayed current was still present and hyperpolarization removed its inactivation, outward currents were still larger from hyperpolarized than from depolarized holding potentials, and again transmitter release was reduced by hyperpolarization. Thus inactivation of both the early and the delayed K^+ channels contributes to the increase in the duration and amplitude of the presynaptic spike with depolarization and the resulting enhancement of transmitter release. This raised the question: Do other currents also contribute? Is all the modulation of the Ca^{2+} current secondary to changes in K^+ current, or is there also a direct modulation of the Ca^{2+} channel?

To explore this question, we blocked all K^+ channels pharmacologically in solutions containing TEA, 4-AP, and Ba^{2+} and found that membrane potential could still modulate transmitter release. These experiments yielded one additional result. Direct observation of inward divalent ion currents showed that membrane potential was not modulating the transient Ca^{2+} current directly. This current was unchanged from hyperpolarized or depolarized holding potentials (Fig. 4A). Thus although transient Ca^{2+} currents were not being directly modulated by membrane potential, membrane potential was able to modulate transmitter release (Fig. 4B).

Depolarization Activates a Steady State Ca^{2+} Current. Although the regulation of transmitter release by membrane potential does not result from direct modulation of the transient Ca^{2+} current, we found that changes in membrane potential lead to changes in the steady state activation of the Ca^{2+} channels (Fig. 5). We have found that in cell L10, steady state activation – the contribution of Ca^{2+} current to the resting current – is greater at depolarized than at hyperpolarized

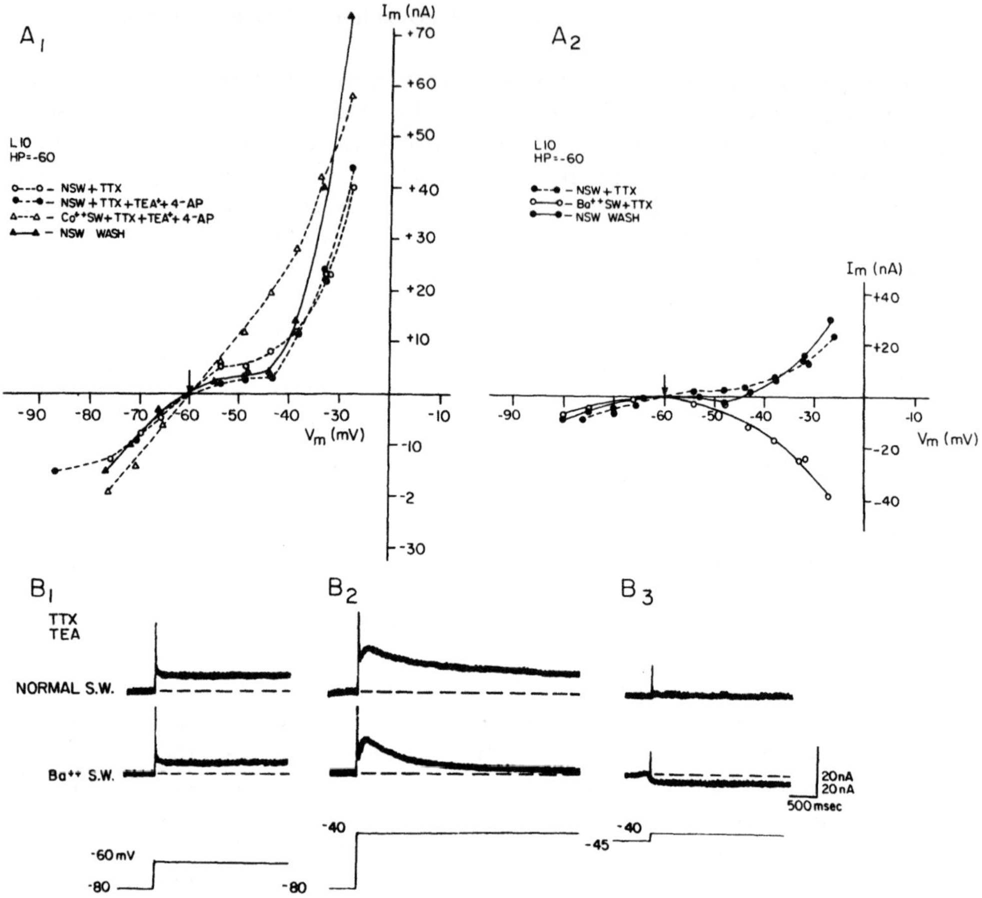

Fig. 5A,B. Presence of steady state Ca^{2+} and Ca^{2+}-dependent K^+ currents in L10. *A* Steady state current-voltage curves of L10. In graphs A_1 and A_2, L10 was voltage clamped to a holding potential (*HP*) of -60 mV and 5-s clamp steps were delivered every 30 s to various hyperpolarized and depolarized membrane potentials. The current at the end of each 5-s voltage step was plotted against membrane voltage. In normal seawater (*NSW*) and TTX the steady state current-voltage curve shows a pronounced region of reduced positive slope in the range of the normal resting potential of L10 (-60 to -40 mV). In A_1 application of TEA^+ and 4-AP have little effect on the steady state current-voltage curve, since transient K^+ channels normally inactivate during the prolonged (5-s) voltage steps. Replacement of extracellular Ca^{2+} with 10 mM Co^{2+} ions, however, linearizes the current-voltage curve by blocking a steady state inward Ca^{2+} current. When normal seawater is washed back the normal steady state current-voltage curve is restored. When, in A_2, Ba^{2+} ions are substituted for Ca^{2+} ions, the normally observed area of steady state reduced positive slope is converted into a region of inward-going rectification. Ba^{2+} carries the steady state inward current through the Ca^{2+} channel, but does not activate an opposing Ca^{2+}-dependent current. V_m, membrane voltage; I_m, membrane current. (Klein et al. 1980)
B Ba^{2+} substitution. Presynaptic neuron L10 voltage clamped in artificial seawater (*S.W.*) containing 30 μM TTX and 50 mM TEA (*upper traces*) and solution containing 30 μM TTX and 50 mM TEA with Ba^{2+} substituted for Ca^{2+} (*middle traces*). The current elicited by a step from -80 mV to -60 mV is only slightly reduced by substitution of Ba^{2+} for Ca^{2+} (*trace* B_1), but the steady state current elicited by steps from -80 to -40 mV is more inward in the presence of Ba^{2+} (*trace* B_2), indicating the presence of a steady state inward current at -40 mV, and blockage of the normally present Ca^{2+}-dependent K^+. The steady state activation of net inward current can be observed directly with smaller steps from -45 to -40 mV, which avoids recruiting the unchanged outward leakage current seen to be elicited by steps from -80 mV to -60 mV (*trace* B_3). (Shapiro et al. 1980a)

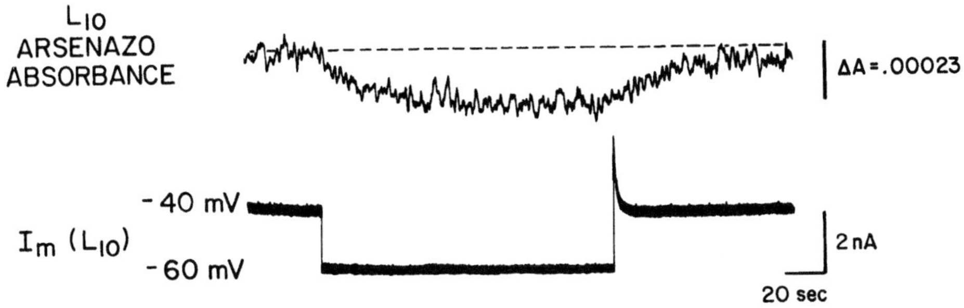

Fig. 6. Steady state changes in Arsenazo absorbance at different L10 membrane potentials. L10 was injected with the Ca^{2+}-sensitive indicator dye Arsenazo III and voltage clamped at -40 mV. A high-gain recording of Arsenazo absorbance (660–690 nm) (*top trace*) shows a slow decrease in dye absorbance when L10 is hyperpolarized to -60 mV, indicating a decrease in intracellular free Ca^{2+}. The dye signal slowly increases when L10 is again depolarized to -40 mV. *Lower trace* shows L10 membrane current. L10 voltage trace is not shown. I_m, membrane current. (Shapiro, E., Connor, J., and Kretz, R., unpublished data)

levels. Similar steady state activation of Ca^{2+} channels has been previously described in molluscan somata by Eckert and Lux (1976), Akaike et al. (1978), and Gorman and Thomas (1978).

We obtained evidence for steady state Ca^{2+} channel activation from examination of the steady state membrane conductance of L10 at different holding potentials in the range of membrane voltages where modulation of release was occurring. We found that the steady state conductance showed a region of reduced positive slope that could be reversibly blocked by Co^{2+}, a Ca^{2+} channel blocker (Fig. 5A). By contrast, Ba^{2+} ions, which flow through Ca^{2+} channels but do not activate the Ca^{2+}-dependent K^+ current (see Figs. 4B, 5B) cause a large inward shift in the steady state current-voltage curve (Fig. 5A_2).

In the presence of Ba^{2+}, a voltage-dependent steady state inward current flows through the Ca^{2+} channels at depolarized levels of holding potential (Fig. 5A_2, 5B); this inward current is not present at hyperpolarized levels (Fig. 5B). In normal seawater solutions the inward current activates a Ca^{2+}-dependent K^+ current that also contributes to the resting steady state conductance at depolarized holding potentials. To activate this K^+ conductance, the concentration of intracellular Ca^{2+} near the membrane must be significantly larger at more depolarized than at more hyperpolarized membrane potentials.

Independent and direct evidence for the presence of a steady state Ca^{2+} current has been obtained with the Ca^{2+}-sensitive dye Arsenazo III. After L10 is injected with Arsenazo, a small increase in dye absorbance indicating an increase in free Ca^{2+} may be seen when L10 is depolarized 20 mV from -60 to -40 mV (Fig. 6; see also Gorman and Thomas 1978).

Free Ca^{2+} Level and Transmitter Release. The increased concentration of intracellular Ca^{2+} at more depolarized holding potentials could enhance transmitter release in several ways. It could add to the Ca^{2+} influx during the action potential, resulting in a higher total intracellular concentration of Ca^{2+} available for transmitter release. In addition, increased intracellular Ca^{2+} could increase the

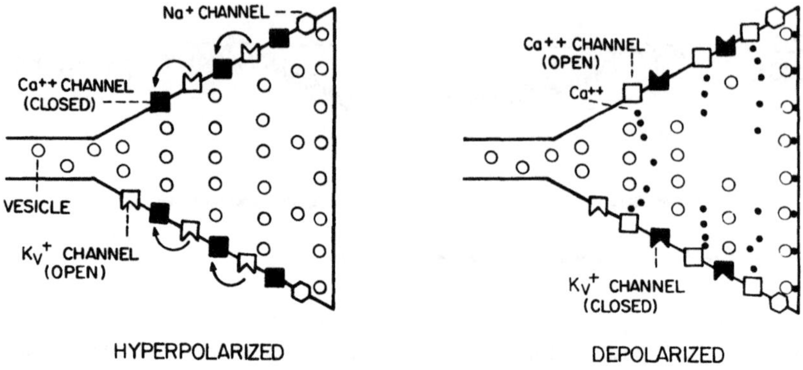

Fig. 7. Diagrammatic summary of effects of resting membrane potential on presynaptic currents and transmitter release. In all the diagrams of presynaptic terminals (see also Figs. 15 and 25), *filled symbols* represent closed channels and *open symbols* represent open channels. The *filled dots* represent the calcium concentration. *Arrows* indicate the fact that opening of K^+ channels repolarizes the membrane potential and in turn closes Ca^{2+} channels.
When the neuron is held at a hyperpolarized resting level two currents are altered. First, the steady state calcium current is reduced, and consequently the resting intracellular calcium concentration is reduced. As a result the influx of calcium that occurs with an action potential may not by itself be sufficient to cause maximal transmitter release, and in the limit, to cause any release. Secondly, the voltage-sensitive potassium currents ($K_V{}^+$) are activated to a greater degree when an action potential occurs from a hyperpolarized than when it occurs from a depolarized membrane potential. The resulting spike is decreased in both duration and amplitude and therefore allows a smaller calcium influx to occur. Thus, at depolarized membrane potentials, increased steady state calcium currents as well as potassium current inactivation cause a larger amount of calcium to be available for transmitter release, resulting in increased synaptic transmission

load on intracellular Ca^{2+} buffering systems, particularly those close to the membrane that are thought to compete with the release process (Alnaes and Rahamimoff 1975; Rahamimoff 1968). The observation that increased Arsenazo absorbance reaches a steady state value (Fig. 6) supports this possibility. A third possibility is that increased intracellular Ca^{2+} concentration causes changes in screening of surface membrane proteins by changing internal membrane surface charges and therefore alters the efficacy of transmitter release (Bass and Moore 1966; Van der Kloot and Kita 1973; Gorman et al. 1983). In each of these instances, invasion of the terminal by an action potential would provide a concentration of Ca^{2+} that is more effective for release.

The role of steady state Ca^{2+} current in the modulation of transmitter release by the membrane potential is supported by the observations of Nicholls and Wallace (1978b), del Castillo and Katz (1954c), and Liley (1956b) that spontaneous quantal release of transmitter in leech neurons and at the frog and mammalian neuromuscular junctions are increased at depolarized potentials. Our results indicate that an additional synergistic mechanism also operates in cell L10: a decrease in both voltage-dependent (the early and the delayed) K^+ currents with depolarization. In an unclamped cell, the decrease in the K^+ currents accounts for the increase in the height and duration of the spike. In this way the transient Ca^{2+} current can also be modulated, albeit indirectly (Shimahara 1981). The synergistic changes in both transient Ca^{2+} current and intracellular Ca^{2+} concentration exert powerful control over synaptic transmission – ranging from total block to enhanced effectiveness. The cartoon of Fig. 7 illustrates schematically how membrane potential can effect release by altering the steady state Ca^{2+} channel and the voltage-dependent K^+ channel.

Modulation of release by membrane potential can have important physiological roles. As first proposed by Shimahara and Tauc (1975), excitatory postsynaptic potentials (EPSPs) and inhibitory postsynaptic potentials (IPSPs) have a dual function. In addition to controlling the probability that a neuron will fire an action potential, synaptic potentials also set the level for transmitter release in those neurons whose release sites are affected by this potential change. These findings also suggest a possible mechanism for long-term regulation; over time, synaptic output may be regulated by metabolic (homosynaptic) or hormonal (heterosynaptic) actions that affect resting potential (Thomas 1972; Mayeri et al. 1979a, b).

Posttetanic Potentiation

Soon after the initial discovery of the endplate potential by Gopfert and Schaefer (1938), Schaefer et al. (1938) and, subsequently, Feng (1941) found that repeated (tetanic) stimulation increased the amplitude of the endplate potential without producing any obvious changes in the action potential of the presynaptic axon. Because the increase in the endplate potential was not restricted to the period of tetanic stimulation, but persisted for several minutes after the tetanus, Feng called this synaptic enhancement "posttetanic potentiation." He also found that longer tetani produced greater potentiation than did shorter ones. In 1947, Larrabee and Bronk found posttetanic potentiation at a peripheral neuron-to-neuron synapse (between preganglionic and postganglionic cells of the stellate ganglion), and 2 years later Lloyd (1949) discovered similar potentiation in the monosynaptic reflex of the spinal cord, thereby showing that these plastic changes also occur in central neurons. Lloyd also found (as had Larrabee and Bronk) that posttetanic potentiation produced by stimulating one afferent pathway did not increase the response of the postsynaptic cell to synaptic activation via another, unstimulated afferent pathway. These experiments indicated that posttetanic potentiation is homosynaptic: it is restricted to the stimulated pathway and results from a change in the synapse itself.

The Residual Ca^{2+} Hypothesis. Posttetanic potentiation has now been axamined at a variety of vertebrate and invertebrate synapses, including those in *Aplysia* (Gage and Hubbard 1966b; Rosenthal 1969; Magleby and Zengel 1975, 1976a, b; Waziri et al. 1969; Schlaepfer et al. 1976; Weinreich 1971; Erulkar and Rahamimoff 1978; Alnaes and Rahamimoff 1975; Rahamimoff et al. 1980). These studies have documented four key findings:

1. That posttetanic potentiation is presynaptic and involves an enhancement of transmitter relaese (Lloyd 1949; Miledi and Thiess 1971)
2. That Ca^{2+} ions play an important role in posttetanic potentiation (Weinreich 1971; Erulkar and Rahamimoff 1978; Erulkar et al. 1978)

3. That posttetanic potentiation is not dependent on the shape of the presynaptic spike (Gage and Hubbard 1966b; Miledi and Thiess 1971; Rosenthal 1969)
4. That posttetanic potentiation is affected by agents that alter the intracellular buffering capability for Ca^{2+}, perhaps including Na^+ (Alnaes and Rahamimoff 1975; Rahamimoff et al. 1980; Atwood et al. 1983; Misler and Hurlbut 1983).

Partly on the basis of some of these findings, Katz and Miledi (1968) and Rahamimoff (1968) proposed that both the synaptic facilitation during the tetanus and posttetanic potentiation itself may result from residual increases in intracellular Ca^{2+} due to the entry of Ca^{2+} during the tetanus and – perhaps additionally – due to its subsequent release from internal stores by Na^+. Rahamimoff (1968; Alnaes and Rahamimoff 1975) further suggested that the intracellular concentration of Ca^{2+} is controlled by intacellular organelles that buffer Ca^{2+} – the mitochondria and endoplasmic reticulum – and that posttetanic potentiation might depend upon this control. More recently, the effects of presynaptic Na^+ concentration in regulating intracellular free Ca^{2+} have been examined (Rahamimoff et al. 1980; Atwood et al. 1983; Misler and Hurlbut 1983) and shown to play a role in posttetanic potentiation.

Although the residual Ca^{2+} hypothesis of posttetanic potentiation is clearly attractive, it has been difficult to test directly because the presynaptic terminals of neurons are inaccessible to direct electrical recording, so that the buffering capabilities or the level of free Ca^{2+} could not be measured. In addition, there are some conceptual problems with the residual Ca^{2+} hypothesis. In most preparations, posttetanic potentiation does not parallel the presumed increase in free Ca^{2+}. Thus posttetanic potentiation does not reach its peak until some time after the tetanus is completed, at a time when the Ca^{2+} concentration is no longer at its maximum (Fig. 8).

To explore this paradox, we carried out an analysis of the presynaptic ionic mechanisms underlying posttetanic potentiation at the output synapses of L10. To examine Ca^{2+} accumulation we initially relied on measurement of the Ca^{2+}-activated K^+ current in L10. This outward current is activated by intracellular Ca^{2+}, and there is good evidence that this current provides an index of the amount and time course of increases in intracellular Ca^{2+} concentration (Meech 1972, 1978; Gorman and Thomas 1978; Gorman and Hermann 1979). We next blocked the accumulation of Ca^{2+} before and after posttetanic potentiation by injecting the calcium-chelating agent EGTA into the presynaptic neuron, and examined the effects on posttetanic potentiation. Finally, we examined Ca^{2+} accumulation during a tetanus with the Ca^{2+}-sensitive indicator dye, Arsenazo III. The results of these experiments supported the residual Ca^{2+} model for posttetanic potentiation.

Posttetanic potentiation could be elicited in unclamped cells with a tetanus consisting of 100 spikes at 4 Hz (Waziri et al. 1969; Kretz et al. 1982). This tetanus increased the size of EPSPs from L10 up to 250% and produced posttetanic potentiation that lasted 3–5 min. In unclamped cells posttetanic potentiation was accompanied by a slow posttetanic hyperpolarization of the presynaptic cell that was 5–10 mV in amplitude and lasted up to 5 min (Fig. 9A, traces b,c). This hyper-

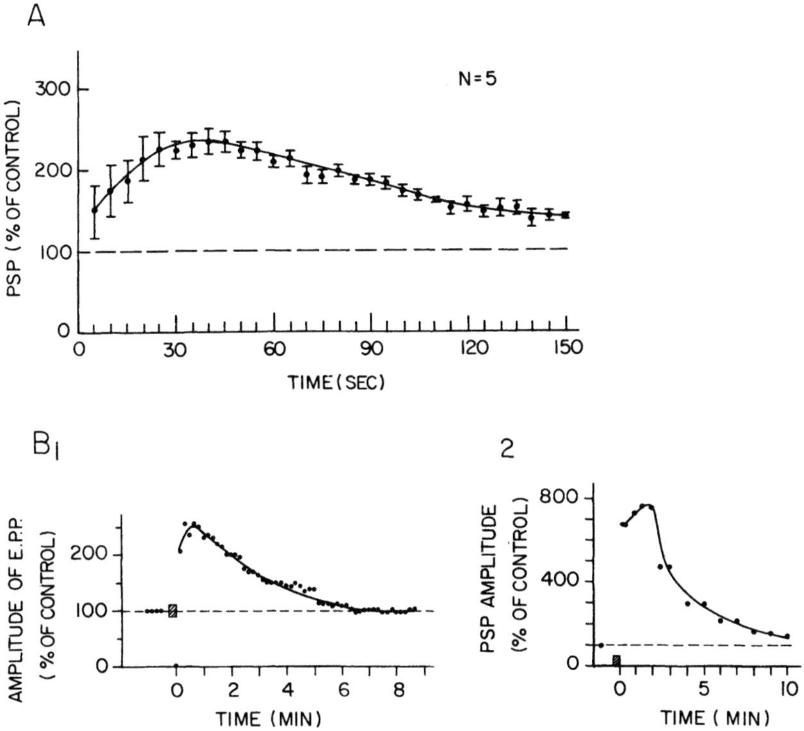

Fig. 8. Time course of posttetanic potentiation in various systems. When amount of potentiation is graphed against time after tetanus the buildup of posttetanic potentiation can clearly be seen. Postsynaptic potentials (PSPs) elicited immediately after the tetanus are not at their maximum level. Rather, the level of transmitter release first increases to a maximum, and then declines. *A* Posttetanic potentiation in *Aplysia* (Kretz et al. 1982); B_1 posttetanic potentiation at the rat neuromuscular junction (Liley 1956a); B_2 posttetanic potentiation in the chick ciliary ganglion (Martin and Pilar 1964). *E.P.P.*, end plate potential

polarization was accompanied by an increased membrane conductance in the presynaptic neuron of 10%–20%. The first few EPSPs after the tetanus were smaller than were later ones, and as in other preparations, posttetanic potentiation in L10 built up to a maximum. Because in *Aplysia* transmitter release is voltage dependent and hyperpolarization reduces transmitter release, we thought that the buildup of posttetanic potentiation might be due to the interaction between the enhanced release mechanism and the membrane potential changes produced by the tetanus.

Controlling Changes in Presynaptic Membrane Potential Reveals an Immediate Maximum in the Posttetanic Potentiation. To control for membrane potential changes, we voltage clamped L10. We elicited control EPSPs with command pulse depolarizations of 50 ms duration to 0 or +10 mV, delivered at 0.2 Hz from a holding potential of -30 to -50 mV. To produce posttetanic potentiation a tetanus was given consisting of 100 pulse depolarizations at 4 Hz. After the tetanus EPSPs were again elicited at the control rate of 0.2 Hz. Under these conditions postte-

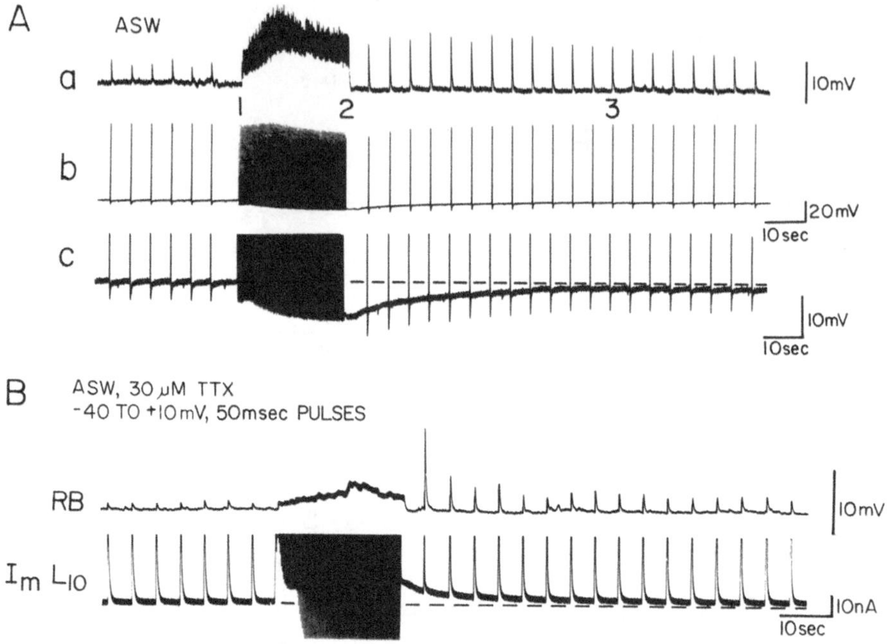

Fig. 9A,B. Posttetanic potantation in *Aplysia*. *A* In unclamped cells. L10 spikes elicited at a control rate of 0.2 Hz (*trace b*) elicit control postsynaptic potential in an RB follower cell (*trace a*). A tetanus of 100 spikes at 4 Hz elicits a large increase in transmitter release (posttetanic potentiation) that builds up to a maximum and then declines. Posttetanic potentiation is accompanied by a presynaptic hyperpolarization (*trace c:* L10 membrane potential at a higher gain). *ASW*, artificial seawater. *B* In voltage-clamped cells. L10 was clamped in a seawater solution containing tetrodotoxin (TTX) at a holding potential of -40 mV. Fifty-millisecond depolarizing pulses to +10 mV at 0.2 Hz elicit control postsynaptic potentials. A tetanus of 100 pulses at 4 Hz elicits posttetanic potentiation, which is at its maximum immediately after the tetanus (*top trace*). The tetanus also recruits a slow outward current (*lower trace*). L10 voltage trace not shown. I_m, membrane current. (Kretz et al. 1982)

tanic potentiation still occurred (Fig. 9B). But now by preventing the hyperpolarization of L10 after the tetanus, we not only completely eliminated the normal buildup of posttetanic potentiation, but also revealed a new fast component (Fig. 10). In clamped cells the EPSPs elicited immediately after the tetanus were at their largest and decayed with a double exponential time course. These two time constants may correspond to what Magleby and Zengel (1975, 1976a, b) have called "augmentation" and "potentiation" respectively. Figure 10 illustrates this point by comparing the average posttetanic potentiation in clamped and unclamped cells.

The voltage clamp now allowed us to examine the ionic conductance changes in the presynaptic neuron. We found that tetanic stimulation caused a buildup of a slow outward current (Fig. 9B), accompanied by an increased membrane conductance. The slow outward current lasted for 3–5 min and decayed with a double exponential time course which correlated well with those for the decay of posttetanic potentiation and posttetanic hyperpolarization (Fig. 10, Table 1).

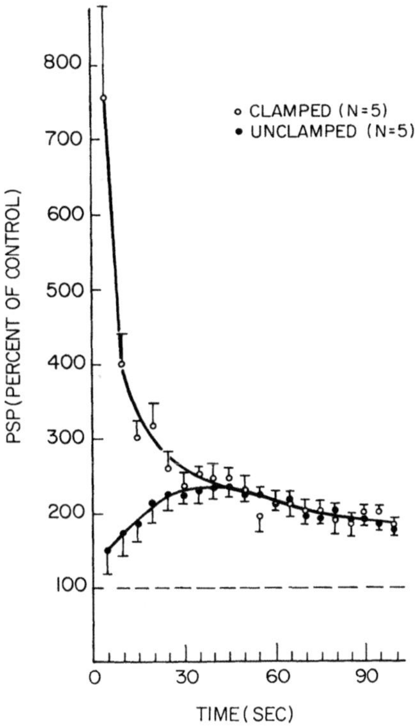

Fig. 10. Comparison of posttetanic potentiation in clamped and unclamped L10. Time course of posttetanic potentiation differs in unclamped and clamped presynaptic cells. Postsynaptic potentials (PSPs) were elicited in clamped (*open circles*) and unclamped (*filled circles*) L10 cells at a frequency of 0.2 Hz. The pretetanic control value is represented by the *broken line* (100%). A standard tetanus was given to elicit posttetanic potentiation. The mean values ($\pm$SEM) of postsynaptic potentials are presented for five separate experiments. The two curves were fitted by eye. L10 was clamped at either -40 or -30 mV and 50-ms pulses were delivered to 0 or +10 mV. The time course of posttetanic potentiation in unclamped L10 cells shows a buildup region that lasts up to 45 s after the end of the tetanus. No such buildup of posttetanic potentiation was seen when the presynaptic cell was clamped. Therefore, the buildup of posttetanic potentiation results from the posttetanic hyperpolarization of L10 and the voltage dependence of transmitter release. (Kretz et al. 1982)

Table 1. Time constants 'τ' of the four parameters that change posttetanically. (Data from Kretz et al. 1982; Connor et al., to be published)

Parameter	Fast component (τ_f) (s)	Slow component (τ_s) (s)
Hyperpolarization	8.8 ± 2.2 ($n=4$)[a]	69.0 ± 25.3 ($n=4$)[a]
Ca^{2+}-activated K^+ current	7.9 ± 1.8 ($n=6$)[b]	79.8 ± 24.1 ($n=6$)[b]
Postsynaptic potential	7.5 ± 2.6 ($n=3$)[b]	99.9 ± 46.6 ($n=3$)[b]
Arsenazo absorbance	6.2 ± 3.7 ($n=9$)[b]	55.6 ± 18.9 ($n=10$)[b]

[a] Artificial seawater (ASW), unclamped cells
[b] ASW, TTX, clamped cells

The Time Course of Posttetanic Potentiation Parallels the Ca^{2+} Activated K^+ Current. The finding of an outward current that parallels posttetanic potentiation is consistent with the residual Ca^{2+} model. This model would predict that the increased intracellular Ca^{2+} responsible for potentiating transmitter release might also activate the Ca^{2+}-dependent K^+ current in those cells that have this channel. But as we have seen, cell L10 and most other cells have at least two other types of K^+ channels that could produce a similar outward current unrelated to Ca^{2+}. It was therefore essential to explore the Ca^{2+} dependence of this current.

To show that the slow posttetanic outward current in L10 was the Ca^{2+}-dependent K^+ current, we first determined its reversal potential. As the membrane potential of L10 was hyperpolarized, the outward current became smaller and reversed at about -75 mV (Fig. 11A). This shows that current is carried by K^+. In addition, we found that the slow outward current was present in solutions in which the Na^+ channel was blocked by TTX and the early and the delayed K^+ channels were blocked by TEA and 4-AP (Fig. 11A, closed circles). These findings supported the idea that the outward current was the Ca^{2+}-activated K^+ current.

We further tested the identity of the slow outward current pharmacologically by examining the outward current produced by tetanic command pulses in solutions containing Co^{2+} or Ni^{2+}. These Ca^{2+} channel antagonists block Ca^{2+} inflow. These solutions blocked the outward current and the membrane conductance increase (Fig. 11B). We next substituted Ba^{2+} for Ca^{2+} ions. Ba^{2+} flows through the Ca^{2+} channels but does not activate the Ca^{2+}-dependent K^+ channel. Again, the slow outward current produced by tetanic stimulation was blocked (Fig. 11B). Finally, the slow outward current was reduced by injecting EGTA into L10 (Fig. 12A,B), a treatment that blocks the Ca^{2+}-activated K^+ current (Meech 1974; Gorman and Hermann 1979; Connor 1979). These several findings demonstrate that tetanic stimulation leads to an increase in intracellular Ca^{2+} concentration which has two effects: (a) it turns on a Ca^{2+}-activated K^+ conductance on the one hand, and (b) it enhances transmitter release on the other. Both of these Ca^{2+}-activated processes decay with similar time constants and presumably reflect increases in the free Ca^{2+} concentration that are due to the Ca^{2+} influx during the tetanus. Although release of Ca^{2+} from internal stores by Na^+ influx (Rahamimoff et al. 1980; Atwood et al. 1983) may contribute a small amount of posttetanic potentiation, the experiments performed in TTX indicate that good posttetanic potentiation can still be obtained in the absence of Na^+ influx (Fig. 8; Weinreich 1971).

Increasing Intracellular Ca^{2+} Buffering Reduces Posttetanic Potentiation. We then examined the effect of the Ca^{2+}-chelating agent EGTA on posttetanic potentiation and on the Ca^{2+}-activated K^+ current. Injection of EGTA into other molluscan neurons decreases the Ca^{2+}-activated K^+ current and reduces inactivation of the transient Ca^{2+} current (Connor 1979; Tillotson 1979), presumably by increasing the intracellular Ca^{2+}-buffering capacity. The residual Ca^{2+} model of posttetanic potentiation predicts that increasing intracellular Ca^{2+}-buffering capacity will reduce posttetanic potentiation. We found that injecting EGTA into L10 decreased the outward currents and increased the inward Ca^{2+} current produced by each depolarizing command. In addition, EGTA injection reduced the

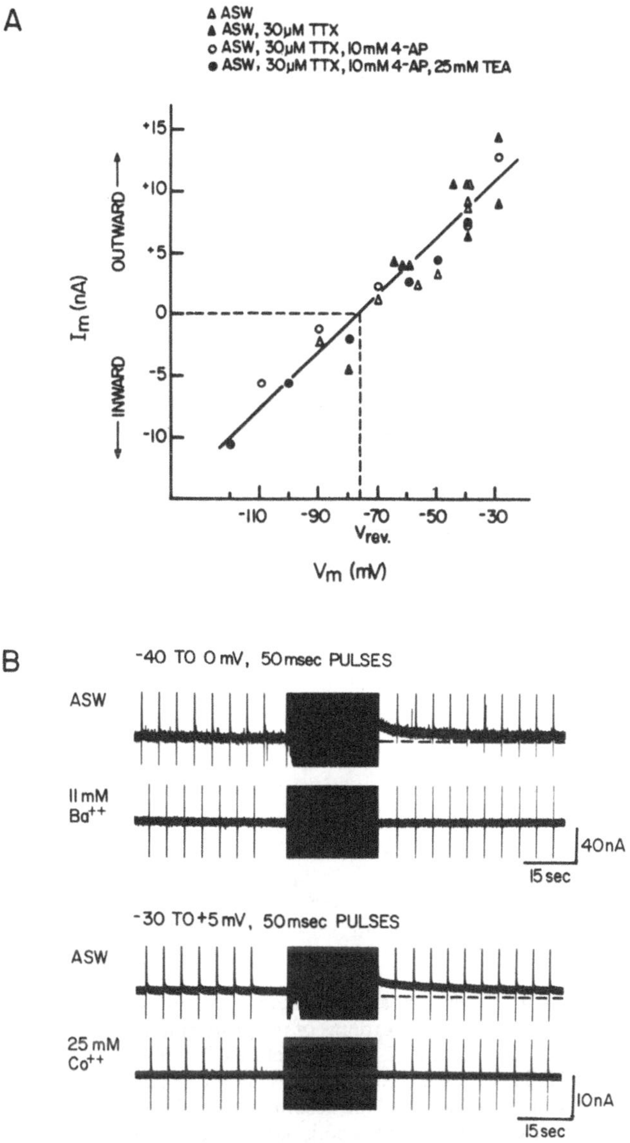

Fig. 11A,B. Properties of slow outward current. *A* Reversal potential of slow outward current. Cell L10 was clamped at various membrane potentials and a tetanus of 100 clamp pulses at 4 Hz was delivered. The size of the peak baseline current shift caused by the tetanus is plotted against holding potential (15 min between runs). The experiment was repeated in the presence of various pharmacologic agents (as indicated in the figure) designed to block voltage-dependent Na^+ and K^+ currents. The reversal potential (V_{rev}) of the slow outward current is about -75 mV. This slow outward current is not blocked by 10 mM 4-AP or 25 mM TEA, suggesting that it is a Ca^{2+}-activated K^+ current. V_m, membrane voltage; I_m, membrane current. *B* Ba^{2+} and Co^{2+} block slow outward current. Control runs show that a tetanus in normal Ca^{2+}-containing artificial seawater (*ASW*) is accompanied by outward baseline current shift. Ba^{2+} substitution for Ca^{2+} eliminates the baseline shift (postsynaptic potential not monitored). In another experiment 25 mM Co^{2+} added to the bath also blocked the slow outward current. (Kretz et al. 1982)

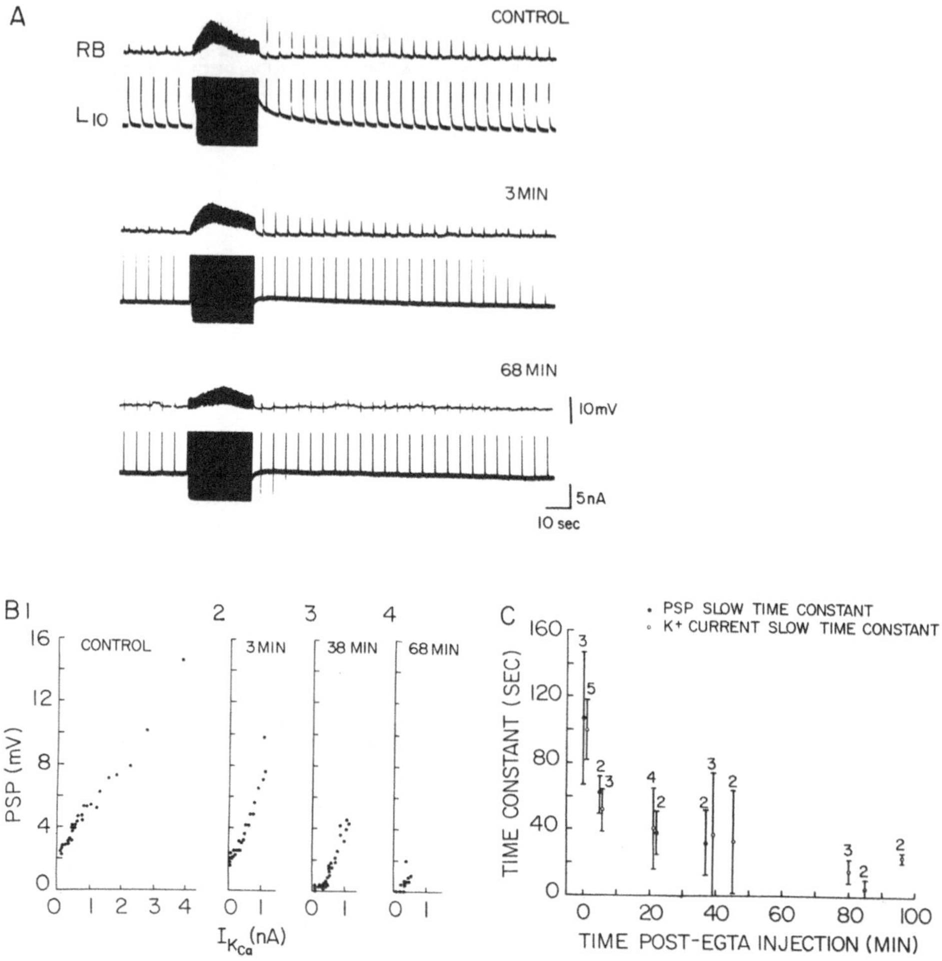

Fig. 12A–C. Time course of posttetanic potentiation and slow outward current before and after EGTA injection. *A* Recordings of the postsynaptic potential (PSP) amplitude (*RB trace*) and the slow outward current (L_{10} *trace*) before (*CONTROL*) and after (*3 MIN and 68 MIN*) the end of 0.5 M EGTA injection into the presynaptic cell L10. Three minutes after the end of EGTA injection, control PSPs and the absolute value of posttetanic potentiation are reduced: 68 min after injection, control PSPs are abolished. However, during the tetanus a buildup of PSPs occurs and posttetanic PSPs are recorded. EGTA injected into the cell body reduces the slow outward current. This effect is more rapid than the changes in PSP amplitude.
B Amplitudes of posttetanic PSPs are plotted as a function of posttetanic Ca^{2+}-activated K^+ current for control (*B1*) and 3 min (*B2*), 38 min (*B3*), and 68 min (*B4*) after EGTA injection. Both posttetanic PSP amplitude and K^+ currents are diminished in parallel by EGTA injection. Note that the baseline voltage clamp current before tetanus has arbitrarily been assigned the value 0 for $I_{K_{Ca}}$, although at holding potential -40 mV a part of the resting leak conductance is a Ca^{2+}-activated K^+ current (see Fig. 5). $I_{K_{Ca}}$, Ca^{2+}-activated K^+ current.
C The slow time constants of decay of posttetanic potentiation (*filled circles*) and posttetanic Ca^{2+}-activated K^+ (*open circles*) currents are plotted as a function of time after EGTA injection. Results are presented as means ±SD with *n* above the error bar. While tetanus is still capable of enhancing transmitter release after EGTA injection, the duration of enhanced release is markedly reduced by EGTA injection. Values at time 0 represent control runs. (Kretz et al. 1982)

slow outward current elicited by the tetanus (Fig. 12A,B). In L10, EGTA also affected the amplitude of the EPSPs as well as the decay time constants of posttetanic potentiation and the slow outward current. Within 5 min after injection of EGTA the control EPSPs were reduced (Fig. 12A). The EPSPs still built up during and after the tetanus, but they did not potentiate to the same size as before EGTA injection (Fig. 12A,B). In parallel with these changes in transmitter release, the Ca^{2+}-activated K^+ current was also reduced (Fig. 12A,B).

These results support the residual Ca^{2+} model. However, they are complicated by the fact that the control EPSP is also reduced by EGTA injection. The clearest indication of the effect of EGTA on posttetanic potentiation, independent of size changes in the control EPSP, was in the changed time course of posttetanic potentiation. As illustrated in Fig. 12C, we found that EGTA caused a progressive reduction in the time constants of posttetanic potentiation that paralleled the reduction in the slow outward current. Within 5 min after EGTA injection the fast time constant of both posttetanic potentiation and the slow outward current was blocked, and both the posttetanic potentiation and outward current could be fitted with a single slow exponential function. With further time after the injection this slow time constant also decreased progressively (Fig. 12C). One to two hours after the injection, tetanic stimulation led to posttetanic potentiation that lasted only seconds.

The Time Course of Posttetanic Potentiation is Paralleled by an Increase in Free Ca^{2+} in the Presynaptic Neuron. To obtain additional evidence for Ca^{2+} accumulation during posttetanic potentiation, we injected cell L10 with Arsenazo III and measured alterations in the level of free Ca^{2+} (Shapiro et al. 1983). Arsenazo absorbance at 660 nm is proportional to the free Ca^{2+} concentration, and this dye has been used to measure intracellular Ca^{2+} concentration in a variety of molluscan preparations (Brown et al. 1975; Ahmed and Connor 1979; Gorman and Thomas 1978; Charlton et al. 1982; Lewis and Wilson 1982). Tetanic stimulation of L10 caused a large increase in the free Ca^{2+} level during the tetanus as indicated by an increase in Arsenazo absorbance (Fig. 13). After the tetanus, the absorbance signal declined, as free Ca^{2+} levels returned to baseline, with a time course similar to those of posttetanic potentiation and the Ca^{2+}-activated K^+ currents (Fig. 13, Table 1).

Posttetanic Potentiation and the Ca^{2+} Accumulation Hypothesis. These results provide direct evidence for the suggestion of Katz and Miledi (1968) and Rahamimoff (1968) that intracellular accumulation of Ca^{2+} during the tetanus contributes importantly to posttetanic potentiation. In addition, our finding indicates that intracellular Ca^{2+} metabolism – the kinetics and activity of Ca^{2+}-buffering systems – contributes to modulation of transmitter release.

Tetanic stimulation could also enhance transmitter release by increasing Ca^{2+} influx through changes in spike configuration (Fig. 14A). Indeed, we found that tetanic stimulation in the presence of TTX caused decreases in the outward currents that could qualitatively account for the posttetanic spike broadening observed in unclamped cells (Fig. 14A,B). After the tetanus, the clamp currents gradually returned to control levels as inactivation wore off. Thus our results do not exclude mechanisms contributing to the full expression of posttetanic potenti-

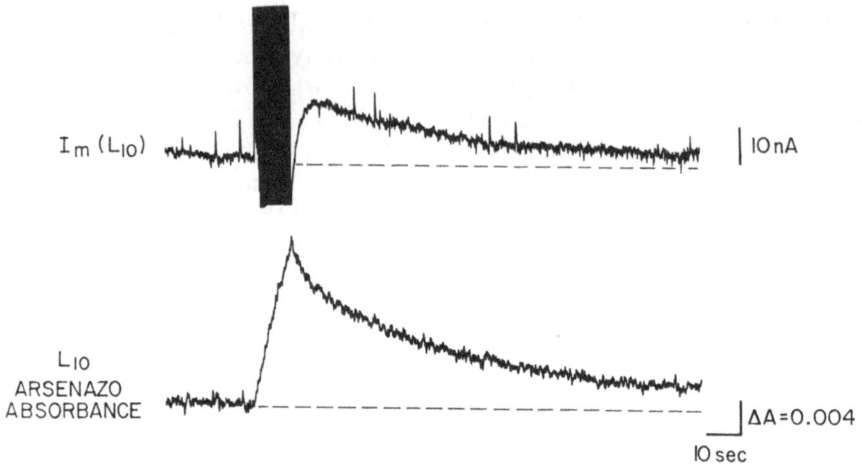

Fig. 13. Tetanic stimulation causes increased Arsenazo absorbance. L10 was injected with Arsenazo III and voltage clamped at -40 mV. Tetanic stimulation (50-ms pulses to +20 mV at 5 Hz for 10 s) elicited a large increase in Arsenazo absorbance during the tetanus. After the tetanus the dye signal decayed with a slow, double exponential time course (*lower trace*). The *upper trace* shows L10 voltage clamp current at high gain. After the tetanus a rapid inward and slow outward tail current is seen. L10 voltage trace not shown. I_m, membrane current. (Shapiro, E., Connor, J., and Kretz, R., unpublished data)

ation, such as change in spike duration. Indeed, these changes do occur and could contribute to posttetanic potentiation. But our demonstration that posttetanic potentiation is present in TTX and can be produced by depolarizing commands of constant duration suggests that Na^+ inflow or changes in spike configuration are not necessary for the expression of posttetanic potentiation.

Pharmacologic blockade of voltage-sensitive K^+ channels leaves residual clamp currents that consist of an inward Ca^{2+} and very small slow outward Ca^{2+}-activated K^+ current (Fig. 14C). Because posttetanic potentiation can still be elicited with voltage clamp commands in solutions containing TTX, TEA, and 4-AP, changes in transient K^+ and Na^+ currents are clearly not essential for posttetanic potentiation (Fig. 14). This treatment also allowed observation of the transient Ca^{2+} current in isolation (Fig. 14C). The Ca^{2+} current was largely inactivated during the tetanus and remained partially inactivated during much of the duration of posttetanic potentiation (Fig. 14C), indicating that the transient Ca^{2+} current is actually decreased, not increased. The increased intracellular

---

Fig. 14A–C. Changes in spike configuration and transient currents. *A* Changes in spike configuration accompanying posttetanic potentiation. After the tetanus, spike width is increased, but the spike returns almost to control levels within 1 min while postsynaptic potential is still enhanced. *Spikes 1, 2, and 3* are from numbered areas of *trace b* in Fig. 9A. *ASW*, artificial seawater. *B* Transient current changes accompanying posttetanic potentiation. L10 is voltage clamped in 30 μM TTX, while an RB follower cell is monitored. Voltage clamp steps from -40 to +5 mV at 0.2 Hz (*bottom trace*) elicit inward Ca^{2+} and outward K^+ currents (*middle trace*) and a stable RB cell postsynaptic potential (*top trace*). After a tetanus of 100 pulses at 4 Hz the outward currents are inactivated and transmitter release is enhanced. Over the course of the next 2 min cur-

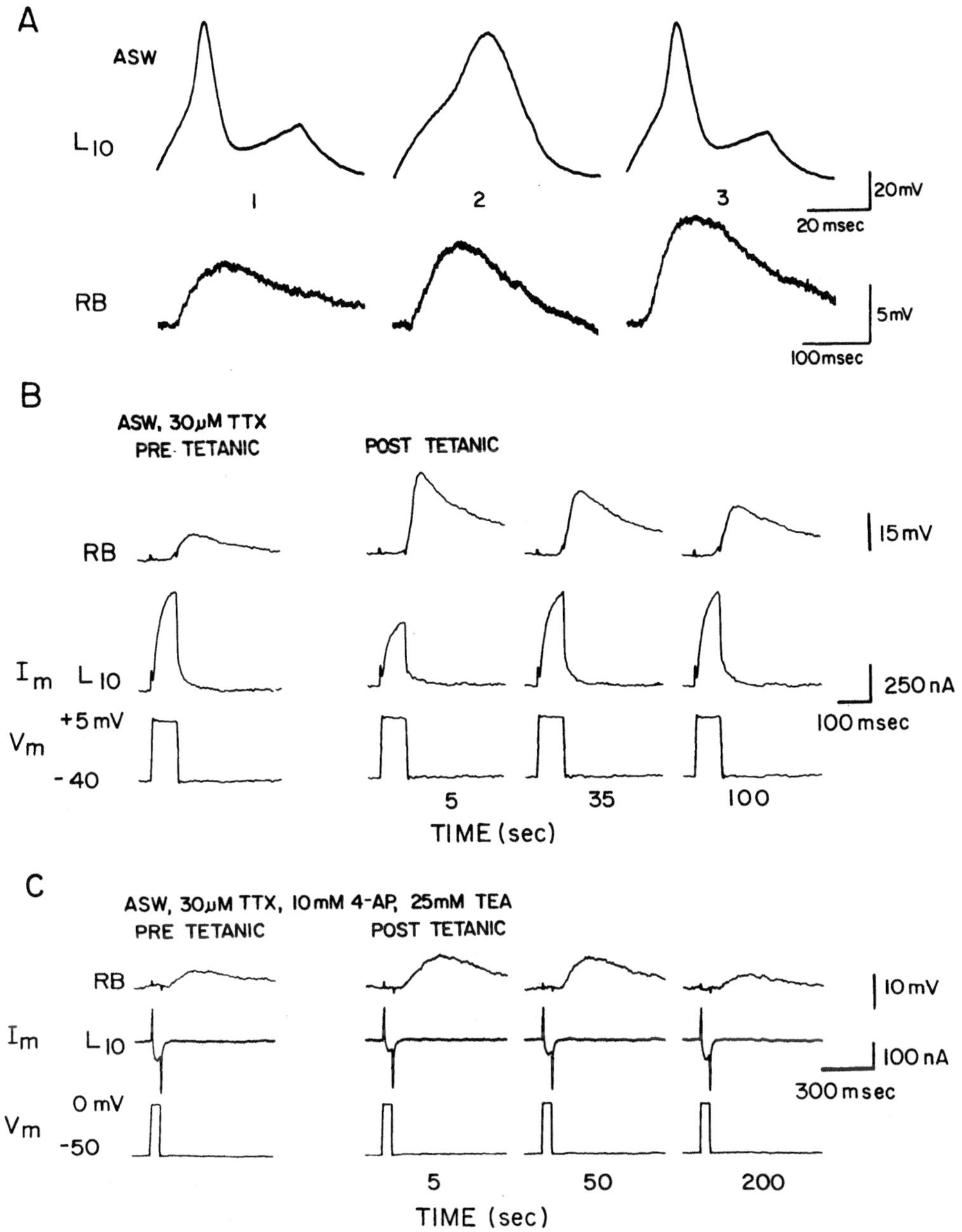

rents return to control levels, while posttetanic potentiation is still seen. V_m, membrane voltage; I_m, membrane current.

C Changes in transient Ca^{2+} current accompanying posttetanic potentiation. When L10 is voltage clamped in solutions containing TTX, 4-AP, and TEA, the transient Ca^{2+} current can be observed. Pretetanic voltage clamp steps elicit inward Ca^{2+} current (*middle trace*) and stable RB cell posttetanic potentiation (*top trace*). After tetanic stimulation of 10 pulses at 4 Hz, the inward Ca^{2+} current is reduced, while transmitter release is enhanced. Over the next 200 s the inward current and transmitter release return to control levels. V_m, membrane voltage; I_m, membrane current. (Kretz et al. 1982)

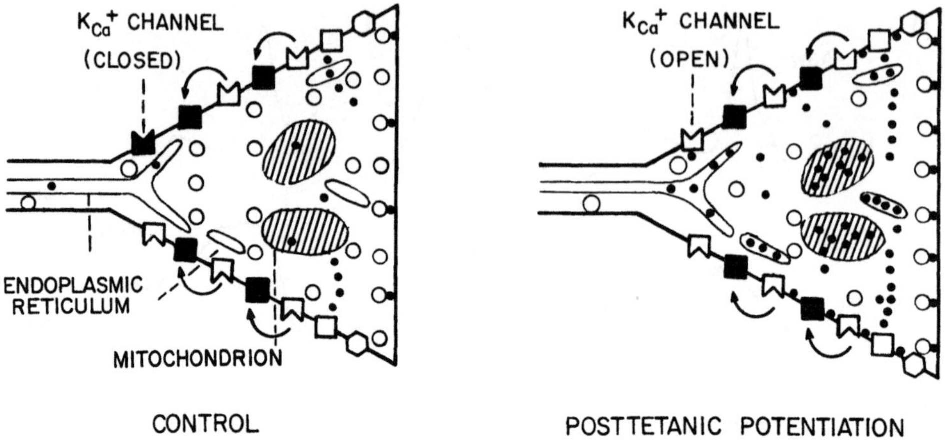

Fig. 15. Mechanism of postetanic potentiation. During tetanic stimulation the intracellular free calcium concentration increases greatly. Intracellular Ca^{2+} buffering systems also become loaded with calcium (mitochondria, endoplasmic reticulum, Ca^{2+}-binding proteins). After the tetanus the greatly increased calcium concentration causes the calcium currents during spikes to be more effective in releasing transmitter. As the increased calcium is gradually removed from the terminal region, the posttetanic increase in intracellular calcium also opens more calcium-activated K^+ channels ($K_{Ca}{}^+$). The opening of these channels reduces membrane excitability, and affects the time course of posttetanic potentiation. (For more details see legend to Fig. 7.)

Ca^{2+} concentration may actually be contributing to transient Ca^{2+} current inactivation (Tillotson 1979). As the posttetanic potentiation was returning to its smaller control value after the tetanus, the transient Ca^{2+} current was increasing as inactivation wore off. These results suggest that in the presence of a high intracellular level of Ca^{2+} (Ca^{2+} accumulation) a reduced transient Ca^{2+} current can still elicit increased transmitter release; they thus emphasize the level-setting role of steady state intracellular Ca^{2+} in the modulation of transmitter release. In addition to actual increases in free Ca^{2+}, tetanic stimulation may also saturate Ca^{2+} buffering systems such as mitochondria and endoplasmic reticulum. Once the intracellular buffering and binding sites are saturated by an increase in intracellular Ca^{2+}, most of the additional Ca^{2+} carried into the terminal by each action potential may be used primarily for release. Figure 15 represents this mechanism of posttetanic potentiation diagrammatically.

In the modulation of release by membrane potential, the changes in transient Ca^{2+} current caused by changes in spike configuration interact synergistically with the change in steady state Ca^{2+}. By contrast, during posttetanic potentiation the posttetanic hyperpolarization causes changes in the transient current during and immediately after the tetanus that counteract the potentiation caused by increased levels of intracellular Ca^{2+}. These results suggest two points: (a) the two counteracting influences – the presynaptic hyperpolarization and Ca^{2+} current inactivation – may be fail-safe devices that prevent an explosive enhancement of transmitter release and the consequent transient depletion of transmitter; and (b) the time course of posttetanic potentiation could be profoundly affected by the membrane potential of the presynaptic cell.

Presynaptic Inhibition

Modulation of synaptic strength may also be heterosynaptic. Following the earlier suggestion of Frank and Fuortes in 1957, Dudel and Kuffler described in 1961 the first clear instance of a heterosynaptic interaction: presynaptic inhibition at the crayfish nerve-muscle synapse. They showed that in the crayfish the inhibitory axon to muscle has a double function: (a) it produces an inhibitory postsynaptic potential in the muscle; and (b) it depresses the excitatory postsynaptic potential produced by the excitatory axon (Dudel and Kuffler 1961a).

Dudel and Kuffler (1961b) found that presynaptic inhibition acts by reducing the number of transmitter quanta released by the excitatory axon without affecting the sensitivity of the receptor molecules. Their experiments provided the first evidence that the membrane of the presynaptic terminals contains receptors to transmitter molecules, and that these receptors can control transmitter release.

Presynaptic inhibition has now been shown to be widely distributed and is found in both vertebrates and invertebrates (Eccles 1964; Schmidt 1971; Burke and Rudomin 1977; Ryall 1978; Nicoll and Alger 1979; Dudel and Kuffler 1961a, b; Tauc 1965; Waziri et al. 1969; Kennedy et al. 1974; Pearson and Goodman 1981; Zengel et al. 1983). As we have seen, one mechanism for the reduction in transmitter output during presynaptic inhibition is due to a synaptically mediated increase in the membrane conductance of the presynaptic terminals (frequently to Cl^-), which either reduces the amplitude of the presynaptic spike (Takeuchi and Takeuchi 1966a, b; Kennedy et al. 1980; Baxter and Bittner 1981), or causes branch-block at bifurcation points on the presynaptic terminal tree (Fuchs and Getting 1980; Halt and Smith 1976). The increased Cl^- conductance may produce either a depolarization or a hyperpolarization of the terminal, depending on the intracellular concentration of Cl^- (Kennedy et al. 1980; Fuchs and Getting 1980; Baxter and Bittner 1981). This mechanism for presynaptic inhibition is generally of short duration, of the order of milliseconds.

A variant of this mechanism is based upon the idea that the change in membrane potential (depolarization) of the presynaptic terminals might itself be responsible for reducing transmitter output. Thus a Cl^--mediated conductance change, or a buildup of extracellular K^+ ions in the restricted spaces around presynaptic release sites due to activity in the inhibitory pathway, would depolarize the presynaptic terminals and reduce transmitter output (as is thought to occur during primary afferent depolarization; for discussion see Schmidt 1971; Burke and Rudomin 1977; Ryall 1978; Nicoll and Alger 1979; Rudomin et al. 1981). This idea is consistent with the release properties described at the squid giant synapse, where depolarization reduces transmitter release (Takeuchi and Takeuchi 1962; Miledi and Slater 1966; Charlton et al. 1982). However, in *Aplysia* and certain other neurons, depolarization enhances rather than depresses transmitter release. At these synapses a different mechanism must account for presynaptic inhibition.

Presynaptic Inhibition Leads to a Hyperpolarization of the Presynaptic Neuron. In Aplysia the synapses of cell L10 that are modulated by membrane potential and posttetanic potentiation can also be modulated by presynaptic inhibition (Waziri et al. 1969). Stimulation of the left pleuroabdominal connective, one of the fiber

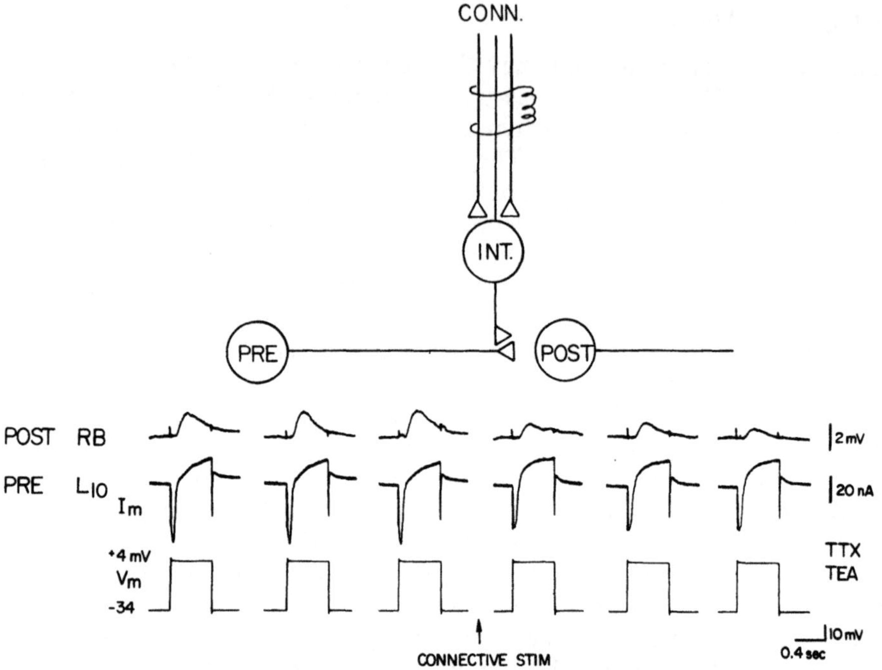

Fig. 16. Presynaptic inhibition with cell L10 under voltage clamp control. In normal seawater containing 30 μM TTX and 50 mM TEA and with the cell held at -34 mV to inactivate the early K^+ current, a step depolarization elicits in the presynaptic neuron an inward Ca^{2+} current and an outward current due to Ca^{2+}-activated K^+. The depolarizing step also elicits an excitatory postsynaptic potential in the RB follower cell. After extracellular stimulation of the pleuroabdominal connectives, a presynaptic inhibitory pathway for 5 s at 6 Hz, the same step depolarization elicits a decreased inward Ca^{2+} current and a reduced postsynaptic potential. V_m, membrane voltage; I_m, membrane current. (Shapiro et al. 1980b)

pathways to the abdominal ganglion, causes a depression of both the monosynaptic excitatory and inhibitory connections made by L10 (Fig. 16; Shapiro et al. 1980b). Recently, Byrne (1980) has discovered a group of cells (the L32 cells) activated by this pathway which, when stimulated, produce presynaptic inhibition independent of any conductance changes in the postsynaptic follower cells of L10. We first concentrated our analysis on identifying the specific ionic conductance changes in the presynaptic neuron associated with presynaptic inhibition, using connective stimulation to elicit presynaptic inhibition. We subsequently repeated our analysis using individual identified presynaptic inhibitory neurons of the L32 cluster.

Each of the two ways of producing presynaptic inhibition usually produces an IPSP in the presynaptic neuron L10. Thus one change in the presynaptic neuron that clearly contributes to presynaptic inhibition is the hyperpolarization that occurs as a result of the IPSP (Fig. 18). As we have seen, this hyperpolarization will reduce the output of transmitter by reducing the steady state Ca^{2+} influx and by removing some of the inactivation of the early and delayed K^+ channels. In the unclamped cells these outward currents will shorten the duration of the action po-

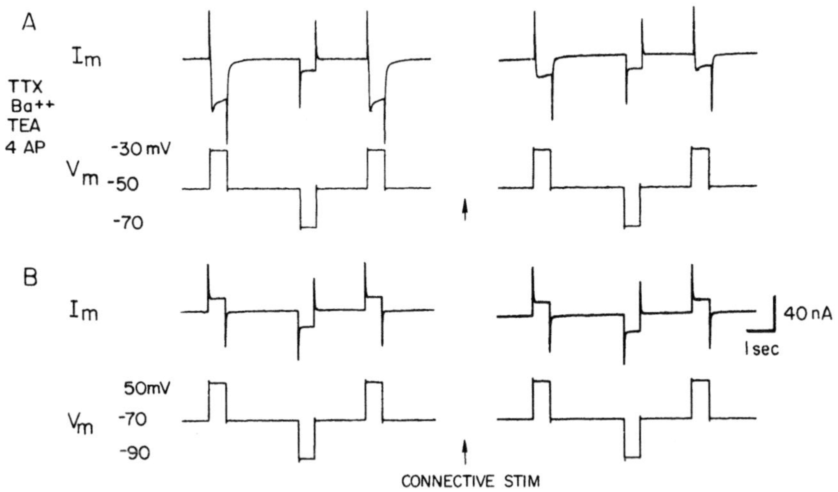

Fig. 17A,B. The conductance changes caused by stimulation of the pleuroabdominal connectives, the presynaptic inhibitory pathway, are voltage sensitive. L10 was voltage clamped in seawater containing Ba^{2+}, TTX, TEA, and 4-AP. Under these pharmacologic conditions 30-min intervals were used between runs. *A* From a holding potential of -50 mV, alternating 20-mV depolarizing and hyperpolarizing voltage steps elicit inward currents. Connective stimulation markedly reduces inward Ba^{2+} current through the Ca^{2+} channel of the depolarizing step and only slightly increases leakage current of the hyperpolarizing step. V_m, membrane voltage; I_m, membrane current. *B* From a holding potential of -70 mV, alternating 20-mV depolarizing and hyperpolarizing voltage steps again elicit inward current with hyperpolarizing steps but elicit primarily outward leakage currents with depolarizing steps. In this voltage range depolarizing command pulses do not significantly activate inward currents. Only small leakage currents are produced by both hyperpolarizing and depolarizing steps. Connective stimulation does not affect these currents. (Shapiro et al. 1980b)

tential and perhaps also change the invasion properties of the presynaptic terminals. However, presynaptic inhibition caused by stimulating each of these two pathways outlasts the hyperpolarization and, perhaps more importantly, presynaptic inhibition can also be observed in cases in which L10 is voltage clamped or in those cases when it is not hyperpolarized (Fig. 16). To explore the additional mechanisms, we examined the Ca^{2+} current of the presynaptic neuron under voltage clamp conditions.

Presynaptic Inhibition Is Correlated with Depression of the Ca^{2+} Current. To analyze the changes in Ca^{2+} current during presynaptic inhibition, we utilized solutions containing TTX and TEA. With the cell held at a membrane potential that inactivates the early outward K^+ current, the inward current produced by a depolarizing clamp step is carried by Ca^{2+} ions, and the residual outward current is mainly due to the Ca^{2+}-dependent K^+ current and the leakage current. Under these circumstances, presynaptic inhibition produced by connective stimulation is correlated with a decrease in transient inward current evoked by the depolarizing command (Fig. 16).

We next examined the current through the Ca^{2+} channel in isolation by blocking the Na^+ channel with TTX and the three known K^+ channels with TEA and

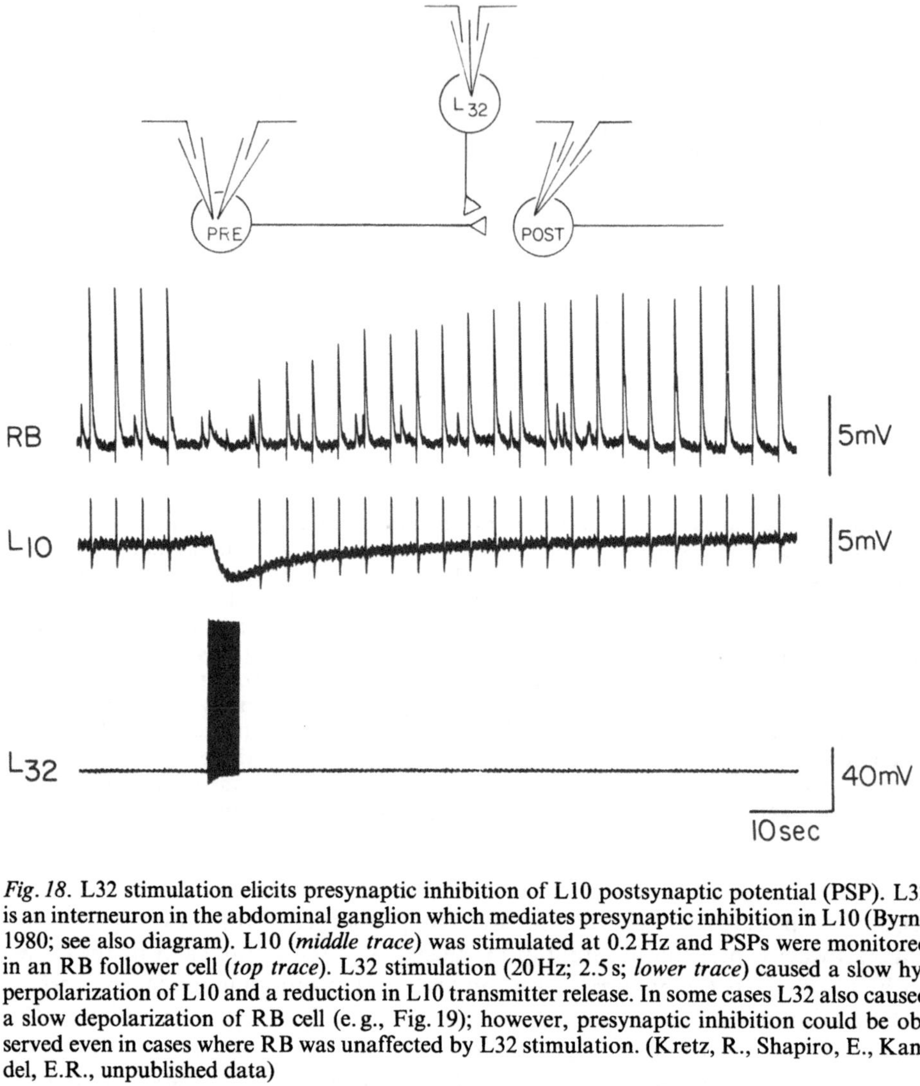

Fig. 18. L32 stimulation elicits presynaptic inhibition of L10 postsynaptic potential (PSP). L32 is an interneuron in the abdominal ganglion which mediates presynaptic inhibition in L10 (Byrne 1980; see also diagram). L10 (*middle trace*) was stimulated at 0.2 Hz and PSPs were monitored in an RB follower cell (*top trace*). L32 stimulation (20 Hz; 2.5 s; *lower trace*) caused a slow hyperpolarization of L10 and a reduction in L10 transmitter release. In some cases L32 also caused a slow depolarization of RB cell (e.g., Fig. 19); however, presynaptic inhibition could be observed even in cases where RB was unaffected by L32 stimulation. (Kretz, R., Shapiro, E., Kandel, E.R., unpublished data)

4-AP, and substituting Ba^{2+} for Ca^{2+}. Again, connective stimulation reduced the inward current (Fig. 17). Since the Ca^{2+} current is voltage dependent, one would expect that under these pharmacologic conditions the synaptic transmitter that mediates presynaptic inhibition would have no effect on ionic currents at hyperpolarized levels of membrane potential where the Ca^{2+} current is not activated. We therefore compared the effects of presynaptic inhibition on currents in the voltage range of Ca^{2+} channel activation with its effects in the range where Ca^{2+} channels are closed. When cell L10 was held in the voltage range in which the Ca^{2+} channel was not activated, no conductance change was seen after connective stimulation (Fig. 17B). These results support the idea that presynaptic inhibition is due to a direct action on the Ca^{2+} channel. A similar mechanism for trans-

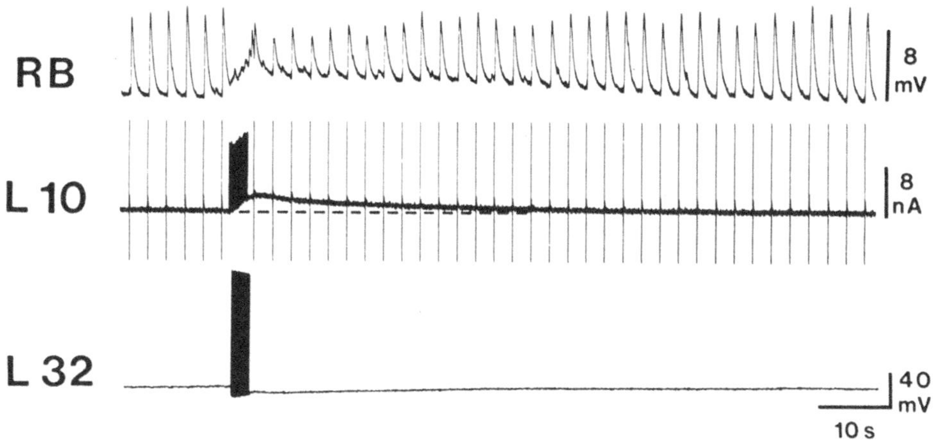

Fig. 19. L32 elicited presynaptic inhibition of voltage-clamped L10. L32 stimulation (20 Hz; 2.5 s; *lower trace*) elicited a slow outward clamp current in L10 (*middle trace*) and reduced transmitter release, as monitored in RB cell (*top trace*). Transmitter release was evoked by depolarizing L10 clamp commands of 50-ms duration from holding potential -40 mV to -20 mV at 0.4 Hz. L10 voltage record not shown (Kretz, R., Shapiro, E., Kandel, E.R., unpublished data)

mitter-mediated inhibition of Ca^{2+} conductance has been discovered in dissociated dorsal root ganglion cells (Dunlap and Fischbach 1978, 1981; Mudge et al. 1979) and in sympathetic ganglion cells (Horn and McAfee 1980; Henon and McAfee 1983). In addition, in cultured dissociated dorsal root ganglion cells, it has been shown that neurotransmitters thought to be involved in presynaptic inhibition, e.g., gamma-aminobutyric acid and enkephalin, inhibit the Ca^{2+} current (Dunlap and Fischbach 1981; Mudge et al. 1979). The observed changes in Ca^{2+} current were voltage sensitive and not associated with a resting membrane potential or resting conductance change (Dunlap and Fischbach 1981).

Presynaptic Inhibition Can Be Produced by Stimulating a Single Presynaptic Inhibitory Neuron. Connective stimulation invariably recruits many inputs onto L10 in addition to those causing presynaptic inhibition. To explore more directly the action of a specified presynaptic inhibitory transmitter, we took advantage of the finding of Byrne (1980) that the L32 cells mediate presynaptic inhibition. We impaled individual L32 cells while simultaneously recording from a follower cell of L10 and examined L10 under voltage clamp. This three-cell arrangement consisting of a presynaptic neuron, a follower cell, and a presynaptic inhibitory neuron has allowed us to examine more selectively the ionic changes responsible for presynaptic inhibition and to determine in more detail the physiological mechanism of presynaptic inhibition. As is the case with stimulating the connective, stimulating an L32 cell causes a slow hyperpolarization of L10 (Fig. 18). But again the presynaptic inhibition does not simply result from this hyperpolarization because the L32 cell causes presynaptic inhibition even when L10 is voltage clamped. Under voltage clamp, presynaptic inhibition is correlated with a slow outward voltage clamp current (Fig. 19). By comparing the amount of inhibition caused by an L32 cell when L10 is clamped with the inhibition caused when it is

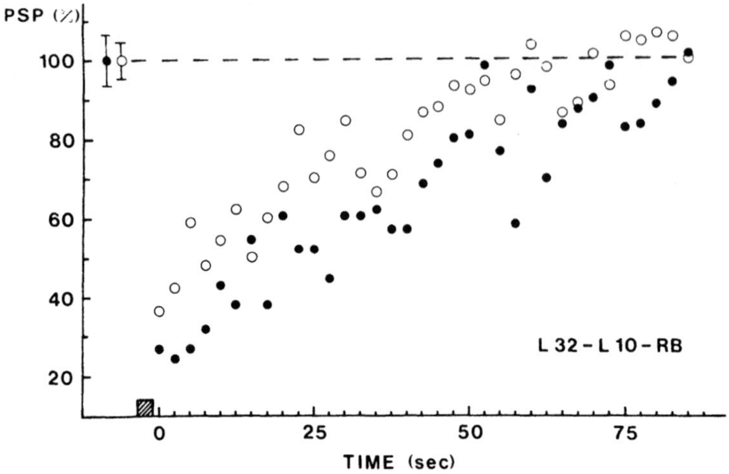

Fig. 20. Comparison between inhibitory effect of L32 stimulation (*hatched bar*) on transmitter release from unclamped (*filled circles*) and clamped (*open circles*) L10. L10 control postsynaptic potential (PSP) size (100%) is average of ten PSPs $\pm$SD, recorded before L32 stimulation. Presynaptic inhibition is still seen when L10 is voltage clamped, although it is partially reduced, indicating that the L32-induced hyperpolarization of L10 contributes to the full expression of L32-induced presynaptic inhibition of L10's output. (Kretz, R., Shapiro, E., Kandel, E.R., unpublished data)

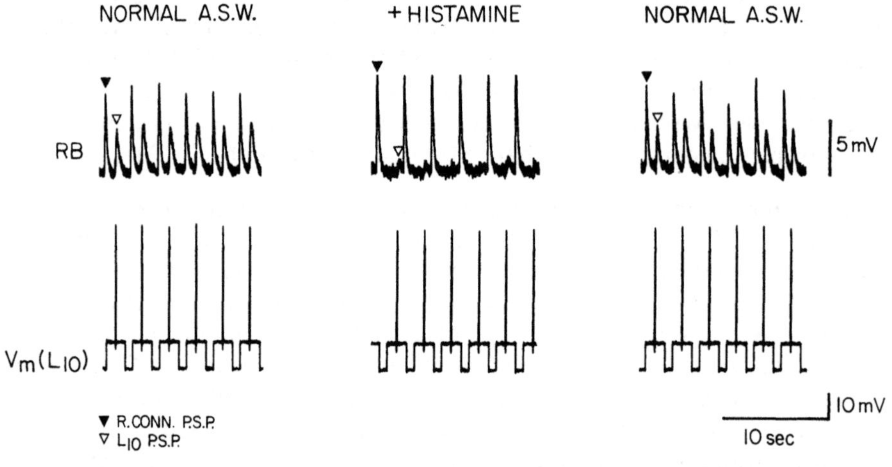

Fig. 21. Bath application of histamine specifically inhibits L10's output. L10 was voltage clamped in a normal seawater solution at holding potential of -40 mV. Depolarizing pulses (at 0.4 Hz) of 50 ms to 0 mV (*lower trace*) elicited excitatory postsynaptic potentials (EPSPs) in RB follower cell (*top traces; open triangles*). These pulses were alternated with brief extracellular stimuli to the right pleuroabdominal connective to elicit a second EPSP onto the RB cell (*closed triangles*). When 5×10^{-4} M histamine was added to the bath, the L10 postsynaptic potential (PSP) was inhibited, while the connective PSP was unaffected. Histamine also depolarized the RB cell (see Fig. 19; not shown in this figure). When the preparation was returned to normal seawater the L10 PSP recovered to control value. *A.S.W.*, artificial seawater; V_m, membrane voltage; *R.CONN.*, right connective PSP. (Kretz, R., Shapiro, E., Kandel, E.R., unpublished data)

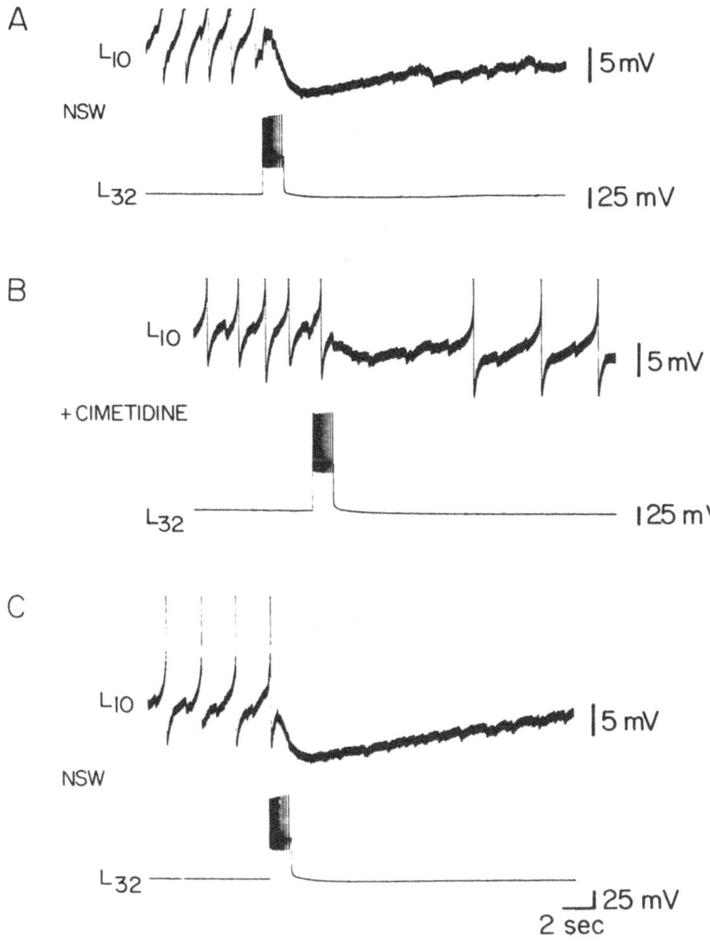

Fig. 22A–C. Cimetidine reduces L32-L10 postsynaptic potential. *A* In a normal seawater (*NSW*) solution L32 stimulation (*lower trace*) causes a slow hyperpolarization of L10 (*top trace*). *B* In 2×10^{-4} M cimetidine, a histamine antagonist, the L32-L10 inhibitory postsynaptic potential (IPSP) is reduced. *C* After the cimetidine was washed out and normal seawater solution restored, L32 stimulation again caused a long-lasting IPSP onto L10. (Kretz et al., to be published)

unclamped, we have determined that the hyperpolarization contributes approximately 20%–30% to the total presynaptic inhibition (Fig. 20).

Histamine Is a Candidate for the Transmitter Used by the L32 Cells for Presynaptic Inhibition. To examine the action of the L32 cells in greater detail, it would be useful to know the transmitter they release to produce their presynaptic inhibitory action. Histamine has been reported to produce presynaptic inhibition in the cerebral ganglion of *Aplysia* (Hinzen and Riehl 1980). We have obtained preliminary evidence that the L32 cluster might also use histamine as its transmitter. This supposition is based on the following findings. First, the vesicle morphology of the L32 cells is similar to that of the known histaminergic cell C2 (Bailey et al.

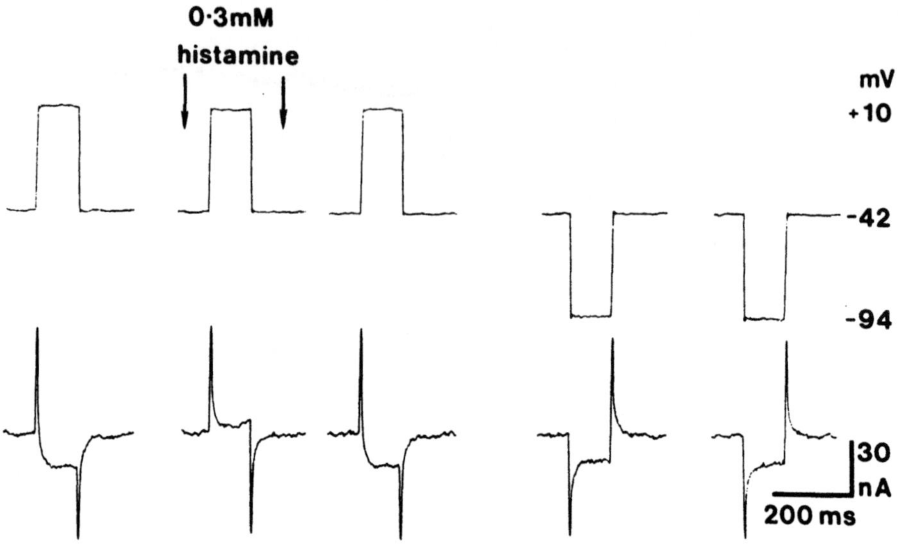

Fig. 23. Histamine reduces L10's Ca^{2+} conductance. L10 was voltage clamped in a solution containing Ba^{2+} (66 mM), TTX, TEA (25 mM), and 4-AP (10 mM) at -42 mV. Under these conditions inward currents (*lower trace*) during depolarizing steps to + 10 mV (*top trace*) are Ba^{2+} currents through Ca^{2+} channels (*first pulse*). When 0.3 mM histamine was added to the bath the inward current was reduced (*second pulse*). After the histamine was washed out and the control solution restored, the inward current recovered (*third pulse*). Hyperpolarizing pulses to measure leakage (-42 to -94 mV) were unchanged by the addition of histamine (compare *fourth and fifth pulses*). (Kretz, R., Shapiro, E., Kandel, E.R., unpublished data) Compare this figure to Fig. 17.

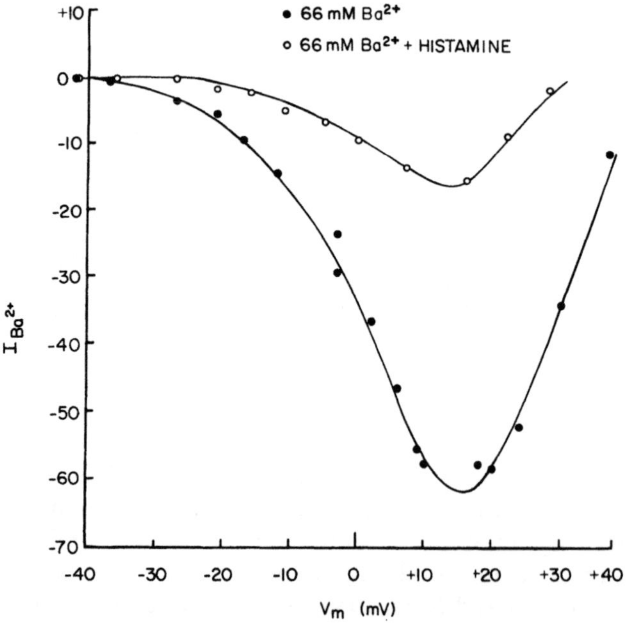

Fig. 24. Current-voltage curve showing that histamine reduces Ba^{2+} current in a voltage-dependent manner. This suggests that histamine is directly reducing the Ca^{2+} conductance of L10. V_m, membrane voltage; $I_{Ba^{2+}}$, Ba^{2+} current. (Same experiment as in Fig. 23.)

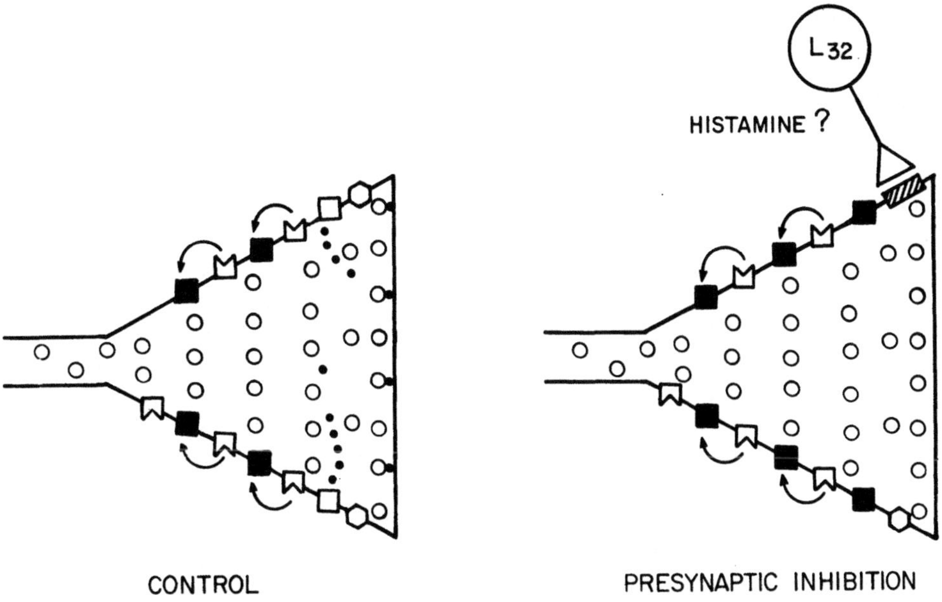

Fig. 25. Mechanism of presynaptic inhibition. A presynaptic inhibitory transmitter, putatively identified as histamine, directly reduces the voltage-dependent calcium current, making less calcium available for transmitter release. A specific cell cluster (L32 cells) can mediate presynaptic inhibition (Byrne 1980). L32 cells cause presynaptic hyperpolarization, which also contributes to presynaptic inhibition (see part *A*). (For more details see legend to Fig. 7.)

1982; Bailey et al. 1983). Secondly, among a number of candidate transmitters which we tested, only histamine simulated the effects of L32 cells both on the conductance of L10 and on transmitter release from its terminals (Fig. 21). Thirdly, the synaptic actions produced by the L32 cells on L10 are blocked by cimetidine, a histamine antagonist (Fig. 22). We are currently trying to obtain more direct evidence for the histaminergic nature of L32 by exploring the conversion of ^{3}H-histidine to histamine in the L32 cells.

Using histamine as a pharmacologic probe, we were able to examine the L10 conductance changes which might underlie presynaptic inhibition under more favorable voltage clamp conditions. In solutions in which K^+ channels were not pharmacologically blocked, histamine application caused a hyperpolarization in L10 due to an increased K^+ conductance (data not shown). We also examined the voltage dependence of the conductance change underlying presynaptic inhibition using histamine. In solutions in which all K^+ currents were blocked, histamine application caused a specific reduction in the inward current through Ca^{2+} channels of L10 (Fig. 23; compare similar experiment done with connective stimulation in Fig. 17). This reduction was voltage dependent (Fig. 24).

These results suggest that presynaptic inhibition in L10 is caused by histamine or a related transmitter acting by means of two synergistic mechanisms: (a) transmitter-mediated increased K^+ conductance leading to a presynaptic hyperpolarization; and (b) transmitter-mediated reduction in presynaptic Ca^{2+} current. This mechanism of presynaptic inhibition is represented diagrammatically in Fig. 25.

An Overall View

Limitations

Ideally, the analysis of the currents underlying transmitter release should be carried out on the membrane of presynaptic terminals of the neuron under study, as has been done in the case of the giant synapse of the squid (Katz and Miledi 1967; Llinás et al. 1981a, b). The analysis we have described here was carried out on the cell body, with the result that most of the total voltage clamp current was recorded across the soma membrane. Although the channels in the soma membrane seem to parallel what happens at the transmitter-releasing membrane area, the channels in the soma do not contribute directly to release. Nevertheless, the terminals are probably electrically quite close to the cell body because we have also been able to modify release and to change the time course of posttetanic potentiation simply by passing current into the cell body, or by injecting EGTA into the cell body. As a result, currents generated in the presynaptic terminals contribute to the current recorded in the cell body. In addition, the recorded voltage clamp currents in the intact cell correlate well with transmitter release as measured directly, and the cell body currents examined in isolated somata, where the space clamp control is optimal, have properties similar to those of the currents recorded in intact cell.

Generality

Despite clear limitations, the ability to analyze directly the Ca^{2+} current and the Ca^{2+}-buffering capability of the presynaptic cell body while simultaneously assaying transmitter release from the terminals has made it possible to begin to explore the mechanisms that underlie a variety of plastic changes in transmitter release (see also Llinás et al. 1981a, b). How likely is it that these mechanisms are general? Although we have shown that even at the same synapse various types of synaptic plasticity have different mechanisms, we believe that the family of mechanisms that operate in *Aplysia* are likely to be general. The ion channels found in L10, including the early K^+ channel, the Ca^{2+}-dependent K^+ channel, and the Ca^{2+} channel, have been described in many other invertebrate and vertebrate nerve cells. Since the TTX-sensitive Na^+ channel and the delayed K^+ channel are highly conserved, it is likely that the other ion channels, including the Ca^{2+} channel and the Ca^{2+}-dependent K^+ channel, will also be shown to share molecular similarities across species (Llinás 1980; Rogart 1981; Meech 1978; Brown et al. 1982; Hagiwara and Byerly 1981, 1983). Indeed, the ability of modulatory neurotransmitters to regulate voltage-dependent and Ca^{2+}-dependent channels is now also being described in other systems (Kehoe and Marty 1980; Hartzell 1981; Madison and Nicoll 1982). These similarities suggest that mechanisms of plasticity that involve changes in voltage-dependent ion channels and regulation of Ca^{2+} influx are also likely to be general.

For example, the striking similarity in time course of posttetanic potentiation in *Aplysia* and in a wide variety of other synapses suggests a common mechanism (Feng 1941; Curtis and Eccles 1960; Martin and Pilar 1964; Lloyd 1949; Liley 1956a, b; Rosenthal 1969; Luescher et al. 1983). That posttetanic potentiation at most synapses correlates with membrane hyperpolarization also supports the generality of the residual Ca^{2+} mechanism for posttetanic potentiation (Lloyd 1949; Braun and Schmidt 1966; Gage and Hubbard 1966a; Kretz et al. 1982). Similarly, transmitter-mediated reduction of voltage-dependent Ca^{2+} current similar to that found in L10 for presynaptic inhibition has also been found in vertebrate neurons in dissociated dorsal root and in sympathetic ganglion cells (Dunlap and Fischbach 1978, 1981; Horn and McAfee 1980; Henon and McAfee 1983). In dorsal root ganglion cells, the reduction in Ca^{2+} current is caused by transmitters that are thought to mediate presynaptic inhibition (Mudge et al. 1979; Dunlap and Fischbach 1981).

Although a mechanism found in *Aplysia* such as that for presynaptic inhibition is likely to be general, the excellent evidence for a Cl^- shunting mechanism in the crayfish neuromuscular synapse clearly suggests that synaptic depression of the Ca^{2+} current is not the only mechanism. Clearly there is no one universal mechanism for presynaptic inhibition. The time course of presynaptic inhibition in crayfish (2–3 ms) is also very different from that in *Aplysia* and in vertebrates (seconds to minutes). Perhaps presynaptic inhibition that lasts only one or a few milliseconds is due to an increased conductance to Cl^- (or some other shunting mechanism) produced by a conventional synaptic action, whereas presynaptic inhibition that lasts from seconds to minutes, such as is seen in *Aplysia* and in the vertebrate spinal cord, is due to modulatory synaptic actions whose time courses are longer. The literature on the correlation between presynaptic inhibition and terminal depolarization or hyperpolarization (or increased and decreased terminal excitability) need not be inconsistent with a modulatory transmitter action on presynaptic Ca^{2+} channels. Decreases in Ca^{2+} conductance can cause depolarization or hyperpolarization (and increased or decreased excitability), depending on the coupling of the Ca^{2+} current and the Ca^{2+}-activated K^+ current (see Grafe et al. 1980; Morita and North 1981; Morita et al. 1982).

Common and Distinctive Features of Short-Term Synaptic Plasticity

The distinction between excitability and plasticity, first outlined by Konorski 40 years ago, has proved consistently useful both conceptually and experimentally. A large number of plastic changes have now been delineated in nervous systems, and these involve not only changes at synapses, of the sort we have reviewed in this paper, but also changes at nonsynaptic sites within the neuron.

Here we have considered three common types of short-term synaptic plasticity: the control of transmitter release by membrane potential, posttetanic potentiation, and presynaptic inhibition. Elsewhere we have also analyzed homosynaptic depression and presynaptic facilitation (see Klein et al. 1980). The five forms of short-term synaptic plasticity have two important features in common: (a) they all involve modulation in transmitter release; and (b) in each case modulation of transmitter release occurs through alterations in the level of free Ca^{2+} in the presynaptic terminals (see Figs. 7, 15, and 25; Table 2). The changes in the level of free Ca^{2+} can come from alterations in influx or alterations in buffering. Alterations in Ca^{2+} influx can in turn result from indirect and direct modulation. Indirect modulation of the Ca^{2+} current is achieved through the regulation of the opposing K^+ curents that control the height and duration of the action potential

Table 2. Modulation of transmitter release through alteration in the level of free Ca^{2+} in the presynaptic terminals

Plastic change	Ca^{2+} current	Direct (Ca^{2+} channel)	Indirect (K^+ channel)	Ca^{2+} accumulation	Homosynaptic	Heterosynaptic
Presynaptic inhibition		+	o	o	o	+
Homosynaptic depression (habituation)		+	o	o	+	o
Presynaptic facilitation (sensitization)		o	+	o	o	+
Depolarization-dependent increase in transmitter release		+	+	o	+	o
Posttetanic potentiation	o	o	o	+	+	o

and the space constant of the neuron, thereby determining the time during which the Ca^{2+} channels are activated. Direct modulation of the Ca^{2+} current, on the other hand, is possibly achieved through actions on the Ca^{2+} channel protein itself. In addition to Ca^{2+} influx, Ca^{2+} buffering and consequently the resting internal Ca^{2+} concentration can also be regulated; and this can also modulate the amount of transmitter released as well as cell excitability.

Thus, in a fascinating and surprising way, the various mechanisms underlying synaptic plasticity represent regulatory steps that act upon the basic processes involved in a cell's excitability: ion channels for K^+ and Ca^{2+}; buffering of Ca^{2+} in the terminals; and the control by Ca^{2+} of transmitter vesicle exocytosis.

For example, the role of Ca^{2+} influx into the terminals with each action potential has been recognized as an essential step for transmitter release for more than a decade. But it has not been clear whether changes in Ca^{2+} influx are involved in the well-known capability of chemical synapses to undergo plastic changes in synaptic effectiveness. The experiments reviewed here provide clear evidence that Ca^{2+} influx into the terminals can be regulated in both directions and that this regulation can contribute to the control of synaptic efficacy lasting minutes to hours. Thus the great capability of chemical synapses for plasticity derives in part from an interesting characteristic of Ca^{2+} and K^+ channel proteins: that they are endowed with several regulatory sites. Sites have been identified on Ca^{2+} channels that are sensitive to (a) membrane voltage (reviewed in Hagiwara and Byerly 1981, 1983); (b) chemical transmitters (Reuter and Scholz 1977; Tsien 1977; Dunlap and Fischbach 1981); (c) cyclic nucleotide action (Reuter and Scholz 1977; Tsien 1977; Irisawa and Kokubun 1983); and (d) intracellular Ca^{2+} concentration (Brehm and Eckert 1978; Tillotson 1979; Eckert and Tillotson 1981; Plant and Standen 1981; Plant et al. 1983). In internally dialyzed cells, the Ca^{2+} current is dependent on the presence of intracellular factors (Kostyuk et al. 1981; Byerly and Hagiwara 1982; Fenwick et al. 1982).

The ability to modulate the Ca^{2+} current indirectly seems to derive from the existence of a similar set of regulatory sites on the K^+ channel. In particular, both the K^+ and the Ca^{2+} channels are capable of being affected by membrane voltage on the one hand and by chemical transmitters (and perhaps by one or more intracellular regulatory signals, including Ca^{2+}) on the other (Klein et al. 1982; Siegelbaum et al. 1982; Marty 1981; Pallotta et al. 1981; Shapiro et al. 1981; Tokimasa et al. 1981; Grafe et al. 1980). Although the Ca^{2+} and K^+ channels have now been shown to be sites of regulation for transmitter release, they are not the only sites. The buffering of Ca^{2+} is also critical in determining the level of free Ca^{2+} and therefore synaptic strength. Moreover, we are just beginning to see how Ca^{2+} acts within the cell to signal vesicle release.

Most models of transmitter release have distinguished two processes regulating transmitter release, vesicle "mobilization" or "depletion," and vesicle "fusion" (e.g., Llinás et al. 1981b). Implicit in this distinction is the recognition of two regions for regulation of release: the presynaptic region at a distance from the membrane release sites, and the presynaptic region directly adjacent to the surface membrane. In studies on chromaffin cells, two independent Ca^{2+}-dependent processes governing the release of granules (vesicles) have been described (Pollard et al. 1981; Fowler and Pollard 1982). The first consists of a Ca^{2+}-dependent dissociation of granules from actin filaments and actin filament depolymerization, which allows granules to approach the surface membrane (Fowler and Pollard 1982). The second is a Ca^{2+}-dependent enhancement of membrane fusion, mediated by the protein synexin, which causes exocytosis (Pollard et al. 1981). Further progress in analyzing the steps in synaptic plasticity will require that the techniques we have outlined here be merged with molecular probes to explore the mobilization and release processes within the cell.

References

Adams D, Smith S, Thompson S (1980) Ionic currents in molluscan soma. Annu Rev Neurosci 3: 141–167

Ahmed Z, Connor JA (1979) Measurement of Ca influx under voltageclamp in molluscan neurones using the metallochromic dye Arsenazo III. J Physiol 286: 61–82

Akaike N, Lee KS, Brown AM (1978) The calcium current of *Helix* neuron. J Gen Physiol 71: 509–531

Alnaes E, Rahamimoff R (1975) On the role of mitochondria in transmitter release from motor nerve terminals. J Physiol 248: 285–306

Atwood HL, Charlton MP, Thompson CS (1983) Neuromuscular transmission in crustaceans is enhanced by a sodium ionophore, monensin, and by prolonged stimulation. J Physiol 335: 179–196

Bailey CH, Chen MC, Weiss KR, Kupfermann I (1982) Ultrastructure of a histaminergic synapse in *Aplysia*. Brain Res 238: 205–210

Bailey CH, Chen M, Kretz R, Shapiro E (1983) Morphology and pharmacology of presynaptic inhibition in *Aplysia* mediated by an identified presynaptic inhibitory neuron. Soc Neurosci Abstr 9: 1025

Bass L, Moore WJ (1966) Electrokinetic mechanism of miniature postsynaptic potentials. Proc Natl Acad Sci USA 55: 1214–1217

Baxter DA, Bittner GD (1981) Intracellular recordings from crustacean motor axons during presynaptic inhibition. Brain Res 223: 422–428

Braun M, Schmidt RF (1966) Potential changes recorded from the frog motor nerve terminal during its activation. Pflügers Arch 287: 56–80

Brehm P, Eckert R (1978) Calcium entry leads to inactivation of calcium channel in *Paramecium*. Science 202: 1203–1206

Brown AM, Camerer H, Kunze KL, Lux HD (1982) Similarity of unitary Ca^{2+} currents in three different species. Nature 299: 156–158

Brown JE, Cohen LB, De Weer P, Pinto LH, Ross WN, Salzberg BM (1975) Rapid changes of intracellular free calcium concentration: detection by metallochromic indicator dye in squid giant axon. Biophys J 15: 1155–1160

Burke RE, Rudomin P (1977) Spinal neurons and synapse. In: Kandel ER (ed) Handbook of Physiology, vol 1. American Physiological Society, Bethesda, pp 877–944

Byerly L, Hagiwara S (1982) Calcium currents in internally perfused nerve cell bodies of *Limnea stagnalis*. J Physiol 322: 503–528

Byrne J (1980) Identification of neurons contributing to presynaptic inhibition in *Aplysia californica*. Brain Res 199: 235–239

Byrne JH, Shapiro E, Dieringer N, Koester J (1979) Biophysical mechanisms contributing to inking behavior in *Aplysia*. J Neurophysiol 42: 1233–1250

Cajal SR (1911) Histologie du systeme nerveux de l'homme et des vertebres, vol 2. Maloine, Paris (Republished Azoulay L (1955) Histologie du systeme nerveux. Instituto Ramon y Cajal, Madrid)

Castellucci VF, Kandel ER (1976) Presynaptic facilitation as a mechanism for behavioral sensitization in *Aplysia*. Science 194: 1176–1178

Charlton MP, Smith SJ, Zucker RS (1982) Role of presynaptic calcium ions and channels in synaptic facilitation and depression at the squid giant synapse. J Physiol 323: 173–194

Connor JA (1979) Calcium current in molluscan neurones: measurement under conditions which maximize its visibility. J Physiol 286: 41–60

Connor JA, Stevens CF (1971a) Inward and delayed outward membrane currents in isolated neural somata under voltage-clamp. J Physiol 213: 1–19

Connor JA, Stevens CF (1971b) Voltage-clamp studies of a transient outward membrane current in gastropod neural somata. J Physiol 213: 21–30

Curtis DR, Eccles JC (1960) Synaptic action during and after repetitive stimulation. J Physiol 150: 374–398

del Castillo J, Katz B (1954a) Quantal components of the end-plate potential. J Physiol 124: 560–573

del Castillo J, Katz B (1954b) Statistical factors involved in neuromuscular facilitation and depression. J Physiol 124: 574–585

del Castillo J, Katz B (1954c) Changes in end-plate activity produced by presynaptic polarization. J Physiol 124: 586–604

Dudel J, Kuffler SW (1961a) The quantal nature of transmission and spontaneous miniature potentials at the crayfish neuromuscular junction. J Physiol 155: 514–529

Dudel J, Kuffler SW (1961b) Presynaptic inhibition at the crayfish neuromuscular juction. J Physiol 155: 543–562

Dunlap K, Fischbach GD (1978) Neurotransmitters decrease the calcium component of sensory neurone action potential. Nature 276: 837–839

Dunlap K, Fischbach GD (1981) Neurotransmitters decrease the calcium conductance activated by depolarization of embryonic chick sensory neurones. J Physiol 317: 519–535

Eccles JC (1964) The physiology of synapses. Springer, Berlin Heidelberg New York

Eckert R, Lux HD (1976) A voltage-sensitive persistent calcium conductance in neuronal somata of *Helix*. J Physiol 254: 129–151

Eckert R, Tillotson DL (1981) Calcium-mediated inactivation of the calcium conductance in caesium-loaded giant neurones of *Aplysia californica*. J Physiol 314: 265–280

Erulkar SD, Rahamimoff R (1978) The role of calcium ions in tetanic and post-tetanic increase of miniature end-plate potential frequency. J Physiol 278: 501–511

Erulkar SD, Rahamimoff R, Rotschenker S (1978) Quelling of spontaneous transmitter release by nerve impulses in low extracellular calcium solutions. J Physiol 278: 491–500

Feng TP (1941) Studies on the neuromuscular junction XXVI. The changes of the end-plate potential during and after prolonged stimulation. Chin J Physiol 16: 341–372

Fenwick EM, Marty A, Neher E (1982) Sodium and calcium channels in bovine chromaffin cells. J Physiol 331: 599–635

Forbes A (1922) The interpretation of spinal reflexes in terms of present knowledge of nerve conduction. Physiol Rev 2: 361–414

Fowler VF, Pollard H (1982) Chromaffin granule membrane-F-actin interactions are calcium sensitive. Nature 295: 336–339

Frank K, Fuortes MGF (1957) Presynaptic and post-synaptic inhibition of monosynaptic reflexes. Fed Proc 16: 39–40

Fuchs PA, Getting PA (1980) Ionic basis of presynaptic inhibitory potentials at crayfish claw opener. J Neurophysiol 43: 1547–1557

Gage PW, Hubbard JI (1966a) The origin of the post-tetanic hyperpolarization of mammalian motor nerve terminals. J Physiol 184: 335–352

Gage PW, Hubbard JI (1966b) An investigation of the post-tetanic potentiation of the end-plate potentials at a mammalian neuromuscular junction. J Physiol 184: 353–375

Geduldig D, Gruener R (1970) Voltage-clamp of the *Aplysia* giant neurone: early sodium and calcium currents. J Physiol 211: 217–244

Gopfert H, Schaefer H (1938) Über den direkt und indirekt erregten Aktionsstrom und die Funktion der motorischen Endplatte. Pflügers Arch 239: 597–619

Goodman CG, Pearson KG (1982) Neuronal development: cellular approaches in invertebrates. Neurosci Res Program Bull 20: 773–942

Gorman ALF, Hermann A (1979) Internal effects of divalent cations on potassium permeability in molluscan neurones. J Physiol 296: 393–410

Gorman ALF, Thomas MV (1978) Changes in the intracellular concentration of free calcium ions in a pacemaker neurone, measured with the metallochromic indicator dye Arsenazo III. J Physiol 275: 357–376

Gorman ALF, Hermann A, Thomas MV (1983) Ionic requirements for membrane oscillations and their dependence on the calcium concentration in a molluscan pace-maker neurone. J Physiol 327: 185–217

Grafe P, Mayer CJ, Wood JD (1980) Synaptic modulation of calcium-dependent potassium conductance in myenteric neurones in the guinea-pig. J Physiol 305: 235–248

Hagiwara S, Byerly L (1981) Calcium channel. Annu Rev Neurosci 4: 69–125

Hagiwara S, Byerly L (1983) The calcium channel. Trends Neurosci 6: 189–193

Hagiwara S, Tasaki I (1958) A study on the mechanism of impulse transmission across the giant synapse of the squid. J Physiol 143: 114–137

Halt H, Smith DO (1976) Synaptic depression related to presynaptic axon conduction block. J Physiol 259: 267–393

Hartzell HC (1981) Mechanisms of slow postsynaptic potentials. Nature 291: 539–544

Henon BK, McAfee DA (1983) The ionic basis of adenosine receptor actions on post-ganglionic neurones in the rat. J Physiol 336: 607–620

Hinzen DH, Riehl J (1980) Presynaptic inhibition in cerebral ganglion of *Aplysia*. J Physiol 305: 57–58P

Horn JP, McAfee DA (1980) Alpha-adrenergic inhibition of calcium-dependent potentials in rat sympathetic neurones. J Physiol 301: 191–204

Horn JP, Miller JJ (1977) A prolonged, voltage-dependent calcium permeability revealed by tetra-ethyl ammonium in the soma and axon of *Aplysia* giant neurons. J Neurobiol 8: 399–415

Horvitz HR (1982) Neuronal cell lineages in the nematode *Caenorhabditis elegans*. In: Garrod D, Feldman J (eds) Development of the nervous system. Cambridge University Press, Cambridge

Hubel DH (1982) Exploration of the primary visual cortex, 1955–78. Nature 299: 515–524

Irisawa H, Kokubun S (1983) Modulation by intracellular ATP and cyclic AMP of the slow inward current in isolated single ventricular cells of the guinea-pig. J Physiol 338: 321–337

Kandel ER (1976) Cellular basis of behavior. Freeman, San Francisco

Kandel ER (1978) A cell-biological approach to learning. Society for Neuroscience, Rockville

Kandel ER, Frazier WT, Waziri R, Coggeshall RE (1967) Direct and common connections among identified neurons in *Aplysia*. J Neurophysiol 30: 1253–1376

Katz B (1962) The transmission of impulses from nerve to muscle, and the subcellular unit of synaptic action. Proc Soc Lond [Biol] 155: 455–477

Katz B, Miledi R (1965) The effect of calcium on acetylcholine release from motor nerve terminals. Proc Soc Lond [Biol] 161: 496–503

Katz B, Miledi R (1967) A study of synaptic transmission in the absence of nerve impulses. J Physiol 192: 407–436

Katz B, Miledi R (1968) The role of calcium in neuromuscular facilitation. J Physiol 195: 481–492

Katz B, Miledi R (1969) Tetrodotoxin resistant electrical activity in presynaptic terminals. J Physiol 203: 459–487

Kehoe J (1972) Three acetylcholine receptors in *Aplysia* neurones. J Physiol 225: 115–146

Kehoe J, Marty A (1980) Certain slow synaptic responses: their properties and possible underlying mechanisms. Annu Rev Biophys Bioeng 9: 437–466

Kennedy DF, Calabrese RL, Wine JJ (1974) Presynaptic inhibition: primary afferent depolarization in crayfish neurons. Science 186: 451–454

Kennedy D, McVittie J, Calabrese R, Fricke RA, Craelius W, Chiappella P (1980) Inhibition of mechanosensory interneurons in the crayfish. I. Presynaptic inhibition from giant fibers. J Neurophysiol 43: 1495–1509

Klein M, Kandel ER (1978) Presynaptic modulation of voltage-dependent Ca^{++} current: mechanism for behavioral sensitization in *Aplysia californica.* Proc Natl Acad Sci USA 75: 3512–3516

Klein M, Kandel ER (1980) Mechanisms of calcium current modulation underlying presynaptic facilitation and behavioral sensitization in *Aplysia.* Proc Natl Acad Sci USA 77: 6912–6916

Klein M, Shapiro E, Kandel ER (1980) Synaptic plasticity and the modulation of the Ca^{2+} current. J Exp Biol 89: 117–157

Klein M, Camardo JS, Kandel ER (1982) Serotonin modulates a specific potassium current in the sensory neurons that show presynaptic facilitation in *Aplysia.* Proc Natl Acad Sci USA 79: 5713–5717

Koester J, Kandel ER (1977) Further identification of neurons in the abdominal ganglion by *Aplysia* using behavioral criteria. Brain Res 121: 1–20

Koester J, Mayeri E, Liebeswar G, Kandel ER (1974) Neural control of circulation in *Aplysia.* II. Interneurons. J Neurophysiol 37: 476–496

Konorski J (1948) Conditioned reflexes and neuron organization. Cambridge University Press, Cambridge

Kostyuk PG, Krishtal OA, Pidoplichko VI (1981) Calcium inward current and related charge movements in the membrane of snail neurones. J Physiol 310: 403–421

Kretz R, Shapiro E, Kandel ER (1982) Post-tetanic potentiation at an identified synapse in *Aplysia* is correlated with a Ca^{++}-activated K^{+} current in the presynaptic neuron: evidence for Ca^{++} accumulation. Proc Natl Acad Sci USA 79: 5430–5434

Larrabee MG, Bronk DW (1947) Prolonged facilitation of synaptic excitation in sympathetic ganglia. J Neurophysiol 10: 139–154

Lewis DV, Wilson WA (1982) Calcium influx and post-stimulus current during early adaptation in *Aplysia* giant neurons. J Neurophysiol 48: 202–216

Liley AW (1956a) An investigation of spontaneous activity at the neuromuscular junction of the rat. J Physiol 132: 650–666

Liley AW (1956b) The effects of presynaptic polarization on the spontaneous activity of the mammalian neuromuscular junction. J Physiol 134: 427–443

Llinás R (1980) Applicability of channel analysis in molloscs to vertebrate central neurons. In: Koester J, Byrne JH (eds) Molluscan nerve cells. Cold Spring Harbor Lab., Cold Spring Harbor

Llinás RR, Steinberg IZ, Walton K (1976) Presynaptic calcium currents and their relation to synaptic transmission: voltage-clamp study in squid giant synapse and theoretical model for the calcium gate. Proc Natl Acad Sci USA 73: 2918–2922

Llinás RR, Steinberg IZ, Walton K (1981a) Presynaptic calcium currents in squid giant synapse. Biophys J 33: 289–322

Llinás RR, Steinberg IZ, Walton K (1981b) Relationship between presynaptic calcium current and post-synaptic potential in squid giant synapse. Biophys J 33: 323–351

Lloyd DPC (1949) Post-tetanic potentiation of response in monosynaptic reflex pathways of the spinal cord. J Gen Physiol 33: 147–170

Lorente de No R (1934) Studies on the structure of the cerebral cortex. II. Continuation of the study of the ammonic stystem. J Phsychol Neurol 46: 113–127
Lorente de No R (1938) Analysis of the activity of the chains of internuncial neurons. J Neurophysiol 1: 207–244
Luescher HR, Ruenzel P, Henneman E (1983) Composite EPSP's in motoneurons of different sizes before and during PTP: implications for transmission failure and its relief in Ia projections. J Neurophysiol 49: 269–289
Lugaro E (1899) I recenti progressi dell'anatomia del sistema nervoso in rapporto alla psicologia ed alla psichiatria. Riv Patol Nerv Ment IV: 11–12 [cited in Cajal (1911)]
Madison DV, Nicoll RA (1982) Noradrenaline blocks accommodation of pyramidal cell discharge in the hippocampus. Nature 299: 636–638
Magleby KL, Zengel JE (1975) A dual effect of repetitive stimualtion on post-tetanic potentiation of transmitter release at the frog neuromuscular junction. J Physiol 245: 163–182
Magleby KL, Zengel JE (1976a) Augmentation: a process that acts to increase transmitter release at the frog neuromuscular junction. J Physiol 257: 449–470
Magleby KL, Zengel JE (1976b) Long term changes in augmentation, potentiation, and depression of transmitter release as a function of repeated synaptic activity at the frog neuromuscular junction. J Physiol 257: 471–494
Martin AR, Pilar G (1964) Presynaptic and postsynaptic events during post-tetanic potentiation and facilitation in the avian ciliary ganglion. J Physiol 175: 17–30
Marty A (1981) Ca-dependent K channels with large unitary conductance in chromaffin cell membrane. Nature 291: 497–500
Mayeri E, Brownell P, Branton WD, Simon SB (1979a) Multiple, prolonged actions of neuroendocrine bag cells on neurons in *Aplysia*. I. Effects on bursting pacemaker neurons. J Neurophysiol 42: 1165–1184
Mayeri E, Brownell P, Branton WD (1979b) Multiple, prolonged actions of neuroendocrine bag cells on neurons in *Aplysia*. II. Effects on beating pacemaker and silent neurons. J Neurophysiol 42: 1185–1197
Meech RW (1972) Intracellular calcium injection causes increased potassium conductance in *Aplysia* nerve cells. Comp Biochem Physiol [A] 42: 493–499
Meech RW (1974) The sensitivity of *Helix aspersa* neurones to injected calcium ions. J Physiol 237: 259–277
Meech RW (1978) Calcium-dependent potassium activation in nervous tissues. Annu Rev Biophys Bioeng 7: 1–18
Miledi R, Slater CR (1966) The action of calcium on neuronal synapses in the squid. J Physiol 184: 473–498
Miledi R, Thiess R (1971) Tetanic and post-tetanic rise in frequency of miniature end-plate potentials in low-calcium solutions. J Physiol 212: 245–257
Misler S, Hurlbut WP (1983) Post-tetanic potentiation of acetylcholine release at the frog neuromuscular junction develops after stimulation in Ca^{2+}-free solutions. Proc Natl Acad Sci USA 80: 315–319
Morita K, North RA (1981) Opiates and enkephalin reduce excitability of neuronal processes. Neuroscience 6: 1943–1951
Morita K, North RA, Tokimasa T (1982) Muscarinic presynaptic inhibition of synaptic transmission in myenteric plexus of guinea-pig ileum. J Physiol 333: 141–149
Mudge AW, Leeman SE, Fischbach GD (1979) Enkephalin inhibits release of substance P from sensory neurons in culture and decreases action potential duration. Proc Natl Acad Sci USA 76: 526–530
Nicholls J, Wallace BG (1978a) Modulation of transmission at an inhibitory synapse in the central nervous system of the leech. J Physiol 281: 157–170
Nicholls J, Wallace BG (1978b) Quantal analysis of transmitter release at an inhibitory synapse in the central nervous system of the leech. J Physiol 281: 171–185
Nicoll RA, Alger BE (1979) Presynaptic inhibition: transmitter and ionic mechanisms. Int Rev Neurobiol 21: 217–258
Palka J (1979) Theories of pattern formation in insect neural development. Adv Insect Physiol 14: 251–349

Pallotta BS, Magleby KL, Barrett JN (1981) Single channel recordings of Ca^{++} activated K^+ currents in rat muscle cell culture. Nature 293: 471–474

Pearson KG, Goodman CS (1981) Presynaptic inhibition of transmission from identified interneurons in the locust central nervous system. J Neurophysiol 45: 501–515

Plant TD, Standen NB (1981) Calcium current inactivation in identified neurones of *Helix aspersa*. J Physiol 321: 273–285

Plant TD, Standen NB, Ward TA (1983) The effects of injection of calcium ions and calcium chelators on calcium channel inactivation in *Helix* neurones. J Physiol 334: 189–212

Pollard HB, Pazoles CJ, Creutz CE (1981) Mechanism of calcium action and release of vesicle-bound hormones during exocytosis. Recent Prog Horm Res 37: 299–332

Rahamimoff R (1968) A dual effect of calcium ions on neuromuscular facilitation. J Physiol 195: 471–480

Rahamimoff R, Lev-Tov A, Meiri H (1980) Primary and secondary regulation of quantal transmitter release: calcium and sodium. J Exp Biol 89: 5–18

Reuter H, Scholz H (1977) The regulation of the calcium conductance of cardiac muscle by adrenaline. J Physiol 264: 49–62

Rogart R (1981) Sodium channel in nerve and muscle membrane. Annu Rev Physiol 43: 711–725

Rosenthal J (1969) Post-tetanic potentiation at the neuromuscular junction of the frog. J Physiol 203: 121–133

Rudomin P, Engberg I, Jimenez I (1981) Mechanisms involved in presynaptic depolarization of group I and rubrospinal fibers in cat spinal cord. J Neurophysiol 46: 532–548

Ryall RW (1978) Presynaptic inhibition. Trends Neurosci 1: 164–166

Schaefer H, Scholmerich P, Haas P (1938) Der Elektrotonus und die Erregungsgesetze des Muskels. Pflügers Arch 241: 310–341

Schlaepfer WT, Tremblay JP, Woodson PBJ, Barondes SH (1976) Frequency facilitation and post-tetanic potentiation of a unitary synaptic potential in *Aplysia californica* are limited by different processes. Brain Res 109: 1–20

Schmidt RF (1971) Presynaptic inhibition in the vertebrate central nervous system. Ergeb Physiol Biol Chem Exp Pharmakol 63: 20–101

Shapiro E, Castellucci VF, Kandel ER (1980a) Presynaptic membrane potential affects transmitter release in an identified neuron in *Aplysia* by modulating the Ca^{2+} and K^+ currents. Proc Natl Acad Sci USA 77: 629–633

Shapiro E, Castellucci VF, Kandel ER (1980b) Presynaptic inhibition in *Aplysia* involves a decrease in the Ca^{2+} current of the presynaptic neuron. Proc Natl Acad Sci USA 77: 1185–1189

Shapiro E, Klein M, Kandel ER (1981) Ionic mechanisms and behavioral functions of presynaptic facilitation and presynaptic inhibition in *Aplysia*: a model system for studying the modulation of signal transmission in sensory neurons. Proc Sensory Physiol 1: 97–137

Shapiro E, Connor J, Kretz R (1983) Presynaptic Ca^{++} accumulation is the mechanism of post-tetanic potentiation in *Aplysia*: evidence from the Ca^{++}-indicator dye Arsenazo III. Soc Neurosci Abstr 9: 1051

Shimahara T (1981) Modulation of synaptic output by the transient outward potassium current in *Aplysia*. Neurosci Lett 24: 139–142

Shimahara T, Peretz B (1978) Soma potential of an interneurone controls transmitter release in a monosynaptic pathway in *Aplysia*. Nature 273: 158–160

Shimahara T, Tauc L (1975) Multiple interneuronal afferents to the giant cells in *Aplysia*. J Physiol 247: 299–319

Siegelbaum SA, Camardo JS, Kandel ER (1982) Serotonin and cyclic AMP close single K^+ channels in *Aplysia* sensory neurones. Nature 299: 413–417

Sperry RW (1963) Chemoaffinity in the orderly growth of nerve fiber patterns and connections. Proc Natl Acad Sci USA 50: 703–710

Sperry RW (1965) Selective communication in nerve nets: impulse specificity vs. connection specificity. Neurosci Res Programm Bull 3: 37–43

Stinnakre J, Tauc L (1973) Calcium influx in active *Aplysia* neurones detected by injected aequorin. Nature 242: 113–115

Takeuchi A, Takeuchi N (1962) Electrical changes in pre- and post-synaptic axons of the giant synapse of *Loligo*. J Gen Physiol 45: 1181–1193

Takeuchi A, Takeuchi N (1966a) A study of the inhibitory action of γ-amino-butyric acid on neuromuscular transmission in the crayfish. J Physiol 183: 418–432

Takeuchi A, Takeuchi N (1966b) On the permeability of the presynaptic terminal of the crayfish neuromuscular junction during synaptic inhibition and the action of γ-amino-butyric acid. J Physiol 183: 433

Tanzi E (1893) I fatti e le induzioni nell'odierna istologia del sistema nervoso. Riv Sper Freniat Med Leg Alien Ment 19: 419–472

Tauc L (1965) Presynaptic inhibition in the abdominal ganglion of *Aplysia*. J Physiol 181: 282–307

Thomas R (1972) Electrogenic sodium pump in nerve and muscle cells. Physiol Rev 52: 563–595

Thompson SH (1977) Three pharmacologically distinct potassium channels in molluscan neurones. J Physiol 265: 465–488

Tillotson D (1979) Inactivation of Ca conductance dependent on entry of Ca ions in molluscan neurons. Proc Natl Acad Sci USA 76: 1497–1500

Tokimasa T, Morita K, North RA (1981) Opiates and clonidine prolong calcium-dependent after hyperpolarization. Nature 294: 162–163

Tsien RW (1977) Cyclic AMP and contractile activity in heart. Adv Cyclic Nucleotide Res 8: 363–420

Tsukahara N (1981) Synaptic plasticity in the mammalian central nervous system. Annu Rev Neurosci 4: 351–379

Van der Kloot W, Kita H (1973) The possible role of fixed membrane surface charges in acetylcholine release at the frog neuromuscular junction. J Membr Biol 14: 365–382

Wachtel H, Kandel ER (1967) A direct synaptic connection mediating both excitation and inhibition. Science 158: 1206–1208

Waziri R (1977) Presynaptic electrical coupling in *Aplysia*: effects on postsynaptic chemical transmission. Science 195: 790–792

Waziri R, Kandel ER, Frazier WT (1969) Organization of inhibition in abdominal ganglion of *Aplysia*. II. Post-tetanic potentiation, heterosynaptic depression, and increments in frequency of inhibitory postsynaptic potentials. J Neurophysiol 32: 509–519

Weinreich D (1971) Ionic mechanism of post-tetanic potentiation at the neuromuscular junction of the frog. J Physiol 212: 431–446

Wiesel TN (1982) Postnatal development of the visual cortex and the influence of environment. Nature 299: 583–591

Zengel JE, Reid SA, Sypert GW, Munson JB (1983) Presynaptic inhibition, EPSP amplitude, and motor-unit type in triceps surae motoneurons in the cat. J Neurophysiol 49: 922–931

Mechanosensibility of Joint Receptors with Fine Afferent Fibers

H.-G. Schaible and R.F. Schmidt, Würzburg

Introduction

In the past 30 years several studies of the mechanosensitivity of afferent units innervating the knee joint of the cat and monkey have been published (Andrew and Dodt 1953; Boyd 1954; Boyd and Roberts 1953; Burgess and Clark 1969; Clark 1975; Clark and Burgess 1975; Eklund and Skoglund 1960; Ferrell 1980; Grigg 1975, 1976; Grigg and Greenspan 1977; Grigg and Hoffman 1982; Grigg et al. 1982; McCall et al. 1974; Skoglund 1956). They were all mainly or exclusively concerned with the properties of receptive units with fast afferent fibers having conduction velocities of more than 20 m/s.

There have been few observations on units with conduction velocities below 20 m/s. Units with conduction velocities below 2.5 m/s, i.e., with unmyelinated fibers, have not been recorded from at all. For instance, Burgess and Clark (1969) recorded in dorsal root filaments from 278 afferent units of the posterior articular nerve (PAN), of which only 17 fibers had conduction velocities of between 10 and 35 m/s. In the study by Clark and Burgess (1975) of 672 medial articular nerve (MAN) units, 104 had conduction velocities of between 4 and 20 m/s, and in the PAN of 713 units only 37 were in this range. In a sample of 331 MAN units Clark (1975) had 41 units with conduction velocities of from 8 to 20 m/s. In the other studies quoted, either conduction velocities were not listed or no slow units were included in the sample.

With regard to the functional properties of the "slow" units (up to 35 m/s) described in the studies quoted above, it was usually stated that they responded either only weakly or not at all to normal joint movements, and that many of them did not even respond to forceful bending, rotating, or pressing of the joint. Consequently, it was assumed that joint receptors with slowly conducting myelinated fibers "probably function as nociceptors" (Burgess and Clark 1969). The same conclusion has been drawn by Clark (1975) for "weakly activated" and "nonactivated" fast myelinated units. The functional properties of the joint receptors with unmyelinated afferent fibers (conduction velocities below 2.5 m/s) remained to be determined. From histological studies of the knee joint innervation it appeared probable to Freeman and Wyke (1967) that the unmyelinated afferents "constitute the pain-sensitive system of the joint tissues."

This situation is particularly unsatisfactory in view of the fact that in joint nerves small-diameter nerve fibers in the groups III and IV range outnumber by far the group II fibers. For instance, in the PAN of the cat, a branch of the tibial nerve which supplies the lateral and posterior structures of the knee joint, of a

Experimental Brain Research, Suppl. 9
© Springer-Verlag Berlin · Heidelberg 1984

total of approximately 1200 nerve fibers only 270 are myelinated afferents (groups II and III), 520 are postganglionic sympathetic efferents, and 410 are unmyelinated afferents. For the MAN, a branch of the femoral nerve which supplies the structures in the medial and anteromedial aspects of the knee joint, the corresponding figures are 1100, 200, 500, and 400 nerve fibers respectively (Langford 1983; Langford and Schmidt 1983). In this latter nerve, of the 200 myelinated afferents not more than 40% fall into the group II range (Langford et al. 1983). Thus the fine group III and IV afferent units with free nerve terminals provide the overwhelming majority of the afferent innervation of the knee joint.

In order to elucidate the functional properties and the physiological role of fine joint afferents in nociception and kinesthesia, single-fiber studies of group III and IV units of the MAN of the cat's knee joint were started in this laboratory some 3 years ago. This short review will sum up the findings related to the mechanosensitivity of such units (cf. also Schaible and Schmidt 1983a, b). Other aspects, such as their structure, their chemosensitivity, and their reactions to joint inflammation are described elsewhere (Langford et al. 1983; Coggeshall et al. 1983; Schaible and Schmidt 1983c; Schmidt and Schaible 1983).

Identification of Fine Joint MAN Units

All experiments were carried out on the right MAN of adult cats anesthetized with α-chloralose and immobilized with pancuronium. The MAN innervates the medial and anteromedial regions of the fibrous capsule of the knee joint, the medial collateral ligament, the medial part of the annular ligament attached to the medial meniscus, the ligamentum patellae, the infrapatellar fat pad, and the medial part of the patella periosteum (Freeman and Wyke 1967). To expose these regions of the joint the cat was placed on its right side and the overlying skin and the sartorius muscle were removed. A threaded bolt was fitted through the femur so that the right lower leg could be moved in a horizontal plane. For electrical stimulation with a fine bipolar electrode, the MAN was dissected free from its surrounding tissue for a length of about 6 mm, starting some 10 mm rostral to the proximal border of the knee joint. A small piece of plastic was placed under the MAN and the electrode in order to isolate it from the underlying tissue. The saphenous nerve was cut about 2 cm distal from the point of origin of the genicular artery to prevent recording of its afferent activity when searching for MAN units in filaments of the saphenous nerve dissected in the inguinal region. Figure 1 shows a schematic view of the experimental situation. To enable close intraarterial injection of algesic and other chemical agents a fine plastic catheter was fitted in the arteria saphena (for the results of such injections see Langford et al. 1983; Schaible 1983; Schaible and Schmidt 1983c).

Whether a filament of the saphenous nerve contained MAN nerve units was tested by stimulating the MAN peripherally. If several units were contained in such a filament (upper record of the specimen in Fig. 2), the filament was progressively split with sharpened watchmaker forceps under a binocular dissecting microscope until only one fiber could be activated (lower record of the specimen in Fig. 2). The fibers were identified by their conduction velocities calculated from

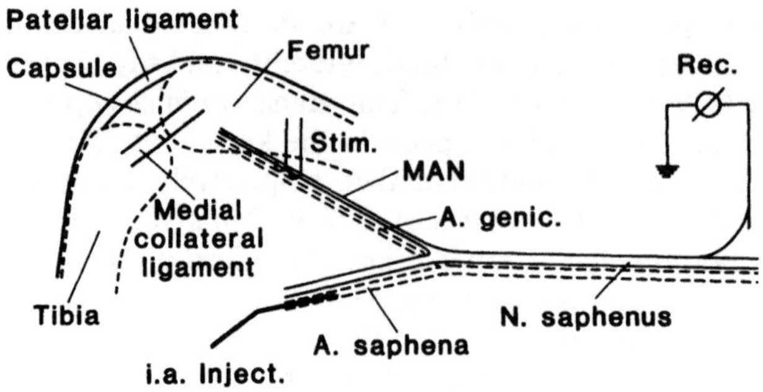

Fig. 1. Schematic view of the medial aspect of the right knee joint and thigh. The medial articular nerve (*MAN*) was stimulated (*Stim.*) via bipolar platinum wire electrodes close to the knee joint (cathode proximal; square pulses of 0.5 ms duration). The nerve was left in continuity. A sheath of plastic insulated the site of stimulation from its surroundings. The distance between stimulating and recording electrodes (*Rec.*) varied from 4 to 7 cm. Chemical stimulation of joint receptors was done by close intraarterial injection via a catheter in the saphenous artery.

the time required for an action potential to cover the distance between the stimulating and recording electrodes (cf. Fig. 1).

Medial articular nerve units conducting at less than 2.5 m/s were classified as group IV fibers, those with conduction velocities of from 2.5 to 20 m/s as group III units. As seen in the histograms of Fig. 2, there is a clear separation between the two fiber groups. The majority of group III units have conduction velocities of between 3 and 12 m/s (Fig. 2A), whereas in the sample of 91 group IV units (Fig. 2B) only eight had conduction velocities above 1.5 m/s. In order to prevent antidromically excited group IV postganglionic sympathetic efferents being mistaken for afferent fibers, a group IV fiber was assumed to be afferent if it was sensitive to local mechanical stimulation (probing with a glass rod) of the joint. As recently reported, with fine afferent fibers the inverse correlation between the electrical threshold and the conduction velocity of their action potential is not very strict for group III units and does not exist for group IV ones (Sato et al. 1983). Furthermore, about 50% of the group III units have electrical thresholds in the same intensity range as most of the group IV units (Sato et al. 1983). Thus the distinct difference in conduction velocity between these two fiber groups (cf. Fig. 2) has no counterpart in the distribution of their thresholds to electrical stimulation.

Receptive Fields and Responses to Local Mechanical Stimulation

One of the surprises of this study was that the receptive fields of mechanosensitive fine afferent units were nearly always small, regardless of their location on the medial aspect of the fibrous capsule and the medial collateral ligament. Examples of such receptive fields are shown in Figs. 3–6. Often units could be activated at one spot only. But we also found units which had two, three, or even four spot-like receptive fields clearly separated and sometimes far apart from each other. Stimulation with von Frey hairs of several times threshold strength less than

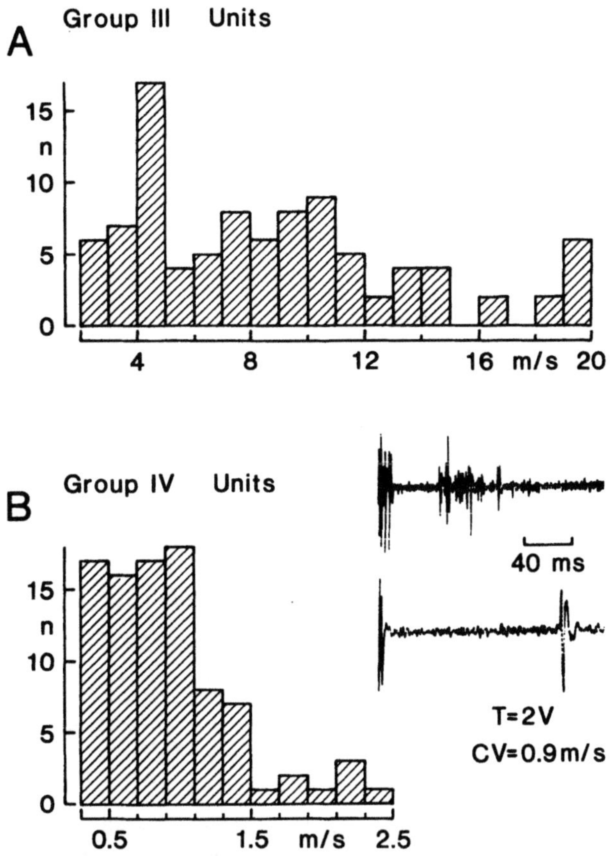

Fig. 2A, B. Conduction velocity histograms of group III (*A*) and IV (*B*) units identified as afferents in the medial articular nerve (MAN). The 95 group III and 91 group IV fibers are from 63 experiments. The histograms include data used to plot Fig. 1B in Schaible and Schmidt 1983a. The *upper specimen record* shows multiunit discharges in a small filament of the saphenous nerve following electrical stimulation of the MAN at the site indicated in Fig. 1 (*Stim.*). Further splitting of the filament resulted in the appearance of one active unit only (*lower record*). Its electrical threshold (*T*) was 2 V and its conduction velocity (*CV*) 0.9 m/s.

0.5 mm outside each receptive field was usually ineffective. In a very few units the spot-like areas were surrounded by a larger field from which the unit could be activated by suprathreshold stimuli. Such receptive fields are frequent in inflamed knee joints (Coggeshall et al. 1983).

A second surprise was that many of the mechanosensitive fine afferent units were very sensitive to stimulation with von Frey hairs. Figure 3A,C shows that of 25 group III units eight (32%) had thresholds below 1 g and a further ten had thresholds below 6 g. Thus 72% of our group III sample have low to moderate von Frey thresholds. With the group IV units the distribution of thresholds clearly shifted to higher values. The cumulative percentages for thresholds up to 1 g, 2 g, and 6 g are 3%, 33%, and 50% respectively (Fig. 3C). Even if it is taken into account that the von Frey thresholds were tested after the removal of the overlying

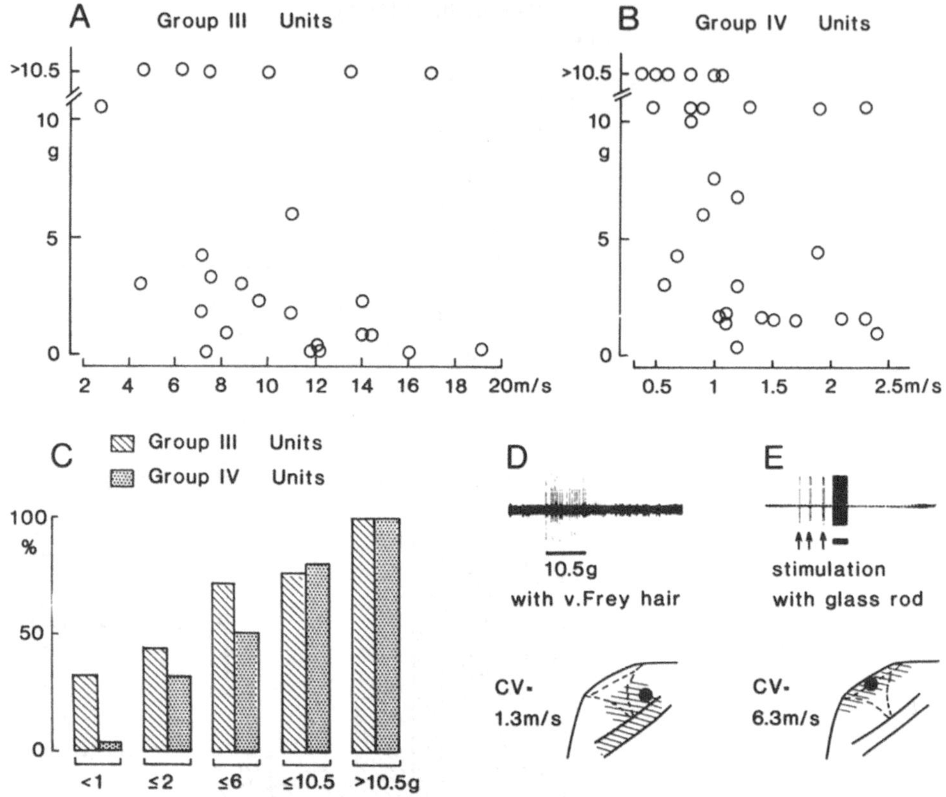

Fig. 3A–E. Relation between the conduction velocity (*CV*) of fine afferent joint units and their von Frey thresholds. *A* Twenty-five group III units. *B* Thirty group IV units. *C* Cumulative diagram showing the percentages of each fiber group for increasing thresholds as indicated. *D* and *E* show the receptive fields (dots) of two units in the medial and anterior areas of the joint respectively. The specimen record in *D* shows that constant pressure with a von Frey hair of 10.5 g gave a tonic response. The unit in *E* was repeatedly probed with a glass rod for brief periods (*arrows*) and for a somewhat longer one (*bar*). Probing outside the receptive fields (hatched areas) did not induce any discharges.

skin, these findings imply that half of the mechanosensitive units with myelinated nerve fibers probably respond to everyday local mechanical stimuli (touch, pressure) of moderate intensity. There is, however, a continuum of von Frey thresholds rather than a clear-cut separation of the low-threshold from the high-threshold ones. Thus for units in the intermediate threshold range it is not possible to decide at this stage whether or not they contribute to nociception.

The relation between von Frey threshold and conduction velocity of group III and IV afferent units is illustrated in Fig. 3A,B. Clearly, in both fiber groups there is a tendency for units with low thresholds to have high conduction velocities and vice versa. For instance, with very few exceptions, the "fast" group III units with conduction velocities above 11 m/s have rather low thresholds, whereas on the other hand most of the "slow" group IV units with conduction velocities below 1 m/s have high to very high thresholds (cf. also the cumulative diagrams of Fig. 3C).

The von Frey threshold does not depend on the location of the receptive field on the medial aspect of the joint (Fig. 4 in Schaible and Schmidt 1983a). Units with two or more receptive fields had similar thresholds in all of them, again either in the high- or in the low-threshold range. Testing the relation between suprathreshold stimuli of increasing strength and peak frequency of these units (Fig. 6 in Schaible and Schmidt 1983a) revealed that quite a few of the group III units with very low thresholds had a steep stimulus-response relationship giving maximum responses of 20–40 Hz at less than 3 g. Higher stimulus strength usually resulted in a decrease in their peak frequency. Most of the other group III units had well-graded responses over the entire stimulus range. With group IV units the stimulus-response relationships were less steep than those of the group III units. Peak frequencies above 16 Hz were rare.

Responses to Passive Joint Movements

In the sleeping as well as in the anesthetized, immobilized cat the knee joint rests in a midrange position, practically halfway between full extension and full flexion. The ankle joint is very similar to that in an awake cat standing quietly on all feet. This natural midrange position has been chosen as the starting and end point of all tests performed to evaluate the responses to limb movements of those fine afferent units having local mechanosensitive receptive fields. The knee was moved manually into flexion and extension and through intermediary angles. At the same time, adduction, abduction, and rotatory movements were carried out. These rotatory movements in the long axis of the tibia were applied in two steps: first, with minimal force taking up the slack of the tissue; and secondly, against definite resistance from the joint structures. The former movements can be applied to the awake cat without any defense reaction by the animal, whereas the latter one will be vividly opposed even by very tame cats. In anesthetized animals such forceful rotations often evoke sharp rises in blood pressure. We therefore regard these forceful movements as noxious ones. Our consistent experience is that the transition between nonnoxious and noxious rotation can be exactly felt by the researcher as a distinct increase in the necessary force which he has to apply to overcome the tissue resistance to rotation.

With such passive innocuous and noxious movements of the knee joint four distinct categories of units were found in both fine fiber groups (Schaible and Schmidt 1983b). Their response behavior is illustrated in Figs. 4–6. The first category, comprising about 20% of our sample of 74 units, responded even to innocuous joint movements, particularly to rotations (Figs. 4A, 6A). For instance, the unit illustrated in Fig. $4A_{1-6}$ discharged to outward rotations from half extension (A_2) and full extension (A_3). The extension movement itself evoked an occasional discharge (A_4). Inward rotation in full extension was also an effective stimulus (A_4), whereas innocuous outward or inward rotations in midposition (A_5 and A_6) resulted in weak activations at best. Units of this type were more frequent in the group III (Fig. 4A) than in the group IV range (Fig. 6A). Practically all of them increased their rate of discharge when the rotations were extended to the noxious range (A_{3-6}) or when extreme flexions or extensions were applied.

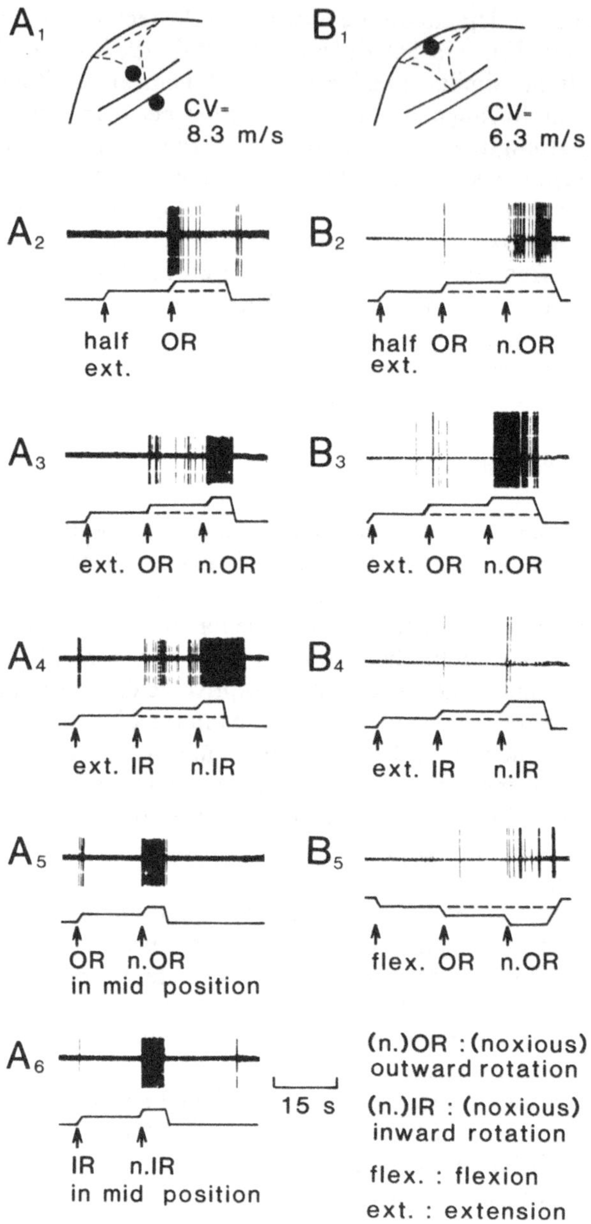

Fig. 4A,B. Responses of group III afferent MAN units to passive movements of the knee joint as indicated. The unit in *A* was more easily activated by innocuous movements than the unit in *B*, which gave only weak responses to such stimuli. Both units responded more vigorously when the movements were extended in the noxious range. The unit in *A* had a receptive field on either side of the medial collateral ligament (A_1). Its von Frey threshold was 0.5 g. During a rest period of 5 min, 13 spontaneous impulses were recorded. The other unit had a receptive field in the anterior patellar region. Due to the experimental situation no von Frey thresholds can be measured in this area. There was no resting discharge. *CV*, conduction velocity.

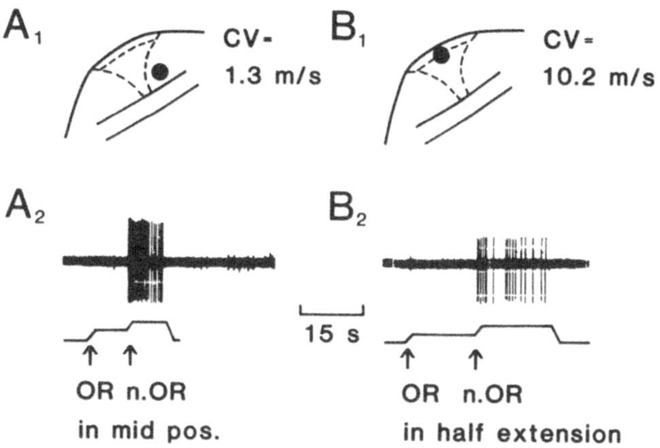

Fig. 5A,B. Receptive fields (A_1, B_1) and responses of a group IV (A_2) and a group III (B_2) MAN afferent unit to the indicated passive movements of the knee joint. None of them showed resting discharges and responses to innocuous movements. The unit in $A_{1,2}$ responded also to inward and outward rotations at flexed and extended positions of the knee joint. The unit in $B_{1,2}$ gave no other responses than the one shown. *CV*, conduction velocity; *(n.)OR*, (noxious) outward rotation.

The second type of unit showed very distinct differences in response rates to innocuous and noxious movements (Figs. $4B_{1-5}$, 6B). These obvious differences were due to the fact that such units were only weakly (Fig. 4B) or occasionally (Fig. 6B) activated by innocuous movements, whereas noxious movements led to pronounced discharges (Figs. $4B_2$, B_3, 6B). Usually, extensions and flexions were ineffective, and outward rotations were more effective than inward ones. Again, this category of units was found in the group III (Fig. 4B) as well as in the group IV range (Fig. 6B). It comprised 16% of all units of our sample.

The third type of unit were those that did not respond to any innocuous joint movements but fired clearly and consistently in response to noxious ones (Figs 5, 6C,D). Some of them responded to a number of outward and inward noxious rotations starting from various joint positions (Fig. 6C; Fig. 5A is also of this type); others were only sensitive to outward rotation from one position, particularly maximum extension (Figs. 5B, 6D). Each kind of response behavior occurred with group III (Figs. 5B, 6C) and with group IV units (Figs. 5A, 6D). All in all, 24 of such units (33% of the total sample) were found.

An example of the fourth type of unit is illustrated in Fig. 6E. Such units had distinct and typical receptive fields, but they did not respond to any joint movement whatsoever. This finding was particularly surprising for units which responded well to high-strength local stimulation, and which, as in Fig. 6E, upon intense local stimulation exhibited distinct and long-lasting afterdischarges. This type of unit was more frequently found in the group IV than in the group III range in our sample (15 against eight units). All together 31% of all units were of this category. In inflamed joint tissue, units of this type are rare (Coggeshall et al. 1983). It may be speculated, therefore, that many of them become sensitized to joint movement by the inflammation.

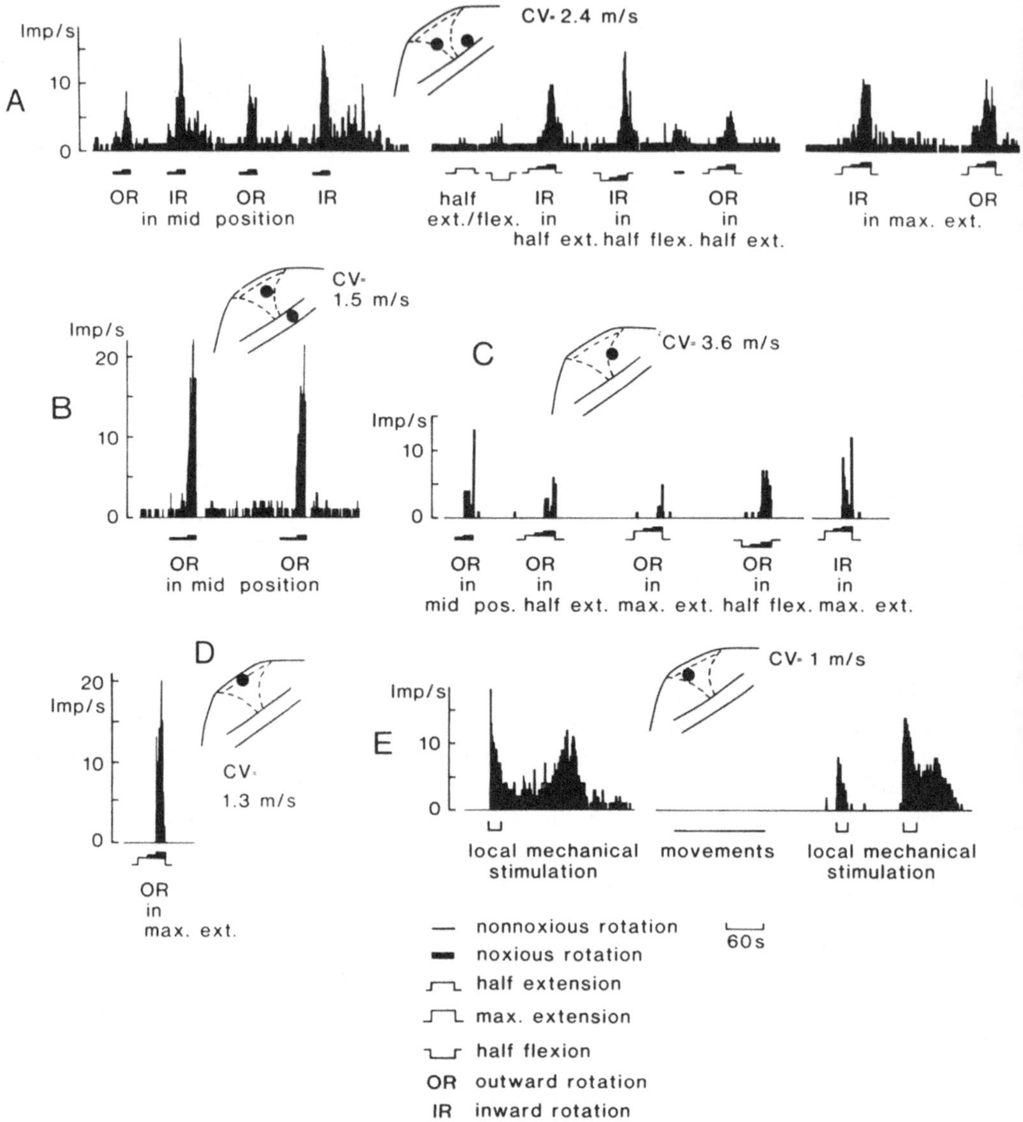

Fig. 6A–E. Response behavior of fine MAN afferent units to passive movements of the knee joint as indicated. The receptive fields of each unit are shown in the *insets*. The conduction velocities (*CV*) are given. Note that the unit in *E* did not respond to any innocuous and noxious movement despite its vigorous responses to local mechanical stimulation.

The units responding to innocuous joint movements quite often discharge throughout the hold period of an effective joint position, i.e., they display a tonic, slowly adapting response behavior (cf. Figs. 4, 6A). In contrast, the phasic responses during movement are usually weak and sometimes absent. From these preliminary findings, which need to be substantiated for a wide range of movement velocities and hold durations, it appears that fine afferent joint units may contribute more to signaling joint position than joint movement.

Relations Between Resting Activity, von Frey Thresholds, Receptive Field Localization, and Movement Behavior

In order to obtain some insight into the transducer mechanisms and into the mode of operation of the joint mechanoreceptors with fine afferent fibers, we have looked for possible correlations between their responses to local stimulation and passive movements, and between these two parameters and their eventual resting activity and the location of their receptive fields on the joint capsule. The results of these investigations are illustrated in Fig. 7–9.

Resting activity, defined as the occurrence of discharges with a frequency of more than one impulse per minute when the joint is in midposition, is found in about 30% of all group III and IV units with local mechanosensitive receptive fields (Schaible and Schmidt 1983a). The discharges are irregular, and their average frequency rarely exceeds 40 impulses per minute. Often it is less than ten impulses per minute. Since resting activity may indicate receptor sensitization due to the experimental manipulation, we have plotted in Fig. 7B the von Frey thresholds of a sample of 25 group III and 29 group IV units in relation to their resting behavior. Of the group III units 16 had no resting activity, 4 displayed a resting activity of less than ten impulses per minute, and 5 had a higher frequency.

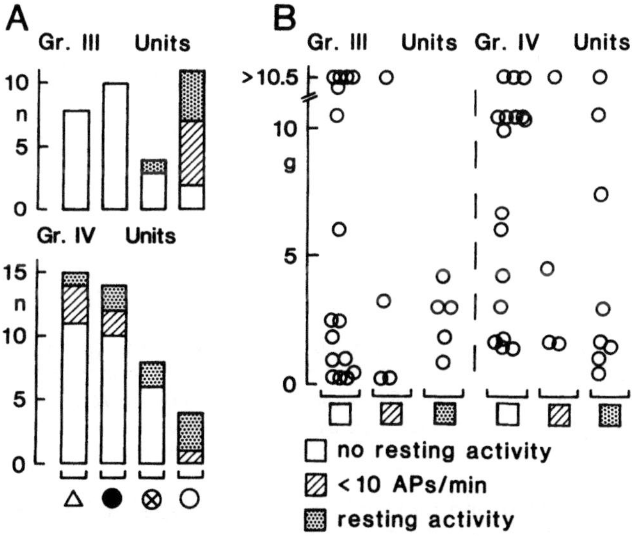

Fig. 7. A The relation between resting behavior in mid-position and response behavior to passive joint movements of 33 group III and 41 group IV MAN afferent units; △ not activated by movement; ● activated only by noxious movements; ⊗ weakly activated by nonnoxious movements; ○ activated by innocuous movements. The symbols for the resting behavior are indicated. *B* Relation between resting behavior and von Frey thresholds for 25 group III and 29 group IV units. *AP*, action potential.

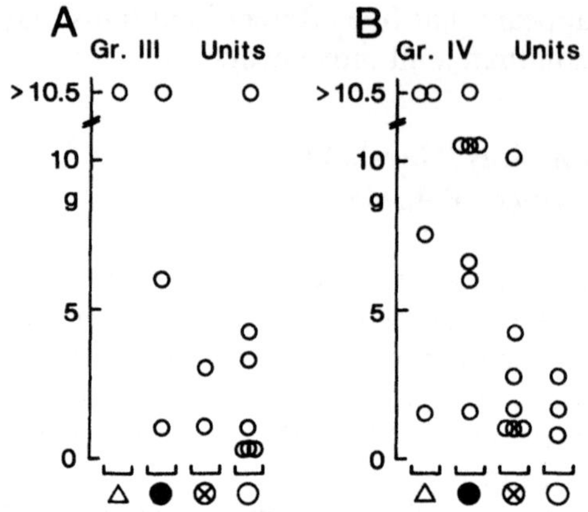

Fig. 8A,B. The relation between von Frey threshold (*ordinate*) and response behavior to passive joint movements (*same symbols as in Fig. 7*) of 13 group III and 21 group IV MAN afferent units. Data from Schaible and Schmidt 1983b and unpublished observations.

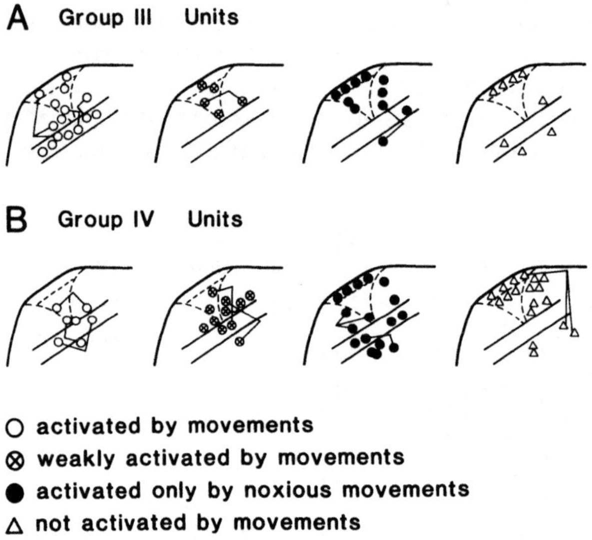

Fig. 9A,B. Location of the receptive fields of the four different types of MAN group III (*A*) and IV (*B*) units as classified by their response behavior to passive joint movements. Data from Schaible and Schmidt 1983b and unpublished observations. *Same symbols as in Fig. 7.* Multiple receptive fields are indicated by *connecting lines.*

The corresponding numbers of the group IV units were 17, 4, and 8, respectively. It can be seen at a glance that not even a weak correlation exists between these two properties. In particular, the majority of the group III units with low thresholds were silent, and very few of the units with resting activity in both fiber groups had high thresholds.

The situation is somewhat different when the relation between resting activity and movement sensitivity is considered (Fig. 7A). Here, indeed, all those group IV units activated by innocuous movements and nearly all group III units of this type display resting activity. In contrast, the group III units only activated by noxious movements and those not activated by any movement have no resting discharges. Similarly, the majority of group IV units of these two types are silent (ten out of 14 and 11 out of 15 units respectively). Thus it may be concluded that resting activity is not a reliable indicator of either the von Frey threshold of a unit or of its response to passive movements. On the other hand, units which respond to innocuous joint movements usually have resting activity. This latter finding, with the additional observation that the responses of such units to innocuous joint movements are well maintained as long as the joint is kept in the new position (Figs. 4, 6A; cf. also Fig. 1 in Schaible and Schmidt 1983b), may be indicative that these "resting discharges" participate in the signaling of joint position. But this matter has to be investigated further, particularly since many discharge features do not favor such a role. These include afterdischarges, changes in resting frequency during long recordings, widely fluctuating impulse intervals, limited sensitivity to small changes in joint position, and increased resting discharges in inflamed joints. Not unexpectedly, local von Frey thresholds and movement sensitivity are closely related to each other, i.e., fine afferent units with von Frey thresholds in the lower range are more easily excited by joint movements than those with higher thresholds (Fig. 8; see also Fig. 5 in Schaible and Schmidt 1983b). In both fiber groups practically all units activated by innocuous movements have von Frey thresholds below 5 g, and those units activated only by noxious movements have higher thresholds. There is no difference in the distribution of thresholds of those units that were activated and those that were only weakly activated by innocuous movements.

Finally, Fig. 9 illustrates the location of the receptive fields of our sample in relation to their responses to passive movements. For group III (A) as well as for group IV units (B) these relations are characterized by the following findings:

1. Units activated or weakly activated by innocuous movements have their receptive fields nearly exclusively on the medial collateral ligament or just anterior to it.
2. The receptive fields of units activated only by noxious movements can be found anywhere in the area innervated by the MAN.
3. The latter statement is also true for units not activated by movements, but the majority of their receptive fields are concentrated in the patellar ligament, and only one field has been found so far on the medial collateral ligament.

Units with more than one receptive field can be found in each category.

Concluding Remarks

The major conclusions to be drawn from these findings have been discussed in detail elsewhere (Schaible and Schmidt 1983a, b). The additional data gathered in the meantime and included in this survey strengthen the notion that many of the group III and IV afferent fibers of the MAN serve as nociceptors, whereas others are sensitive to low to moderate innocuous local mechanical stimulation and to innocuous joint movements, particularly to rotatory movements of the tibia. Therefore, these latter units may play an important role in signaling this type of movement as well as local mechanical events. Since we found only a rough correlation or none at all between joint angle and unitary activity, we propose that these units are mainly engaged in signaling that the joint is about to leave its normal working range, and that these warning signals increase in urgency as units with high thresholds to movements come into play.

A great number of open questions require further investigations. Only three may be mentioned here:

1. In our hands, all mechanosensitive units can be classified with regard to their responses to passive movements as belonging to one of the categories characterized above. But it may still be asked whether this different response behavior indicates distinctly different receptor types, of which the two high-threshold categories are nothing but nociceptors. At this stage the alternative cannot be ruled out that this population of afferent units has a continuum of local and movement thresholds but that we cannot properly recognize this continuum by our modes of stimulation.
2. The functional role of those units which have local mechanosensitive receptive fields on the joint capsule but which are not activated by any joint movement remains a puzzle. In inflamed joints many of these units seem to become very sensitive to joint movements (Coggeshall et al. 1983). It may well be that they have no task to fulfill in normal joints.
3. All mechanosensitive units of our sample can be activated by close intraarterial injection of KCl and of other algogenic agents such as bradykinin and serotonin, i.e., they are polymodal nociceptors (Langford et al. 1983; Schaible 1983; Schaible and Schmidt 1983c). In addition many of them are sensitized to bradykinin by prostaglandins (Schaible 1983; Schaible and Schmidt 1983c). Again we are confronted with the question whether with regard to this chemosensitivity we face a homogeneous population of receptors with slight differences in individual response behavior due to the location in the joint tissue or other such extraneural factors, or whether receptors exist with distinctly different chemoreceptive properties. Our preliminary results favor the second possibility. For instance, prostaglandins E_1 and E_2 increase preferentially, if not specifically, the bradykinin sensitivity of presumed nociceptive joint units, whereas units involved with reception of innocuous mechanical events remain unaffected (Langford et al. 1983; Schaible 1983; Schaible and Schmidt 1983c).

Acknowledgment. This work was supported by grants from the Deutsche Forschungsgemeinschaft and the Senator Kurt and Inge Schuster-Stiftung.

References

Andrew BL, Dodt E (1953) The deployment of sensory nerve endings at the knee joint of the cat. Acta Physiol Scand 28: 287–296

Boyd IA (1954) The histological structure of the receptors in the knee-joint of the cat correlated with their physiological response. J Physiol 124: 476–488

Boyd IA, Roberts TDM (1953) Proprioceptive discharges from stretch receptors in the knee joint of the cat. J Physiol 122: 38–58

Burgess PR, Clark FJ (1969) Characteristics of knee joint receptors in the cat. J Physiol 203: 317–335

Clark FJ (1975) Information signaled by sensory fibers in medial articular nerve. J Neurophysiol 38: 1464–1472

Clark FJ, Burgess PR (1975) Slowly adapting receptors in cat knee joint: can they signal joint angle? J Neurophysiol 38: 1448–1463

Coggeshall RE, Hong KA, Langford LA, Schaible HG, Schmidt RF (1983) Discharge characteristics of fine medial articular nerve afferents at rest and during passive movements of inflamed knee joints. Brain Res 272: 185–188

Eklund G, Skoglund S (1960) On the specificity of the Ruffini like joint receptors. Acta Physiol Scand 49: 184–191

Ferrell WR (1980) The adequacy of stretch receptors in the cat knee joint for signaling joint angle throughout a full range of movement. J Physiol 299: 85–99

Freeman MAR, Wyke B (1967) The innervation of the knee joint. An anatomical and histological study in the cat. J Anat 101: 505–532

Grigg P (1975) Mechanical factors influencing response of joint afferent neurons from cat knee. J Neurophysiol 38: 1473–1484

Grigg P (1976) Response of joint afferent neurons in cat medial articular nerve to active and passive movements of the knee. Brain Res 118: 482–485

Grigg P, Greenspan BJ (1977) Response of primate joint afferent neurons to mechanical stimulation of knee joint. J Neurophysiol 40: 1–8

Grigg P, Hoffman AH (1982) Properties of Ruffini afferents revealed by stress analysis of isolated sections of cat knee capsule. J Neurophysiol 47: 41–54

Grigg P, Hoffman AH, Fogarty KE (1982) Properties of Golgi-Mazzoni afferents in cat knee joint capsule, as revealed by mechanical studies of isolated joint capsule. J Neurophysiol 47: 31–40

Langford LA (1983) Unmyelinated axon ratios in cat motor, cutaneous and articular nerves. Neurosci Lett 40: 19–22

Langford LA, Schaible HG, Schmidt RF (1983) Structure and function of fine joint afferents: observations and speculations. Symposium on sensory receptor mechanisms, Hongkong (in press)

Langford LA, Schmidt RF (1983) Afferent and efferent axons in the medial and posterior articular nerves of the cat. Anat Rec 206: 71–78

McCall Jr WD, Farias MC, Williams WJ, Bement SL (1974) Static and dynamic responses of slowly adapting joint receptors. Brain Res 70: 221–243

Sato, Y, Schaible HG, Schmidt RF (1983) Types of afferents from the knee joint evoking sympathetic reflexes in cat inferior cardiac nerves. Neurosci Lett 39: 71–75

Schaible HG (1983) Bradykinin sensitivity of polymodal fine afferent joint units. Sensitization by prostaglandin E1. Naunyn Schmiedebergs Arch Pharmacol [Suppl] 352: R98

Schaible HG, Schmidt RF (1983a) Activation of groups III and IV sensory units in medial arcticular nerve by local mechanical stimulation of knee joint. J Neurophysiol 49: 35–44

Schaible HG, Schmidt RF (1983b) Responses of fine medial articular nerve afferents to passive movements of knee joint. J Neurophysiol 49: 1118–1126

Schaible HG, Schmidt RF (1983c) Coding of pain information from joints. Proc symp development organization and processing in somatosensory pathways (in press)

Schmidt RF, Schaible HG (1983) Neurophysiology of nociception in normal and inflamed knee joints of the cat. Proc Int U Physiol Sci (in press)

Skoglund S (1956) Anatomical and physiological studies of knee joint innervation in the cat. Acta Physiol Scand 36: [Suppl 124]: 1–101

V. The Physiology of Synapses

Stephan Kuffler, John Eccles, and David Katz in Sydney 1940 on the way to university to lecture, and in Oxford 1972

The Form and Function of Synapses

JOHN C. ECCLES, Contra (Locarno) and Göttingen

An essential feature of the operation of chemically transmitting synapses is that the current flowing in through the opened ionic channels in the postsynaptic membrane has to pass through the very constricted synaptic cleft, measuring about 20 nm across. As shown in Fig. 1 (note scale), it is the flow of this current that causes the excitatory postsynaptic potential (EPSP). As pointed out by Eccles and Jaeger (1958), two conflicting requirements are represented in the dimensions of the synaptic cleft: (a) it should be very narrow so that after its liberation the synaptic transmitter is applied as efficiently and quickly as possible to the receptors of the subsynaptic membrane; (b) it should be wide relative to its area so that the current may flow as freely as possible for maximum generation of the EPSP (cf. Fig. 1). There was a rigorous mathematical treatment of this dilemma of design.

This early theoretical work has been lost sight of, yet it could form the basic explanation of remarkable morphological findings on spine synapses of neocortical pyramidal cells and of the hippocampus during growth, degeneration, and regeneration. Hitherto no explanation has been produced for these findings on Gray type I excitatory synapses.

A small spine synapse is a disc of less than 1 μm in diameter (Fig. 2) with an accurate apposition of the postsynaptic density (PSD) to the presynaptic dense protrusions. Pioneer studies by Akert (1973) showed that active presynaptic sites were indicated by freeze etching electron microscopy, and appeared in four shapes: compact disc, annulus, horseshoe, and multifocal. Many comparable observations have since been made using techniques that revealed the shapes of the active components of both the presynaptic and postsynaptic elements of the spine

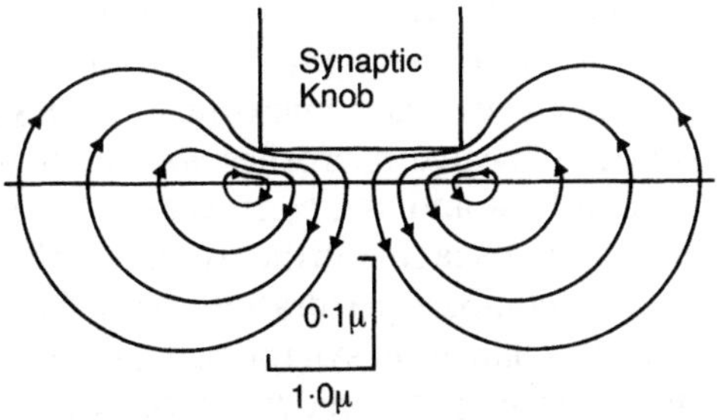

Fig. 1. Diagram of an excitatory synapse showing the lines of postsynaptic current flow when the subsynaptic membrane is activated by the transmitter substance. Note that, as shown by the scales, the width of the synaptic cleft is increased by a factor of 10 relative to its length.

Experimental Brain Research, Suppl. 9
© Springer-Verlag Berlin · Heidelberg 1984

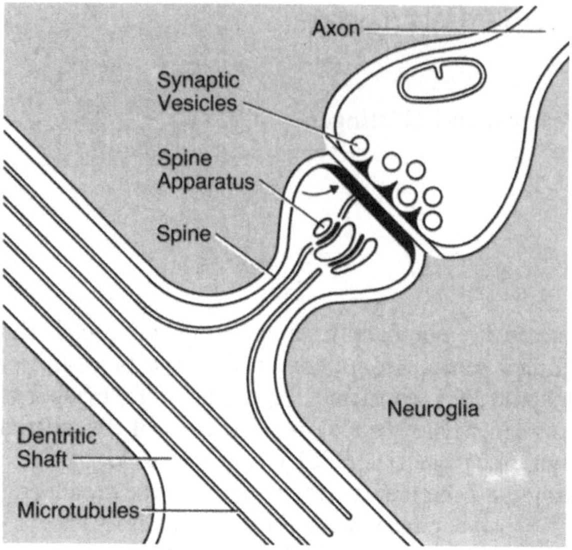

Fig. 2. Drawing of spine synapse on a dentate granule cell showing dendritic microtubules going into a spine up to the spine apparatus. The postsynaptic density is indicated by the *arrow*. From the presynaptic membrane there are dense projections up to and between the synaptic vesicles. (Gray 1982)

synapses (Vrensen et al. 1980; Vrensen and Nunes Cardozo 1981; Nieto-Sampedro et al. 1982; Carlin and Siekevitz 1983). There is a general agreement that in all synaptic shapes there is close matching of the active presynaptic elements (the dense protrusions) with the active postsynaptic elements, the PSDs. It is believed that the sequence from small round, to larger annular, to horseshoe or complex perforation, to multifocal or fragmented (Fig. 3) represents an evolutionary process (cf. Nieto-Sampedro et al. 1982), but no statement is made with respect to the responsible causal factors.

It is proposed that an explanation can be derived from the earlier mathematical treatment of the synapse (Eccles and Jaeger 1958). That treatment was simplified by making the calculation for "efficiency" of the inward current through the activated postsynaptic membrane as if it were uniformly distributed, which is unrealistic. The synaptic cleft resistance builds up progressively from the periphery. As a consequence with a simple disc-shaped PSD there will be stronger inward current through the channels of the peripheral receptor sites than of those more central. The calculated overall efficiency may be as high as 95%, but most of this would be for the peripheral zone and the central zone would be quite low.

The hypothesis is that inward current flow through the channels is essential for maintenance of each microsite of the PSD. As a disc-shaped synapse enlarges, the centrally located microsites will suffer from a progressive diminution of inward current flow and as a consequence regress, so forming the central perforation or the annular shape. Further results of this regression would be the observed horseshoe or complex perforations with the eventual fragmentation. It is to be noted that the changes in synaptic shape are initiated postsynaptically, the presynaptic matching shapes being due to some secondary trophic action across the synaptic cleft.

The hypothesis is that the inward current through a channel in the PSD influences only the immediately adjacent microsites, possibly only that associated with

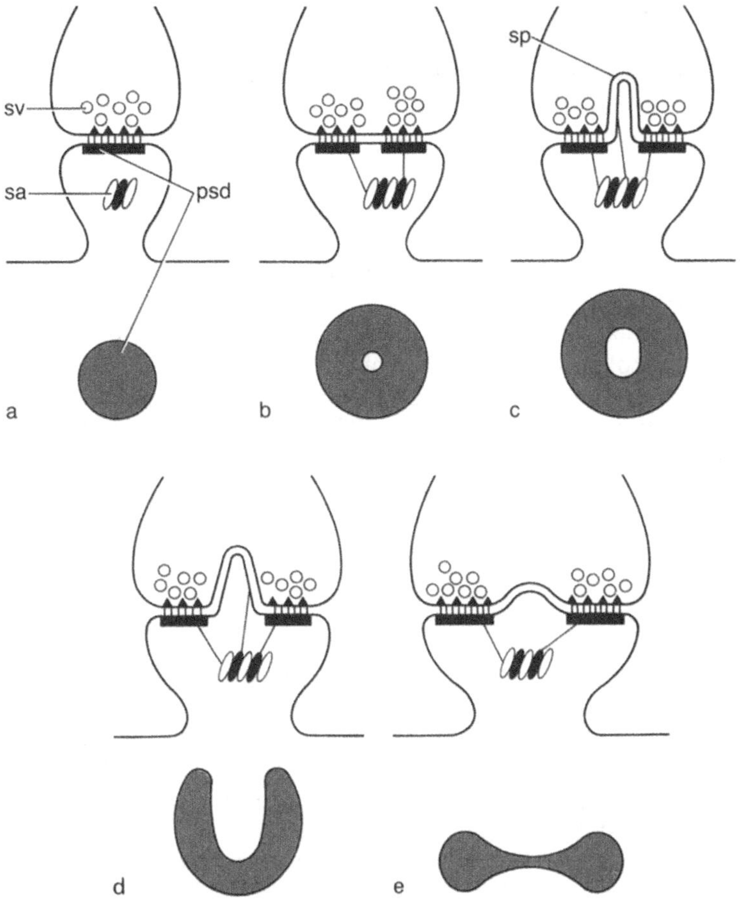

Fig. 3a–e. Drawing of spine synapses to show evolution of shape from a small disc to a larger annulus to a horseshoe, and finally to the fragmentation. *psd*, postsynaptic density; *sv*, synaptic vesicles; *sa*, spine apparatus; *sp*, spine. (Carlin and Siekevitz 1983)

that channel. It is therefore essential to enquire further into the way in which ionic currents through a channel could exert this maintenance. Almost all of the ionic current is carried by Na^+ and K^+ (Takeuchi and Takeuchi 1960; Gage 1976), but there is a very small inward Ca^{2+} current also (Takeuchi 1963; Gage 1976). Some 5%–10% of the protein composition of PSDs is the special receptor protein for calcium, calmodulin (Grab et al. 1980; Matus 1981), so there is opportunity for the inward flux of Ca^{2+} ions to combine with calmodulin, four Ca^{2+} ions to each calmodulin molecule. The Ca^{2+}-calmodulin complex has powerful metabolic actions (Cheung 1980, 1982), particularly on proteins, phosphorylating them and organizing their structural relationships, and these have been studied in the PSD by special electron microscopic techniques (Matus 1981). Furthermore, it has been shown that calcium increases the synaptic transmitter (glutamate) receptors in the hippocampus, presumably by action on the PSDs (Lynch et al. 1983), and that this is also effected by repetitive electrical stimulation (Baudry et al 1980).

Thus in summary the observed shape changes in spine synapses can be explained by the hypothesis that each microsite of the PSD is maintained by the local action of the inward Ca^{2+} currents during synaptic activation. As a disc-shaped synapse grows, the central zone will suffer from a progressive decline in Ca^{2+} input with the consequent disorder resulting from a decrease in the Ca^{2+}-calmodulin metabolic support. It will atrophy to give the central perforation or annular shape and the eventual fragmentation. By trophic influence across the cleft the active presynaptic sites will match the postsynaptic. A corollary of the hypothesis is that recovery follows automatically in the small pieces of the fragmented synapse.

An extension of the hypothesis is that the atrophy of disused synapses would also be explained by the failure of the Ca^{2+}-calmodulin metabolic support. However, there does not seem to be a study of disused synapses with a sufficiently discriminative electron microscopy. The Golgi studies of Valverde (1967, 1971) disclosed that, under conditions of disuse, synapses did not develop, but there appears to be no study of disuse atrophy of already formed synapses using electron microscopy.

References

Akert K (1973) Dynamic aspects of synaptic ultrastructure. Brain Res 49: 511–518

Baudry M, Oliver M, Greager R, Wieraszko A, Lynch G (1980) Increase in glutamate receptors following repetitive electrical stimulation in hippocampal slices. Life Sci 27: 325–330

Carlin RK, Siekevitz P (1983) Plasticity in the central nervous system: do synapses divide? Proc Natl Acad Sci USA 80: 3517–3521

Cheung WY (1980) Calmodulin plays a pivotal role in cellular regulation. Science 207: 19–27

Cheung WY (1982) Calmodulin. Sci Am 246,6: 48–56

Eccles JC, Jaeger JC (1958) The relationship between the mode of operation and the dimensions of the junctional regions at synapses and motor end-organs. Proc. R Soc Lond [Biol] 148: 38–56

Gage PW (1976) Generation of end-plate potentials. Physiol Rev 56: 177–247

Grab DJ, Carlin RK, Siekevitz P (1980) The presence and functions of calmodulin in the postsynaptic density. Ann NY Acad Sci 356: 55–72

Gray EG (1982) Rehabilitating the dendritic spine. Trends Neurosci 5: 5–6

Lynch G, Halpain S, Baudry M (1983) Structural and biochemical effects of high frequency stimulation in hippocampus. In: Seifert W (ed) Molecular, cellular and behavioral neurobiology of the hippocampus. Academic, New York pp 253–262

Matus A (1981) The postsynaptic density. Trends Neurosci 4: 51–53

Nieto-Sampedro M, Hoff SW, Cotman CW (1982) Perforated postsynaptic densities: probable intermediates in synapse turnover. Proc Natl Acad Sci USA 79: 5718–5722

Takeuchi A, Takeuchi N (1960) On the permeability of the end-plate membrane during the action of transmitter. J Physiol (Lond) 154: 52–67

Takeuchi N (1963) Effects of calcium on the conductance change of the end-plate membrane during the action of transmitter. J Physiol (Lond) 167: 141–155

Valverde F (1967) Apical dendritic spines of the visual cortex and light deprivation in the mouse. Brain Res 3: 337–352

Valverde F (1971) Rate and extent of recovery from dark rearing in the visual cortex of the mouse. Brain Res 33: 1–11

Vrensen G, Nunes Cardozo J (1981) Changes in size and shape of synaptic connections after visual training: an ultrastructural approach of synaptic plasticity. Brain Res 218: 79–97

Vrensen G, Nunes Cardozo J, Müller L, van der Want J (1980) The presynaptic grid: a new approach. Brain Res 184: 23–40

Norepinephrine and Acetylcholine Block a Calcium-Activated Potassium Hyperpolarization in Hippocampal Pyramidal Cells

R.A. Nicoll, A.E. Cole, D.V. Madison, and N.R. Newberry, San Francisco

Summary. We have examined with intracellular recording the effects of norepinephrine (NE) and acetylcholine (ACh) on pyramidal cells in the hippocampal slice preparation. The primary action of NE was to reduce spike frequency adaptation or accommodation, which occurs during long duration current pulses. Thus in the presence of NE cells fired many more spikes to the depolarizing stimulus. NE produced this effect by blocking calcium-activated potassium conductance [$G_{K(Ca)}$], which is largely responsible for the accommodation. This effect was mediated by beta-receptor activation and was mimicked by cyclic adenosine monophosphate (cAMP). Since NE and cAMP did not reduce calcium spikes, it appeares that $G_{K(Ca)}$ is blocked at a step subsequent to calcium entry into the neuron. ACh also blocked $G_{K(Ca)}$ and accommodation. In addition it depolarized neurons and increased the input resistance of neurons. All actions were blocked by the muscarinic antagonist atropine. Stimulation of sites in the slice known to contain cholinergic fibers mimicked all of these actions of ACh. Furthermore, these evoked synaptic responses were enhanced by the cholinesterase inhibitor eserine, and were blocked by atropine. Possible roles of these nonclassical synaptic actions of NE and ACh are discussed.

Key words: *hippocampus – norepinephrine – acetylcholine – accommodation – calcium-activated potassium conductance*

Introduction

The traditional and most firmly established view of the action of neurotransmitters is that they bind to receptors, resulting in the brief opening of ion channels. Depending on the ion involved, this change in conductance will either increase or decrease the excitability of the cell. Virtually all of the synaptic responses analyzed in the central nervous system (CNS) thus far conform to this view (Eccles 1964). However, recent studies performed primarily in heart muscle, and in neurons of invertebrate and vertebrate ganglia, suggest that neurotransmitters can also act on ion channels that are controlled by voltage (Nicoll 1982). Such an action can result in marked changes in firing patterns of the neuron in the absence of large changes in resting excitability. In the CNS, acetylcholine (ACh), acting on muscarinic receptors, produces a slow depolarization associated with a decrease in potassium conductance (Benardo and Prince 1982; Dodd et al. 1981; Halliwell and Adams 1982; Krnjevic et al. 1971). It has also been noted that this

Experimental Brain Research, Suppl. 9
© Springer-Verlag Berlin · Heidelberg 1984

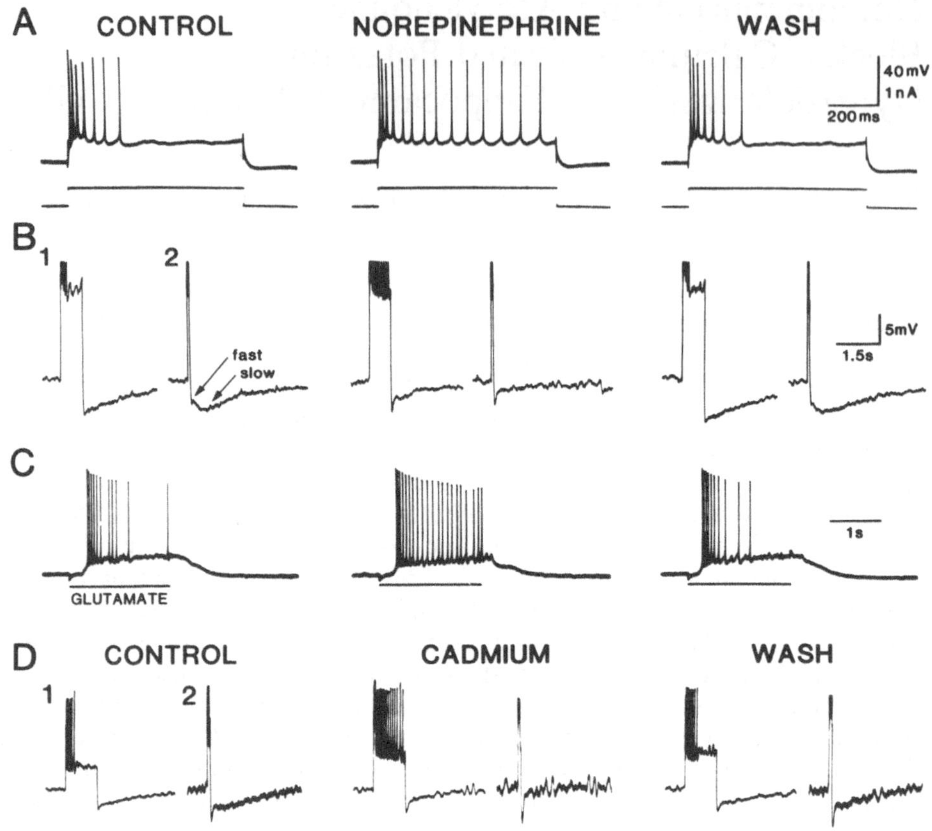

Fig. 1A–D. Norepinephrine (NE) and cadmium block calcium-activated potassium responses and reduce accommodation in hippocampal pyramidal cells. The responses in *A–C* are all from the same pyramidal cell. *A* Film records of the response of the cell to a depolarizing current pulse. The *current trace* is positioned below the *voltage trace*. The response in the *middle column* was obtained 4 min after the addition of 10 μM NE to the bathing solution. The membrane potential did not change during the application of NE. The records in the *right-hand column* were obtained 30 min after returning to the control solution. *B* Simultaneously recorded chart records of the response shown in *A* (B_1), and response to a short (100-ms) depolarizing current pulse (B_2). *C* Film records of response to 30 nA iontophoretically applied glutamate. Membrane potential -58 mV. The gain of *C* is the same as in *A*. *D* Chart records from another cell showing responses to a long (D_1) and short (D_2) depolarizing current pulse recorded at different gains. The calcium antagonist cadmium (0.1 mM) was applied for 8 min. The records in the right-hand column were obtained 32 min after returning to the control bathing solution. The spikes were attenuated by the pen recorder. The time calibration is the same as in *B*. The gain for D_2 is the same as in *B*, while D_1 is half the gain of *B*. Membrane potential -57 mV. (Madison and Nicoll 1982).

action of ACh facilitates repetitive cell discharge (Krnjevic et al. 1971). However, it has yet to be shown that these effects can be reproduced by synaptically released ACh.

A number of observations with extracellular recording have suggested that the action of norepinephrine (NE) in the CNS may not conform to the traditional view of transmitter action (Woodward et al. 1979). However, with the available evidence, it is difficult to propose concrete mechanisms that could account for these observations. We have been studying the action of ACh (Cole and Nicoll 1983) and NE (Madison and Nicoll 1982) on CA1 pyramidal cells in the hippocampal slice preparation. These cells are believed to be innervated by fibers containing these transmitters.

In our initial studies with NE we were disappointed by the variable and meager effects that this transmitter had on the membrane potential of pyramidal cells. The responses were rarely over a few millivolts in size. Since NE had been reported to enhance the action of glutamate in other systems (Rogawski and Aghajanian; Woodward et al. 1979), we looked for an interaction between these two transmitters on pyramidal cells (Madison and Nicoll 1982). The response of pyramidal cells to iontophoretically applied glutamate typically consisted of an initial high- frequency discharge which then slowed during the glutamate depolarization. NE (1–10 μM) markedly increased the number of spikes evoked by glutamate (Fig. 1C), often in the absence of any change in membrane potential. To determine whether this effect was due to a specific interaction between these two transmitters or to a more general effect of NE on the responsiveness of the cell, we excited neurons with direct depolarizing current pulses applied through the recording electrode instead of with glutamate. We found that NE also caused an increase in the number of spikes elicited by depolarizing the cell with direct current. In control conditions the action potential discharge rapidly accommodated (adapted) during the depolarizing pulse and the cell typically fell silent after about six to eight spikes. NE strongly depressed this accommodation (Fig. 1A). Since the NE augmentation of the response appeared to be independent of the mode of depolarization, it seems unlikely that NE exerts a specific action on glutamate responses.

The mechanism underlying this accommodation appears to be due largely to a calcium-activated potassium conductance [$G_{K(Ca)}$], since bath application of the calcium antagonist cadmium (Fig. 1D) or intracellular injection of the calcium chelator ethyleneglycol-bis(β-aminoethyl ether)N, N'-tetraacetic acid (EGTA) reduced the accommodation. The slow hyperpolarization which follows the depolarizing pulse, known to be due to $G_{K(Ca)}$ (Hotson and Prince 1980), is also reduced. NE appears to block accommodation by a reduction in $G_{K(Ca)}$, since it also blocked the afterhyperpolarization (Fig. 1B). A beta-adrenergic receptor appears to be involved in this effect because the response is blocked by the beta-antagonist propranolol (10 μM), but not by the alpha-antagonist phentolamine (20 μM). Furthermore, the beta-agonist isoproterenol was more potent than NE, while the alpha-agonist phenylephrine was ineffective at 10 μM, which is a near-maximal concentration for the action of NE. Other investigators have also observed a beta-adrenergic blockade of $G_{K(Ca)}$ (Gruol and Siggins 1982; Haas and Konnerth 1983). Histamine also blocks this response in hippocampal pyramidal cells (Haas and Konnerth 1983) and FMRF amide peptides have a similar action on an identified snail neuron (Cottrell 1982b).

Cyclic adenosine monophospate (cAMP) has been found to mediate most of the effects of beta-receptor activation in other systems. We therefore bath applied the membrane-permeable analog of cAMP, 8-bromoadenosine-3',5'-cyclic monophosphate (1 mM) and found that it blocked both the afterhyperpolarization

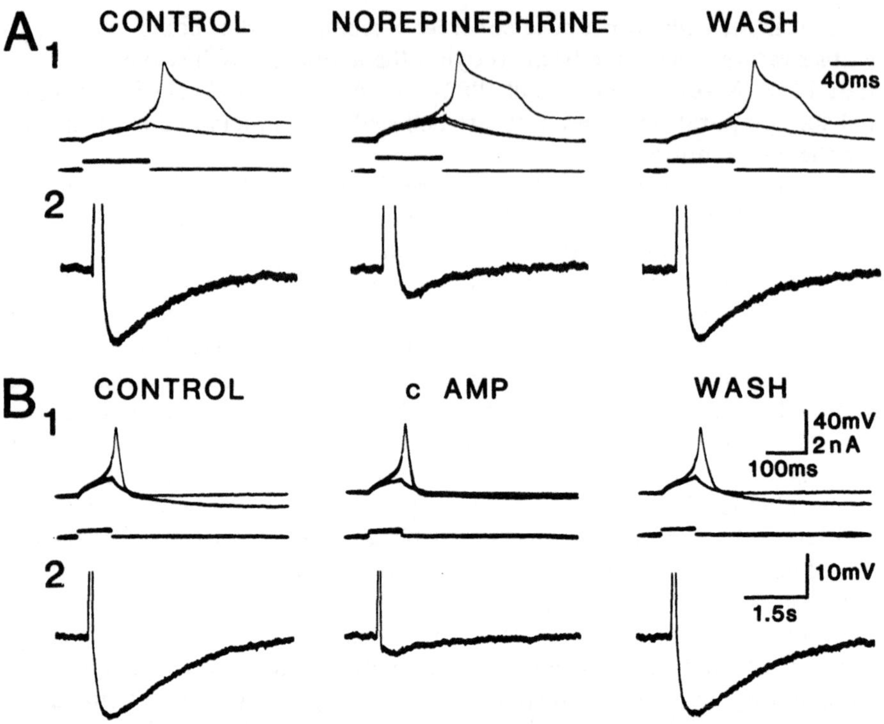

Fig. 2A,B. Norepinephrine and cyclic adenosine monophosphate (*cAMP*) block calcium-activated potassium responses without reducing calcium spikes. *A* The *top traces* (A_1) are film records of the Ca^{2+} spike evoked in a pyramidal cell and the corresponding current pulses. The *bottom traces* (A_2) are the simultaneously recorded chart records. The preparation was bathed with a solution containing 1 μM tetrodotoxin and 5 mM tetraethylammonium. The response in the *middle column* was obtained 7 min after starting the iontophoretic application of NE (200 nA). In this cell the threshold for spike initiation was slightly increased in NE but this was not a typical finding. The response in the *right-hand column* was obtained 15 min after termination of NE iontophoresis. Membrane potential -56 mV. *B* shows similar responses to those in *A*, but from another cell. The record in the *middle column* was obtained 9 min after the addition of 1 mM 8-bromoadenosine-3',5'-cyclic monophosphate (sodium salt) to the superfusate. The record in the *right-hand column* was obtained 66 min after returning to the control superfusate. Membrane potential -62 mV. The calibrations in *A* are the same as in *B* except for the time calibration in A_1. (Madison and Nicoll 1982)

(Fig. 2B) and accommodation. A similar blockade was observed when cAMP was injected into neurons via the recording microelectrode. These results suggest that cAMP mediates the action of NE in hippocampal pyramidal cells. cAMP appears to block $G_{K(Ca)}$ in the bag cells of *Aplysia* (Kaczmarek and Strumwasser 1981).

The next question concerned the site at which NE and cAMP exert their antagonism of $G_{K(Ca)}$. In sympathetic ganglia NE blocks the $G_{K(Ca)}$ by blocking calcium entry into the cell (Galvan and Adams 1982; Horn and McAfee 1980). Calcium currents in embryonic sensory neurons are also reduced by NE (Dunlap and Fischbach 1981). We have examined the effects of NE on calcium spikes in pyramidal cells to see if a similar mechanism might explain our results. Tetrodotoxin (1 μM) and tetraethylammonium (5 mM) were added to the medium to increase

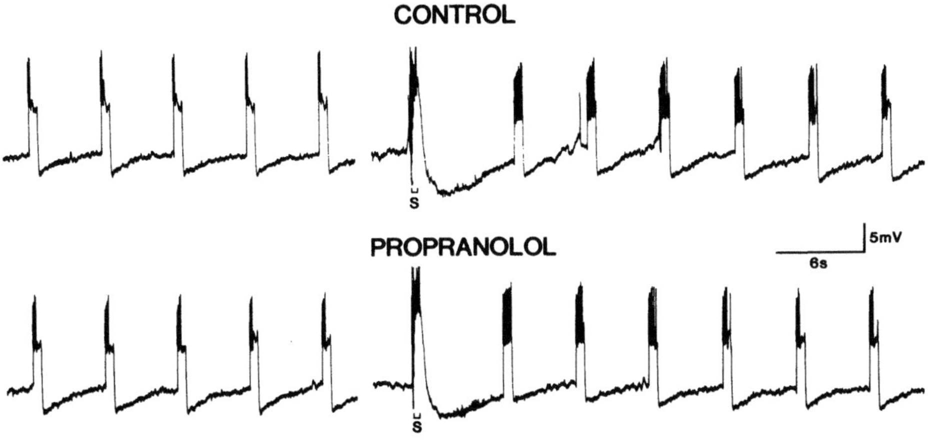

Fig. 3. Electrical stimulation of stratum radiatum and stratum oriens blocks accommodation of pyramidal cell discharge. A 50-Hz stimulus (*S*) for 0.5 s duration caused an initial depolarization followed by a hyperpolarization. In addition there was a prolonged block in the ability of the cell to accommodate during depolarizing pulses. The addition of the beta-adrenergic antagonist propranolol (20 μM) had little effect on the stimulus-induced block in accommodation, suggesting that norepinephrine was not involved.

the size of calcium spikes. Although NE and cAMP severely depressed the afterhyperpolarization that followed calcium spikes, they did not reduce the size of the calcium spikes. Rather the spikes were often prolonged, and in some cells repetitive spiking occurred. Therefore, NE and cAMP do not appear to reduce calcium entry and must be acting to reduce $G_{K(Ca)}$ at some step that occurs after calcium entry into the cell.

We attempted to activate NE-containing fibers by stimulating in stratum radiatum and stratum oriens. Following repetitive stimulation at these sites, a prolonged block of the afterhyperpolarization and the accommodation of spike discharge in pyramidal cells occurred (Fig. 3). However, propranolol, which effectively blocked the action of NE, had little effect on the stimulus-induced block of accommodation. Therefore, we considered the possibility that the stimulus might be releasing another substance which had effects similar to NE. By altering the position of the stimulating electrode it was found that the stimulation in stratum oriens usually resulted in the strongest blockade of pyramidal cell accommodation and this was often associated with a slow depolarization. Since the cholinergic input to the CA1 region of the hippocampus lies mainly in stratum oriens (Lewis and Shute 1967; Mosko et al. 1973), the action of ACh (10–100 μM) and carbachol (0.1–10 μM) on pyramidal cells was studied. In agreement with others (Benardo and Prince 1982; Dodd et al. 1981; Halliwell and Adams 1982; Krnjevic et al. 1971) we found that ACh and carbachol depolarized pyramidal cells and that there was a concomitant increase in the cell input resistance. This is shown in Fig. 4 for two concentrations of carbachol. The resistance increase is most clearly seen when the membrane potential is hyperpolarized back to the control level (see bar beneath trace). At the higher concentration there was some desensitization of the depolarizations (see two arrows in Fig. 4B). In addition, we found

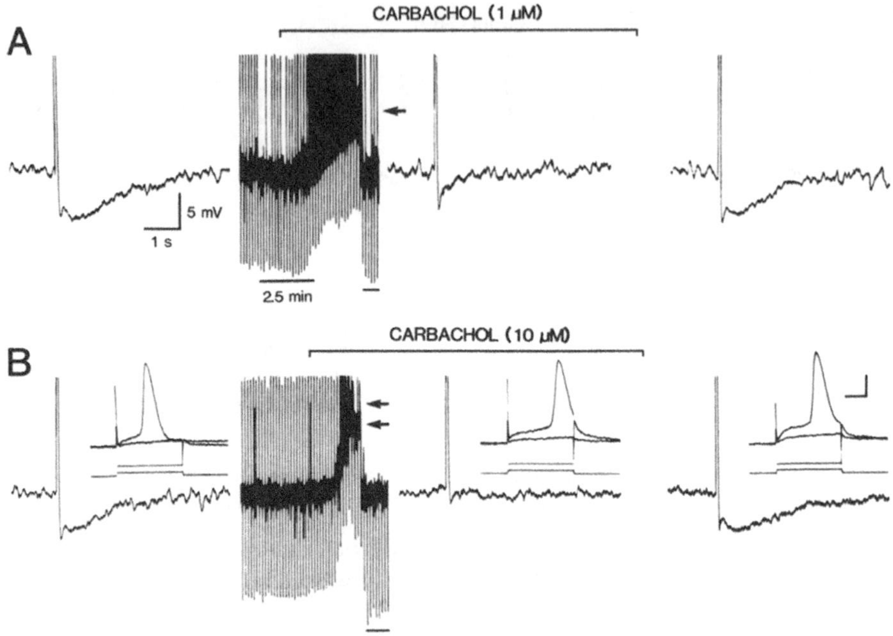

Fig. 4A,B. Effect of carbachol on hippocampal pyramidal cell. *A* Carbachol was bath applied during the *bar*. The *trace on the left* is the control afterhyperpolarization elicited by a depolarizing current pulse. After the depolarization had stabilized, the membrane was hyperpolarized back to the resting level by injecting current (see *bar beneath record*). The afterhyperpolarization was then repeated. The *record on the right* was obtained 10 min after washing out carbachol. *B* shows response in the same cell to a tenfold higher concentration of carbachol. In addition the shape of a single action potential evoked by depolarizing current is shown in the *film inset*. The depolarization to the higher concentration is much more rapid and shows some desensitization. The *response on the right* was obtained 20 min after washing out carbachol. The calibration for the *inset* is 1 nA and 20 mV for the gain and 2 ms for time.

that ACh and carbachol blocked $G_{K(Ca)}$ (cf. Benardo and Prince 1982), as can be seen by comparing the afterhyperpolarization before and after the addition of carbachol in Fig. 4. It can also be seen that at a time when $G_{K(Ca)}$ is blocked, there is little effect on the repolarization of the action potential (see inset in Fig. 4B). This suggests that carbachol does not affect the delayed rectifier K current. The block of $G_{K(Ca)}$ could be seen at concentrations of ACh or carbachol that had only small effects on the membrane potential. In the presence of tetrodotoxin and tetraethylammonium, ACh and carbachol reduced the afterhyperpolarization with little reduction in the size of calcium spikes. Thus we believe that ACh, like NE, reduces $G_{K(Ca)}$ at some step subsequent to calcium entry into the cell. ACh also greatly reduced the accommodation of pyramidal cell discharge (not shown). All of these effects on pyramidal cells were mimicked by the muscarinic agonists muscarine and pilocarpine (0.1–10 μM), but not by nicotine (10 μM). These effects were also reversed by low concentrations of atropine (0.1–1.0 μM), indicating that they are muscarinic in nature. With this information on the action of ACh we returned to the pathway stimulation experiments to see if we could duplicate these actions by activating cholinergic fibers.

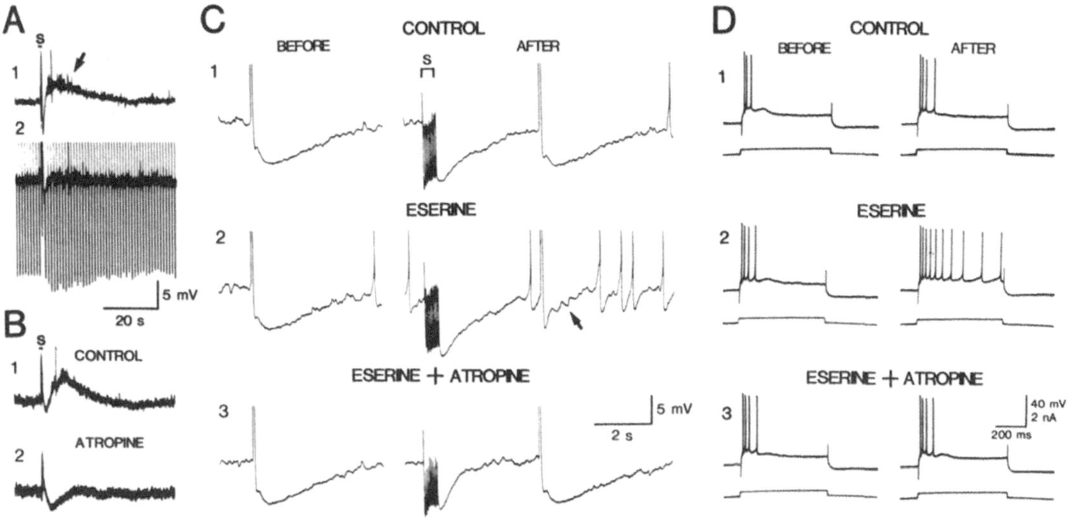

Fig. 5A–D. Stimulation on stratum oriens mimics the effect of exogenous ACh. *A–D* are chart records from four different cells. *A* Immediately after a train of stimuli (*S*) the cell hyperpolarizes (condensed on this time scale) and then depolarizes (see *arrow in* A_1). In A_2, the slow depolarization was manually voltage clamped, which clearly shows the increase in size of the hyperpolarizing current pulses. Resting membrane potential = -53 mV. In another cell, slow depolarization (B_1) was completely blocked by 1 μM atropine (B_2). Resting membrane potential = -59 mV. Calibration in *A* also applies to *B*. *C* shows the afterhyperpolarizations before pathway stimulation (*left-hand column*) and 4 s after the stimulus (*S*) (*right-hand column*). In control solution (C_1) the afterhyperpolarization is slightly depressed after the stimulus; following the addition of 1 μM eserine (C_2) the afterhyperpolarization is greatly reduced by the stimulus (see *arrow*); the afterhyperpolarization is restored to control amplitude by 1 μM atropine (C_3). Resting membrane potential = -56 mV. *D* Same experimental protocol as in *C* except showing the accommodation during a 600-ms pulse before and after pathway stimulation (*S*). Resting membrane potential = -64 mV. (Cole and Nicoll 1983)

Electrical stimulation in stratum oriens ("S" in Fig. 5A$_1$) produced a noncholinergic hyperpolarization lasting 2–3 s, followed by a slow cholinergic depolarization lasting 20–30 s (see arrow in Fig. 5A$_1$). This depolarization was associated with an increase in the cell's input resistance. This can be seen in Fig. 5A$_2$, where the size of the voltage defelections to hyperpolarizing current pulses is increased when the slow depolarization was manually voltage clamped. The addition of eserine increased the size and duration of this depolarization, resulting in spike discharge. Atropine blocked the depolarization (Fig. 5B). The depolarization, along with the earlier noncholinergic hyperpolarization, was blocked by the calcium antagonist cadmium (0.1 mM), and also by tetrodotoxin (1 μM). The finding that this slow depolarization increases excitability and is dependent on calcium and propagated action potentials indicates that this potential is a slow excitatory postsynaptic potential (sEPSP).

In addition to generating a sEPSP, the stimulus also reduced the hyperpolarization following activation of the cell by depolarizing current pulses, indicating a reduction in $G_{K(Ca)}$. In Fig. 5C an afterhyperpolarization was elicited before and 4 s after pathway stimulation (S), which caused a small decrease in the amplitude of the response. In the presence of eserine (Fig. 5C_2) the same stimulus caused a much larger reduction (see arrow). Atropine entirely blocked the stimulus induced depression of the afterhyperpolarization (Fig. 5C_3). Pathway stimulation also attenuated accommodation of spike discharge during long direct depolarizing current pulses, and this effect was markedly enhanced by eserine and blocked by atropine (Fig. 5D). Both the afterhyperpolarization and the accommodation provided very sensitive measures of cholinergic activity, since dramatic effects on these parameters could be seen in the absence of any change in membrane potential.

Discussion

We have shown in the slice preparation that both NE and ACh block calcium-activated potassium afterhyperpolarizations in CA1 hippocampal pyramidal cells. Since this effect could occur without any decrease in the size of calcium spikes, both transmitters appear to exert their blockade at a site after calcium entry into the cell. This effect can occur in the absence of any large changes in resting membrane potential. $G_{K(Ca)}$ exerts a strong braking influence on pyramidal cell discharge and, therefore, in the presence of NE or ACh, many more spikes are elicited in response to a constant depolarizing stimulus. In the case of ACh, we have been able to mimic its actions precisely by electrically stimulating cholinergic afferent fibers. The actions described here for NE and ACh are nonclassical in that the responses do not result from an increase in ion conductance. Rather, these transmitters act on a process (i.e., accommodation) that is only brought into play when the neuron is activated by a depolarizing input.

In addition to ACh's action on $G_{K(Ca)}$, which has also been observed in myenteric neurons (Morita, North and Tokimasa 1982), ACh has long been known to depolarize neurons by blocking a resting potassium conductance (Adams et al. 1982; Benardo and Prince 1982; Dodd et al. 1981; Halliwell and Adams 1982; Krnjevic et al. 1971; Morita et al. 1982; Weight and Votava 1970). In sympathetic and hippocampal neurons this effect is largely due to the blockade of a time- and voltage-dependent potassium current, termed I_M (Adams et al. 1982; Halliwell and Adams 1982). Both the block in $G_{K(Ca)}$ and I_M would be expected to contribute to the blockade of accommodation.

Catecholamines have been shown to have nonclassical actions at other sites. One of the first indications that neurotransmitters could act on voltage-sensitive ion channels was the finding that epinephrine enhances inward calcium currents in heart muscle (Reuter and Scholz 1977). Interestingly, in sympathetic and dorsal root ganglion cells, norepinephrine reduces calcium influx (Dunlap and Fischbach 1981; Galvan and Adams 1982; Horn and McAfee 1980). These nonclassical actions are not restricted to NE and ACh. Serotonin, which can depolarize neurons by a reduction in a resting potassium conductance (Cottrell 1982a; Gerschenfeld and Paupardin-Tritsch 1974; Klein et al. 1982; VanderMaelen and Aghajanian 1980; Wood and Mayer 1979), can also block $G_{K(Ca)}$ (Wood and Mayer 1979) and accommodation (Cottrell 1982a; Wood and Mayer 1979).

Norepinephrine and ACh were the first two substances to be established as neurotransmitters in the peripheral nervous system and, furthermore, ACh was the first transmitter identified in the CNS (Eccles et al. 1954). Ever since their presence in brain tissue was established, a great deal of speculation has been generated on the possible roles of these two compounds. It has been proposed on numerous occasions that both of these transmitters might be involved in a variety of complex behaviors such as arousal, attention, memory, and learning (Aston-Jones et al. 1983; Coyle et al. 1983). One reason for suspecting a role in such global behaviors is that, although the neurons containing NE and ACh are few in number, they have extremely divergent patterns of cortical innervation. Although the neuronal mechanisms involved in these behaviors are largely unknown, it is conceivable that a transmitter-mediated increase in neuronal responsiveness to excitatory inputs could play such a role. For instance, neurons in the visual cortex respond more briskly to visual stimuli during arousal than during sleep (Livingstone and Hubel 1981), even though the spontaneous activity of these cells is unchanged or even decreased. Mountcastle et al. (1981) have found that attentive fixation exerts a powerful facilitation of the response of parietal neurons to visual stimuli. This facilitation, which plays an important role in visually guided behavior, occurs in the absence of any consistent change in spontaneous activity. The modulatory effects that we have described here for NE and ACh are strikingly similar to those associated with changes in the behavioral state of the animal, since these transmitters can act to facilitate the responsiveness of neurons to excitatory inputs with little change in their resting potential. The use of specific antagonists in these behavioral studies would be of interest in establishing a role for NE and ACh in the modulation of neuronal responsiveness associated with behavioral state.

Acknowledgments. This work was supported by NIH predoctoral fellowships GM-07449 to D.V.M. and grants NS 15764, NS 16485, RCDA MH 00437, and the Klingenstein Fund to R.A.N. N.R.N. thanks the Wellcome Trust for a travel grant.

References

Adams PR, Brown DA, Constanti A (1982) Pharmacological inhibition of the M-current. J Physiol (Lond) 332: 223–262

Aston-Jones G, Foote SL, Bloom FE (1983) Anatomy and physiology of locus coeruleus neurons: functional implications. In: Ziegler MG, Lake CR (eds) Norepinephrine: clinical aspects. William and Wilkins, Baltimore (to be published)

Benardo LS, Prince DA (1982) Ionic mechanisms of cholinergic excitation in mammalian hippocampal pyramidal cells. Brain Res 249: 333–344

Cole AE, Nicoll RA (1983) Acetylcholine mediates a slow synaptic potential in hippocampal pyramidal cells. Science 221: 1299–1301

Cottrell GA (1982a) Physiological role of a slow voltage-sensitive, synaptic response mediated by an identified serotonin-containing neurone. J Exp Physiol 67: 179–183

Cottrell GA (1982b) FMRF amide neuropeptides simultaneously increase and decrease K^+ currents in an identified neurone. Nature 296: 87–89

Coyle JT, Price DL, DeLong MR (1983) Alzheimer's disease: a disorder of cortical cholinergic innervation. Science 219: 1184–1190

Dodd J, Dingledine R, Kelly JS (1981) The excitatory action of acetylcholine on hippocampal neurones of the guinea pig and rat maintained in vitro. Brain Res 207: 109–127

Dunlap K, Fischbach GD (1981) Neurotransmitters decrease the calcium conductance activated by depolarization of embryonic chick sensory neurones. J Physiol (Lond) 317: 519–535
Eccles JC (1964) The physiology of synapses. Springer, Berlin Heidelberg New York
Eccles JC, Fatt P, Koketsu K (1954) Cholinergic and inhibitory synapses in a pathway from motor-axon collaterals to motoneurones. J Physiol (Lond) 126: 524–562
Galvan M, Adams PR (1982) Control of calcium current in rat sympathetic neurons by norepinephrine. Brain Res 244: 135–144
Gerschenfeld HM, Paupardin-Tritsch D (1974) Ionic mechanisms and receptor properties underlying the responses of molluscan neurones to 5-hydroxytryptamine. J Physiol (Lond) 243: 427–456
Gruol DL, Siggins GR (1982) Norepinephrine (NE) and adrenergic agonists in low concentrations alter spontaneous activity and membrane potential of CA1 pyramidal neurons of the rat hippocampus *in vitro*. Neurosci Abst 8: 797
Haas HL, Konnerth A (1983) Histamine and noradrenaline decrease calcium-activated potassium conductance in hippocampal pyramidal cells. Nature 302: 432–434
Halliwell JV, Adams PR (1982) Voltage-clamp analysis of muscarinic excitation in hippocampal neurons. Brain Res 250: 71–92
Horn JP, McAfee DA (1980) Alpha-adrenergic inhibition of calcium-dependent potentials in rat sympathetic neurones. J Physiol (Lond) 301: 191–204
Hotson JR, Prince DA (1980) A calcium-activated hyperpolarization follows repetitive firing in hippocampal neurons. J Neurophysiol 43: 409–419
Kaczmarek LK, Strumwasser F (1981) Net outward currents of bag cell neurons are diminished by a cAMP analogue. Neurosci Abst 7: 932
Klein M, Camardo J, Kandel ER (1982) Serotonin modulates a specific potassium current in the sensory neurons that show presynaptic facilitation in *Aplysia*. Proc Natl Acad Sci USA 79: 5713–5717
Krnjevic K, Pumain R, Renaud L (1971) The mechanism of excitation by acetylcholine in the cerebral cortex. J Physiol (Lond) 215: 247–268
Lewis PR, Shute CCD (1967) The cholinergic limbic system: projections to hippocampal formation, medial cortex, nuclei of the ascending cholinergic reticular system, and the subfornical organ and supra-optic crest. Brain 90: 521–540
Livingstone MS, Hubel DH (1981) Effects of sleep and arousal on the processing of visual information in the cat. Nature 291: 554–561
Madison DV, Nicoll RA (1982) Noradrenaline blocks accommodation of pyramidal cell discharge in the hippocampus. Nature 299: 636–638
Morita K, North RA, Tokimasa T (1982) Muscarinic agonists inactivate potassium conductance of guinea-pig myenteric neurones. J Physiol 333: 125–140
Mosko S, Lynch G, Cotman CW (1973) The distribution of septal projections to the hippocampus of the rat. J Comp Neurol 152: 163–174
Mountcastle VB, Andersen RA, Mottes BC (1981) The influence of attentive fixation upon the excitability of the light-sensitive neurons of the posterior parietal cortex. J Neurosci 1: 1218–1235
Nicoll RA (1982) Neurotransmitters can say more than just "yes" or "no." Trends in Neuroscience 5: 369–374
Reuter H, Scholz H (1977) The regulation of the Ca conductance of cardiac muscle by adrenaline. J Physiol (Lond) 264: 49–62
Rogawski MA, Aghajanian GK (1980) Modulation of lateral geniculate neurone excitability by noradrenaline microiontophoresis or locus coeruleus stimulation. Nature 287: 731–734
VanderMaelen CP, Aghajanian GK (1980) Intracellular studies showing modulation of facial motoneurone excitability by serotonin. Nature 287: 346–347
Weight F, Votava Z (1970) Slow synaptic excitation in sympathetic ganglion cells: evidence for synaptic inactivation of potassium conductance. Science 170: 755–758
Wood JD, Mayer CJ (1979) Serotonin activation oftonic-type enteric neurons in guinea-pig small bowel. J Neurophysiol 42: 582–593
Woodward DJ, Moises HC, Waterhouse BD, Hoffer BJ, Freedman R (1979) Modulatory actions of norepinephrine in the central nervous system. Fed Proc 38: 2109–2116

Evidence for Both Pre- and Postsynaptic Mechanisms During Long-Term Potentiation in Hippocampal Slices

P. ANDERSEN, M. AVOLI, and Ø. HVALBY, Oslo

Introduction

The plastic changes that occur in certain hippocampal synapses during long-term potentiation (LTP) have attracted considerable interest (Swanson et al. 1982). The phenomenon requires short periods of afferent activation, at a frequency and for a duration which are comparable to those which can be met in physiological conditions (Ranck 1975). Furthermore, the changes are often considerable and may last for several hours or days in intact preparations after a single tetanic priming (Bliss and Lømo, 1973; Bliss and Gardner-Medwin 1973). Even in isolated preparations like brain slices, the phenomenon has been shown to occur (Schwartzkroin and Wester 1975) although here the process is usually of shorter duration (Andersen et al. 1977; Deadwyler et al. 1975, 1976).

There is suggested evidence of a presynaptic control mechanism, partly because of an enhanced excitatory postsynaptic potential (EPSP) without any change in the membrane resistance or potential as measured across the soma membrane (Andersen et al. 1980), and partly because an increased liberation of glutamate has been observed to parallel LTP development (Skrede and Malthe-Sørenssen 1981; Dolphin et al. 1982). However, it is far from certain whether an increased transmitter release gives a complete explanation. First, the enhanced EPSP could be due to an increased efficiency of the same amount of transmitter, provided postsynaptic sensitivity changes took place. As such, unmasking of glutamate receptors has been proposed (Baudry and Lynch 1980). Admittedly, glutamate excitability testing is difficult because of the extreme sensitivity for the localization of the test electrode (Schwartzkroin and Andersen 1975). However, the experiments that have been performed have turned out to show no change, or even a decrease, in glutamate sensitivity (Lynch et al. 1976). The second reason for questioning the idea that only presynaptic mechanisms are involved in LTP is the paradoxical discrepancy between the small or moderate increase in the EPSP and the very large increases observed in the population spike amplitude, and in the probability of single-unit discharges and of the latency of individual spikes. Thirdly, although most investigators agree that the LTP effect is homosynaptic to the activated synapses, several heterosynaptic control mechanisms are also operating. Both the necessity of activating a certain number of afferent fibres, a phenomenon called "cooperativity" (McNaughton et al. 1978), and the role of monoamines in LTP (Bliss et al. 1983) could be compatible with both a presynaptic and/or a postsynaptic locus. However, since both procedures cause greater changes in the amplitude and latency of the population spike (dis-

Experimental Brain Research, Suppl. 9
© Springer-Verlag Berlin · Heidelberg 1984

charge of pyramidal neurons) than in the amplitude of the synaptic potentials alone (Andersen et al. 1977, 1980; Douglas et al. 1982), a postsynaptic mechanism appears a possibility. Finally, since calcium ions are required for the development of LTP (Dunwiddie et al. 1978; Dunwiddie and Lynch 1979; Wigström et al., 1979), beyond the tetanic stimulation, a postsynaptic involvement seems plausible.

For these reasons, we have undertaken experiments in order to show whether postsynaptic changes could occur in parallel with the presynaptic changes suggested by the evidence referred to above. We have further studied the behaviour of the participating neurons during the tetanization in order to describe more precisely any neuronal changes that may be associated with subsequent development of LTP. For this purpose, we have employed isolated hippocampal slices, and recorded intracellularly from CA1 pyramidal cells and extracellularly from ensembles of such cells that were similarly treated. Such recordings were made both during the tetanic stimulation and for several tens of minutes during short-lasting potentiation and the first part of the LTP.

Methods

Adult rats weighing 220–280 g were anaesthetized by ether, the brain was removed and divided in two, and the hippocampus was taken out. Slices of 400–500 μm, were cut transversely to the longitudinal axis and transferred to an incubation chamber. Here the slices were maintained in the interphase between 95% oxygen, 5% CO_2 and an artificial cerebrospinal fluid, consisting of (in mM): NaCl 124, KCl 3, K_2HPO_4 1.25, $CaCl_2$ 2, $MgCl_2$ 1.25, $NaHCO_3$ 26, glucose 10. Stimulation electrodes, made of glass- or lacquer-insulated tungsten needles, were placed under visual control. Separate stimulation electrodes, spaced about 300 μm apart along the same bundle of afferent fibres, were employed for tetanization and for testing to avoid polarization changes at the test stimulus location. Recordings were made with micropipettes filled with 4 M potassium acetate from the layer of the activated dendritic synapses (stratum radiatum) and from the layer of the cell bodies (stratum pyramidale) or intracellularly from pyramidal cells in the CA1 region. The synapses we tested belonged to fibres from the commissural and the Schaffer collateral pathways, terminating in the outer half of the apical dendritic tree in the stratum radiatum.

Results

Short periods of tetanic stimulation caused an increase in the transmission between fibres running in the stratum radiatum and the CA1 pyramidal cells. Figure 1 shows an example in which the lower traces in A–D show intracellular recording from a CA1 pyramidal cell (two trials), whereas the upper trace in each pair shows the recording from the dendritic layer, where the synapses in question are located.

A and B show control records to two stimulus strengths. Following McNaughton et al. (1978) we have standardized the stimuli by using a constant

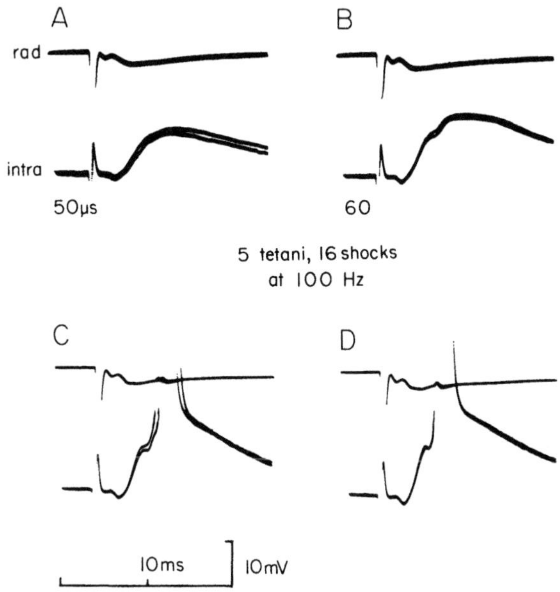

Fig. 1A–D. Long-term potentiation of transmission across apical dendritic synapses of CA1 pyramidal cell. *Upper and lower traces* are extra- and intracellular records respectively. The first is taken from the level of the activated dendritic synapses. *A* and *B* are responses to stimulation of radiatum fibres (100 μA) for durations of 50 and 60 μs respectively. *C* and *D* show similar responses, taken between 8 and 10 min after five periods of tetani, each consisting of 16 shocks at 100 Hz.

stimulation current and varying the pulse duration. The intracellular records (intra) show an EPSP which did not give rise to an action potential at either strength (A,B). The extracellular record from the layer of the activated synapse (rad) shows an initial small deflection indicating the activity of the presynaptic afferent fibres followed by the extracellular negative (downward) deflection corresponding to the EPSP. After a series of five tetani, each comprising 16 shocks at 100 Hz, had been given to the afferent fibres, test pulses were delivered at various times after the tetanization. The examples shown in C and D were taken between 8 and 10 min after the end of the last tetanus, and at stimulation strengths corresponding to those in A and B. When the records before and after the tetani are compared, the enhanced synaptic transmission for this cell appears evident. There is an EPSP increase and the occurrence of spike discharges on all records, indicating a greatly enhanced probability of discharge. There was also a clear latency decrease for the same shocks. The presynaptic volley, however, did not change; it must be remembered that its later part is riding on the enhanced extracellular EPSP.

By doing similar experiments with field recordings, the effect is substantiated for an assemblage of cells. In Fig. 2, the records in B show responses to a constant stimulus taken before (Control) and 1 and 5 min after a series of five tetani of 16 shocks at 100 Hz. The 1-min record shows a clear potentiation with an increased population spike (upper trace), including one single unit, but with a remarkably

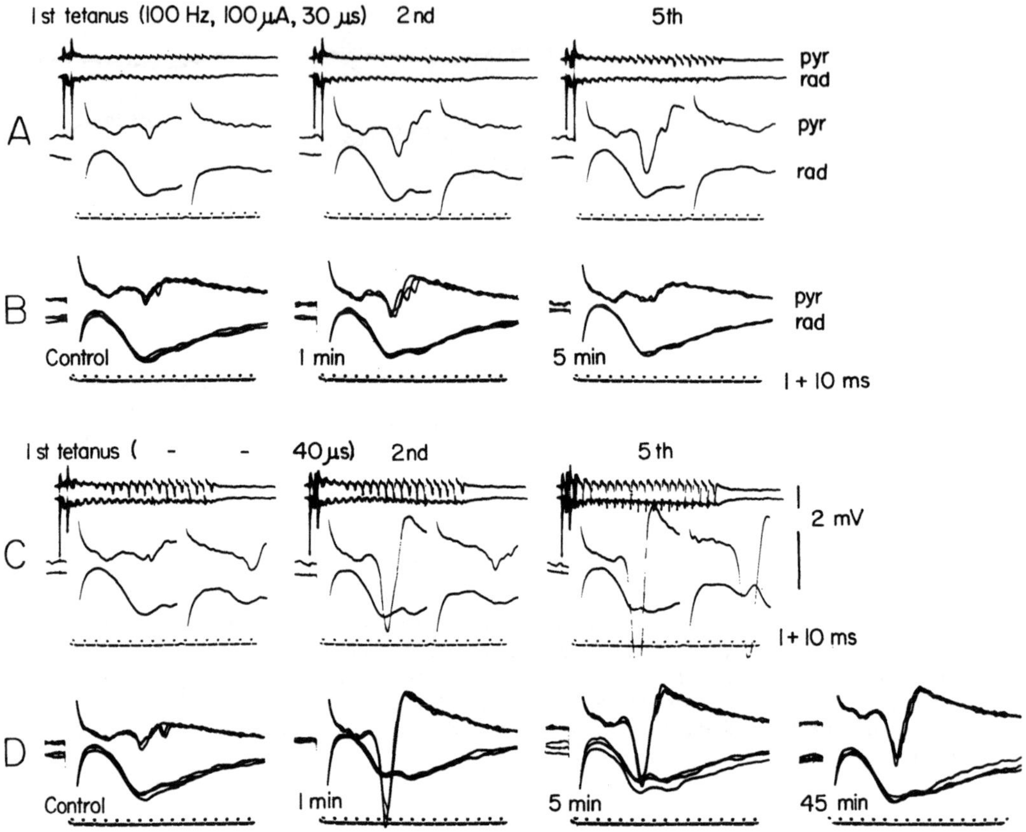

Fig. 2. A shows extracellular CA1 responses during the first, second and fifth of five tetani, delivered 5 s apart. Each period was composed by a constant test stimulus followed by 15 shocks at 100 Hz (100 μA, 30 μs duration). The *two upper traces* are from the pyramidal (*pyr*) and radiatum (*rad*) layers. Their initial parts are shown on expanded sweep in the *third and fourth traces*. *B* shows the responses from the same locations to a constant test stimulus, before (*left*) and 1 and 5 min after the series of five tetani. *C* and *D* show corresponding responses but with the tetanic pulse duration increased from 30 to 40 μs. The tetani show more recruitment of population spikes and the synaptic transmission is greatly increased and lasts for more than 45 min.

small increase in the extracellular EPSP (lower trace). After 5 min, there is little left of the enhancement of the population spike, apart from a slightly increased width. This followed the tetanization, which is indicated in row A. Here, the weak tetanic stimuli (100 μA, 30 μs duration) were able to set up small field potentials which summed to a hardly discernible plateau recorded from the dendritic electrode (second trace) and small potentials from the pyramidal cell electrode (first trace). The second tetanus appeared largely similar, but the fifth tetanus (right column) showed a recruitment of population spikes in its latter half. The expanded sweeps in the third and fourth traces show how there was an enhanced response to the first (test) stimlus indicating not only that the facilitatory effect lasted for 5 s after the previous tetanus but also that this enhancement was cumulative for the five tetani.

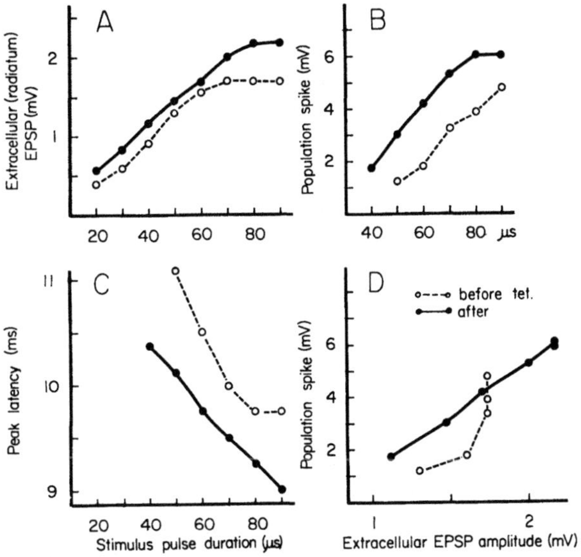

Fig. 3. A. Size of extracellular dendritic EPSP plotted against stimulus strength (pulse duration) before and after (*open* and *closed symbols* respectively) five tetani (each 16 shocks at 100 Hz, 650 μA, 50 μs). *B* and *C*. Similar data for the amplitude and peak latency of the population spike respectively. *D*. Graph of the population spike amplitude plotted against the radiatum EPSP amplitude. The left-shift of the initial part indicates a greater effect on the population spike than can be explained solely by an EPSP increase. Note also in the control curve that the population spike may increase, even if the EPSP does not (*three upper open circles*), and that both the EPSP and the population spike may increase after the tetani.

Following a series of five slightly stronger tetani (100 μA, 40 μs duration), indicated in row C of Fig. 2, there was greater enhancement during the tetanization with a slow negative DC change in the dendritic layer during the latter half of the tetanus. In addition, there was a clear growth of the population spikes in the latter half of the tetanus, building up from one tetanic period to the next. The lower traces in C indicate how the test response was enhanced because of the previous tetanus and the cumulative nature of this facilitation, which particularly affects the population spike. The consequences of this tetanization are seen in D. The left-hand trace is the control record, which is similar to the response in B. One minute after the tetanus there was a large increase in the population spike, which also had a decreased latency, but there was only a small increase in the rate of rise of the extracellular EPSP. The inflection near its peak is due to the reversed population spike. After 5 min there was still a large population spike enhancement and it lasted for at least 45 min. Throughout, there was a clear discrepancy between the changes in amplitude and latency of the population spike on the one hand and the moderate increase in the extracellular EPSP on the other.

The discrepancy between the EPSP and the population spike changes is indicated in Fig. 3. Here, the open symbols indicate the situation before a series of five tetani, whereas the full symbols indicate the post-tetanic situation. Panel A shows the increase in the extracellular EPSPs recorded from the layer of the activated synapses in response to increasing stimulus strength. There is a small indication of EPSP increase. Panel B indicates the change in population spike during the same procedure. Here, the amplitude increase is more pronounced. Correspondingly, panel C shows the large and consistent change in the peak spike latency. When the increase of the population spikes is plotted as a function of the corresponding amplitude of the extracellular EPSP, the curve shifts to the left. This indicates that a considerably larger fraction of the cells are brought to discharge at a given size of the extracellular EPSP. Since the intracellular EPSP develops in parallel to the dendritic field EPSP, this phenomenon implies a postsynaptic excitability change which is added to the possible presynaptic changes shown by the EPSP increase.

In order to study the conditions that induce the LTP, we have also recorded the events taking place at the level of the activated synapses during the tetanization and related these events to the degree of synaptic transmission enhancement. When a series of shocks were delivered to the radiatum fibres at 20 Hz, the second radiatum EPSP was potentiated, and the potentiated level was maintained for the remainder of the tetanus. In the pyramidal records there was a marked enhancement of the population spike. At higher frequencies, the picture was more complicated, however. Characteristically, at 50 and 100 Hz the radiatum EPSP showed a facilitation of the second response but then a decline of the summated trace. The intracellularly recorded EPSP behaved similarly. Provided the stimulus was sufficiently strong and long-lasting, there was a subsequent growth of the summated potential, in spite of the fact that the individual EPSPs were depressed. The procedure, therefore, elicits a slow process which appears as a negative DC shift at the activated synapses in addition to the summated EPSPs. The same picture was seen with intracellular recording.

In the pyramidal layer, the 50-Hz record showed a considerably smaller potentiation of the second population spike and a transient depression of the following four spikes, compared to the 20-Hz responses. In the latter half of the 50-Hz tetanus the population spikes grew, in spite of the persisting small amplitude of the radiatum EPSPs. This tendency was even more marked for the 100-Hz tetanus. The late population spike growth was parallelled by a slowly developing negativity in the dendritic record, while the individual EPSPs were very small indeed. This late developing slow dendritic DC shift is associated with depolarization of the cell membrane as shown by intracellular recording.

The degree of potentiation during a tetanus also increased when the tetani were repeated at 5 s intervals (Fig. 4). In the first tetanus, an action potential was produced by EPSP number 2 only (Fig. 4A). Five seconds later, action potentials appeared on EPSP numbers 2 and 3 (Fig. 4B). During the fifth tetanus, spikes occurred on all EPSPs and a few spikes even occurred after the tetanus (Fig. 4C). This enhanced excitability was associated with a slowly growing depolarization, which was seen particularly well during the latter two thirds of the tetanic period (note the reference lines at a fixed DC level). The growth was not due to increased EPSPs per se.

The build-up of excitability is also seen by the responses to the first shock in each tetanic group, which was delivered by the test cathode. They are shown in Fig. 4D. Here, the response elicited 5 s after the first tetanus shows an increased

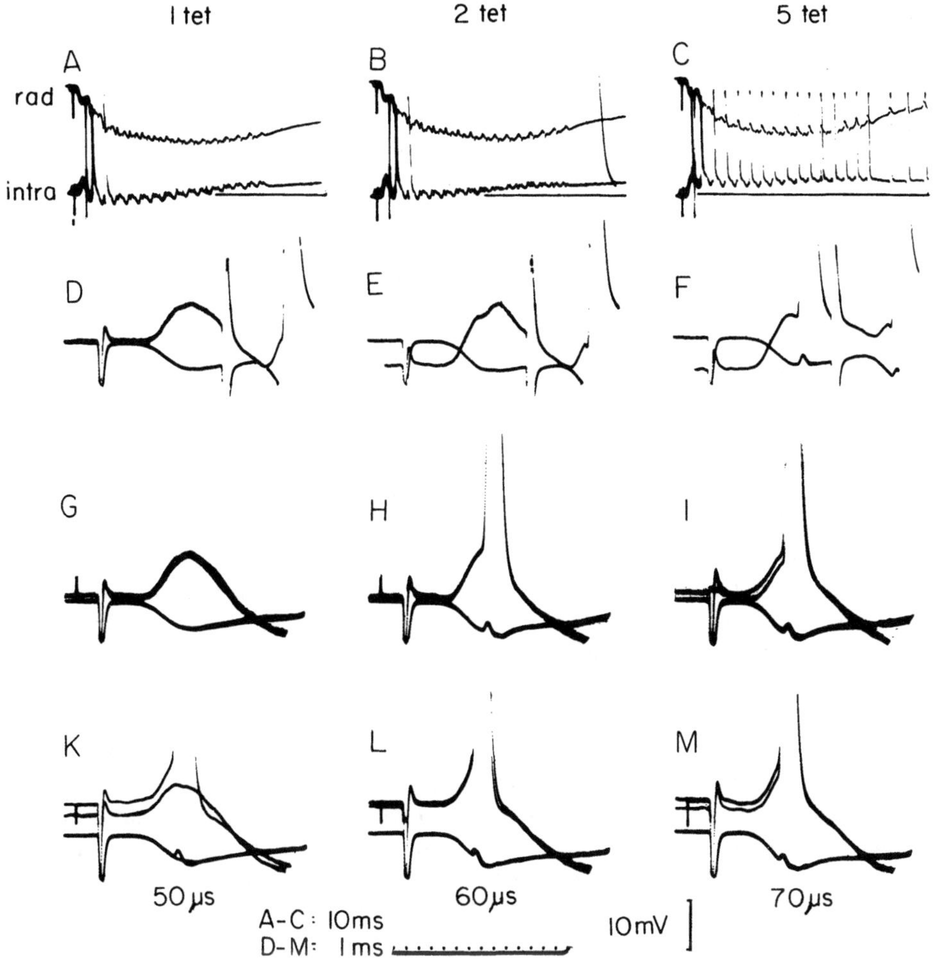

Fig. 4. A–C Extracellular dendritic (*rad*) and intracellular (*intra*) responses to the first, second and fifth tetani in a series of five (16 shocks at 100 Hz) delivered 5 s apart. The initial responses are shown on expanded sweep in *D–F*. *G–I* Responses from the same sites to three test stimuli (140 μA; 50, 60 and 70 μs duration) delivered to radiatum fibres before the first tetanus seen above. *K–M* show the responses to the same stimulus strengths 8 min after the last tetanus.

rate of rise of the EPSP (Fig. 4E). The effect was progressive, as shown by the even steeper EPSP and the occurrence of an action potential in the response taken before the fifth tetanus, i.e., after four previous tetani (Fig. 4F). The effect of the synaptic transmission is demonstrated by excerpts from input/output curves taken before (G–I) and after (K–M) the series of tetani. The responses in K–M illustrate the greater probability of discharge or a reduced latency when compared to the appropriate records in G–I.

We believe that the accretion of a slow depolarization during the later part of subsequent tetani is causally related to the subsequent increased transmission, seen as LTP. Our main reason is that the duration of LTP was observed to be longer when such slow build-up was seen during the tetani, and it was absent when plain plateau-like potentials were found.

We also postulate that the difference in the pattern of potentials seen during the tetanization with weak and stronger stimuli has a similar basis (see Fig. 2). During weak stimulation, the subsequent field EPSPs summate to a plateau which remains constant, not only during the first tetanus but also during the four subsequent tetani. In contrast, the potentials produced by the stronger tetanic stimuli gave a slowly increasing negativity in the dendritic recording and population spikes of increasing magnitude in the last half of the tetanus. The slow build-up of dendritic negativity associated with intracellularly recorded depolarization, depended not only upon the stimulus strength but also upon the frequency and duration of the tetanus.

In conclusion, during the latter part of a tetanus there may be a slow depolarizing tendency, seen as a change in the average membrane potential in impaled cells and as a negative DC shift in the record from the activated synapses. The more developed this late depolarization, the more clear and long-lasting was the following LTP. Furthermore, a useful index for the later acquisition of LTP seems to be that the tetanus is able to create an enhancement of a test response within the 5 s interval between subsequent tetani. We further propose that both the locus of control and the locus of enhancement for the postsynaptic changes are in the dendritic membrane. This conclusion is based upon the observation that the slow depolarization during the tail of the priming tetani was reflected as a slow negative shift of the local potential in the dendritic region where the activated synapses are located. Furthermore, the responses showing increased excitability of the cells often showed clear indications of dendritic involvement in the form of an increased occurrence of dendritic spikes (incrementing or fractionated spikes) and of spikes which occurred after the peak of the EPSP. Strikingly, under such conditions a spike could occur with very long latency to stimulation currents which were completely ineffective before the tetanization.

Discussion

The main aim of this presentation has been to point out the clear occurrence of postsynaptic excitability changes during the production of LTP of hippocampal synapses. We take the large amplitude increase of the population spike as the point of departure. This increase was not accompanied by a correspondingly large increase of the extracellular or intracellular EPSP. The most important evidence for this conclusion is the shift of the EPSP/population spike curve. Similar relations can be produced between intracellular EPSP and spike occurrence or spike latency. In some intracellularly impaled cells, the absence of EPSP change in spite of an increased excitability clearly indicates that the change not only resides postsynaptically but may take place in a part of the cell which lies remotely from the sodium spike trigger point, which in these cells is likely to be the initial segment of the axon.

Possible candidates for a postsynaptic change in the dendritic territory would be slow changes in the ionic conductances in this region. Of particular interest seem to be slow calcium or sodium currents. By blocking sodium currents with tetrodotoxin, Schwartzkroin and Slawsky (1977) were able to record slow potentials; they also found that these potentials were blocked by manganese or cobalt. Slow calcium potentials and spikes have been recorded intradendritically (Wong and Prince 1978; Wong et al. 1979). The postulated conductance changes are either moderately strong or located remotely in the dendritic tree, since no conductance changes were detected by hyperpolarizing test pulses across the soma membrane.

We regard the slow depolarizing event in the dendrites as important for a subsequent LTP development. This probably requires a certain input condition, most often appearing as a minimal stimulus strength. However, the cooperativity required for the LTP development (McNaughton et al. 1978; Goddard 1982) is not necessarily the result of a minimum number of activated afferent fibres, but may be set by the degree of depolarization of the recipient dendrites. This could be done by a sufficiently large number of activated fibres, but also through modulating neurons or other depolarizing processes coupled to the dendritic membrane.

Although postsynapic excitability changes seem to be present, one should not forget the clear indications of the presynaptic change (Skrede and Malthe-Sørenssen 1981; Dolphin et al. 1982). As indicated by Andersen et al. (1980), there is usually a clear EPSP increase, but without any change in membrane resistance or membrane potential. This suggests a presynaptic increased transmitter release, although it is also possible that an enhanced number of receptors or unmasking of already present receptors could account for the same increase in the EPSP (Baudry and Lynch 1980). In the latter case, it will be necessary to explain the rapid onset of the LTP process. If the enhanced reaction, which follows as quick as 5 s after a tetanus, is substantiated as essential for LTP production, this possibility appears less likely. A changed spine configuration as proposed by Bliss and Lømo (1973) and reported by Fifkova and van Harreveld (1977) will by itself probably be of little importance, since models have given little change in the soma EPSP even after drastic changes of the chosen spine parameters (Turner and Andersen, unpublished observations).

Acknowledgement. This research was supported by the Norwegian Research Council for Science and the Humanities (NAVF) and by the Medical Research Council of Canada.

References

Andersen P, Sundberg SH, Sveen O, Wigström H (1977) Specific long-lasting potentiation of synaptic transmission in hippocampal slices. Nature 266: 736–737

Andersen P, Sundberg SH, Swann JN, Wigström H (1980) Possible mechanisms for long-lasting potentiation of synaptic transmission in hippocampal slices from guinea pigs. J Physiol 302: 463–482

Baudry M, Lynch G (1980) Hypothesis regarding the cellular mechanisms responsible for long-term synaptic potentiation in the hippocampus. Exp Neurol 68: 202–204

Bliss TVP, Gardner-Medwin AR (1973) Long-lasting potentiation of synaptic transmission in the dentate area of the unanaesthetized rabbit following stimulation of the perforant path. J Physiol (Lond) 232: 357–374

Bliss TVP, Lømo T (1973) Long-lasting potentiation of synaptic transmission in the dentate area of the anaesthetized rabbit following stimulation of the perforant path. J Physiol (Lond) 232: 331–356
Bliss TVP, Goddard GV, Riives M (1983) Reduction of long-term potentiation in the dentate gyrus of the rat following selective depletion of monoamines. J Physiol 334: 475–491
Deadwyler SA, Dudek FE, Cotman CW, Lynch G (1975) Intracellular responses of rat dentate granule cells *in vitro*: post-tetanic potentiation of perforant path stimulation. Brain Res 88: 80–85
Deadwyler SA, Gribkoff V, Cotman CW, Lynch G (1976) Long-lasting changes in the spontaneous activity of hippocampal neurons following stimulation of the entorhinal cortex. Brain Res Bull 1: 1–7
Dolphin AC, Errington ML, Bliss TVP (1982) Long-term potentiation of the perforant path *in vivo* associates with increased glutamate release. Nature 297: 496–498
Douglas RM, Goddard GV, Riives M (1982) Inhibitory modulation of long-term potentiation: evidence for a postsynaptic locus of control. Brain Res 240: 259–272
Dunwiddie TV, Lynch G (1979) The relationship between extracellular calcium concentration and the induction of hippocampal long-term potentiation. Brain Res 169: 103–110
Dunwiddie T, Madison D, Lynch G (1978) Synaptic transmission is required for initiation of long-term potentiation. Brain Res 150: 413–417
Fifkova E, van Harreveld A (1977) Long-lasting morphological changes in dendritic spines of dentate granular cells following stimulation of the entorhinal area. J Neurocytol 6: 211–230
Goddard GV (1982) Heterosynaptic control of LTP. In: Swanson LW, Teyler TJ, Thompson RF (eds) Hippocampal long-term potentiation: mechanisms and implications for memory. Neurosci Res Prog Bull 20: 676–680
Lynch GS, Gribkoff VK, Deadwyler SA (1976) Long-term potentiation by a reduction in dendritic responsiveness to glutamic acid. Nature 263: 151–153
McNaughton BL, Douglas RM, Goddard GV (1978) Synaptic enhancement in fascia dentata: cooperativity among coactive afferents. Brain Res 157: 277–293
Ranck JB (1975) Behavioral correlates and firing repertoires of neurons in the dorsal hippocampal formation and septum of unrestrained rats. In: Isacsson RL, Pribram KH (eds) The hippocampus. Plenum, New York, pp 207–244
Schwartzkroin PA, Andersen P (1975) Glutamic acid sensitivity of dendrites in hippocampal slices *in vitro*. In: Kreutzberg GW (ed) Advances in Neurology. Raven, New York, pp 45–51
Schwartzkroin PA, Slawsky M (1977) Probable calcium spikes in hippocampal neurons. Brain Res 135: 157–161
Schwartzkroin PA, Wester K (1975) Long-lasting facilitation of a synaptic potential following tetanization in the *in vitro* hippocampal slice. Brain Res 89: 107–119
Skrede KK, Malthe-Sørenssen D (1981) Increased resting and evoked release of transmitter following repetitive electrical tetanization in hippocampus: a biochemical correlate to long-lasting synaptic potentiation. Brain Res 208: 436–441
Swanson LW, Teyler TJ, Thompson RF (1982) Hippocampal long-term potentiation: mechanisms and implications for memory. Neurosci Res Prog Bull 20: 611–769
Wigström H, Swann JW, Andersen P (1979) Calcium dependency of synaptic long-lasting potentiation in the hippocampal slice. Acta Physiol Scand 105: 126–128
Wong RKS, Prince DA (1978) Participation of calcium spikes during intrinsic burst firing in hippocampal neurons. Brain Res 159: 385–390
Wong RKS, Prince DA, Basbaum AI (1979) Intradendritic recordings from hippocampal neurons. Proc Natl Acad Sci USA 76: 986–990

Long-Term Potentiation and Inhibition in CA3 Neurons

ULRICH MISGELD and MANFRED R. KLEE, Frankfurt-Main

Summary. The changes which inhibitory postsynaptic potentials (IPSPs) undergo during long-term potentiation (LTP) were investigated in vitro in CA3 neurons of hippocampal slices obtained from guinea pigs.

By means of parallel intra- and extracellular recording it was found that IPSP amplitudes have decreased in some neurons but increased in others during the time the population spike of the field potential is increased after a brief burst of tetanic stimulation to the mossy fibers. The decrease in IPSP amplitude is accompanied by a reduction of the inhibitory shunt.

The results are discussed in respect to the following conclusions:

1. Inhibitory synapses and/or neurons participate in LTP after mossy fiber stimulation.
2. Activation of only a fraction of CA3 neurons is sufficient to cause an increase in the amplitude of the population spike during LTP.
3. The majority of CA3 neurons is under the influence of enhanced inhibition during LTP.

Introduction

Long-term potentiation (LTP) is defined as an increase in efficacy of synaptic transfer lasting minutes to days after a fiber input to a group of cells has been activated by a brief burst of high-frequency stimulation (see reviews in Bliss 1979; Bliss and Dolphin 1982; Eccles 1979a, b). In contrast to posttetanic potentiation, LTP develops and decays slowly after tetanic stimulation. LTP has been described and defined mainly on the basis of observations made at pathways of the mammalian hippocampus. The locus where the plastic changes take place is generally believed to be the synapse. There is an ongoing discussion as to whether the change occurs at the pre- or postsynaptic site (for references see above and Andersen et al. 1980; Lynch et al. 1979).

Not related to this problem, but in our opinion also of significance for the understanding of LTP, is the question of the fate of inhibition during LTP. Surprisingly, this question has not often been explored, although hippocampal pyramidal cells respond with inhibitory postsynaptic potentials (IPSPs) or excitatory postsynaptic potential (EPSP)/IPSP sequences to stimulation of their inputs (Eccles et al. 1977; Kandel et al. 1961). There are probably two pathways through which pyramidal cells can be inhibited: (a) the recurrent pathway, and (b) the pathway conveying the so-called feed-forward inhibition. The term "recurrent in-

Experimental Brain Research, Suppl. 9
© Springer-Verlag Berlin · Heidelberg 1984

hibition" denotes the circuit formed by collaterals of the pyramidal cell axons terminating on basket cells, which in turn send their axons to pyramidal cells (Andersen et al. 1964; Kandel et al. 1961). The pathway of feed-forward inhibition has been postulated on the basis of electrophysiological (Alger and Nicoll 1982) and immunocytochemical evidence (Ribak et al. 1978, 1981). The assumption is that afferents to pyramidal cells form direct contacts with interneurons also, which in turn inhibit the pyramidal cells. The existence of such a circuit has been decisively demonstrated for the commisural input by the combination of electron microscopic degeneration techniques with Golgi staining (Frotscher and Zimmer 1983). Although we have to await similar investigations on the other afferents, such a circuit should be taken into account when interpreting responses of pyramidal neurons to stimulation of their fiber inputs.

Many studies on LTP have taken advantage of the slice preparation which offers high stability for long-duration recording. CA1 neurons were most thoroughly investigated in this respect. However, IPSPs were found to be not very prominent in CA1 pyramidal cells in vitro (Andersen 1981), except in cells in slices which were maintained at low temperature (Alger and Nicoll 1982) or cut obliquely to the septotemporal axis of the hippocampus (Dingledine and Langmoen 1980). We therefore opted to investigate LTP in CA3 neurons which in vitro as in situ exhibit prominent IPSPs to stimulation of their mossy fiber input (Yamamoto 1972; Misgeld et al. 1979).

Methods

Details of slice preparation and maintenance are described elsewhere (Misgeld et al. 1979). The 200–400 μm thick slices were obtained from guinea pigs. The composition of the Krebs-Ringer solution in which the slices were bathed was (in mM): NaCl 124, KCl 5, $MgSO_4$ 1.3, KH_2PO_4 1.25, $NaHCO_3$ 26, $CaCl_2$ 2, glucose 10. Temperature was kept at 36°C during preincubation and recording period. Background stimulation (5–20 V, 0.1 ms) was delivered at a frequency of 0.5 Hz. For tetanization mossy fibers were stimulated for 5–20 ms with frequencies ranging from 5 to 300 Hz. Stimulus intensities were adjusted for tetanization to evoke IPSPs or EPSP/IPSP sequences subthreshold for spike generation before tetanization. The bipolar mossy fiber stimulation electrode was placed in the dentate hilus; in some experiments an additional bipolar electrode was placed in the stratum radiatum of field CA1. Double stimulations through these two electrodes demonstrated that separated synaptic inputs were activated (Misgeld et al. 1979). For intracellular recording 2 M potassium methylsulfate and for field potential recording 4 M NaCl electrodes were used.

Results

Of 73 CA3 neurons recorded intracellularly, 28 were tested for changes induced by tetanic stimulation. The mean resting membrane potential (RMP) of the neurons was -60 mV (SD $\pm$9.4). Electrical stimulation of mossy fibers in the den-

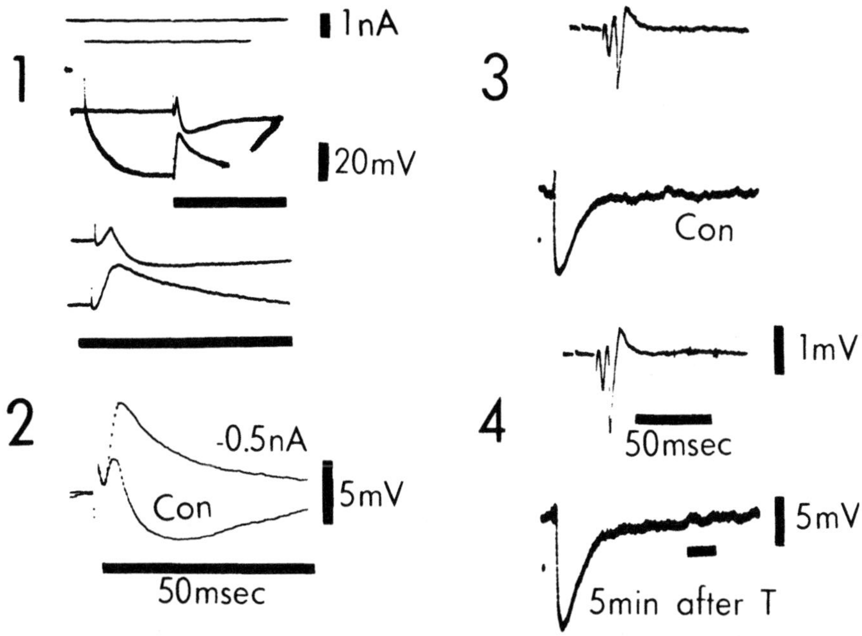

Fig. 1. 1,2 Reversal of IPSPs and estimation of IPSP latency. *1* In the middle traces an EPSP/IPSP sequence evoked by mossy fiber stimulation is shown with and without concomitant injection of an inward current (current monitor: *upper traces*). The *bottom traces* show the pair of responses at higher sweep speed. *2* Eight responses evoked before and eight responses evoked during injection of -0.5 nA were averaged and superimposed. The *point of divergence of two curves* denotes the onset of the IPSP. *3,4* Potentiation of the synaptically triggered population spike (*upper trace*) and of an intracellularly recorded IPSP (*lower trace*). The intracellular and the extracellular electrode were positioned side by side in the CA3 pyramidal layer. *3* shows the responses before (*Con*) and *4* the responses 5 min after a tetanic stimulation (*T*) (50 Hz, 10 s) to the mossy fibers. Calibration in *4* applies also to *3*.

tate hilus elicited EPSPs, IPSPs, and orthodromic and, sometimes, antidromic action potentials (APs). "Pure" IPSPs or IPSPs several times larger in amplitude than the preceeding EPSPs were evoked by low stimulus intensities. The field electrode side by side with the intracellular electrode did not reveal a population spike when stimulations of this low intensity were applied. Higher stimulus intensities were required to evoke EPSP/IPSP sequences of which the EPSPs could generate an AP. IPSPs could be reversed to depolarizing potentials when the cell membrane was hyperpolarized by the injection of inward current. The reversal potential for the IPSPs was 10–20 mV more negative than the RMP (Fig. 1.1). When the reversed IPSPs obtained by displacement of the membrane potential in a negative direction were averaged and superimposed on the hyperpolarizing IPSPs recorded and averaged at the RMP, the latency of the IPSPs could be estimated from the point of divergence between the superimposed traces. In this way the IPSPs were found to have their onset 1.5 ms after the beginning of the EPSPs, i. e., they commenced when the EPSPs were still in their rising phase and had not yet reached firing threshold (Fig. 1. 1,2).

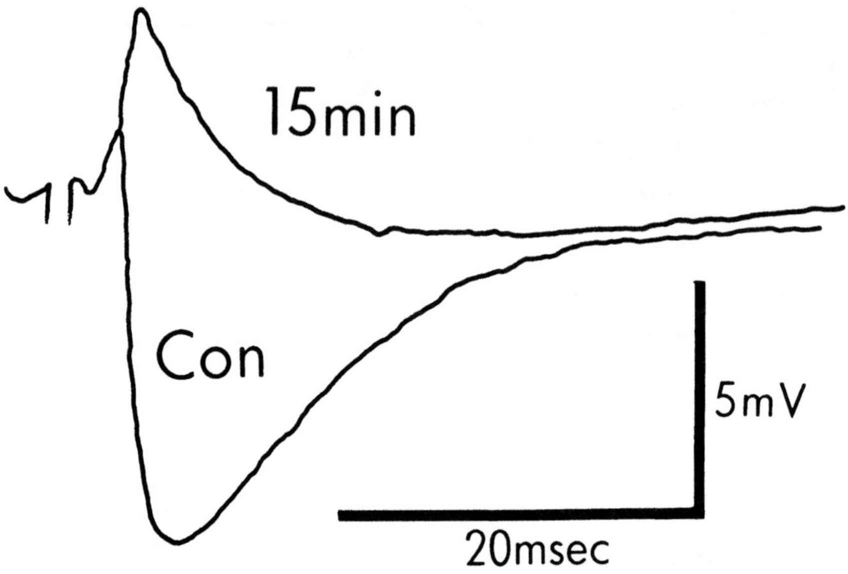

Fig. 2. Change of a predominantly inhibitory PSP to a predominantly excitatory PSP during LTP. Before tetanization (*Con*) the neuron responded with a small EPSP followed by a large IPSP. The same stimulation intensity evoked an increased EPSP 15 min after tetanus (20 Hz, 10 s). The inhibitory component was strongly reduced.

Single or repetitive bursts of high-frequency stimulation changed the postsynaptic potentials (PSPs) but not the RMP or the input resistance. The time course of development of the PSP changes is described elsewhere (Misgeld et al. 1979). In this paper only changes are described and discussed which persisted longer than 5 min after tetanization and hence can be considered to represent LTP. Because of the variability of the PSP changes (see below) it turned out to be necessary to pursue the occurrence of LTP by means of extracellular recording through a separate electrode. As in other extracellular studies, an increase in the amplitude of the synaptic population spike lasting up to several hours was taken as the paradigm to indicate the occurrence of LTP (upper traces of Fig. 1.3,4). Parallel to the increase in amplitude of the synaptic population spike the following observations obtained by intracellular recording were made:

1. In most cases (ca. 70%) an increase in IPSP amplitudes was monitored (Fig. 1.3,4).
2. In about 20% of the observations an increase in the amplitude of the depolarizing component and a decrease in the hyperpolarizing component of the PSP sequence was seen (Figs. 2 and 3).
3. In about 10% no change could be ascertained.

Only the change directed toward an enhancement of the EPSP and attenuation of the IPSP (Fig. 2) is compatible with the extracellular observation of an enhanced population spike in the field potential during LTP. The observation of enhanced IPSPs or the lack of a change indicate that the potentiation observed in the field potential is brought about by a subgroup of neurons only, whereas

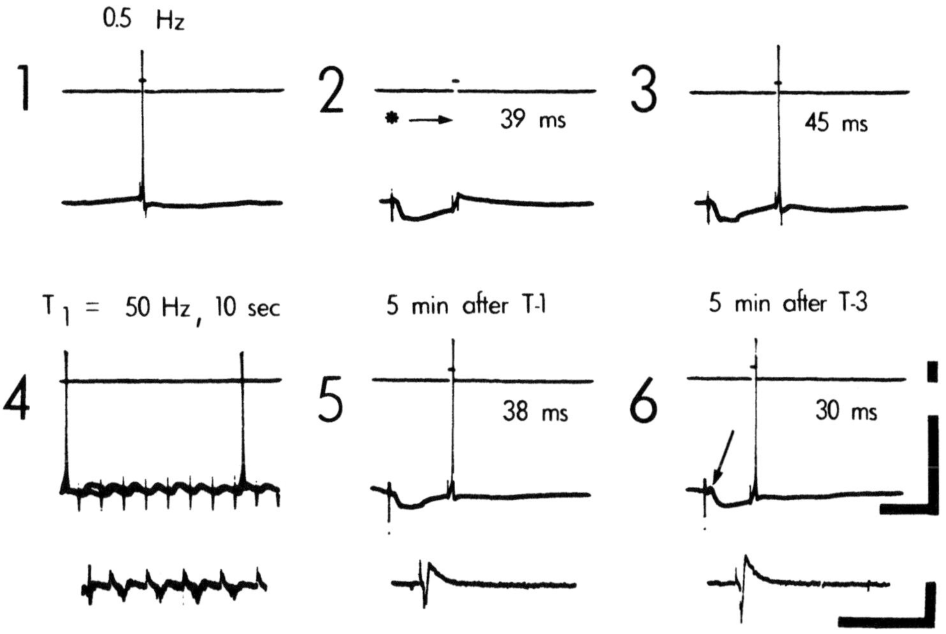

Fig. 3. Shortening of the inhibitory shunt during LTP. An intracellularly injected current pulse was set to trigger a spike (*1*). In control, the IPSP prevented this pulse generating a spike (*2*) if it was not at least 45 ms delayed after the mossy fiber stimulus (*3*). Parallel to the increase in the population spike (*bottom traces 4–6*), after three repetitive tetani (*T1–T3*) an EPSP (*arrow, 6*) appeared and the time shortened during which the IPSP prevented the current pulse from triggering a spike (*5,6*). *Top horizontal lines* represent the outside potential and the current monitor.

others do not participate in the generation of enhanced output. Repetition of tetanization augmented the probability that the EPSP increased and the IPSP decreased in the particular neuron subjected recording (Fig. 3). Neurons could even show enhanced IPSPs after the first tetanization and display enhanced EPSPs accompanied by reduced IPSPs after further repetition of tetanization. It could be assumed that the change of the slope of the IPSP is caused solely by an increase in the EPSP amplitude (Haas and Rose 1982). For a test we checked the inhibitory conductance increase, which is less likely to be influenced by EPSP changes. It was found that the period of conductance increase shortened (Fig. 3) when the IPSP was attenuated, indicating that it was a true diminution of the IPSP.

Discussion

Ample evidence has been accumulated suggesting that frequency stimulation of most hippocampal pathways induces LTP by the enhancement of the efficacy of excitatory synapses. A set of complex changes at the pre- and postsynaptic site has been established to secure the potentiation of synaptic transfer (for references see Introduction). However, we found that complicated changes also take place in the inhibitory circuits. On the first glance, this may seem to complicate the understanding of LTP, but in fact it offers the possibility of casting some light upon various obscurities.

Increase in the efficacy of postsynaptic inhibition during LTP can have two causes:

1. LTP is commonly measured as an increase in the synaptic population spike, in other words, as an increased output of pyramidal cells. As the consequence of an enhanced output of pyramidal cells an intensification of the inhibitory bombardment via the recurrent pathway is bound to occur. Increased IPSP amplitudes are therefore displayed by neurons which are not themselves potentiated but are neighbored by potentiated cells.
2. Excitatory synapses formed between axon collaterals of axons projecting to pyramidal cells and inhibitory interneurons or between these neurons and axons intermingled with the fibers to pyramidal neurons are activated by the high-frequency stimulation. Thus increased IPSP amplitudes may even reflect LTP of synapses on interneurons. In fact, a potentiation of these synapses is also easier to understand than a lack of potentiation of IPSPs. For the latter, one would have to postulate that terminals, even if they arise from the same cell, can be "LTP-negative" at one location and "LTP-positive" at the other, or, if a postsynaptic origin of LTP is assumed, that inhibitory interneurons do not demonstrate plasticity whereas pyramidal cells do. In extracellular studies, however, a potentiation of interneurons only was observed when low stimulus intensities were used, which activated and potentiated interneurons before pyramidal neurons (Buzsaki and Eidelberg 1982).

Increased inhibition, on the other hand, is a paradox with respect to enhanced neuronal output during LTP. The observation of enhanced EPSPs accompanied by attenuated IPSPs is therefore significant. It is the intracellular correlate of the extracellularly monitored enhanced output. For the observation of an increase in a population spike, it is sufficient that only a few neurons are more efficiently activated. Besides the increase in the EPSP amplitude, the decrease in the IPSP amplitude is also critical in our opinion. A parallel increase in excitation and inhibition will probably not result in an intensified output because the IPSPs curtail the EPSPs already during their rising phase, but a decrease in inhibition, even alone, will. The significance of the diminution of IPSP amplitudes is further underlined by the observation of heterosynaptic LTP in field CA3 (Misgeld et al. 1979; Yamamoto and Chujo 1978). Histochemical staining of glutamic acid decarboxylase (GAD) suggests that the somata of pyramidal cells are densely covered by gamma-aminobutyric acid (GABA)-ergic terminals (Ribak et al. 1978, 1981). If an inhibition of such a topical arrangement is reduced, the neuron will very likely respond with increased output to synaptic activation, not only through the activated pathway but also through other afferents. Since we suggested that IPSPs increase during LTP because of an enhanced output of pyramidal cells or because of potentiation of synapses on inhibitory interneurons, another mechanism must account for the decrease in IPSP amplitude. Recently, evidence has been provided (Wong and Watkins 1982) that responses of pyramidal neurons to local GABA application are changed from hyperpolarizing to depolarizing responses during and after frequent stimulation. A similar mechanism may affect the GABAergic IPSPs of some neurons during LTP.

According to the present description, LTP is characterized not only by enhanced excitatory synaptic transfer but also by a changed inhibitory regulation of neuronal activity. Inhibition in some cells is enhanced, in others reduced. This mixture may provide an explanation of why in extracellular studies increases in population spikes are not necessarily associated with an equal increase in the amplitude of the population excitatory postsynaptic potential, epsp, (Andersen et al. 1980; Haas and Rose 1982). It may be well that the population epsp is composed not only of excitatory but also of inhibitory synaptic current. In any case, increased excitatory transfer in some neurons surrounded by increased inhibition in others may very well bear functional meaning: it may ensure that LTP does not result in seizure activity, as it occurs after kindling. It has been argued (Andersen et al. 1980; Bliss 1979; Bliss and Dolphin 1982) that the specifity of LTP is guaranteed by confinement to the synapse activated. Activated neurons surrounded by enhanced inhibition, however, could also provide a high degree of specifity, yet on a different level of integration. For the theory of learning (Eccles 1979a; Marr 1969, 1970) neuronal specifity allowing heterosynaptic potentiation may be considered even more significant than synaptic specifity.

Acknowledgment. The authors thank C. Ziegler for technical assistance, M. Duesmann, H. Thomas, and L. Kannenbrock for preparing the figures, and S. Thel for typing of the manuscript.

References

Alger BE, Nicoll RA (1982) Feed-forward dendritic inhibition in rat hippocampal pyramidal cells studied in vitro. J Physiol 328: 105–123

Andersen P (1981) Brain slices – a neurobiological tool of increasing usefulness. Trends in Neuroscience 4: 53–56

Andersen P, Eccles JC, Løyning Y (1964) Location of postsynaptic inhibitory synapses on hippocampal pyramids. J Neurophysiol 27: 592–607

Andersen P, Sundberg SH, Sveen O, Swann JW, Wigström H (1980) Possible mechanisms for long-lasting potentiation of synaptic transmission in hippocampal slices from guinea pigs. J Physiol 302: 463–482

Bliss TVP (1979) Synaptic plasticity in the hippocampus. Trends in Neuroscience 2: 42–46

Bliss TVP, Dolphin AC (1982) What is the mechanism of long-term potentiation in the hippocampus? Trends in Neuroscience 5: 289–290

Buzsaki G, Eidelberg E (1982) Direct afferent excitation and long-term potentiation of hippocampal interneurons. J Neurophysiol 48: 597–607

Dingledine R, Langmoen IA (1980) Conductance changes and inhibitory actions of hippocampal recurrent IPSPs. Brain Res 185: 277–287

Eccles JC (1979a) Synaptic plasticity. Naturwissenschaften 66: 147–153

Eccles JC (1979b) Hippocampal plasticity. Prog Brain Res 51: 133–138

Eccles JC, Nicoll RA, Oshima T, Rubia FJ (1977) The anionic permeability of the inhibitory postsynaptic membrane of hippocampal pyramidal cells. Proc R Soc Lond [Biol] 198: 345–361

Frotscher M, Zimmer J (1983) Commissural fibers terminate on non-pyramidal neurons in the guinea pig hippocampus – a combined Golgi-EM degeneration study. Brain Res 265: 289–293

Haas HL, Rose G (1982) Long-term potentiation of excitatory synaptic transmission in the rat hippocampus: the role of inhibitory processes. J Physiol 329: 541–552
Kandel ER, Spencer WA, Brindley FJ Jr (1961) Electrophysiology of hippocampal neurons: I. Sequential invasion and synaptic organization. J Neurophysiol 24: 225–242
Lynch G, Browning M, Bennett WF (1979) Biochemical studies of long-term synaptic plasticity. Fed Proc 38: 2117–2123
Marr D (1969) A theory of the cerebellar cortex. J Physiol 202: 437–470
Marr D (1970) A theory for cerebral neocortex. Proc R Soc Lond [Biol] 176: 161–234
Misgeld U, Sarvey JM, Klee MR (1979) Heterosynaptic postactivation potentiation in hippocampal CA3 neurons: long-term changes of the postsynaptic potentials. Exp Brain Res 37: 217–229
Ribak CE, Vaughn JE, Saito K (1978) Immunocytochemical localization of glutamic acid decarboxylase in neuronal somata following colchicine inhibition of axonal transport. Brain Res 140: 315–332
Ribak CE, Vaughn JE, Barber RP (1981) Immunocytochemical localization of GABAergic neurones at the electron microscopical level. Histochem J 13: 555–582
Wong RKS, Watkins DJ (1982) Cellular factors influencing GABA response in hippocampal pyramidal cells. J Neurophysiol 48: 938–951
Yamamoto C (1972) Activation of hippocampal neurons by mossy fiber stimulation in thin brain sections in vitro. Exp Brain Res 14: 423–435
Yamamoto C, Chujo T (1978) Long-term potentiation in thin hippocampal sections studied by intracellular and extracellular recordings. Exp Neurol 58: 242–250

Synaptic Plasticity in the Red Nucleus

N. TSUKAHARA, Osaka

Introduction

The Red Nucleus (RN) has attracted the attention of many neurobiologists because of its role in motor control (Allen and Tsukahara 1974; Massion 1967). The large size of the RN along with the capacity for unique synaptic organization makes it a particularly favorable target for electrophysiological investigation of synaptic plasticity. Of particular importance is the segregation of the two major types of synaptic inputs: those from the nucleus interpositus (IP) of the cerebellum, terminating on the somatic portions of the cell membranes, and those from the cerebral cortex, forming synapses on the distal dendrites (Fig. 1A). We found that the latter are characterized by a variety of synaptic plasticity, such as sprouting, long-term potentiation (LTP) paired pulse potentiation, or posttetanic potentiation, whereas the former lack any of the types of plasticity listed above.

Sprouting of Corticorubral Input

The most remarkable long-term plasticity in the brain is the formation of new synaptic connections, a process called "sprouting." Sprouting is now recognized as a widespread phenomenon in the mammalian brain. In 1958, Liu and Chambers using the light microscope demonstrated that dorsal root fibers can sprout in the cat spinal cord. A decade later Raisman (1969) and Raisman and Field (1973) showed that new synapses are formed by sprouting of fibers in the rat brain. Now there is a growing list of investigations of sprouting (see Cotman and Lynch 1976; Tsukahara 1981 for reviews) and it is clear that the phenomenon is common in the mammalian brain. Our question is whether we can demonstrate sprouting by physiological techniques which are the only means of deciding whether the newly formed synapses are functional or not.

More specifically, our question is whether, following the destruction of one of the inputs to the RN, IP for example, the sprouting of physiologically active corticorubral inputs can be demonstrated. If sprouting takes place at the proximal portion of the somadendritic membrane, then we can expect that the time course of the corticorubral EPSPs might change, because the electrotonic distortion would be much smaller due to its proximal location. So we would have a fast-rising EPSP superimposed on the preexisting corticorubral EPSPs. It turned out that this is indeed the case.

After chronic lesions of the IP in adult cats, a new fast-rising EPSP appears superimposed on the slow-rising corticorubral EPSPs, as shown in Fig. 1B. The

Experimental Brain Research, Suppl. 9
© Springer-Verlag Berlin · Heidelberg 1984

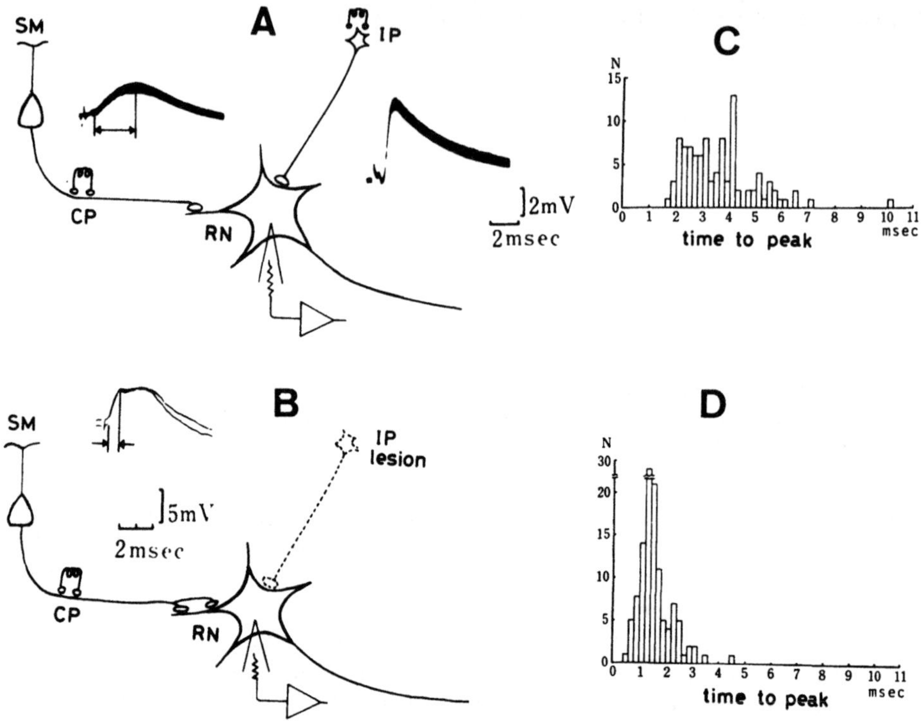

Fig. 1A–D. Lesion-induced sprouting in the adult feline red nucleus (*RN*). *A* Synaptic organization of the normal RN. Monosynaptic excitatory input from ispilateral sensorimotor cortex (*SM*) through the cerebral peduncle (*CP*) impinges on the distal dendrites and that from the contralateral nucleus interpositus (*IP*) of the cerebellum on the soma. Stimulation of CP produces a slow-rising EPSP in an RN cell and stimulation of IP produces a fast-rising EPSP, as shown in the *inset*. *B* After IP lesion a fast-rising component appears superimposed on the slow-rising CP-EPSPs, as shown in the *inset*. *C* Frequency distribution of the time to peak of the CP-EPSPs of normal cats measured as in the *inset of A*. *D* Same as *C* but CP-EPSPs of cats with chronic IP lesions measured as in the *inset of B*.

histogram of Fig. 1C illustrates the frequency distribution of CP-EPSPs induced by stimulating the corticorubral fibers at the cerebral pendunclе (CP) in normal cats. The inset diagram shows how we measured the time to peak. Figure 1D shows the frequency distribution of CP-EPSPs in cats with lesions of IP. It is noted that the time to peak is shorter in D than in C. These data were taken to indicate that new, functionally active synapses are formed at the proximal portion of the somadendritic membrane of the RN cells. It should be noted that a change of cable properties of RN neurons following IP lesions accounts for only a minor portion, less than 5%, of the observed change in time to peak of the corticorubral EPSPs (Tsukahara et al. 1975).

The next question would be whether there are any morphological correlates for the physiological findings? Electron microscopic studies (Nakamura et al. 1974; Hanaway and Smith 1978; and Murakami et al. 1982) have provided evidence in support of the physiological conclusion of sprouting. Previous electron

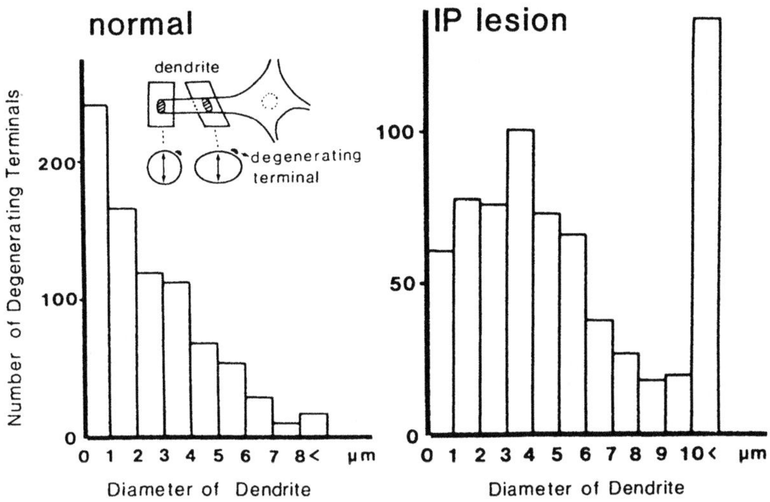

Fig. 2. Frequency distribution of the dendritic diameters where degenerating terminals of corticorubral synapses were found. *Left:* normal cat. *Right:* cats with chronic destruction of the nucleus interpositus. *Ordinates*, number of degenerating presynaptic terminals; *abscissae*, minor diameters of dendrites on which the corticorubral degenerating terminals were found. (Modified from Murakami et al. 1982)

microscopic studies of the RN were not able to survey the whole neuronal surface because of the extended nature of the dendritic tree of neurons from the RN region. However, intracellular horseradish peroxidase (HRP) injection with microelectrodes has made it possible to survey whole labeled neurons electron microscopically. This technique, combined with techniques involving degeneration of the postsynaptic terminals, has enabled us to locate the site of termination of the two major inputs of RN: inputs from the cerebral cortex and those from the IP of the cerebellum.

Quantitative measurement of the dendritic tree by a computer-assisted image-processing system developed in our laboratory (Maeda and Tsukahara 1980) revealed that the diameters of dendritic branches decrease monotonically with distance. Therefore, it is possible to estimate the position of synapses on RN cell dendritic branches in electron micrographs of HRP-stained RN cells. Furthermore, corticorubral synapses can be distinguished by the degeneration of their terminals after lesions of the cerebral sensorimotor cortex.

The degenerating terminal boutons after destruction of the cerebral sensorimotor cortex were found mainly on dendritic branches with small diameters, i.e., 1–4 μm. Figure 2 shows the frequency distribution of the diameters of the dendritic cross sections of HRP-stained RN neurons. We used the smallest diameter of the cross section when the shape of the cross section was not round. These histological findings therefore support the conclusion reached on the basis of electrophysiological studies. They show that cerebral inputs terminate on the distal dendrites of RN neurons, whereas inputs from the IP terminate on the soma.

Degenerated corticorubral terminals were also studied in cats in which the IP had been destroyed about 3 months previously. In these animals, in accordance with the physiological findings, the cerebral synapses were found to terminate not only on the distal dendrites but also on the proximal dendrites and on the somatic portion of the RN cells (Fig. 2).

Sprouting at the Early Developmental Stages

In the neonate a unique feature of lesion-induced sprouting is that the individual sprouts can extend over a considerable distance. For example, cerebral lesions which destroy the ipsilateral corticorubral inputs induced sprouting from the contralateral cerebral cortex (Tsukahara and Fujito 1976; Tsukahara et al. 1983). The abnormal crossed corticorubral fibers cross the midline and reach the RN from the contralateral cerebral cortex. Judging from the slow rise time of the EPSPs produced by these newly formed connections, the site of their synaptic contact is on the distal dendrites (Fig. 3). The relative insensitivity of the EPSPs to membrane hyperpolarization also supports this view.

The degree and extent of formation of the contralateral corticorubral projection depends on the age of the animal at the time of the cerebral lesions. In order to evaluate the age dependency of sprouting of the contralateral corticorubral fibers, recordings were made from kittens operated upon at different times after birth. It was found that the probability of appearance of detectable contralateral CP-EPSPs is highest in kittens operated on at 17 days of age and gradually decreases.

Sprouting Without Lesion

Sprouting and synaptic reorganization are not limited to cases in which direct synaptic inputs are removed. For example, we have studied evidence for synaptic reorganization in the RN after cross-innervation of forelimb flexor and extensor nerves (Tsukahara and Fujito 1976; Tsukahara et al. 1982; Fujito et al. 1982). Sperry (1942, 1947) investigated functional compensation in movement disorders after cross-innervation of flexor and extensor nerves of various animals and found that in monkeys motor relearning occurs, an observation which has never been made in similarly treated rats. Eccles et al. (1962) examined the possible change in the synaptic connections of spinal motoneurons after cross-innervation of hindlimb flexor and extensor nerves. Although they found evidence suggesting the formation of new synapses on motoneurons, sprouting was restricted to a group of motoneurons which were axotomized. We found that after cross-innervation, a fast-rising component appears in both compound and unitary corticorubral EPSPs. This is not due to concomitant changes in the cable properties of the RN neurons and therefore suggests that synapse formation is occurring on the proximal dendrites, just as was the case following lesions of the IP. However, since none of the synaptic inputs to the RN were destroyed in this case, we suggested that sprouting can occur in the absence of lesions to the synaptic inputs and seems to be a general reaction to environmental changes.

For cross-innervation of the forelimb nerves, the musculocutaneous, median, and ulnar nerves were united with the peripheral stumps of the radial nerve by

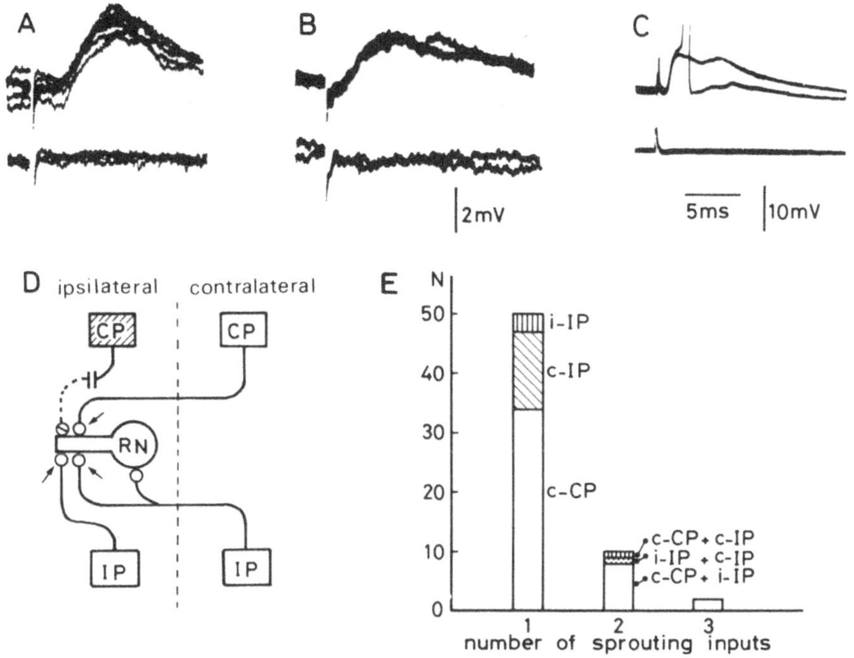

Fig. 3A–E. Newly appeared EPSPs in the kitten red nucleus (*RN*) after destruction of the ipsilateral cerebrorubral input. *A* Newly appeared EPSPs induced in a RN cell by stimulation of the contralateral cerebral peduncle (*c-CP*) (*upper trace*). Corresponding extracellular records are illustrated by the *lower traces in A–C*. *B* EPSPs induced by stimulating the contralateral IP (*c-IP*) of the cerebellum. *C* Dual-peaked EPSPs induced by stimulation of the c-IP. Voltage calibration in *B* also applies to *A*. Time calibration in *C* also applies to *A* and *B*. *D* Diagram illustrating formation of new synapses in the kitten RN after destruction of the ipsilateral cerebral cortex. *Arrows* indicate newly formed synapses. *Dashed line* shows the midline. *E* Frequency distribution of numbers of synaptogenesis sources in each RN cell. *Ordinate*, number of RN neurons. *Left column* indicates RN cells where only one source of synaptogenesis was found. *Open column*, RN cells where only c-CP-EPSPs were recorded; *obliquely lined column*, RN cells where only double-peaked c-IP-EPSPs were recorded; *vertically lined column*, RN cells where only i-IP-EPSPs appeared. *Middle column* shows RN cells in which two sources were simultaneously observed. *Open column*, RN cells where c-CP and i-IP-EPSPs appeared; *obliquely lined column*, RN cells where i-IP-EPSPs and c-IP-EPSPs were recorded; *vertically lined column*, RN cells where c-CP-EPSPs and c-CP-EPSPs and c-IP-EPSPs were recorded. *Right column* RN cells where all three new inputs were simultaneously recorded. (Modified from Tsukahara et al. 1983)

suturing the nerve sheaths. A similar procedure was applied to the central stumps of the radial nerve, which were united with the peripheral stumps of the musculocutaneous, median, and ulnar nerves.

These studies showed that 2–6 months after cross-innervation the rising time course of the corticorubral EPSPs induced by stimulating the CP became shorter than in normal cats. A typical slow EPSP evoked by stimulating the CP in a normal cat is illustrated in Fig. 4C. In contrast, the EPSPs shown in Fig. 4A are from a cat with cross-innervation of the forelimb nerves. Recordings are from cells innervating the upper spinal segments (C cells). These EPSPs had a much faster ris-

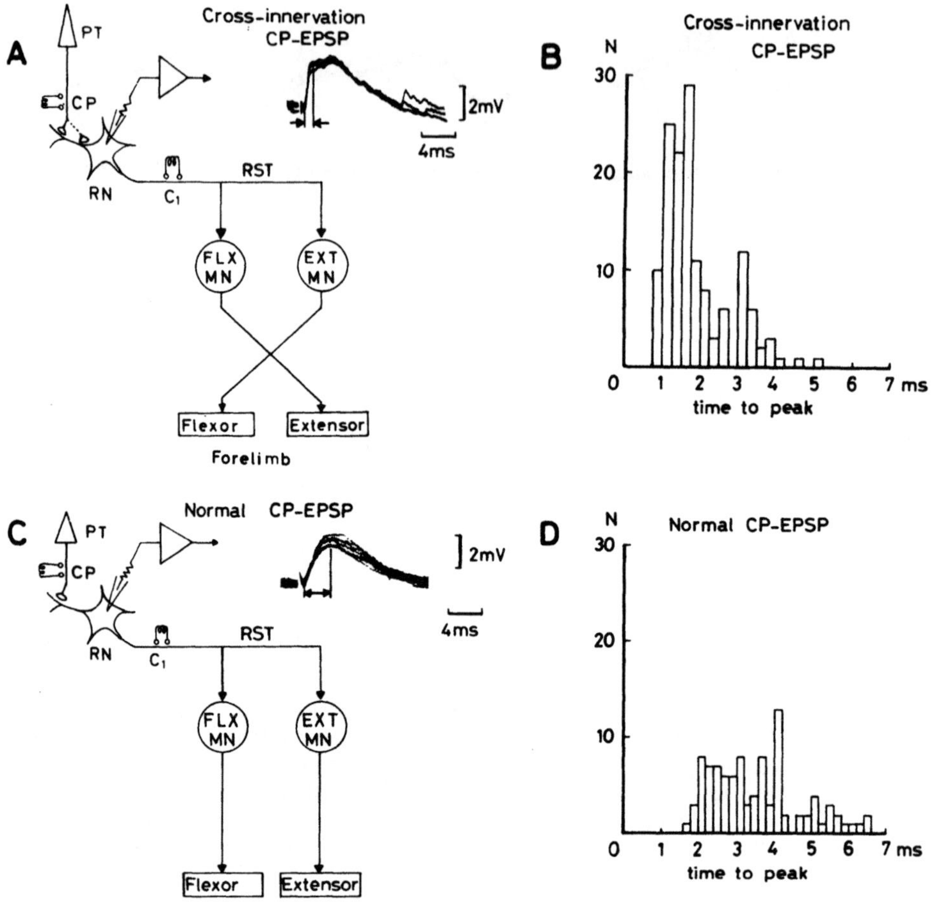

Fig. 4A–D. Corticorubral EPSPs in cross-innervated cats. *A Traces* are intracellular responses in a RN neuron, from cross-innervated cat 176 days before acute experiments. *C Traces* are intracellular responses in a RN cell in normal cat. *A,C* Diagram illustrates the experimental arrangements. *PT*, pyramidal tract cell; *RST*, rubrospinal tract; *FLX MN*, flexor motoneuron; *EXT MN*, extensor motoneuron; C_1, C1 spinal segment; *RN*, red nucleus; *CP*, cerebral peduncle. *B* The frequency distribution of time to peak of CP-EPSPs in cross-innervated cats. Number of cells is shown on the *ordinate* and the time to peak of CP-EPSPs in milliseconds on the *abscissa*. *D* The same as *B* but with CP-EPSPs in normal cats.

ing time course than those in normal cats. In these experiments, C cells were identified by the antidromic excitation produced by stimulating the C1 spinal segment. The frequency distribution of the time to peak of the CP-EPSPs of C-cells in cross-innervated cats is shown in Fig. 4B. The mean time to peak of CP-EPSPs of C cells in cats cross-innervated more than 2 months previously was significantly shorter (1.9 ± 0.9 ms, $n = 160$) than that in the normal cats (3.6 ± 1.4 ms, $n = 100$). The slower time to peak in C cells was also found when recordings were performed less than 2 months after cross-innervation. In contrast, the EPSPs of RN cells in normal cats had a much slower rise time than those of C cells in cross-innervated cats, as shown in Fig. 4D.

Plasticity of Synaptic Transmission – Long-Term Potentiation

The corticorubral input is also characterized by its short-term plasticity, such as paired pulse potentiation and posttetanic potentiation (Murakami et al. 1977). Recently, LTP has also been found in corticorubral synaptic transmission.

A slice preparation has been used advantageously for the study of plasticity of synaptic transmission in several loci of the brain. Experiments were therefore performed on a slice preparation of kitten midbrain containing the RN. The animals (180–1100 g at ages of 2–14 weeks postnatally) were anesthetized with pentobarbitone sodium (35 mg/kg), the brain was removed, and the RN was dissected out. Frontal sections of the midbrain at the rostral and the caudal ends of the RN were obtained using the boundary of the midbrain and pons as a landmark. The brain stem block thus obtained was then sectioned at the midline. Furthermore, a pair of midbrain blocks containing the RN was cut horizontally at a thickness of about 600 μm. The block was laid on the stage of a binocular microscope and a handmade chopper was used to make slices. Several slices thus obtained were transferred to a glass dish perfused with Krebs-Ringer solution (see below) with 95% O_2 and 5% CO_2. After incubation of about 1 h, the slice was transferred to the plastic recording chamber. Shielded bipolar tungsten electrodes insulated except for 300 μm at the tip were used for stimulation. The stimulating electrodes were placed on the rostrolateral part of the RN in order to stimulate the corticorubral fibers.

Figure 5A–C illustrates examples of EPSPs. Frequently, EPSPs were followed by another peak of EPSPs, as shown in Fig. 5C. Figure 5A,B illustrates two kinds of EPSPs, with fast and slow time courses. Figure 5D illustrates the frequency dis-

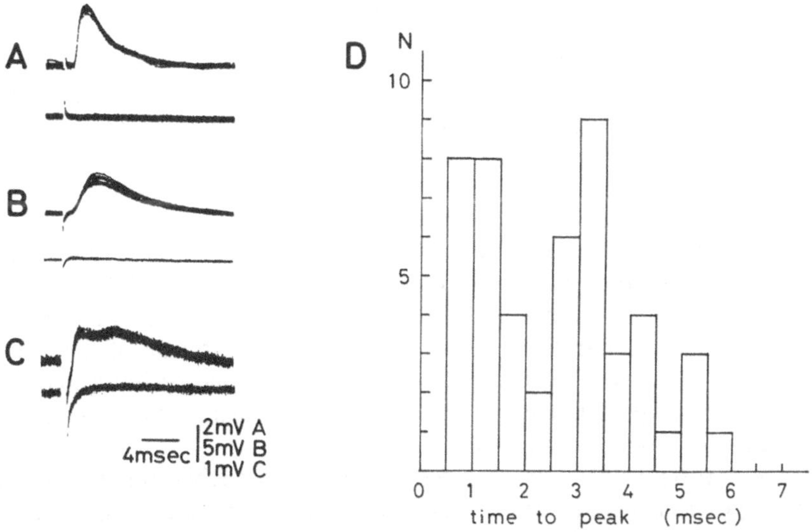

Fig. 5A–D. EPSPs induced in RN cells in in vitro slice preparation. *A* Fast-rising EPSPs. *B* Slow-rising EPSPs. *C* Fast-rising EPSPs superimposed on the slow-rising ones. *D* Frequency distribution of the time to peak of the EPSPs. *Ordinate*, number of EPSPs; *abscissa*, time to peak of the EPSPs.

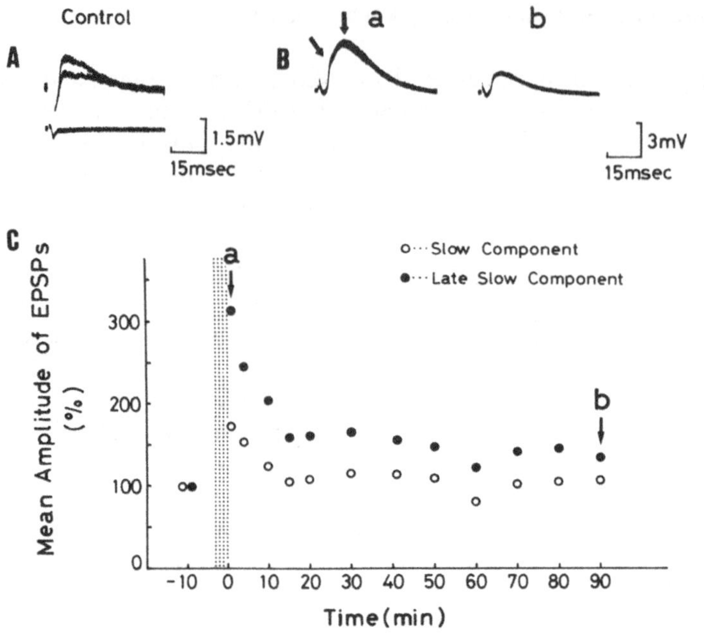

Fig. 6A–C. Long-term potentiation (LTP) of the EPSPs. *A,B* Examples of the EPSPs recorded *A* before and *B* after tetanic stimulation of 50 Hz for 3 min. *C Ordinate*, amplitudes measured at 5 ms and 14 ms after the onset of the stimulation as the percentage of those of the EPSPs recorded 10 min before tetanus.

tribution of the time to peak of these two kinds of EPSPs. It is noted that there are two peaks in the frequency distribution of the time to peak, with mean time to peak of 1.2 ms and 3.3 ms respectively. Since in the in vivo RN preparation of kitten, cerebral EPSPs have an average time to peak of 3.4 ms and interpositorubral EPSPs have a time to peak of 0.75 ms, it is suggested that the main component of the fast-rising EPSPs is the interpositorubral EPSP and the slow-rising EPSP is a corticorubral EPSP. EPSPs disappeared after Ca was removed from the perfusing solution.

In contrast to the tetanic stimulation of 200 Hz for 30 s, which induces post-tetanic potentiation lasting for only several minutes in RN cells, tetanic stimulation of 50 Hz for 3 min induces potentiation lasting longer than an hour or so in the RN slice preparation. In view of the longer duration of the potentiation, it could be termed a LTP. Furthermore, tetanic stimulation induced a later slow potential with a peak latency of about 15–20 ms, which persisted several tens of minutes after tetanic stimulation.

Figure 6 illustrates an example of the LTP in a RN cell in which slow and late slow potentials were recorded simultaneously. The latter appeared after the tetanic stimulation. The amplitudes of the potentials were measured at two points which corresponded to the peak of the slow and the late slow potential, and these amplitudes were plotted as a percentage of the control, as shown in Fig. 6C. It is noted that the potentiation lasted for 30 min and was larger for the late slow po-

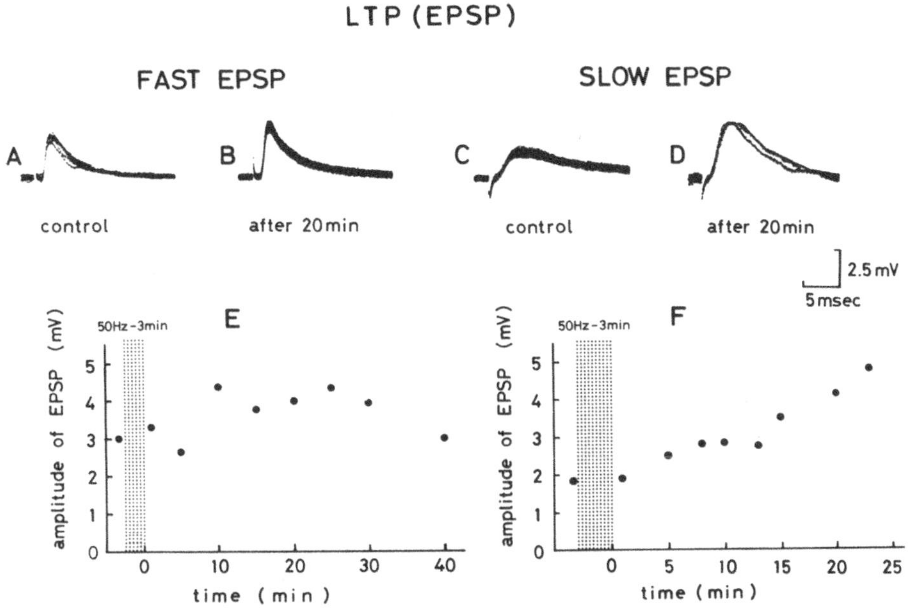

Fig. 7A–F. Differential behavior of fast- and slow-rising EPSPs after tetanic stimulation. ***A,B*** Fast-rising EPSP. ***C,D*** Slow-rising EPSP. ***A,C*** Before tetanus; ***B,D*** after tetanus. ***E,F*** ***Ordinates,*** amplitudes of the EPSPs; *abscissae,* time in minutes. *Dotted bars* represent the tetanic stimulation. *E* Fast-rising EPSPs; *F* slow-rising EPSPs.

tential than for the fast ones. Figure 7 shows other examples of LTP from two RN cells in which fast EPSP or slow EPSPs were recorded relatively in isolation. The time course of the potentiation is illustrated in the lower graph (Fig. 7E,F). The potentiation was greater for the slow potential than for the fast one. In some cases, the potentiation was preceded by a transient depression.

Several investigations have shown that gamma-aminobutyric acid (GABA) is the inhibitory transmitter of corticorubral inhibition. First, application of GABA induces membrane hyperpolarization with associated conductance increase in both in vivo and in vitro preparations (Altman et al. 1976; Sakaguchi, H., Kubota, M., Nakamura, M., and Tsukahara, N., personal communication). Secondly, GABA is released in the extracellular fluid of the RN slice preparation by several depolarizing procedures (Kubota et al. 1983). Thirdly, glutamic acid decarboxylase (GAD) is present in the presynaptic terminals. Electron microscopic staining using GAD antibody showed that GAD positive grains were found in nerve terminals around the somas of the RN neurons (Murakami et al. 1983). On the basis of these observations, it is likely that GABA is the inhibitory transmitter of RN cells. Since bicuculline blocks the IPSPs mediated by GABA, we used bicuculline to block the IPSP and tested whether LTP could be observed in the absence of IPSPs. The LTP of the slow-rising EPSPs still persisted in the preparation where the IPSPs were blocked by applying bicuculline. Therefore, it is concluded that potentiation is due to a genuine increase in excitatory synaptic transmission and not due to the depression of IPSPs.

Comments

Corticorubral synapses are characterized by various types of plasticity, sprouting as well as potentiation of synaptic transmission. Sprouting occurs not only after lesions of the synaptic inputs, but also under more natural conditions of the brain such as after cross-innervation. This phenomenon, sprouting without lesions, might provide a neuronal basis for adaptive behavior such as learning and memory. Recent investigations have shown that classical conditioning can be synthesized by using corticorubral synapses as modifiable elements: a conditioning stimulus was applied by electrical stimulation of the corticorubral fibers and electrical stimulation of the forelimb skin was used as an unconditioned stimulus (Tsukahara et al. 1981). It would be an interesting future study to decide what kinds of mechanisms are indeed responsible for the establishment of the conditioned response in this experimental paradigm.

References

Allen GI, Tsukahara N (1974) Cerebrocerebellar communication systems. Physiol Rev 54: 957–1006

Altman H, Bruggencate GT, Pickelman P, Steinberg R (1976) Effects of GABA, glycine, picrotoxin and bicuculline methochloride on rubrospinal neurones in cat. Brain Res 111: 337–345

Cotman CW, Lynch GS (1976) Reactive synaptogenesis in the adult nervous system. The effects of partial deafferentation on new synapse formation. In: Barondes S (ed) Neuronal recognition. Plenum, New York, pp 69–108

Eccles JC, Eccles RM, Shealy CN, Willis WD (1962) Experiments utilizing monosynaptic excitatory action on motoneurons for testing hypothesis relating to specificity of neuronal connection. J Neurophysiol 25: 559–579

Fujito Y, Tsukahara N, Oda Y, Yoshida M (1982) Formation of functional synapses in adult cat red nucleus from the cerebrum following cross-innervation of forelimb flexor and extensor nerves. II. Analysis of newly-appeared synaptic potentials. Exp Brain Res 45: 13–18

Hanaway J, Smith J (1978) Sprouting of corticorubral terminals in the cerebellar deafferented cat red nucleus. Soc Neurosci Abstr 4: 1507

Kubota M, Sakaguchi H, Tsukahara N (1983) Release of endogenous GABA from the cat red nucleus slices. Brain Res 270: 190–192

Liu CN, Chambers WW (1958) Intraspinal sprouting of dorsal root axons. Arch Neurol 79: 46–61

Maeda J, Tsukahara N (1980) Digital image processing system for measuring the dendritic diameters of HRP-stained neurons. In: Ito M, Tsukahara N, Kubota K, Yagi K (eds) Integrative control function of the brain. III. Kodansha/Elsevier, Tokyo, pp 368–370

Massion J (1967) The mammalian red nucleus. Physiol Rev 47: 383–436

Murakami F, Tsukahara N, Fujito Y (1977) Properties of the synaptic transmission of he newly-formed corticorubral synapses after lesion of the nucleus interpositus of the cerebellum. Exp Brain Res 30: 245–258

Murakami F, Katsumaru H, Saito K, Tsukahara N (1982) A quantitative study of synaptic reorganization in red nucleus neurons after lesion of the nucleus interpositus of the cat: an electron microscopic study involving intracellular injection of horseradish peroxidase. Brain Res 242: 41–53

Murakami F, Katsumaru H, Wu JY, Matsuda T, Tsukahara N (1983) Immunocytochemical demonstration of GABAergic synapses on identified rubrospinal neurons. Brain Res 267: 357–360

Nakamura Y, Mizuno N, Konishi A, Sato M (1974) Synaptic reorganization of the red nucleus after chronic deafferentation from cerebellorubral fibers. An electron microscopic study in the cat. Brain Res 82: 298–301

Raisman G (1969) Neuronal plasticity in the septal nuclei of the adult rat. Brain Res 14: 25–48
Raisman G, Field PM (1973) A quantitative investigation of the development of the collateral innervation after partial deafferentation of septal nuclei. Brain Res 50: 241–264
Sperry RW (1942) Transplantation of motor nerves and muscles in the forelimb of the rat. J Comp Neurol 76: 283–321
Sperry RW (1947) Effect of crossing nerves to antagonistic limb muscles in the monkey. Arch Neurol 58: 452–473
Tsukahara N (1981) Synaptic plasticity in the central nervous system. Annu Rev Neurosci 4: 351–379
Tsukahara N, Fujito Y (1976) Physiological evidence of formation of new synapses from cerebrum in the red nucleus neurons following cross-union of forelimb nerves. Brain Res 106: 184–188
Tsukahara N, Hultborn H, Murakami F, Fujito Y (1975) Electrophysiological study of formation of new synapses and collateral sprouting in red nucleus neurons after partial denervation. J Neurophysiol 38: 1355–1372
Tsukahara N, Oda Y, Notsu T (1981) Classical conditioning mediated by the red nucleus in the cat. J Neurosci 1: 72–79
Tsukahara N, Fujito Y, Oda Y, Maeda J (1982) Formation of functional synapses in the adult cat red nucleus from the cerebrum following cross-innervation of forelimb flexor and extensor nerves. Exp Brain Res 45: 1–12
Tsukahara N, Fujito Y, Kubota M (1983) Specificity of the newly-formed corticorubral synapses in the kitten red nucleus. Exp Brain Res 51: 45–56

VI. The Understanding of the Brain

John Eccles explaining the brain (Schloß Ringberg/Oberbayern 1983)

Neocortical Neuron Circuit Models Reconsidered in the Light of the New Techniques

J. SZENTÁGOTHAI, Budapest

Summary. Earlier neuron circuit models are reconsidered in the light of recent studies using the new "hard criteria" for light and electron microscopic cross-identification of neurons and synapses defined by physiological, anatomic, fine structural, and biochemical (immunocytochemical) methods and the respective new tracing methods (horseradish peroxidase, autoradiography of tritiated compounds). Although certain naivities of the earlier circuit models (especially those due to general overestimation of the "target specificity" (type and site of neuron contacted) and gross underestimation of the axonal span of some inhibitory interneurons) have now to be rectified, the earlier models are holding their place remarkably well even after close to 15 years.

A study of one "global" architecture of inhibitory intracortical connections, made possible by selective labeling of aspinous stellate cells (and neuropil) through uptake of ^{3}H-gamma-aminobutyric acid, shows a remarkable number of specific vertical (inhibitory and/or disinhibitory) connections between closely neighboring layers (sublayers), for example IVc, deep III, IVa; more remote layers, e.g., V and superficial III and II; and distant layers: VI and II–I. The tangential spread of inhibitory axons may reach 350–250 μm distances as defined by this retrograde uptake method. This procedure cannot, though, reveal connections formed by cells (e.g., large basket cells) with very widespread axonal arborizations. The span of intracortical inhibitory connections exceeds the width of the corticocortical columns very considerably by invading as many as five neighboring columns lined up in one direction.

These general principles of intracortical circuitry can most probably be applied without major adjustments to the somatomotor cortex.

Introduction

It is now exactly 30 years since Eccles attempted in his book *The Neurophysiological Basis of Mind* (Eccles 1953, Fig. 89) to visualize the dynamic path of excitation in the cortex on the basis of the scant data then available. This diagram was extremely simple, and Eccles tried to elaborate on it in some of his later publications, but even his most recent versions using arrays of open and closed modules had their roots in these very early ideas. The dynamic concept underlying this attempt to visualize neural functions was the trigger stimulus for me at least to take up again the earlier histological studies that had ended in frustration.

The cause of this frustration was the same as that which defeated Ramón y Cajal (1899, 1911, 1935), who in spite of the still unmatched richness of his Golgi

Experimental Brain Research, Suppl. 9
© Springer-Verlag Berlin · Heidelberg 1984

observations on the neocortex could never make up his mind about what ought to be considered a synapse in the cortex. In his last paper (Ramón y Cajal 1935) he spoke of the intertwinements of axonal and dendritic arborizations, apart from the basket axon terminals, which he illustrated as terminating on the cell bodies in the usual form of terminal knobs or enlargements of passage. When Lorente de Nó (1938) assembled his ingenious diagrams of the vertically ascending and descending neuronal chains of the neocortex, he had to make the bold step of introducing a single bouton as an abstraction of a cortical synapse. But this did not render it possible to overcome the difficulty of not knowing what the cortical synapse was in terms of classical neurohistology, and of interpreting correctly the signs of axonal degeneration that could already be clearly seen under the light microscope.

It was not until the dendritic spines were shown under the electron microscope by Gray (1959) as being one of the main synaptic targets – although direct synaptic boutons on smooth dendrites or interspine dendrite portions or somata were also identified – that synapses of the cortex were realized to be essentially the same as in most other parts of the nervous system.

The sudden reappearance of Golgi studies on cerebral cortex – which had all but fallen into oblivion after the great classical period of neurohistology, i.e., after the publication of Ramón y Cajal's summary in 1911 – during the 1960s and their increasing momentum during the 1970s was prompted by this new understanding of the applicability of all basic principles of synaptology to the cortical organization.

It is intended as a tribute to Eccle's far-sighted and imaginative contribution to the study of the cerebral cortex that the various earlier models of cortical neuron circuits are reconsidered in the light of the recent, almost revolutionary, explosive development of new histological methods. A large number of new tracing techniques have contributed to a radical change in our structural understanding: second-generation Golgi techniques (defining physiologically identified nerve cells by microinjection with horseradish peroxidase (HRP), cobalt procedures, etc.), immunocytology, and cross-identification between light and electron microscopy of the same neurons and synapses defined by various methods.

How Much Is Still Valid of the Early Circuit Models?

The basic assumption, already made in the eariest generalized diagrams of Ramón y Cajal (1911), has been shown to be correct in principle. It is the arciform through-conduction of the supposed close "reflex" loop bridging the gap between an "afferent" (or input) chain of neurons and an "efferent" (or output) chain of neurons. Only very exceptionally does this mean that specific afferents would be routed through a local chain into a corresponding motor path, as in the case studied in area 4 by Rosén and Asanuma (1972). The pyramidal tract fibers arising from somatosensory area S I have also to be considered. No major tangential connections inside the cortex can be assumed to exist for any immediate sensory motor transfer even in the primate brain, where there is a certain proximity and topical isomorphy between sensory and motor projection. However, since cytoarchitectonic area 3a, the closest neighbor of area 4, does not seem to belong to the real "core" of the somatosensory cortex (Dreyer et al. 1974), a tangential intracortical connectivity would be of little help in direct sensory motor transfer. What is characteristic instead, for all specific sensory cortical areas, is the preferential termination of the specific thalamic (geniculate) afferents on a specific type of excitatory interneuron, the spiny stellate cells, with additional direct termination on spineless interneurons that are probably inhibitory. There is also an unknown but probably small proportion of monosynaptic connections of specific sensory afferents to pyramidal cells. These features were all included in my first general cortical circuit models (rather vaguely in that of Szentágothai 1967 and more specifically in that of 1969). The spiny stellate cells as the putative main targets of specific sensory afferents have been established by Golgi studies on the visual cortex (Lund 1973), the somatosensory cortex (Jones 1975b), and the auditory cortex (Szentágothai 1975), although conclusive direct proof of this assumption was furnished only later by serial electron microscopic section studies on Golgi-stained material combined with experimental degeneration of the specific afferent pathway (Somogyi 1978). The next link in the main neuron chain was visualized over predominantly ascending axonal branches of the spiny stellates (Szentágothai 1967, 1969, 1973, 1975; Jones 1975b), a fact well documented by Golgi material but not so far ascertained by sufficient extensive studies on Golgi-stained spiny stellates (or type 7 nonpyramidal cells of Jones 1975b). Whether, and if so how far, the vertically ascending axons of the spiny stellates could explain the physiologically experienced "columnar" arrangement of cells with similar receptive field properties is still a matter of speculation.

Although direct information concerning the direct input-output (excitatory) neuron chain of the cortical basic circuit is still meager, it seems reasonably established that specific afferents are mainly di- but also trisynaptically connected with lamina III–VI pyramidal cells, which would mean that input-output connections might be established within a minimum distance in the tangential direction of the cortex, but might use almost the entire depth of the cortex. In this respect the neuron circuit model of Jones (1975b) and my own (Szentágothai 1967, 1969, 1973, 1975) did not show major differences, and may still appear valid today. However, since the pyramidal cells of laminae III and II (except for those with purely local axonal termination) of the primary sensory areas project mainly to other ipsilateral cortical areas (some at shorter distances, for example into secondary sensory areas, others at greater distances), and since many lamina II–VI pyramidal cells project to contralateral cortical areas (i.e., are callosal), the onward conduction from the sensory receiving areas is mainly toward other cortical regions. The majority of the sub-Gennarian lamina V and VI pyramidal cells (some of the latter are not pyramidal but fusiform cells) project to subcortical targets, in most cases to the sensory thalamic and geniculate relay nuclei. (This aspect, however, being beyond the immediate concern of the present summary, will not be pursued here.)

Consequently, the chief output of the cortex and of the specific sensory areas – and as far as is known, of all other neocortical areas too – is corticocortical. This shifted attention from the cortical circuit models toward the corticocortical connectivity system. The radical turn came with the fundamental paper of Goldman and Nauta (1977) showing a hitherto unexpected degree of discreteness and specificity of the corticocortical columns. These columns of 200–300 μm diameter were an anatomic reality in contrast to physiologically experienced receptive field columns, the structural basis of which remained elusive. Further studies showed that this basic architectural principle is probably general for most mammals and most neocortical regions. A general definition of the corticocortical columns and how they may emerge from their anatomic constituents was given in my Ferrier Lecture of 1977 (Szentágothai 1978), but the whole problem is better known from several of Eccles' papers in which many of my diagrams were used and suitably combined in order to fit into a wider conceptual framework (Eccles 1979, 1980, 1981). What seems to emerge also from more recent studies is that corticocortical columns appear to be invariant structural units of the neocortex over a wide phylogenetic range and over most of the neocortex in any mammalian species.

This leads us to accept reentrance or recursivity as the true overall connectivity principle of the neocortex (most clearly spelled out by Mountcastle in 1978). In this new view the whole cortex has to be envisaged as an immense system of reentrant loops – in fact, multiple and often reciprocal loops – arranged in an infinite richness of loops in succession and loop superimposed on loop ad infinitum. This connectivity principle has been suggested already by Ramón y Cajal as a general principle of "avalanche conduction," and an early model of it was proposed in an (as we now know) oversimplified form by Braitenberg (1978), which nevertheless seems to represent the truth in the end. All my circuit models dating after 1977 were built on this fundamental principle (Szentágothai 1978, 1981, 1983).

However, realizing the fundamental importance of the corticocortical connectivity system did not in itself change our views on local connectivity. The major turnabout came with the new techniques for the complete reconstruction of physiologically identified and HRP-injected neurons, both on the level of the light microscope and on the level of subsequent electron microscopic series made from the same cell. It would obviously be impossible to go into any detail in a short paper, so that only some of the main well-established facts can be listed.

Some specific inhibitory interneurons have now been identified according to the above criteria:

1. The axoaxonic cell (earlier "chandelier cell"), which is found everywhere where pyramidal cells are present (Somogyi 1977, 1979; Somogyi et al. 1982), even in the hippocampus (Somogyi et al. 1983b)
2. The original "*cellule à double bouquet*" (double bouquet cell) of Ramón y Cajal (Somogyi and Cowey 1981)
3. One type of large basket cell (Somogyi et al. 1983c)
4. Other types of smaller, more compact basket cells ("clutch cells"), which are under study (Somogyi et al., to be published).

All these identified putative inhibitory (or in the case of the double bouquet cells partially disinhibitory) interneurons and all so far studied smooth dendrite interneurons of the cortex seem to be gamma-aminobutyric acid (GABA)-ergic (Ribak 1978; Somogyi et al. 1981a, 1981b; Freund et al. 1983).

The specificity of the well-identified interneurons is much higher in certain respects and much lower in others than was assumed earlier. However, "specificity" has to be defined more strictly and analyzed, because neurons may be "specific" or "less specific" in various respects:

1. A(n) (inter)neuron may be absolutely "target specific" (a term coined by Somogyi) in the sense that its axon terminals may be addressed exclusively to a specific site on other specific neurons. This is the case in the cell type (the axoaxonal cell) cited above.
2. Other (inter)neurons may show a strict geometric specificity, i. e., their axon (dendrite) arborizations may be restricted to geometrically strictly defined (and circumscript) spaces. This "geometric specificity" of axonal arborizations was used as the leading architectural parameter in my earlier neuron circuit models. It turned out that specificity of axonal geometry had been greatly overestimated in the basket cell. Basket cells (or basket cell axons) that had been intracellularly labeled (Somogyi et al. 1983c) showed total spans of their axon arborization up to 1500 μm in the anteroposterior and up to 900 μm in the lateral direction.
3. A third type of specificity is "specificity of addressing," i. e., certain neurons may have axons – or more often axon branches – that have terminal arborizations of high density confined to such small specific sites (geometrically not always well defined) that this suggests specific addressing to one or a few neighboring elements. This type of specificity is probably more general in pyramidal cell collaterals arising in lower strata of the cortex, and occurs also in basket cells (Somogyi et al. 1983c) and in the axoaxonic cells (Freund et al. 1983).

Various kinds of specificities need not be linked. Target specificity may be very high in some cells and relatively high in others, which suggests the predominantly disinhibitory nature of the double bouquet cell. There may be relative preponderance for one specific type of target, for example in the basket cells, where only about one-third of all identified synapses are in contact with pyramidal cell bodies, in contrast to the earlier assumption. Conversely, geometric specificity does not necessarily also mean target specificity; for example, only half of the double bouquet cell axon terminals identifiably contact smooth nonspiny dendrites (i. e., putative inhibitory interneurons), so that their disinhibitory nature is at best a predominant but by no means a unique feature of these cells. Addressing specificity is not yet sufficiently well defined, due to a deficiency of direct information. We know it to exist in pyramidal cells, in double bouquet cells, and in basket cells, but it applies only to a tiny fraction of the total axonal arborization. This means that the local cortical neuron circuit might be redrawn by removing all not strictly identified connections, as tentatively shown in Fig. 1.

Some New Features

An entirely new aspect of cortical connectivity was discovered quite recently. It was known much earlier from classical Golgi studies that certain cells of the upper cortical layers project to the deeper layers and that conversely, cells of the deeper cortical layers project to the upper strata. The latter were even labeled "Martinotti cells" with reference to the author who mentioned them first (Martinotti

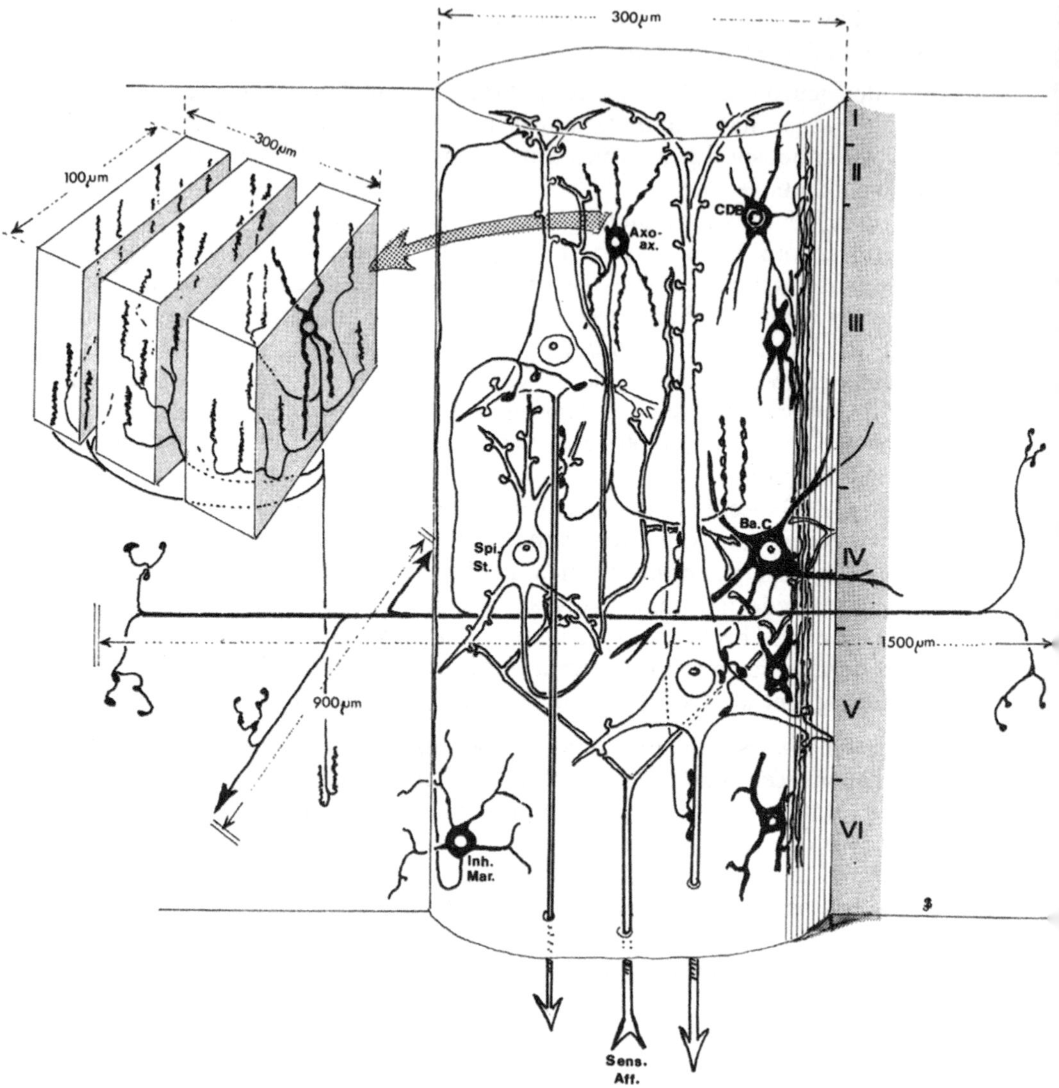

Fig. 1. A neocortical neuron circuit model stripped down to include exclusively neurons and connections that have been cross-correlated on the basis of the new "hard criteria" between physiological, anatomic, fine structural, and/or biochemical (immunocytochemical) identification of the same elements (either or both in neurons and synapses). The excitatory input line of sensory afferents (*Sens.Aff.*), the onward conduction over spiny stellate cells (*Spi.St.*), and the output lines (pyramidal cells) are indicated in *outline drawings*. Of the inhibitory interneurons so far identified and all defined as GABAergic (on the basis of containing glutamic acid decarboxylase), a large basket cell (*Ba.C*) and an axoaxonic ("chandelier") (*Axo-ax.*) cell are drawn *in full black* with indication of the dimensions of their axonal arborizations (in the case of the axoaxonic cell this had to be done in a separate *inset drawing in upper left*). An inhibitory cell projecting vertically to layers I and II has also been identified and is assumed to be identical with the hitherto elusive Martinotti cells (*Inh.Mar.*). The double bouquet cells (*CDB*), one of them indicated at *far right*, are assumed to be predominantly disinhibitory on the basis of their numerous synaptic contacts on smooth dendrite stellate cells and the fine structural criteria of the synapses.

1890). The very fact that the tangential axonal plexus all but disappeared from lamina I of chronically isolated cortical slabs, while a fairly rich plexus presisted whenever the isolated slab contained more than lamina IV (Szentágothai 1962, 1971), suggested the existence of cells in laminae V–VI with a strong projection to lamina I (and II).

By injection technique using 30–50 μm diameter micropipettes it was possible to study vertical (and horizontal) connectivity of local inhibitory interneurons (Somogyi et al. 1983a). The Golgi type II cells with aspinous dendrites, which gave synapses of symmetrical membrane structure and contained flattened vesicles, could be defined by specific uptake of ^{3}H-GABA and by retrograde labeling of their axons. Following small 0.1–0.05 μl injections of the labeled GABA 1 or 0.5 mm apart along surface-parallel micropipette penetrations, there was a consistent labeling of perikarya in layers IVc–VI vertically below the pipette track in layers I–IVc. The closer the injection track was to the pia, the deeper the peak density of labeled cells appeared. After injection in layers IVa and lower III the highest number of labeled cells was found in layer IVc. After injection into the upper strata of lamina III, most labeled neurons were in layer V, and after injection in laminae I–II the proportion of labeled neurons increased in laminae V and VI. Golgi staining of the material before it was processed for autoradiography made it possible to identify some of the labeled cells as aspinous stellate cells in layer V. It had been shown already that stellate cells mainly in lamina II and upper III can be labeled from small injection foci at the border between layer VI and the white matter (Somogyi et al. 1981b). It was conjectured that these calls were double bouquet cells retrogradely labeled over their vertical axon strands.

Although the studies of this kind of global intracortical connectivity are still in their beginnings and do not, so far, offer significant insight into connections in the tangential direction, some important conclusions on vertical connectivity can be made already. These connections are illustrated diagrammatically in Fig. 2 for the visual cortex in the monkey, where these experiments were performed. But it stands to reason when the basic similarities between the main sensory receiving areas (especially what concerns the several cell types are considered), that the situation may be essentially similar also in the somatosensory area. The diagram includes – with gross simplifications and generalizations – the general routes of excitation both entering through specific (thalamic and/or geniculate) sensory channels and nonsensory (yet specific) corticocortical routes (open arrows in Fig. 2). In the descending inhibitory connections two different types can be distinguished: one originating mainly in lamina III, causing a massive labeling of the neuropil sharply at the border of layer IVc and extending into layer V (see Fig. 6A in Somogyi et al. 1983a), and a vertically descending longitudinally oriented neuropil labeling, which may originate from layer II and upper III double bouquet cells (Somogyi et al. 1981b). Since this cell type appears to be predominantly disinhibitory (as judged from the large portion of double bouquet axon terminals in contact with aspinous neurons), the corresponding arrow in Fig. 2 is not drawn in full black but densely stippled. While in lamina IVc the inhibitory connections are mainly tangential, they become increasingly vertical as one moves up into lamina IVa and into the bottom of III, from where a massive labeling, expecially in lamina IVc, is induced with a tangential spread of 250 μm. Vertical connections

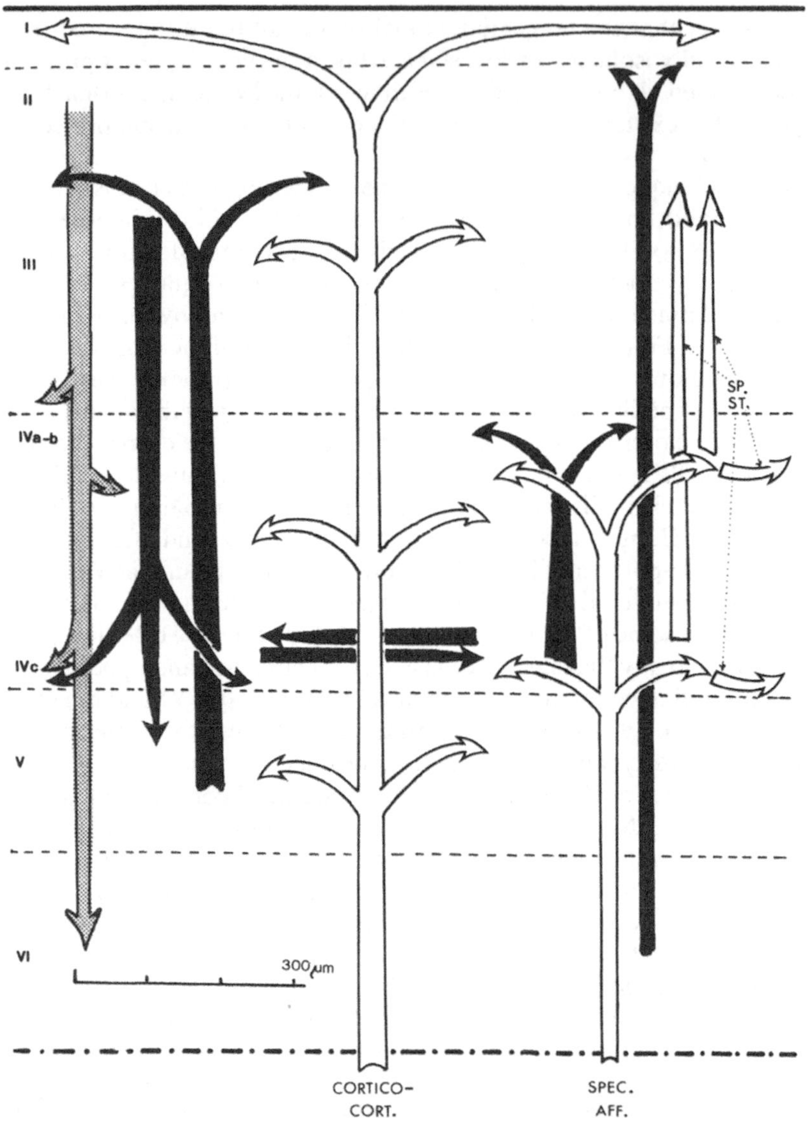

Fig. 2. Overall local connectivity (mainly in the vertical direction with indication of the tangential span) by inhibitory neuron systems, as identified on the basis of selective uptake of ^{3}H-GABA from microinjections given with surface parallel micropipette penetrations (*black arrows*). For comparison the main excitatory input lines are indicated by *open arrows* for the general corticocortical (*CORTICOCORT.*) and the sensory afferent (*SENS.AFF.*) and the forward conduction over spiny stellate (*SP.ST.*) cells. Inhibitory connections reaching far beyond the assumed corticocortical columns of 200 μm diameter (as indicated in Fig. 1) are not shown here.

arising from lamina V inhibitory interneurons are directed into the upper part of lamina III (probably also II) with a large tangential spread of 350 μm in any direction. The deepest inhibitory interneurons of the cortex (in lamina VI) project massively to laminae I and II, but the tangential spread of these connections cannot be judged for technical reasons (Somogyi et al. 1983a). These latter cells and their connections are apparently what used to be referred to as Martinotti cells. It is very likely that these inhibitory connections are (at least partly) those that may interfere with the transmission from the excitatory axonal network fed by the corticocortical fiber systems (see in more detail in my earlier models and particularly in Szentágothai 1975, 1978). As described in more detail there (Szentágothai 1975), the deeper part of lamina I abounds in multiple axon-spine synapses of which one or two are of symmetrical and one (or two) of asymmetrical character. The latter were interpreted as mainly the terminals of corticocortical excitatory fibers contacting apical dendritic spines of pyramidal neurons. The former were thought to originate mainly from local inhibitory interneurons (the axonal tuft cells of Ramón y Cajal 1899; see photographs in Szentágothai 1975), but now it looks as if we must assume many to originate from lamina VI inhibitory interneurons.

The overall picture that emerges from these considerations and Fig. 2 may still be rather vague for the time being, but it nevertheless gives some impression of how vertical connectivity of the cortical network has to be envisaged within the boundaries of a 300 μm wide corticocortical column. However, all the new data discussed in this chapter indicate that inhibitory connections reach far beyond the boundaries of the classical corticocortical (meso)column, not to mention minicolumns of much smaller diameter (see Mountcastle 1978). I have always emphasized that anything smaller in diameter (in the tangential direction) than 300–200 μm cannot have any direct anatomic substrate (apart, perhaps, from the double bouquet cells); smaller architectonic units, for example orientation columns, do exist, but they are the results of functional (dynamic) operations of the cortical network.

Applications to the Somatomotor System

What has so far been discussed was based in the "hard" new factual material mainly on the visual cortex of the monkey. However, we have sufficient reason (both direct application of the new strict identification criteria to the visual cortex of the rat and the cat and to the hippocampus of the monkey and conventional neurohistological observations, including degeneration and retrograde labeling, on other areas of he cortex) to assume that the basic structural principles apply equally to all other specific sensory areas. The detailed electron microscopic analysis of the somatosensory cortex of the cat under normal conditions and with degenerating afferent systems by Jones and Powell (1970), and the studies both of distribution patterns of thalamic afferents (Jones 1975a) and for cell types (Jones 1975b) in the squirrel monkey, offer sufficient basis for making direct comaprisons with the visual cortex.

This is not so simple with area 4, from which the absence of any compact granular layer is apparent in the cytoarchitectonic picture. And yet, as known from the studies of Ramón y Cajal (1899), one finds in the Golgi picture an abundance of various types of stellate interneurons. The ones studied most extensively are the large basket cells (Marin-Padilla 1969, 1974; Marin-Padilla and Stibitz 1974), and even if our confidence in the completeness of stellate cell arborizations, as seen

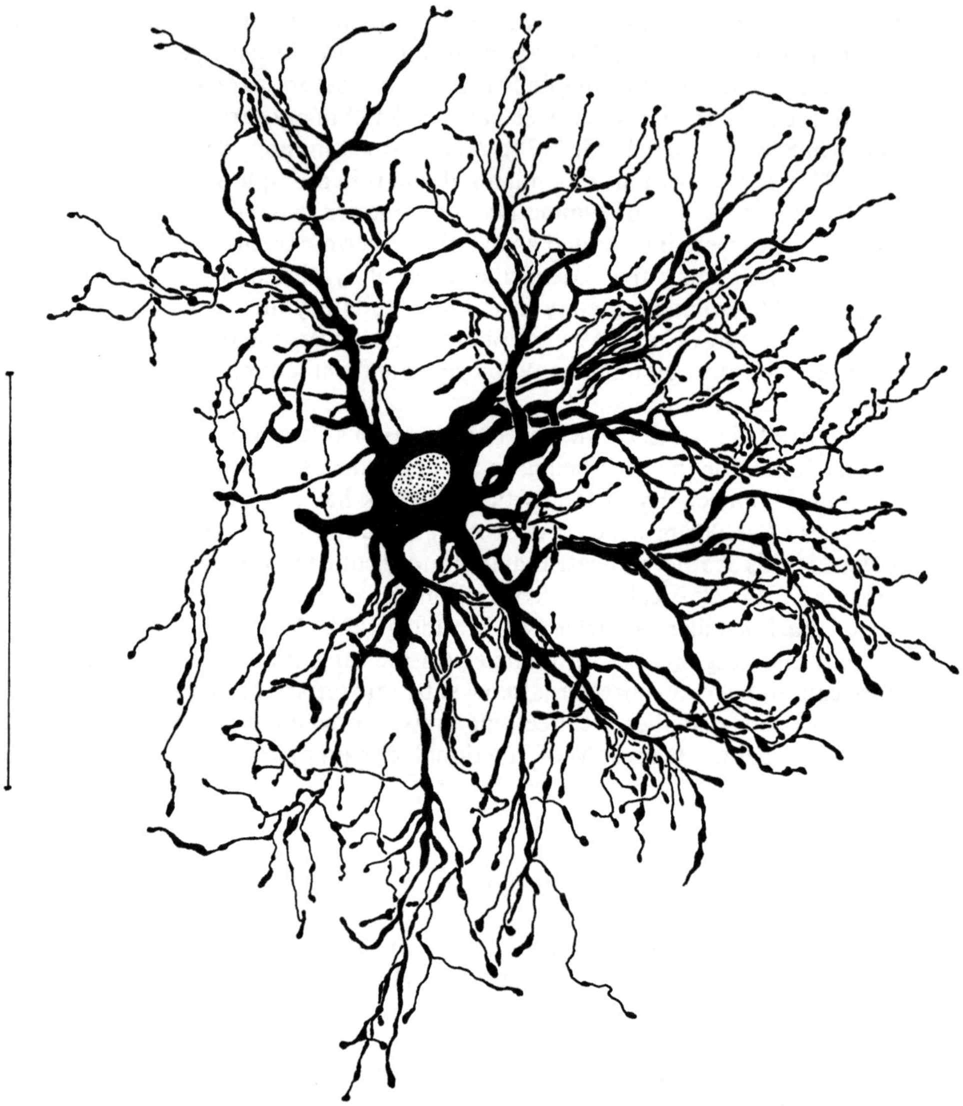

Fig. 3. An example of a microgliform cell from area 4 (border of layers V–VI) in *Macaca mulatta*. *Scale*: 50 μm. (The origin of the axon was not traceable, but many of the finer arborizations belong to the axon of this cell.)

in the Golgi picture, has been considerably reduced by the discovery of many myelinated stretches of the axons, their existence in the motor cortex is well established. From some of my own Golgi observations on the precentral gyrus of *Macaca mulatta* it appears that probably all important aspinous stellate cell types are present. Double bouquet cells certainly do occur, and even so-called "microgliform" cells (see Fig. 3). The presence of axoaxonal cells is practically certain, also simply on the basis of the fact that all pyramidal cell initial axons not only in the motor cortex but also in the hippocampus are supplied liberally with the characteristic rows of axoaxonic contacts, wherever an initial axon segment can be unequivocally identified.

In conclusion, therefore, there is no reason whatever to doubt that the general principles of intracortical neuron circuits and connectivity are also applicable to the somatomotor cortex.

References

Braitenberg V (1978) Cortical architectonics: general and areal. In: Brazier MAB, Petsche H (eds) Architectonics of the cerebral cortex. Raven, New York, pp 443–465

Dreyer DA, Schneider RJ, Metz CB, Whitsel BL (1974) Differential contributions of spinal pathways to body representation in postcentral gyrus of *Macaca mulatta*. J Neurophysiol 37: 119–145

Eccles JC (1953) The neurophysiological basis of mind. The principles of neurophysiology. Clarendon, Oxford, p 314

Eccles JC (1979) The human mystery. Springer, Berlin Heidelberg New York, p 225

Eccles JC (1980) The human psyche. Springer, Berlin Heidelberg New York

Eccles JC (1981) The modular operation of the cerebral neocortex considered as the material basis of mental events. Neuroscience 6: 1839–1856

Freund TF, Martin KAC, Smith AD, Somogyi P (1983) Glutamate decarboxylase-immunoreactive terminals of Golgi-impregnated axo-axonic cells and of presumed basket cells in synaptic contact with pyramidal neurons of the cat's visual cortex. J Comp Neurol 221: 263–278

Goldman PS, Nauta WJH (1977) Columnar distribution of corticocortical fibers in the frontal association, limbic and motor cortex of the developing rhesus monkey. Brain Res 122: 393–413

Gray EG (1959) Axo-somatic and axo-dendritic synapses of the cerebral cortex: an electron microscope study. J Anat [Lond] 93: 420–433

Jones EG (1975a) Lamination and differential distribution of thalamic afferents within the sensory-motor cortex of the squirrel monkey. J Comp Neurol 160: 167–204

Jones EG (1975b) Varieties and distribution of non-pyramidal cells in the somatic sensory cortex of the squirrel monkey. J Comp Neurol 160: 205–268

Jones EG, Powell TPS (1970) Electron microscopy of the somatic sensory cortex of the cat. Philos Trans R Soc Lond [Biol] 257: 1–62

Lorente de Nó (1938) The cerebral cortex: architecture, intracortical connections and motor projections. In: Fulton JF (ed) Physiology of the nervous system. Oxford University Press, London, pp 291–321

Lund JS (1973) Organization of neurons in the visual cortex area 17, of the monkey (*Macaca mulatta*). J Comp Neurol 147: 455–496

Martinotti C (1890) Beitrag zum Studium der Hirnrinde und dem Centralursprung der Nerven. Int Mschr Anat Physiol 7: 69–90

Marin-Padilla M (1969) Prenatal and early postnatal ontogenesis of the human motor cortex. II. The basket pyramidal system. Brain Res 14: 633–646

Marin-Padilla M (1974) Three-dimensional reconstruction of the pericellular nests (baskets) of the motor (area 4) and visual (area 17) areas of the human cerebral cortex. A Golgi study. Z Anat Entwickl-gesch 144: 123–135

Marin-Padilla M, Stiblitz GR (1974) Three-dimensional reconstruction of the basket cell of the human motor cortex. Brain Res 70: 511–514

Mountcastle VB (1978) An organizing principle for cerebral function. In: Edelman GM, Mountcastle VB (eds) The mindful brain. MIT Press, Cambridge, pp 7–50

Ramón y Cajal S (1899) Estudios sobra la cortezza cerebral humana. Trab Inst Cajal Invest Biol 4: 1–63

Ramón y Cajal S (1911) Histologie du système nerveux de l'homme et des vertébrés II. Maloine, Paris, p 997

Ramón y Cajal S (1935) Die Neuronenlehre. In: Bumke O, Foerster O (eds) Handbuch der Neurologie. Springer, Berlin Heidelberg New York, pp 887–994
Ribak CE (1978) Aspinous and sparsely-spinous stellate neurons in the visual cortex of rats contain glutamic acid decarboxylase. J Neurocytol 7: 461–478
Rosén I, Asanuma H (1972) Peripheral afferent inputs to the forelimb area of the monkey motor cortex: input-output relations. Exp Brain Res 14: 257–273
Somogyi P (1977) A specific axo-axonal interneuron in the visual cortex of the rat. Brain Res 136: 345–350
Somogyi P (1978) The study of Golgi stained cells and of experimental degeneration under the electron microscope: a direct method for the identification in the visual cortex of three successive links in a neuron chain. Neuroscience 3: 167–180
Somogyi P (1979) An interneuron making synapses specifically on the axon initial segment (AIS) of pyramidal cells in the cerebral cortex. J Physiol (Lond) 296: 18–19
Somogyi P, Cowey A (1981) Combined Golgi and electron microscopic study on the synapses formed by double bouquet cells in the visual cortex of the cat and monkey. J Comp Neurol 195: 547–566
Somogyi P, Freund TF, Halász N, Kisvárday ZF (1981a) Selectivity of neuronal ^{3}H-GABA accumulation in the visual cortex as revealed by Golgi staining of the labelled neuron. Brain Res 225: 431–436
Somogyi P, Cowey A, Halász N, Freund TF (1981b) Vertical organization of neurons accumulating ^{3}H-GABA in the visual cortex of the rhesus monkey. Nature 294: 761–763
Somogyi P, Freund T, Cowey A (1982) The axo-axonic interneuron in the cerebral cortex of the rat, cat and monkey. Neuroscience 7: 2577–2608
Somogyi P, Cowey A, Kisvárday ZF, Freund T, Szentágothai J (1983a) Retrograde transport of γ-amino [^{3}H]butyric-acid reveals specific interlaminar connections in the striate cortex of monkey. Proc Natl Acad Sci USA 80: (to be published)
Somogyi P, Nunzi MG, Gorio A, Smith AD (1983b) A new type of specific interneuron in the monkey hippocampus forming synapses exclusively with the axon initial segments of pyramidal cells. Brain Res 259: 137–142
Somogyi P, Kisvárday ZF, Martin KAC, Whitteridge D (1983c) Synaptic connections of morphologically identified and physiologically characterized large basket cells in the striate cortex of cat. (to be published)
Szentágothai J (1962) On the synaptology of the cerebral cortex. In: Sarkissov SA (ed) Structure and function of the nervous system. Medgiz, Moscow, pp 6–14 (in Russian)
Szentágothai J (1967) The anatomy of complex integrative units in the nervous system. In: Lissák K (ed) Recent development in neurobiology in Hungary, vol 1. Akad Kiadó, Budapest, pp 9–45
Szentágothai J (1969) Architecture of the cerebral cortex. In: Jasper HH, Ward AA, Pope A (eds) Basic mechanisms of the epilepsies. Little Brown, Boston, pp 13–28
Szentágothai J (1971) Some geometrical aspects of the neocortical neuropil. Acta Biol Acad Sci Hung 22: 107–124
Szentágothai J (1973) Synaptology of the visual cortex. In: Jung R (ed) Handbook of sensory physiology, vol 7, 3B. Springer, Berlin Heidelberg New York, pp 269–324
Szentágothai J (1975) The "module concept" in cerebral cortex architecture. Brain Res 95: 475–496
Szentágothai J (1978) The neuron network of the cerebral cortex: a functional interpretation. The Ferrier Lecture 1977. Proc R Soc Lond [Biol] 201: 219–248
Szentágothai J (1981) Principles of neural organization. In: Szentágothai J, Palkovits M, Hámori J (eds) Regulatory functions of the CNS vol 1. Pergamon, Akadémiai Kiadó Budapest, pp 1–16 (Adv physiol sci, vol 1)
Szentágothai J (1983) The "modular" architectonic principle of neural centers. In: Jung R (ed) Rev Physiol Biochem Pharmacol, Vol. 98, 11–61

The Sensorimotor Integration in Area 3a of the Cat*

S. LANDGREN, H. SILFVENIUS, and K.Å. OLSSON, Umeå

Introduction

A new hypothesis concerning the functional organization of the postcentral somatosensory cortex is accepted in the contemporary literature. The idea of one topographical representation of the body in the postcentral gyrus – the SmI of Woolsey (1958) – is no longer tenable. We can now see SmI subdivided into at least four zones with different input and output connections and different functional characteristics. These zones extend in mediolateral direction in parallel with the central sulcus of the primates. They are somatotopically organized and there is a relative correspondence in extent between the functional zones and the cytoarchitectonically defined areas 3a, 3b, 1, and 2 originally described by Brodman (1903) and by Vogt and Vogt (1919).

The evidence of a proprioceptive zone corresponding to area 3a in the depth of the central sulcus has played an important role in the establishment of the new hypothesis. Two excellent reviews of the literature concerning area 3a were recently published by Jones and Porter (1980) and by Wiesendanger and Miles (1982). The emphasis in these reviews is, however, placed on the evidence obtained in primates. A recollection of the results found in the cat may therefore be of some comparative interest.

Localization Within the Cerebral Cortex

When observations on function obtained from different species in different laboratories are compared, a reference frame for localization is required. Three such frames are available for localization in the cerebral cortex. They involve reference to gyri and sulci, to cytoarchitectonic areas, or to cortical action potentials evoked by the stimulation of primary projection paths. All the three reference systems have limitations. The pattern of the sulci varies between animals. So does the extent of the cytoarchitectonic areas. Borders between the areas are sometimes difficult to define. Maps based on the distribution of evoked potentials have arbitrary borders because they have to be defined in terms of a certain decrease in amplitude of the maximal response. The three methods do, however, support one another, and used together in the individual animals investigated they allow a reasonably reliable localization.

* This work was supported by the Mecidal Research Council (proj.no. B84-14X-00045-20C).

Experimental Brain Research, Suppl. 9
© Springer-Verlag Berlin · Heidelberg 1984

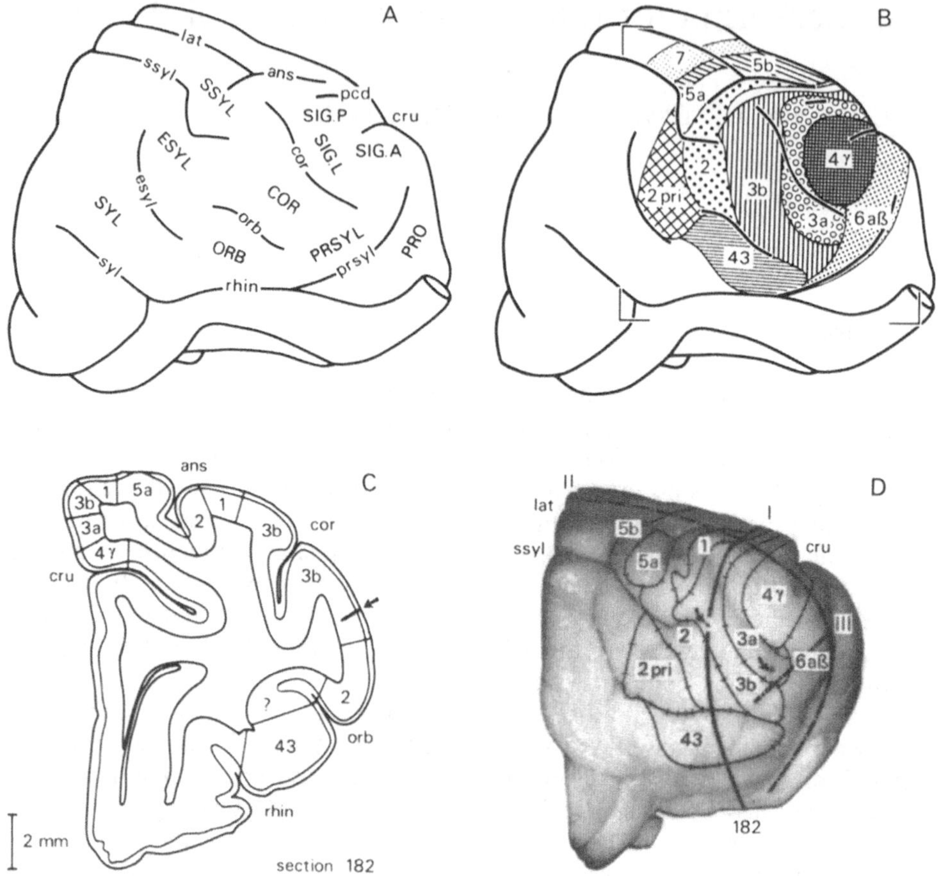

Fig. 1A–D. Diagrams summarizing the method used for identifying cytoarchitectonic areas of the frontal part of the right hemisphere. *A* Nomenclature of sulci and gyri according to Hassler and Muhs-Clement (1964). *B* Cytoarchitectonic areas. The boundaries represent typical features observed in the individual experiments (cf. *D*). The frame locates the enlarged diagrams of Fig. 2. *C* Outline of a serial section (cf. sectioning plane *I*, section number 182 in *D*) with classified cytoarchitecture as determined according to the criteria of Hassler and Muhs-Clement (1964). The position of an india ink puncture is given by the *arrow*. *D* Photograph of the perfused hemisphere from an individual experiment. Borders of the cytoarchitectonic areas (*thin lines*) are given. *Coarse lines (I, II, III)* indicate the different planes of sectioning. *Curved arrows* denote india ink punctures made at points of interest. *ans*, sulcus ansatus; *cor*, sulcus coronalis; *cru*, sulcus cruciatus; *esyl*, sulcus ectosylvius; *lat*, sulcus lateralis; *orb*, sulcus orbitalis; *pcd*, postcruciate dimple; *prsyl*, sulcus presylvius; *rhin*, sulcus rhinalis; *ssyl*, sulcus suprasylvius; *syl*, sulcus sylvius; *COR*, gyrus coronalis; *ESYL*, gyrus ectosylvius; *ORB*, gyrus orbitalis; *PRO*, gyrus proreus; *PRSYL*, gyrus presylvius; *SIGA*, gyrus sigmoideus anterior ; *SIGL*, gyrus sigmoideus lateralis; *SIGP*, gyrus sigmoideus posterior; *SSYL*, gyrus suprasylvius; *SYL*, gyrus sylvius. (Landgren and Olsson 1980)

The cytoarchitectonic areas in the cat's rostral hemisphere were described by Hassler and Muhs-Clement (1964). When studying the somatosensory cortex of the cat we had the opportunity of confirming their observations (Landgren and Olsson 1980). Our results are summarized in Fig. 1B, which shows a standardized diagram generalizing the observation from several animals. The borders of the

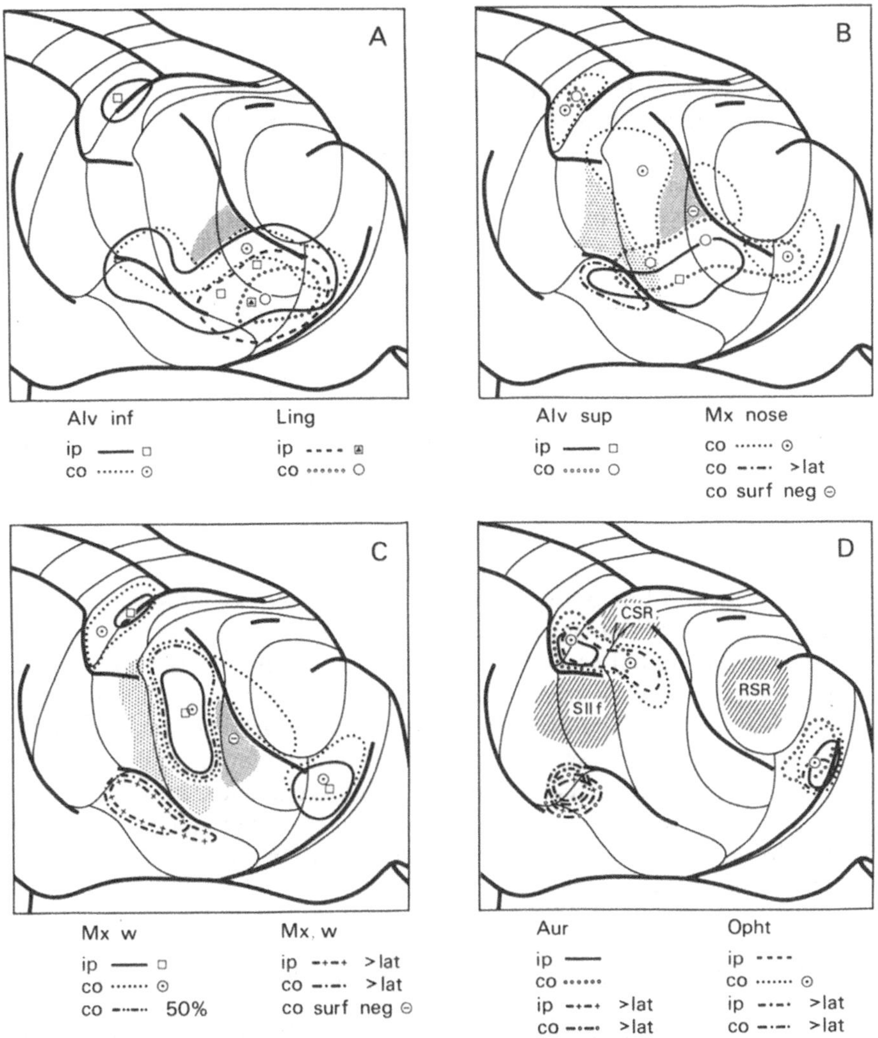

Fig. 2A–D. Diagrams showing the typical location of oral (*A* and *B*) and face projections (*B, C,* and *D*) related to the anatomy of the sulci and gyri and to the borders of the cytoarchitectonic areas (*thin lines drawn in full*). *Symbols* of borderlines and maximum points of the fields are explained in the figures. Borders at 20% of maximal amplitude are given for the initially surface-positive short-latency potentials unless otherwise indicated (*50% or > lat*). *> lat* = responses of longer latency in projection to the orbital gyrus. *Dense stippling*, field of initially negative or diphasic positive-negative potentials. ⊖, maximum point. *Thin stippling*, field of diphasic negative-positive potentials. *Oblique hatching*, the caudal (*CSR*) and rostral (*RSR*) projection fields of low-threshold superficial contralateral radial afferents (*SR*). *SIIf*, the forelimb subdivision of the second somatosensory projection area. (Landgren and Olsson 1980)

areas may be compared with those obtained in an individual experiment shown in Fig. 1C and D.

Area 3a is a mediolateral band starting on the medial surface of the hemisphere near the cruciate sulcus and extending over the posterior sigmoid gyrus near the postcruciate dimple (Pcd). It dives into the coronal sulcus and follows the course of this sulcus in its rostral direction. The medial border toward area 4 is generally found below the cortical surface in the medial bank of the coronal sulcus. Laterally, area 3a extends into the coronal gyrus and it ends near the rostral end of the coronal sulcus. It is useful to note that the Pcd in the cat corresponds to the depth of the central sulcus in the primate. The cruciate sulcus, on the other hand, is found within the motor cortex (area 4).

The main cytoarchitectonic criteria used by Hassler and Muhs-Clement (1964) to characterize area 3a were the appearance of a layer IV, a differentiation of layer III, and a breakdown of the band of giant pyramids in layer V to nests of a few giant cells alternating with smaller pyramids. Jones and Porter (1980) have reviewed this evidence, and they stress the importance of the layer IV criterion because the observation of occasional giant pyramids in layer V in area 3a only holds true for the hand area but not for the foot and leg in the monkey. In our experience the border between area 4 and 3a is relatively easy to establish. Also, the border toward area 3b may be found with relative accuracy due to the much less laminated appearance of area 3b compared to area 3a. It is, however, obvious that very sharp borders between the areas cannot be drawn.

The use of evoked potential mapping as an aid to localization is illustrated in Fig. 2. The localization of the surface-positive evoked potential with maximal amplitude and minimal latency is indicated by symbols. It was required that the surface-positive potential should change phase to a negative potential of maximal amplitude near layer IV when recorded along a microelectrode track normal to the cortical surface. The borders of the fields are arbitrarily chosen at a point where the amplitude is reduced to 20% of the maximum. This commonly used method adds somatotopical detail to the cytoarchitectonic areas. It locates the dense termination of the thalamocortical input from a chosen primary projection path. The validity of this conclusion is confirmed by repeated sampling of action potentials from single neurons.

The Projection Paths from the Primary Muscle Spindle Endings to Area 3a

Area 3a is functionally characterized by its input from muscle spindle afferents. Figure 3 summarizes the course of the primary projection paths from large muscle spindle afferents through the spinal cord, medulla oblongata, and thalamus to the cerebral cortex. The paths from the contralateral forelimb and hind limb muscles are drawn separately because certain differences between their course and connections exist in the cat.

The Forelimb Path

The forelimb path from primary muscle spindle afferents (group Ia) to area 3a enters the dorsal column at the level of C4–C5 and proceeds to the main cuneate nucleus (MCN). It enters the nucleus from its dorsolateral aspect and terminates ventrally in MCN deep to the cutaneous relay neurons. The cuneate relay is found between the obex and 7 mm caudal to this landmark.

The organization of the cuneate relay was described by Rosén (1969a, b). The receptive field of the relay cells was restricted to one muscle in 50% of the ob-

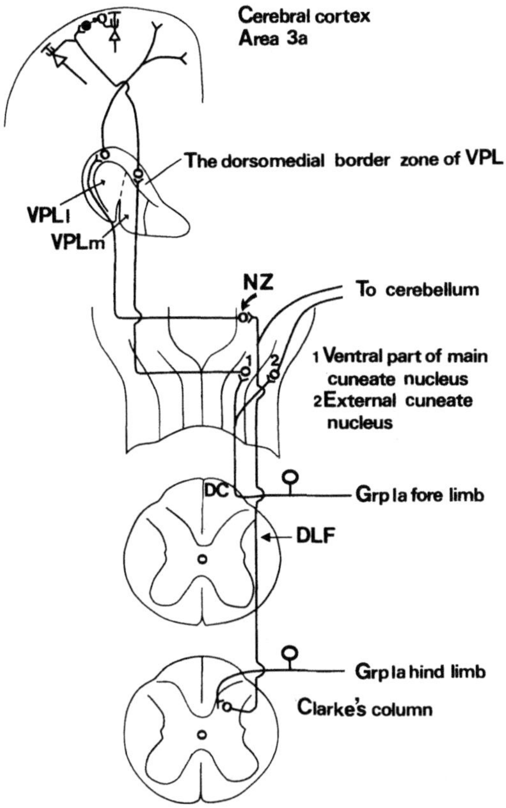

Fig. 3. The projection paths from the primary muscle spindle afferents to area 3a of the cerebral cortex of the cat. *DC*, dorsal column; *DLF*, dorsolateral fascicle; *NZ*, nucleus Z.

served units, but convergence from two or more muscles was observed. The convergence from other types of afferent was limited. Some convergence from group II muscle afferents and from low-threshold cutaneous afferents was, however, found. The population of neurons in the group I nucleus of MCN may be divided into relay cells and nonprojecting interneurons, as was also found by Andersen et al. (1964b) in their study of the cutaneous relay of MCN.

No group Ia relays to the thalamus were observed by Rosén in the cat's external cuneate nucleus. This nucleus relays group I impulses in the cuneocerebellar tract (cf. Holmqvist et al. 1963). It is, however, of interest to note that Boivie and Boman (1981) have described a projection from the external cuneate nucleus to the rostral part of the caudal subdivision of the ventroposterolateral nucleus (VPLc) in the monkey's thalamus. When considering the strict separation between the group I paths to the cerebellum and those to the thalamus, it is also relevant to note that Johansson and Silfvenius (1977a) found that many relay cells in the bulbothalamic group I path from the hind limb were activated by collaterals of axons in the dorsal spinocerebellar tract (DSCT).

The second-order neurons in the group Ia path from the forelimb muscles send their axons across the midline into the medial lemniscus. The axons terminate in

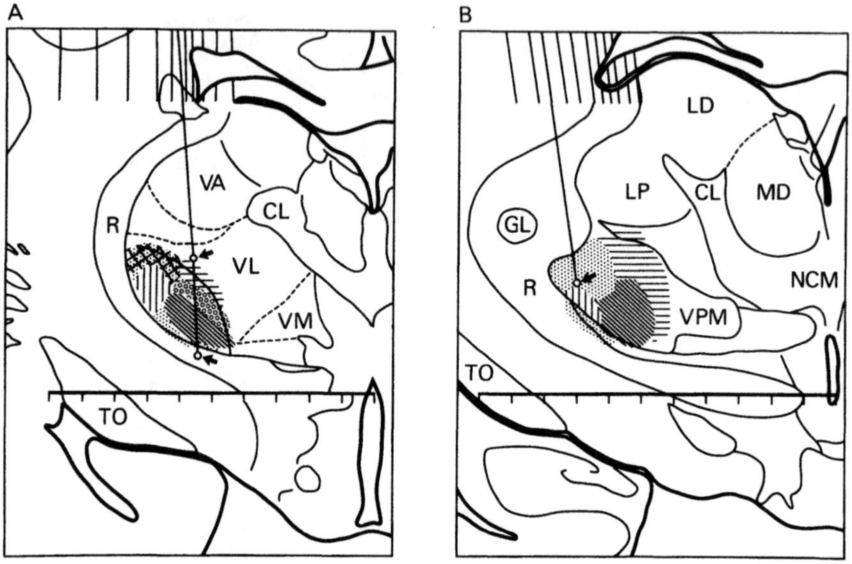

Fig. 4A,B. Tracings of histological sections with microelectrode tracks passing through VPL of the cat. *A* Grid of tracks through the middle third of VPL (640 μm rostral to the rostral pole of GL). *B* Grid of tracks through the caudal third of VPL (40 μm caudal to the rostral pole of GL). The *thin lines at top of diagram* indicate the course of identified recording tracks. The *heavy arrows* mark the Prussian blue reference points used for dorsoventral localization. The *scales* give millimeters in the living preparation. The *ring symbols* show the location of the responses evoked by group I muscle afferents from the contralateral forelimb. *Cross-hatching*, responses evoked by group I muscle afferents from the contralateral hind limb; *oblique hatching*, short-latency responses from contralateral superficial radial nerve; *vertical hatching*, short-latency responses from contralateral dorsal columns at L1; *horizontal hatching*, responses from contralateral superficial radial nerve and group II and III muscle afferents; *stippling*, convergent responses from the dorsal column-lemniscal and the spinocervicolemniscal paths (the zone of convergence of Landgren et al. 1965).
CL, n. centralis lateralis; *GL*, corpus geniculatum laterale; *LD*, n. lateralis dorsalis; *LP*, n. lateralis posterior; *MD*, n. medialis dorsalis; *NCM*, n. centralis medialis; *R*, n. reticularis; *TO*, tractus opticus; *VA*, n. ventralis anterior; *VL*, n. ventralis lateralis; *VM*, n. ventralis medialis; *VPL*, n. ventralis posterolateralis; *VPM*, n. ventralis postéromedialis. (Modified from Andersson et al. 1966, and Landgren and Silfvenius 1970)

a dorsomedial border zone within the rostral two-thirds of the ventroposterolateral thalamic nucleus (VPL) in the cat (Mallart 1964; Andersson et al. 1966). This border zone (see Fig. 4) is part of a shell of neurons surrounding the specific cutaneous relay nucleus in the center of the ventrobasal complex (VB)

The shell has afferent and efferent connections that differ from those of the centrally located cutaneous relay. The dorsomedial and dorsolateral parts of the border zone receive proprioceptive afferents and the lateral and ventral parts convergence between the spinocervicothalamic path and the dorsal column lemniscal paths from both fore- and hind limbs (cf. Landgren et al. 1965 and Andersen et al. 1966). The physiological observations are in agreement with the anatomic analysis of the afferent termination in the VPL described by Boivie (1970) and Berkley (1980).

The neurons in the dorsomedial border zone of the medial subdivision of the ventroposterolateral nucleus (VPLm) were analyzed by single-unit recording technique by Andersson et al. (1966) and Rosén (1969c). In the population described by Rosén (1969c), 75% of the neurons were thalamocortical relay cells. The spatial convergence onto these cells was moderate. One-third of them were activated only by one muscle and two-thirds by at least two synergist muscles. A convergence from afferents other than group Ia was observed. Fifteen percent of the relay cells were activated by group I and group II muscle afferents, often from the same nerve; 39% showed a weak convergence from cutaneous afferents.

Responses from neurons discharged by electrical stimulation of the elbow joint nerve were recorded by Andersson et al. (1966). No convergence between group I and low-threshold joint afferents was, however, found on the same cell. Neurons in the dorsolateral part of the border zone of the lateral subdivision of the ventroposterolateral nucleus (VPLl) responding to knee joint capsule were observed in the cat by Yin and Williams (1976).

The existence of a proprioceptive border zone in VPLc, or in a zone overlapping the border between VPLc and the oral subdivision of the ventroposterolateral nucleus (VPLo), was for a long time questioned in studies on the monkey. The findings of Poggio and Mountcastle (1963), that responses to deep stimuli were found dorsally in VB of the monkey, suggested a subdivision of the cutaneous and deep inputs. As the dorsal VB was dominated by the proximal parts of the limb, where deep receptors could be expected to form a larger proportion, the question of a dorsal proprioceptive zone was left open. Recent inverstigations by Friedman and Jones (1981), Jones and Friedman (1982), and Maendly et al. (1981) have, however, fully confirmed the observations previously made in the cat. The dorsal and rostral shell of VPLc contains proprioceptive relays to the cerebral cortex. Some disagreement still exists concerning the overlap into VPLo, probably mainly due to difficulties in defining the cytoarchitectonic criteria of the two nuclei. An overlapping termination in the rostral VPL and adjacent caudal parts of the ventrolateral thalamic nucleus (VL) of the cat was also described by Grant et al. (1973) for the nucleus Z axons. It is possible that the VPL (VPLc) and the VL (=VPLo in the monkey) relays in the group Ia path differ in their cortical projection, the latter contributing a direct input to area 4.

The thalamocortical group I neurons project to the Pcd of the cat's cerebral cortex and thus to area 3a (Andersson et al. 1966; Rosén 1969c). In addition, there is a projection to the lower bank of the anterior suprasylvian sulcus (cf. Andersson et al. 1966; Landgren et al. 1967a). The projection to area 3a in the cat was confirmed by Larsen and Asanuma (1979), who injected horseradish peroxidase in areas 3a and 4. The 3a projections had their cell somas in the dorsomedial shell of VPL, whereas the cells of the 4 projection were found in this area but also spread out over a larger part of VL.

There is now general agreement that the proprioceptive border zone of VPLc of the monkey projects to area 3a (Kalil 1981; Friedman and Jones 1981; Maendly et al. 1981). In addition, Friedman and Jones (1981) and Jones and Friedman (1982) have demonstrated a projection to area 2 mainly from the posterior part of the VPLc border zone. Convergence between a short-latency vestibular input and deep somatosensory afferents onto neurons in the VPLc-VPLo border region of the monkey has been reported by Liedgren et al. (1976). Some of their thalamic neurons projected to area 3a.

The Hind Limb Path

The projection path from group I hind limb muscle afferents to area 3a of the cerebral cortex of the cat has its secondary neuron in Clarke's column at the level of L1–L3 (Landgren and Silfvenius 1969, 1971). The secondary axons enter the ipsilateral dorsolateral fascicle (DLF) and ascend to the ipsilateral medulla oblongata.

The bulbar relay in the path is located in nucleus Z of Brodal and Pompeiano (1957) in rostral parts of the main cuneate nucleus and in nearby parts of the descending vestibular nucleus and nucleus X of Brodal and Pompeiano (1957) (cf. Landgren and Silfvenius 1971; Johansson and Silfvenius 1977a–c). This nuclear complex receives axon collaterals from DSCT. The group I relay cells send their axons via the medial lemniscus to the thalamus.

The bulbar group I cells were analyzed by Johansson and Silfvenius (1977a–c). They found that 52% of 34 bulbothalamic relay cells located in nucleus Z were discharged by DSCT axon collaterals. The relay cells exhibited a restricted spatial convergence. Convergence between group I and II muscle afferents and group I cutaneous afferents was observed in some relay cells. When the convergence in all nucleus Z cells, relay as well as non relay cells, was considered (Johansson and Silfvenius 1977b), it was found that 60% were excited exclusively by group I muscle afferents from the hind limb, 18% were discharged only be cutaneous afferents, and 7% by knee joint afferents. Only 15% received a convergent input between two or more of these receptor types.

The axons of the tertiary neurons in the group I hind limb path terminate in the dorsolateral border zone of VPLl and in adjacent parts of VL, as shown in anatomic degeneration studies by Grant et al. (1973) and with neurophysiological recordings by Landgren and Silfvenius (1970).

Distribution of the Group Ia Muscle Afferent Projections in the Cerebral Cortex

Projections to Area 3a

Responses from low-threshold muscle afferents in the posterior sigmoid gyrus of the cat were first recorded by Amassian and Berlin (1958). A more widespread interest in this new cortical projection path was, however, aroused by the careful investigations of Oscarsson and Rosén (1963, 1966). They studied the group Ia path from the contralateral forelimb muscles in the cat. Efforts were made to search for the muscle receptors from which the path originated. There is now general agreement that the primary muscle spindle afferents discharge neurons in area 3a. All receptor types responding to movement of the limb do, however, contribute to the 3a input.

The group Ia projection was located in the posterior sigmoid gyrus near the Pcd, and referring to Hassler and Muhs-Clement (1964) it was suggested that the location was within area 3a. The analysis of responses from single neurons in the

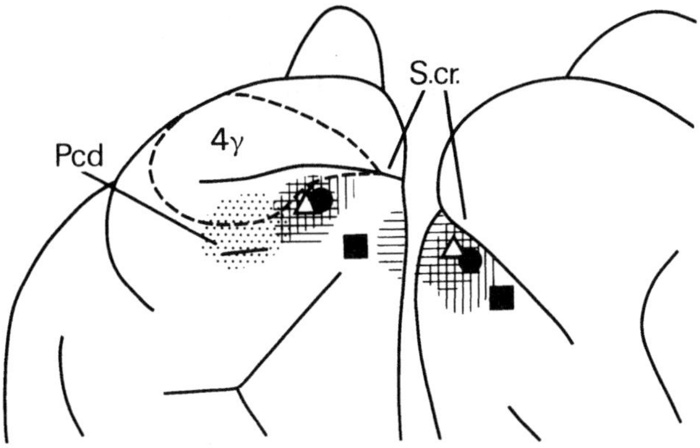

Fig. 5. Dorsal and medial aspect of the rostral pole of the cat's cerebral hemisphere. Location of the projection areas of the group I muscle afferents from the contralateral forelimb (*stippling*) and hind limb (*hatching*). *Horizontal hatching*, area of quadriceps (Q). *Vertical hatching*, area of posterior biceps-semitendinosus (PBSt). Maximum points: for Q, △; for PBSt, ●; for sural projection, ■. *Pcd*, postcruciate dimple. *S.cr.*, cruciate sulcus; *Su*, sural nerves. (Landgren and Silfvenius 1969)

forelimb group Ia projection was carried out by Oscarsson et al. (1966). The population of neurons sampled near the Pcd was located between 500 and 1500 μm below the cortical surface (layers III–V). Compared to the thalamic and cuneate populations the cortical neurons showed an increasing spatial convergence. Convergence from synergistic muscles was observed and also from antagonistic muscles and muscles acting at different joints. Some neurons did, however, preserve a restricted spatial input.

Convergence from other types of afferents was also found at the cortical level. Some group Ia discharged neurons received input from group II muscle afferents, as was the case with the thalamic and cuneate populations. Convergence from cutaneous afferents was often found (cf. Swett and Bourassa 1967). Area 3a also receives afferent input from low-threshold joint (Clark et al. 1973), pacinian (Silfvenius 1970), and vestibular (Ödqvist et al. 1975) afferents. As evidenced by evoked potential records, the 3a projections were separated from the projection maxima located in areas 3b and 1. This evidence suggests reception of joint, pacinian, and vestibular impulses in the area (zonal convergence) but does not allow conclusions concerning convergence onto single cortical neurons. Such convergence is, however, likely to take place at some of the stages in the cortical integration.

The question whether afferents from Golgi tendon organs contribute to the area 3a input is controversial. Evidence based on cortical surface potentials evoked by electrical stimulation of group Ia and Ib afferents from the contralateral hind limb muscles does, however, suggest that impulses from the Golgi afferents also reach area 3a (cf. Landgren and Silfvenius 1969).

The group I projections from the hind limb muscles to area 3a was mapped by Landgren and Silfvenius (1969). They found partially overlapping projections from the gastrocnemius, quadriceps, and biceps semitendinosus muscles medial to the Pcd in the posterior sigmoid gyrus of the cat. A separate projection was recorded on the medial surface of the hemisphere, just caudal to the cruciate sulcus. Preaxial as well as postaxial muscles were somatotopically represented on the dorsal as well as on the medial aspect of the hemisphere (see Fig. 5).

A correlation between the location of group I evoked potential maxima and cytoarchitectonic criteria in the individual experiment was carried out and the localization within area 3a was established (cf. Clark et al. 1973 and Grant et al. 1975).

Projections to Other Parts of the Sensorimotor Cortex

The group Ia muscle afferents also project to the cerebral cortex outside area 3a in the cat. A projection to the lower bank of the anterior suprasylvian sulcus was thus described by Landgren et al. (1967a), and the properties of the neurons were analyzed by Silfvenius (1972a). This projection borders onto the forelimb area of the second somatosensory projection (SmII) (Carreras and Andersson 1963). Further caudally in the anterior suprasylvian sulcus it overlaps with short-latency vestibular, auditory, and visual inputs (cf. Landgren et al. 1967b).

Cortical focal potentials evoked by group Ia muscle afferents and by low-threshold pacinian afferents were recorded from a locus buried in the rostral bank of the lateral ansate sulcus (Silfvenius 1968, 1970). Neurons in the ansate sulcus responding to joint movements and to cutaneous stimuli were recently described by Waters et al. (1982). These observations confirm the early reports of Mountcastle (1957), who found an increasing number of units responding to deep stimuli when the recording position approached the ansate sulcus.

A separate projection of the group Ia afferents to area 4 was suggested by Silfvenius (1972b). He found short-latency group I evoked potentials near the lateral end of the cruciate sulcus. The responsive area was clearly separated from the Pcd projection. Compatible evidence was presented by Asanuma et al. (1979a).

Projections from Primary Muscle Spindle Afferents to Area 3a of the Monkey

An input from low-threshold muscle afferents to area 3a of the baboon was first demonstrated by Phillips et al. (1971). The group I forelimb responses were further analyzed by Lucier et al. (1975) and Heath et al. (1976). The muscle spindle input from hind limb muscles to areas 3a and 4 was investigated by Hore et al. (1976).

There is a good agreement between the observations made on primates and those on cats. The evidence of the origin in primary muscle spindle afferents has been refined in the experiments on the primates and further evidence of convergence from the secondary spindle afferents has accumulated. A considerable convergence from cutaneous afferents on group I cells is also seen in the monkey.

The existence of a direct group I input to area 4 is still a controversial issue. In the monkey the group I input seems to be transmitted to area 4 through area 3a. In the cat, however, there are some indications of a direct path, although the responses to group I afferents are small compared to those evoked by group II (Asanuma et al. 1979b).

Another open question concerns the directness of the cutaneous path to the group I cells of area 3a. The latencies of action potentials evoked by the cutaneous

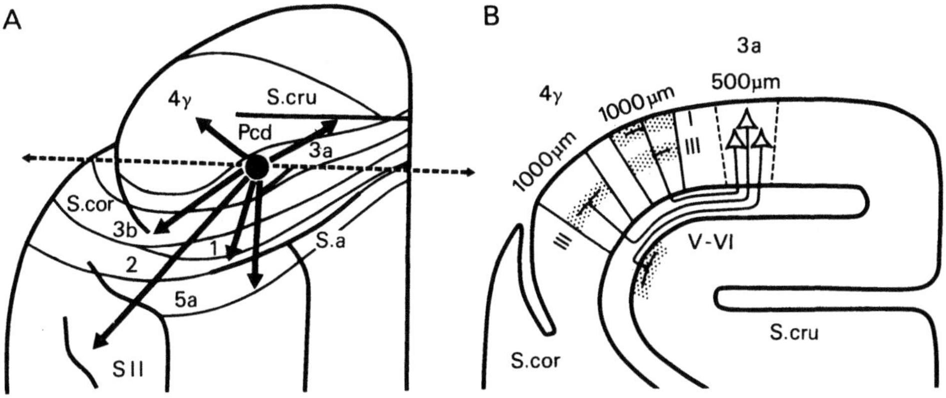

Fig. 6. A Diagram summarizing the observed U fiber connections from the lesion in area 3a (*Pcd*) to area 4γ, to the coronal (*S.cor.*) and the ansate (*S.a.*) sulci, and to SII. *S.cru.*, cruciate sulcus. The *broken line* approximates the plane of the transverse section shown in *A*. (Modified from Grant et al. 1975) *B* Diagram showing the three types of columns with degeneration in cortical layers I and III, III, and V–VI respectively. U fibers from pyramidal cells in the lesion of area 3a are drawn schematically.

input fall into two groups. The responses with longer latency may be transmitted through cutaneous relays in area 3b (cf. Heath et al. 1976). It is further likely that the type of cutaneous receptors which project to area 3a differ from those projecting to area 3b. The available evidence suggest that all inputs to area 3a are generated by movements. If so, the afferents from the direction-sensitive Ruffini endings (the SAII receptors) should contribute to area 3a (cf. Hulliger et al. 1979).

The Microorganization of Area 3a

The transmission of impulses in the group Ia path to area 3a is fast and characterized by a high degree of synaptic efficiency. Oscarsson et al. (1966) thus recorded sharply rising unitary excitatory postsynaptic potentials (EPSPs) from group I neurons, which were monosynaptically activated by the thalamocortical input. Most of the group I excited 3a neurons do not send their axons into the pyramidal tract (non-PT cells). Excited PT cells were, however, found by Swett and Bourassa (1967) and by Zarzecki et al. (1978). The PT cells in area 3a were often disynaptically inhibited by the group I volley. Grampp and Oscarsson (1968) therefore suggested that the group I afferents modulated the excitatory input from other paths by inhibition of the PT cells.

Area 3a sends corticocortical fibers via the subcortical white matter (U fibers) to area 4, as shown by Jones and Powell (1968). The U fibers terminate in discrete cortical cylinders of columnar shape (Grant et al. 1975). Needle penetrations in the forelimb area of 3a resulted in columnar degeneration in the forelimb as well as in the hind limb areas of 4γ. There was also a columnar projection to areas 2 and 5, 3b, and SmII, as indicated in the diagram of Fig. 6.

In agreement with the anatomical findings, Zarzecki et al. (1978) have described group I activated neurons in area 3a, which were antidromically discharged by microstimulation in area 4γ. One of the 4γ projecting neurons was a

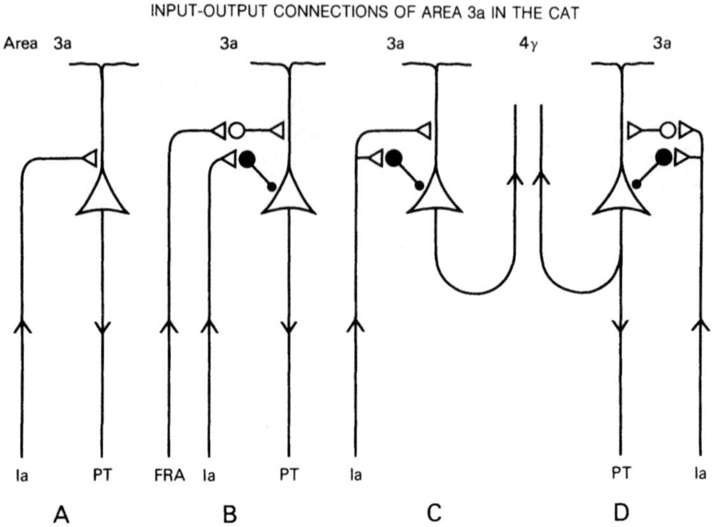

Fig. 7. Observed connections between the thalamocortical group I afferents and the corticospinal (*PT*) cells or the 4γ projecting cells in area 3a. *A*, direct PT cell excitation (Swett and Bourassa 1967); *B*, inhibitory Ia modulation of an excitatory input from flexor reflex afferents (*FRA*) (Grampp and Oscarsson 1968); *C*, excitation or inhibition of 3a cell projecting to area 4γ; *D*, group Ia excitation or inhibition of PT cell with axon collateral to area 4γ. (*C* and *D* after Zarzecki et al. 1978)

PT cell. The neuronal circuits suggested for the transmission of group Ia impulses through area 3a are shown in the diagram of Fig. 7.

The observations suggest that proprioceptive impulses to area 3a are transmitted to the motor cortex via excitatory links. They do, however, also subserve an important inhibitory function within area 3a itself. The proprioceptive input may further excite the PT cells of area 3a and thus contribute to the output through the pyramidal tract. Area 3a also sends axons to the proprioceptive area 2 and to the parietal association cortex, which is involved in the initiation of movement in the immediate extracorporeal space (Hyvärinen and Poranen 1974; Mountcastle et al. 1975).

The Output from Area 3a to the Spinal Cord and the Trigeminal Nucleus

The pyramidal cells in layer V of area 3a send axons to the cervical and lumbar spinal cord in the cat. This was demonstrated by Groos et al. (1978) by experiments with retrograde transport of horseradish peroxidase from the cord to the 3a pyramids. The output to the spinal segments was also observed in the monkey by Coulter and Jones (1977).

Effects on spinal motoneurons and interneurons evoked by electrical stimulation of the posterior sigmoid gyrus have been described by Lundberg and Voorhoeve (1962) and Hongo and Jankowska (1967). The effects were evoked by pyramidal and rubrospinal paths. Cytoarchitectonic localization of the cortical origin of these effects was not offered, but the reports of the location of the point where they were evoked by minimal stimulus strength is compatible with a 3a origin.

Electrical excitation of area 3a may, however, be transmitted via the output from area 4γ. The observation of Olsson and Landgren (1980) is of some interest in this connection. They studied the influence of electrical stimulation of the cerebral cortex on the monosynaptic jaw-closing and the disynaptic jaw-opening reflexes in the cat. The largest and most complex effects on these reflexes were evoked from the projection fields of the whiskers and the periodontal afferents of area 3a, and not from area 4γ. Stimulation of the 3a projections discharges interneurons in the intertrigeminal nucleus, which is located between the main sensory trigeminal nucleus and the trigeminal motor nucleus. The interneurons were excited with a latency indicating monosynaptic corticotrigeminal connections, and they send their axons to the motor nucleus (Landgren and Olsson, to be published). These findings suggest that the trigeminal reflexes may be powerfully influenced from area 3a.

Descending Control of the Group I Path to Area 3a

The relay cells of the group I path to the cerebral cortex receive a descending corticofugal influence. The thalamic group I relay cells were antidromically as well as transsynaptically discharged by electrical stimulation of the cerebral cortex. The transsynaptic response was evoked from the Pcd (Rosén 1969c) or from the Pcd and SIIA (Andersson et al. 1966). In the case of the thalamic relay, the possibility that the transsynaptic discharge was evoked antidromically via axon collaterals cannot be excluded. Axon collaterals of ventrobasal relay cells do, however, evoke inhibition as shown by Andersen et al. (1964a).

Corticofugal neurons localized in area 3a and antidromically discharged from nucleus gracilis and cuneatus were described by Brech et al. (1977). In his investigation of cuneothalamic relay cells in the group I path, Rosén (1969b) found that only 7% of these cells were excited by cortical stimulation. Cortical excitation of group I excited nonrelay cells in the cuneate nucleus was observed in 75% of these neurons. Rosén also found evidence of a presynaptic inhibition of the transmission through the cuneate group I relay, which was evoked from the cortex near the lateral end of the cruciate sulcus, thus in area 4γ. The cortically excited nonrelay cells may thus be engaged in the presynaptic inhibition of the cuneate group I relay. Pre- as well as postsynaptic inhibition of cuneate neurons evoked from the cerebral cortex of the cat has been described by Andersen et al. (1964b, c). In the nucleus gracilis Gordon and Jukes (1964) also found neurons which were excited or inhibited by electrical stimulation of the posterior sigmoid gyrus. A corticofugal influence of the medullary relay nuclei is thus established. Part of this influence originates in area 3a.

As described above, the spinal relay in the group I path from the hind limb is located in Clarke's column and some of the axons ascending to nucleus Z are DSCT axons. It is therefore of interest to note that Hongo and Okada (1967) found that postsynaptic inhibition of group I discharged DSCT neurons could be evoked from the contralateral posterior sigmoid gyrus in the cat. The best point for stimulation was located 2–4 mm lateral to the midline and 1–3 mm caudal to the cruciate sulcus. This location is similar to that of the area 3a projection of the hind limb group I path. A presynaptic inhibition of DSCT evoked from the same cortical area was described by Hongo et al. (1967). These observations indicate that the spinal relay of the group I paths may also be influenced from area 3a.

Functional Considerations

The function of the proprioceptive projections to area 3a can only be considered in the most general terms. Behavioral experiments involving lesions of SmI and II (Norrsell 1978, 1980) have been carried out. They indicate a role of SmI in "active touch," in tasks involving tactile perception in combination with movements. Randolph and Semmes (1974) have observed impairment of discrimination of angles in experiments on monkeys with lesions in area 2. No comparable behavioral experiments on area 3a have come to our knowledge. Because of the connections between areas 3a and 2 it nevertheless seems reasonable to expect area 3a to play a role in perception of active touch and in stereognosis.

The most obvious functional suggestion made for the 3a system is that it may play a role in motor control. Area 3a extends along the border of the motor cortex. It receives detailed and integrated signals from receptors discharged by movements. The signals are delivered through a fast path with efficient synaptic transmission. They are transmitted to area 4γ of the motor cortex, bypassing the cerebellum. The contribution of sensory signals to the motor performance is an established fact, and the role of the proprioceptive 3a system in motor control is thus not a farfetched assumption. The control mechanisms are, however, largely unknown.

The 3a system could be part of long loop reflexes, as suggested by Phillips (1969). The arguments for and against this hypothesis were considered by Wiesendanger and Miles (1982). The observed input and output connections of area 3a are compatible with a long loop arc from muscle spindle afferents via the cerebral cortex to the motoneurons. Attempts to demonstrate long loop function were, however, not convincing, because they could be explained by segmental mechanisms.

Another attractive suggestion is that area 3a may take part in the design of voluntary motor programs. The proprioceptive system of area 3a has remarkable features in terms of rapid communication from and to the sensory relay nuclei. It does also influence interneurons of reflex arcs. As exemplified by Olsson and Landgren (1980) with trigeminal reflexes, there is plenty of time for the passage of signals up and down between area 3a and the lower motor centers during the development of the readiness potential of Deecke et al. (1969). The 3a system may therefore take part in the updating of motor programs.

References

Amassian VE, Berlin L (1958) Early cortical projection of group I afferents in the forelimb muscle nerves of cat. J Physiol (Lond) 143: 61

Andersen P, Eccles JC, Sears TA (1964a) The ventrobasal complex of the thalamus: types of cells, their responses and their functional organization. J Physiol (Lond) 174: 370–399

Andersen P, Eccles JC, Schmidt RF, Yokota T (1964b) Identification of relay cells and interneurones in the cuneate nucleus. J Neurophysiol 27: 1080–1095

Andersen P, Eccles JC, Oshima T, Schmidt RF (1964c) Mechanisms of synaptic transmission in the cuneate nucleus. J Neurophysiol 27: 1096–1116

Andersen P, Andersson SA, Landgren S (1966) Some properties of the thalamic relay cells in the spino-cervico-lemniscal path. Acta Physiol Scand 68: 72–83

Andersson SA, Landgren S, Wolsk D (1966) The thalamic relay and cortical projection of group I muscle afferents from the forelimb of the cat. J Physiol (Lond) 183: 576–591

Asanuma H, Larsen KD, Yumiya H (1979a) Receptive fields of thalamic neurons projecting to the motor cortex in the cat. Brain Res 172: 217–228

Asanuma H, Larsen KD, Zarzecki P (1979b) Peripheral input pathways projecting to the motor cortex in the cat. Brain Res 172: 197–208

Berkley KJ (1980) Spatial relationships between the terminations of somatic sensory and motor pathways in the rostral brain stem of cats and monkeys. J Comp Neurol 193: 283–317

Boivie J (1970) The termination of the cervicothalamic tract in the cat. An experimental study with silver impregnation methods. Brain Res 19: 333–360

Boivie J, Boman K (1981) Termination of a separate (proprioceptive?) cuneothalamic tract from external cuneate nucleus in monkey. Brain Res 224: 235–246

Brech AJ, Gordon G, Powell TPS (1977) Corticofugal cells responding antidromically to stimulation of the cuneate or gracile nuclei of the cat. Brain Res 128: 39–52

Brodal A, Pompeiano O (1957) The vestibular nuclei in the cat. J Anat 91: 438–454

Brodman K (1903) Beiträge zur histologischen Lokalisation der Grosshirnrinde. Erste Mitteilung; die Regio Rolandica. J Psychol Neurol 2: 79–107

Carreras M, Andersson SA (1963) Functional properties of neurones of the anterior ectosylvian gyrus of the cat. J Neurophysiol 26: 100–126

Clark FJ, Landgren S, Silfvenius H (1973) Projections to the cat's cerebral cortex from low threshold joint afferents. Acta Physiol Scand 89: 504–521

Coulter JD, Jones EG (1977) Differential distribution of corticospinal projections from individual cytoarchitectonic fields in the monkey. Brain Res 129: 335–340

Deecke L, Scheid P, Kornhuber HH (1969) Distribution of readiness potential, pre-motion positivity and motor potential of the human cerebral cortex preceding voluntary finger movements. Exp Brain Res 7: 158–168

Friedman DP, Jones EG (1981) Thalamic input to areas 3a and 2 in monkeys. J Neurophysiol 45: 59–85

Gordon G, Jukes MGM (1964) Descending influences on the exteroceptive organizations of the cat's gracile nucleus. J Physiol (Lond) 173: 291–319

Grampp W, Oscarsson O (1968) Inhibitory neurones in the group I projection area of the cat's cerebral cortex. In: von Euler C, Skoglund S, Söderberg U (eds) Structure and function of inhibitory neuronal mechanisms. Proceedings of the fourth international meeting of neurobiologists, Stockholm. Pergamon, Oxford, pp 351–356

Grant G, Boivie J, Silfvenius H (1973) Course and termination of fibres from the nucleus Z of the medulla oblongata. An experimental light microscopical study in the cat. Brain Res 55: 55–70

Grant G, Landgren S, Silfvenius H (1975) Columnar distribution of U-fibres from the postcruciate cerebral projection area of the cat's group I muscle afferents. Exp Brain Res 24: 57–74

Groos WP, Ewing LK, Carter CM, Coulter JD (1978) Organization of corticospinal neurons in the cat. Brain Res 143: 393–419

Hassler R, Muhs-Clement K (1964) Architektonischer Aufbau des sensomotorischen und parietalen Cortex der Katze. J Hirnforsch 6: 377–420

Heath CJ, Hore J, Phillips CG (1976) Inputs from low threshold muscle and cutaneous afferents of hand and forearm to areas 3a and 3b of baboon's cerebral cortex. J Physiol (Lond) 257: 199–227

Holmqvist B, Oscarsson O, Rosén J (1963) Functional organization of the cuneocerebellar tract in the cat. Acta Physiol Scand 58: 216–235

Hongo T, Jankowska E (1967) Effects from the sensorimotor cortex on the spinal cord in cats with transected pyramids. Exp Brain Res 3: 117–134

Hongo T, Okada Y (1967) Cortically evoked pre- and postsynaptic inhibition of impulse transmission to the dorsal spinocerebellar tract. Exp Brain Res 3: 163–177

Hongo T, Okada Y, Sato M (1967) Corticofugal influences on transmission to the dorsal spinocerebellar tract from hindlimb primary afferents. Exp Brain Res 3: 135–149
Hore J, Preston JB, Durkovic RG, Cheney PD (1976) Responses of cortical neurones (areas 3a and 4) to rampstretch of hindlimb muscles in the baboon. J Neurophysiol 39: 484–500
Hulliger M, Nordh E, Thelin A-E, Vallbo ÅB (1979) The responses of afferent fibres from the glabrous skin of the hand during voluntary finger movements in man. J Physiol (Lond) 291: 233–249
Hyvärinen J, Poranen A (1974) Function of the parietal associative area 7 as revealed from cellular discharges in alert monkeys. Brain 97: 673–692
Johansson H, Silfvenius H (1977a) Axon-collateral activation by dorsal spino-cerebellar tract fibres of group I relay cells of nucleus Z in the cat medulla oblongata. J Physiol (Lond) 265: 341–369
Johansson H, Silfvenius H (1977b) Input from ipsilateral proprio- and exteroceptive hind limb afferents to nucleus Z of the cat medulla oblongata. J Physiol (Lond) 265: 371–393
Johansson H, Silfvenius H (1977c) Connections from large, ipsilateral hindlimb muscle and skin afferents to the rostral main cuneate nucleus and to the nucleus X region in the cat. J Physiol (Lond) 265: 395–428
Jones EG, Friedman DP (1982) Projection pattern of functional components of thalamic ventrobasal complex on monkey somatosensory cortex. J Neurophysiol 48: 521–544
Jones EG, Porter R (1980) What is area 3a? Brain Res 2: 1–43
Jones EG, Powell TPS (1968) The ipsilateral cortical connections of the somatic sensory areas in the cat. Brain Res 9: 71–94
Kalil K (1981) Projections of the cerebellar and dorsal column nuclei upon the thalamus of the rhesus monkey. J Comp Neurol 195: 25–50
Landgren S, Olsson KÅ (1980) Low threshold afferent projections from the oral cavity and the face to the cerebral cortex of the cat. Exp Brain Res 39: 133–147
Landgren S, Silfvenius H (1969) Projection to cerebral cortex of group I muscle afferents from the cat's hindlimb. J Physiol (Lond) 200: 353–372
Landgren S, Silfvenius H (1970) The projection of group I muscle afferents from the hindlimb to the contralateral thalamus of the cat. Acta Physiol Scand 80: 10A
Landgren S, Silfvenius H (1971) Nucleus Z, the medullary relay in the projection path to the cerebral cortex of group I muscle afferents from the cat's hind limb. J Physiol (Lond) 218: 551–571
Landgren S, Nordwall A, Wengström C (1965) The location of the thalamic relay in the spino-cervico-lemniscal path. Acta Physiol Scand 65: 164–175
Landgren S, Silfvenius H, Wolsk D (1967a) Somato-sensory paths to the second cortical projection area of the group I muscle afferents. J Physiol (Lond) 191: 543–559
Landgren S, Silfvenius H, Wolsk D (1967b) Vestibular, cochlear and trigeminal projections to the cortex in the anterior suprasylvian sulcus of the cat. J Physiol (Lond) 191: 561–573
Larsen KD, Asanuma H (1979) Thalamic projections to the feline motor cortex studied with horseradish peroxidase. Brain Res 172: 209–215
Liedgren SRC, Milne AC, Rubin AM, Schwarz DWF, Tomlinson RD (1976) Representation of vestibular afferents in somatosensory thalamic nuclei of the squirrel monkey (*Saimiri scirueus*). J Neurophysiol 39: 601–612
Lucier GE, Rüegg DC, Wiesendanger M (1975) Responses of neurones in motor cortex and in area 3a to controlled stretches of forelimb muscles in cebus monkeys. J Physiol (Lond) 251: 833–853
Lundberg A, Voorhoeve P (1962) Effects from the pyramidal tract on spinal reflex arcs. Acta Physiol Scand 56: 201–219
Mallart A (1964) Projection des afférences musculaires de la patte antérieure au niveau du thalamus chez le chat. Cr Seances Acad Sci 295: 1215–1218
Maendly R, Rüegg DG, Wiesendanger M, Wiesendanger R, Lagowska J, Hess B (1981) Thalamic relay for group I muscle afferents of forelimb nerves in the monkey. J Neurophysiol 46: 901–917
Mountcastle VB (1957) Modality and topographic properties of single neurons of cat's somatic sensory cortex. J Neurophysiol 20: 408–434

Mountcastle VB, Lynch JC, Georgopoulos A, Sakata H, Acuna A (1975) Posterior parietal association cortex of the monkey: command function for operations within extrapersonal space. J Neurophysiol 38: 871–908
Norrsell U (1978) Sensory defects caused by lesions of the first (SI) and second (SII) somatosensory areas of the dog. Exp Brain Res 32: 181–195
Norrsell U (1980) Behavioral studies of the somatosensory system. Physiol Rev 60: 327–354
Ödqvist L, Larsby B, Fredrickson JM (1975) Projection of the vestibular nerve to the SI arm field in the cerebral cortex of the cat. Acta Otolaryngol (Stockh) 79: 88–95
Olsson KÅ, Landgren S (1980) Facilitation and inhibition of jaw reflexes evoked by electrical stimulation of the cat's cerebral cortex. Exp Brain Res 39: 149–164
Oscarsson O, Rosén I (1963) Projection to cerebral cortex of large muscle spindle afferents in fore limb nerves of the cat. J Physiol (Lond) 169: 924–945
Oscarsson O, Rosén I (1966) Short-latency projections to the cat's cerebral cortex from skin and muscle afferents in the contralateral forelimb. J Physiol (Lond) 182: 164–185
Oscarsson O, Rosén I, Sulg I (1966) Organization of neurones in the cat cerebral cortex that are influenced from group I muscle afferents. J Physiol (Lond) 183: 189–210
Phillips CG (1969) Motor apparatus of the baboon's hand. Proc R Soc [Biol] 173: 141–174
Phillips CG, Powell TPS, Wiesendanger M (1971) Projection from low threshold muscle afferents of hand and forearm to area 3a of baboon's cortex. J Physiol (Lond) 217: 419–446
Poggio FG, Mountcastle VB (1963) The functional properties of ventrobasal thalamic neurons studied in unanesthetized monkeys. J Neurophysiol 26: 775–806
Randolph M, Semmes J (1974) Behavioral consequences of selective subtotal ablation in the postcentral gyrus of *Macaca mulatta*. Brain Res 70. 55: 70–
Rosén I (1969a) Localization in caudal brain stem and cervical spinal cord of neurones activated from forelimb group I afferents in the cat. Brain Res 16: 55–71
Rosén I (1969b) Afferent connections to group I activated cells in the main cuneate nucleus of the cat. J Physiol (Lond) 205: 209–236
Rosén I (1969c) Excitation of group I activated thalamocortical relay neurones in the cat. J Physiol (Lond) 205: 237–255
Silfvenius H (1968) Cortical projections of large muscle afferents from the cat's forelimb. Acta Physiol Scand 74: 25–26A
Silfvenius H (1970) Projections to the cerebral cortex from afferents of the interosseous nerves of the cat. Acta Physiol Scand 80: 196–214
Silfvenius H (1972a) Properties of cortical group I neurons located in the lower bank of the anterior suprasylvian sulcus of the cat. Acta Physiol Scand 84: 555–577
Silfvenius H (1972b) Projections to the cat cerebral cortex from fore- and hindlimb group I muscle afferents. Umea Univerity Medical Dissertations No. 4
Swett JE, Bourassa CM (1967) Short latency activation of pyramidal tract cells by group I afferent volleys in the cat. J Physiol (Lond) 189: 101–117
Vogt C, Vogt O (1919) Allgemeine Ergebnisse unserer Hirnforschung. J Psychol Neurol (Lpz) 25: 277–462
Waters RS, Favorov O, Asanuma H (1982) Physiological properties and pattern of projection of cortico-cortical connections from the anterior bank of the ansate sulcus to the motor cortex, area 4, in the cat. Exp Brain Res 46: 403–412
Wiesendanger M, Miles TS (1982) Ascending pathway of low-threshold muscle afferents to the cerebral cortex and its possible role in motor control. Physiol Rev 62: 1234–1270
Woolsey CN (1958) Organization of somatic sensory and motor areas of the cerebral cortex. In: Harlow HF, Woolsey CN (eds) Biological and biochemical bases of behaviour. University of Wisconsin Press, Madison, pp 63–81
Yin TCT, Williams WJ (1976) Dynamic response and transfer characteristics of joint neurones in somatosensory thalamus of the cat. J Neurophysiol 39: 582–600
Zarzecki P, Shinoda Y, Asanuma H (1978) Projection from area 3a to the motor cortex by neurones activated from group I muscle afferents. Exp Brain Res 33: 269–282

Neuronal Organization of Cat Motor Cortex

TOMOKAZU OSHIMA, Tokyo

Summary. Spatiotemporal patterns of neuronal activities within the cat motor cortex were analyzed during EEG arousal by the intracellular recording technique. A cascade transmission model was proposed to interpret different neuronal activities as a functional ensemble. The neuronal organization of the motor cortex was discussed in relation to the behavioral context of an initial motor arrest followed by orienting and searching movements.

Key words: *cortical arousal – motor cortex – cascade transmission model*

Introduction

Application of the intracellular recording technique has been extended to relatively small-sized neurons acting under various physiological conditions of the mammalian central nervous system. Nowadays, all the constituent neurons of the cerebral cortex can be examined in a given physiological state. For a systematic analysis we chose the cortical EEG arousal reaction. It is an expression in the brain of behavioral arousal or the orienting response, perhaps involving a primitive form of animal awareness. We can repeatedly induce it by means of natural or electrical stimulation in transected brain preparations suitable for an intracellular study. Behaviorally, a typical arousal reaction is initiated by a transient phasic phase characterized by arrest or diminution of ongoing activities in response to a novel stimulus, and is followed by a later tonic phase of orienting and searching in the environment. These two phases are manifested also in various neuronal responses, in which we have become interested as the first neuronal sign triggering a voluntary act.

Methods

Torii and his sleep research group, who had been working for years with chronically prepared cats, transected the brain under ether anesthesia at the midpontine level (pretrigeminal preparation) or the upper cervical cord [*encéphale isolé* (isolated encephalus) preparation] (Inubushi et al. 1978a, b). The head of the cat was fixed in a stereotaxic apparatus atraumatically with a device originally designed for the chronic experiment. After ether was discontinued, the preparation was immobilized by gallamine triethiodide, artificially ventilated, and carefully main-

Experimental Brain Research, Suppl. 9
© Springer-Verlag Berlin · Heidelberg 1984

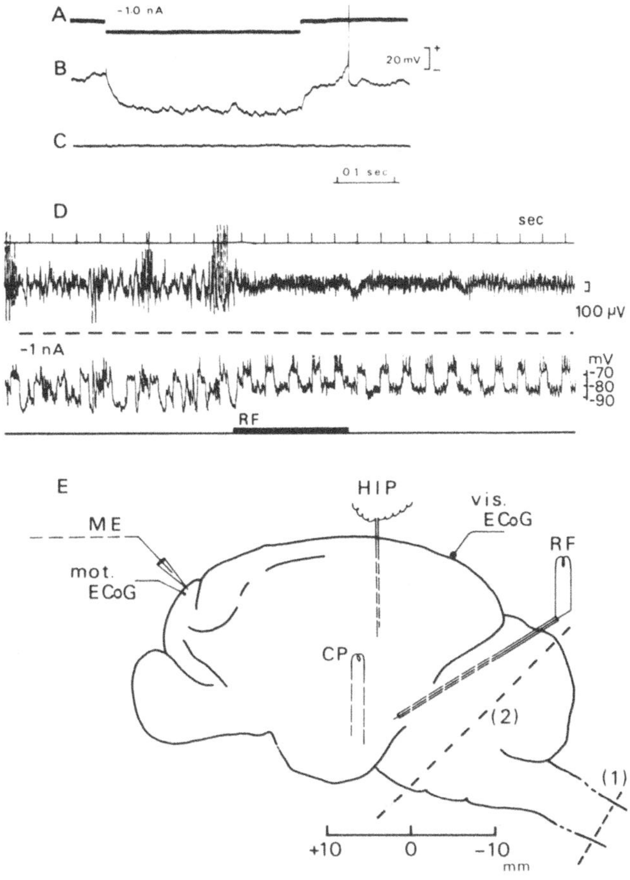

Fig. 1A–E. Measurement of effective membrane impedance by *A* passing a hyperpolarizing current step and *B* recording deflections in membrane potential with *C* a control in extracellular space. *D* illustrates an episode of reticular-induced EEG arousal in a polygraphic record including, *from top to bottom*, a time scale in seconds, electrocorticogram (ECoG) in motor area, marks of current passage, membrane potential, and reticular stimulation indicated by a *thick bar (RF)*. *E* shows experimental setups of recording electrodes of motor ECoG, visual ECoG, hippocampal activity (*HIP*), microelectrode (*ME*), and stimulating electrodes in cerebral peduncle (*CP*) and midbrain reticular formation (*RF*). *Scale* for anteroposterior axis is given. (*1*) and (*2*), transections for *encéphale isolé* and pretrigeminal preparations respectively. (Modified from Inubushi et al. 1978a, b)

tained in a dimly lit, sound-attenuated, shielded room. The encéphale isolé preparation exhibited spontaneous alternation of the well-known sleep and waking EEG patterns, and was very sensitive to external stimuli. The pretrigeminal preparation was more stabilized and suitable for the stimulation experiment.

While EEG activities were being monitored at several sites (Fig. 1E), a glass micropipette filled with 2 M KCH_3SO_4 or 2 M potassium citrate solution was inserted perpendicularly into the precruciate gyrus to record from as many cells as possible in all the cortical layers in single trackings. Since the physiological state was well maintained, the intracellular technique was most successful in recording natural fluctuations of the membrane potential. The effective membrane impedance was measured by injecting hyperpolarizing current steps every second (Fig. 1A–C), so that the membrane potential record showed two levels which thus oscillated (D). Four response types, excitation (E), inhibition (I), disfacilitation (DF), and disinhibition (DI), were identified by combining changes in the membrane potential and the impedance. In addition, we counted the rate of occurrence of unitary synaptic potentials to assess the identification.

Results

In the case illustrated in Fig. 1D we stimulated the midbrain reticular formation with a high-frequency train of pulses. The membrane potential was then depolarized with decreases of the membrane impedance. This response was identified as synaptic excitation. The arousal effect outlasted the period of stimulation, while this cell exhibited the excitatory response for a long time.

Double phase responses were first demonstrated in large-sized fast pyramidal tract cells or the so-called Betz cells. Their membrane potential was initially hyperpolarized and later depolarized. The membrane impedance was increased first, revealing response by disfacilitation, and was decreased later, exhibiting excitation (Inubushi et al. 1978b; Ezure and Oshima 1981a).

To sum up, in a series of experiments (Ezure and Oshima 1981a, b, c), a total of 164 cells were classified according to their responses in two phases of EEG arousal. These 164 cells showed either E, I, DF, or DI responses during the phasic arousal. We identified 71 E, 49 I, 38 DF, and 6 DI cells. In the subsequent tonic arousal 32 cells failed to respond, being termed "late unresponsive." Other cells were identified as 102 late E (+E), 16 late I (+I), 12 late DF (+DF), and 2 late DI (+DI) cells. There were nine types of combination of the initial and late responses. Forty E+E, 10 E+I, 10 E+DF, 22 I+E, 6 I+I, 2 I+DF, 2 I+DI, 34 DF+E, and 6 DI+E cells were identified. These various response patterns were complicated, but their depth distributions showed unique topographical patterns.

In phasic arousal (Fig. 2A) we found only E cells in the uppermost laminae, I–II, both E and I cells in the upper half of lamina III (III_a), and E, I, DF, and DI cells from the lower half of the lamina III (III_b) to V and VI. Together with other evidence of the response time courses, we postulated a simple model circuit (Fig. 2B) in which cascade transmissions are initiated from the uppermost layer cells excited by the nonspecific afferent inputs and are processed toward the deep layers. Excitatory and inhibitory cells in this model are connected with all possible combinations, and the more complex responses are produced in the course of downward processing. This cascade transmission model may also involve additional positive feedback loops, which intensify the basic cascade pattern even after the afferent impulses are discontinued (Inubushi et al. 1978b).

One problem with these oligosynaptic relays was the time domain of processing arousal, which took as long as a few seconds. Then we asked: when we apply the single electrical pulse stimulation directly at a certain site on the surface of the cortex, can we observe a quick rehearsal of the cascade transmissions through

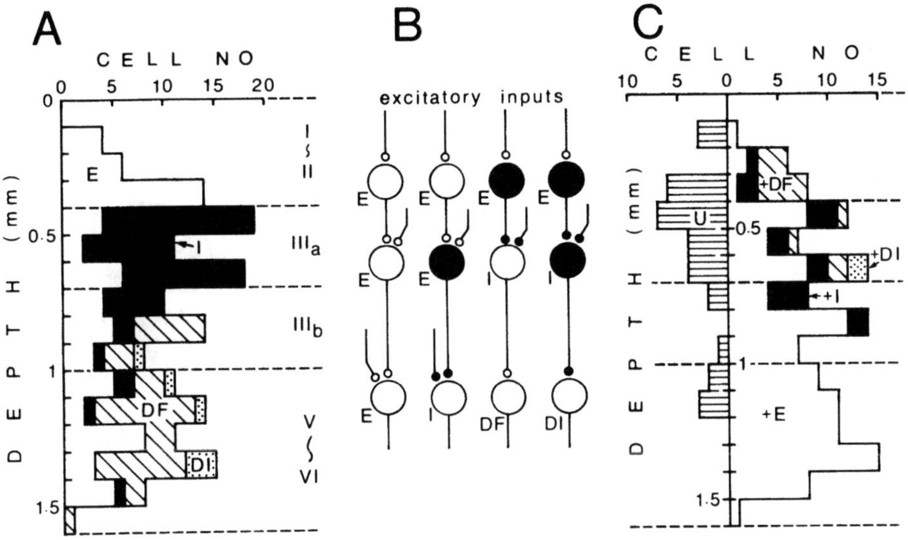

Fig. 2A–C. Depth distributions of cells with identified responses: E (*open columns*), I (*filled*), DF (*hatched*), DI (*stippled*), and U (*horizontally shaded*) in *A* phasic and *C* tonic arousal. *B* A cascade model of neuron relays with excitatory (*open symbols*) and inhibitory cells (*filled*). (Modified from Ezure and Oshima 1981c)

these relay connections? And are the responses to this epicortical stimulation identical in each cell with the type of response in the phasic arousal? Affirmative results have been obtained in another series of experiments (Ezure, Oguri and Oshima, unpublished observations).

During tonic arousal the laminar organization of cellular responses was drastically changed (Fig. 2C). There were +E cells in laminae V and VI, +E and +I cells in III_b, and cells of any of the four response types more superficially, so that the responses became more complex toward the cortical surface. Late unresponsive cells (U) were concentrated more superficially. We postulated that the cascade transmissions were brought about by the afferent inputs impinging deeply, and processed them from deep to superficial layers (Ezure and Oshima 1981c). A small number of late unresponsive cells indicated that the nonspecific arousal process was then taken over by a more specific process in the selected cell groups. From phasic to tonic arousal, we thus saw a dynamic reversal of the direction in the cascade transmissions. This transition would be determined by a balance in synaptic bombardments from two separate afferent systems with more terminal branches in the superficial and deep layers, as well as by the intracortical mechanisms involving refractoriness and recycling of impulses in the neuron relays (Ezure and Oshima 1981c).

Discussion

The functional implications have been variously understood (Inubushi et al. 1978b; Ezure and Oshima 1981c). For example, disfacilitation of fast pyramidal

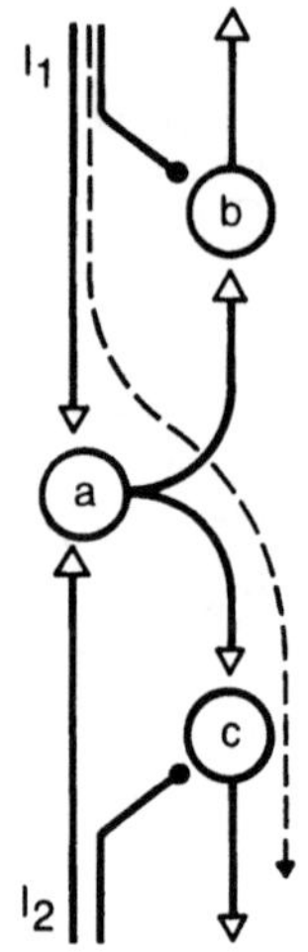

Fig. 3

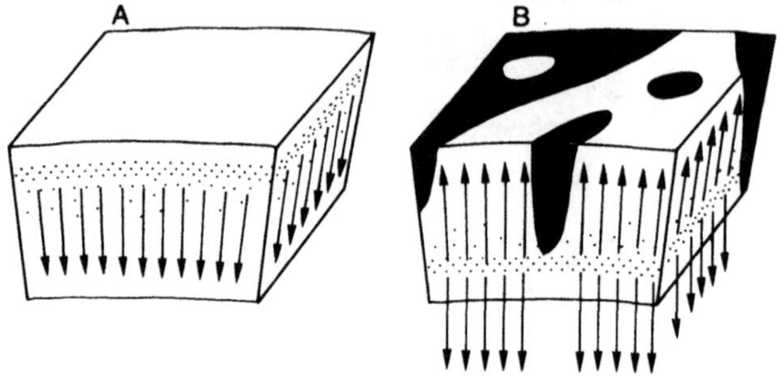

Fig. 4

Fig. 3. A model neuron (*a*) with two inputs (I_1, I_2) and two target neurons (*b*, *c*) is illustrated as participating in either downward (directed by a *broken line*) or upward neural processing. Excitatory and inhibitory axon terminals are indicated by *open triangles* and *filled circles* respectively.

Fig. 4A,B. Dynamic cerebral states in *A* phasic and *B* tonic arousal. (Oshima 1981)

tract cells during phasic arousal was interpreted as a "resetting" of their activity. The increased membrane impedance would be effective in building up greater depolarization to the subsequent excitatory drive, and elimination of ongoing excitatory synaptic noise would improve the signal-to-noise ratio. This disfacilitation showed a peculiar "all-or-none" property in response to the reticular stimuli with different intensities (Ezure and Oshima 1981a), perhaps in the manner of releasing an innately fixed central program of motor arrest.

The cascade transmission model of arousal provides a good example of a cooperative system in which various constituent neurons show different activity patterns in a single behavioral context. Furthermore, each of these neurons behaves differently in two phases of arousal. For example, most superficial layer cells are excited as the first recipient of nonspecific afferents on phasic arousal, but later exhibit one of the four response types with possibly more specific nature during tonic arousal. Figure 3 illustrates a situation of a neuron (a) in our model that can play double roles in processing neural information from an upper input (I_1) to lower output (neuron c) and other information in a reversed direction ($I_2 \rightarrow$ b). Therefore, even in the cortical network with a rigidly fixed anatomic framework, physiologists might see a broad capacity of functioning because each cell could be utilized in a polyfunctional fashion. The laminar structure with multiple inputs may enable this capacious mode of operation. Referring to his earlier n-dimensional network, Eccles maintained in the book *The Self and Its Brain*: "There are relatively limited numbers of cells in the brain, but the opportunities for pattern, the potentialities for pattern are enormously greater than the cell numbers. Theories of pattern will have to be developed especially for this purpose" (p. 466 in

Popper and Eccles 1977). Eccles' suggestion was not taken seriously by physiologists as a working hypothesis because it was not technically possible to test it. Although we wonder if an example of cortical arousal is inadequate to prove this hypothesis, the function of single neurons should be viewed differently in our future studies according to various behavioral contexts.

To emphasize the developmental aspect of arousal, we propose a stereoscopic view (Fig. 4). The phasic arousal starts with excitation in the superficial layer cells represented by dots (Fig. 4A). Excitation spreads horizontally and downward (arrows). All the cortical neurons participate in arousal by one of four different actions. The tonic arousal (Fig. 4B) starts with excitation of deep layer cells (dots). The cascade transmissions occur upward within the cortex, a part of which, as illustrated with filled blocks with late unresponsive cells, remains inactive. Simultaneously, excitation of deep layer cells is processed as the cortical outputs down to the subcortical target nuclei. Development from phasic to tonic phases might be further advanced to a more specific pattern in which the minimum functional units such as the cortical columns or modules take part, as Eccles has illustrated in many of his books and papers (cf. Popper and Eccles 1977). We analyzed a cortical network only in a confined area of the cat motor cortex. However, it is to be hoped that various dynamic patterns of neuronal activities will be found in different cerebral states by further application of the structuralistic approach.

Acknowledgments. The author wishes to thank Professor Shizuo Torii, Mr. Shikio Inubushi, Dr. Toshinori Kobayashi, Dr. Kazuhisa Ezure, and Miss Takako Sato for their cooperation in this study. This research was supported by the Ministry of Education, Science and Culture of Japan.

References

Ezure K, Oshima T (1981a) Dual activity patterns of fast pyramidal tract cells and their family neurones during EEG arousal in the cat. Jpn J Physiol 31: 717–736

Ezure K, Oshima T (1981b) Excitation of slow pyramidal tract cells and their family neurones during phasic and tonic phases of EEG arousal. Jpn J Physiol 31: 737–748

Ezure K, Oshima T (1981c) The structure of EEG arousal as a dynamic ensemble of neuronal activities in cat motor cortex. Jpn J Physiol 31: 809–829

Inubushi S, Kobayashi T, Oshima T, Torii S (1978a) Intracellular recordings from the motor cortex during EEG arousal in unanaesthetized brain preparations of the cat. Jpn J Physiol 28: 669–688

Inubushi S, Kobayashi T, Oshima T, Torii S (1978b) An intracellular analysis of EEG arousal in cat motor cortex. Jpn J Physiol 28: 689–708

Oshima T (1981) Cortical neurones in arousal. Jpn J EEG EMG [Suppl]: 73–78

Popper KR, Eccles JC (1977) The self and its brain. Springer, Berlin Heidelberg New York, pp 235–244, 466

The Supplementary Motor Area in the Light of Recent Investigations

M. WIESENDANGER and R. WIESENDANGER, Fribourg

Summary. The supplementary motor area (SMA), according to the hypothesis of Eccles (1982), plays a crucial role in movement initiation. The proposition is based mainly on cerebral blood flow measurements and brain potential recordings in man. A number of single-unit recordings in monkeys are also compatible with the hypothesis. We here adduce evidence from anterograde and retrograde tracing studies in monkeys that the SMA has far-reaching connections with cortical and subcortical structures. Among the former are inputs from the frontal eye field, the superior temporal and intraparietal sulci and the cingulate cortex and outputs to these same structures, as well as to the prefrontal cortex. The SMA also projects heavily to the caudate nucleus and, less importantly, to the pontine nuclei. Among the "non-motor" inputs (in the traditional sense) are the nuclear complex of the amygdala and the thalamic mediodorsal nucleus. All these connections may play a significant role in movement initiation.

On the other hand, the SMA is also directly connected with the spinal cord. The presence of a fair number of corticospinal cells and the demonstration of positive microstimulation effects underscores the relatively potent control of the SMA on the motor apparatus. It is therefore inferred that the SMA, in addition to its proposed role in movement initiation, also participates in the immediate control of movement execution. In particular, we advance the speculation that the motor control function of the SMA via direct and indirect pathways to the spinal cord might be utilized for anticipatory postural adjustments during volitional movements. The command signals for such anticipatory adjustments must be released at an early stage of movement initiation.

Key words: *Supplementary motor area – voluntary movements – motor cortex – movement initiation*

Introduction

Three lines of research have considerably changed our views on the possible functions of the supplementary motor area (SMA): first, anatomical and electrophysiological investigations of its connections; secondly, studies of single neurons in behaving monkeys; and thirdly, cerebral blood flow measurements and slow potential recordings in man. On the basis of the main results of these investigations, Eccles (1982) developed the hypothesis that "in all voluntary movements the initial neural event is generated in the supplementary motor area of both sides". The arguments are as follows:

Experimental Brain Research, Suppl. 9
© Springer-Verlag Berlin · Heidelberg 1984

1. Cerebral blood flow shows local increases bilaterally in the region of the SMA when patients perform non-trivial, self-paced movements or even during movement "mentation" (Roland et al. 1980).
2. Readiness potentials occurring before movement onset are found to be prominent over the region of the SMA even in cases of reduced motor cortical function (Deecke and Kornhuber 1978).
3. Many neurons of the SMA show changes in their firing rate well before motor cortical cells in a complex arm movement (Brinkman and Porter 1979).

In 1973 we briefly reviewed the then available knowledge about the SMA (Wiesendanger et al. 1973). Since the detailed electrical stimulation experiments performed by Woolsey et al. (1952), the SMA had been viewed essentially as a second, parallel motor cortex with a detailed somatotopically organized outflow to the spinal cord, a concept which was pictorially represented with the well-known simiusculus on the medial wall of the cerebral cortex. Lesion experiments in animals had not contributed any new concept about the function of the SMA. Some, mainly clinical, observations suggested to us that the SMA might be implicated in postural control. However, we furthermore advanced the speculation, also on the basis of observations in humans, that the SMA, together with the premotor cortex, may also play a role in programming movements. The possibility was therefore considered that the SMA might be involved at a higher level as well as at a lower level of motor control. In the detailed account of the structure and function of the SMA in the *Handbook of Physiology* (Wiesendanger 1981), the proposed double function of the SMA was developed further. In the following, we wish to summarize more recent anatomical and microelectrophysiological results which are compatible with the proposition that the SMA may be implicated in the programming of movements as well as in a direct control of spinal circuits.

The Direct Relationship of the SMA with the Spinal Cord

Previous degeneration and autoradiographic studies have yielded equivocal results concerning a direct projection to the spinal cord, and the number of corticospinal neurons in medial area 6 has been considered to be of minor importance (cf. Wiesendanger 1981). However, introduction of retrograde labelling techniques has revealed beyond any doubt the presence of a considerable number of corticospinal cells in the SMA (Fig. 1; cf. also Biber et al. 1978; Murray and Coulter 1981; Macpherson et al. 1982a, b). Furthermore, and contrary to previous negative findings, microstimulation, performed in minute detail in the awake monkey, revealed a zone in the SMA from which positive effects were observed at stimulation intensities below 30μA and as low as 10μA (Macpherson et al. 1982a). These anatomical and electrophysiological observations suggested to us that at least part of the SMA has a relatively tight coupling with the spinal cord. This seems to confirm results reported by Woolsey et al. (1952). However, some notable differences in the organization of the descending neurons were observed: like Penfield and Welch (1951), we were unable to detect a clear somatotopy with the head representation rostrally and the legs in area 4 beneath the primary leg representation as had been described by Woolsey et al. (1952). Instead, lumbar

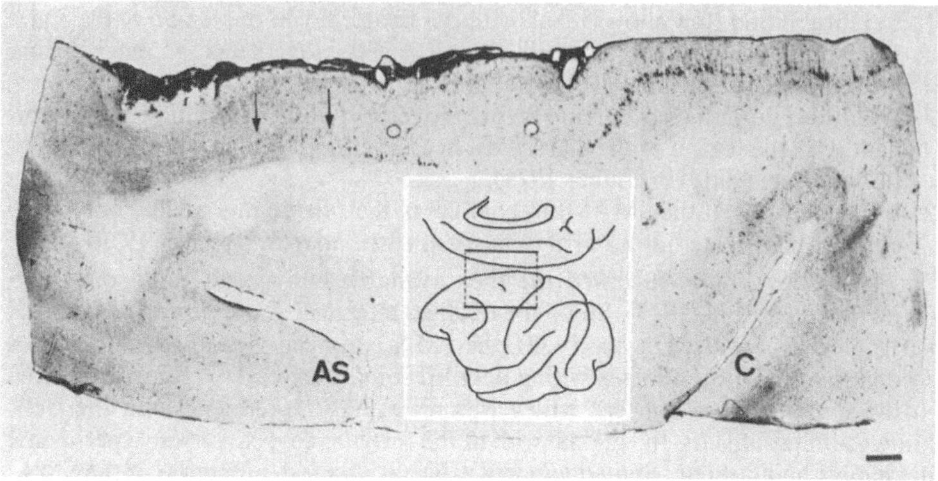

Fig. 1. Labelling of corticospinal neurons following injections of the retrograde tracer HRP into the cervical cord of *Macaca fascicularis.* Note labelling of cells in medial area 6 of the SMA, the most rostral corticospinal cells are marked by arrows. *Inset* shows the orientation of the horizontal section. TMB-treated horizontal section counterstained with thionin. *Calibration bar,* 1 mm; *AS,* arcuate sulcus; *C,* central sulcus.

cord injections with the retrograde marker horseradish peroxidase (HRP) labelled cells in the entire rostro-caudal extent of medial area 6 (Macpherson et al. 1982a). Microstimulation sites producing hind limb effects were intermingled with sites producing forelimb effects. Similarly, neurons responding antidromically to stimulation of the dorsolateral funiculus of the upper lumbar cord were found in the caudal as well as in the rostral part of medial area 6 (Macpherson et al. 1982b). Surprisingly (but in accord with Penfield and Welch 1951), we were unable to elicit any face movements. Such a negative finding must, of course, be regarded with caution; however, recent anterograde studies of the SMA with the sensitive marker HRP conjugated to the lectin wheat germ agglutinin (WGA) equally failed to reveal a projection to the facial nucleus or to its immediate surroundings (unpublished observations). It may be that the motor effects observed

---

Fig. 2A–F. Some SMA projection patterns to the caudate nucleus and thalamic nuclei. *A* Schematic drawing of the projection site in the caudate nucleus. *B* Field of ipsilateral anterograde labelling. A nearly symmetrical contralateral projection field was also seen (not shown). *C* Schematic drawing of thalamic nuclei VA/VLo to indicate the place where the photomicrograph shown in *D* was taken (*arrow*). *D* Small island of thalamic cells in the border zone of VA and VLo containing terminal fibre labelling as well as retrogradely labelled cells. Edge of a larger field of heavy anterograde labelling on the right. *E* Projection territory shown in *F*: typical thalamic lamella containing anterograde labelling as well as retrogradely filled cells. The lamella extends from dorsal VLc to turn medially along the border of area X. Small injection of HRP, conjugated to the lectin WGA, injected into the SMA (*Macaca fascicularis*). TMB-treated sections counterstained with neutral red. *Calibration bars,* 1 mm (*A–E*), 0.1 mm (*F*). *Cd,* nucleus caudatus; *GP,* globus pallidus; *P,* putamen; *NR,* nucleus ruber; *STh,* nucleus subthalamicus; *VA,* nucleus ventralis anterior thalami; *mc,* pars magnocellularis; *pc,* pars parvocellularis; *VLm,* nucleus ventralis lateralis thalami, pars medialis; *VLo,* nucleus ventralis lateralis thalami, pars oralis; *VPLo,* nucleus ventralis posterior lateralis thalami, pars oralis; *X,* area X thalami.

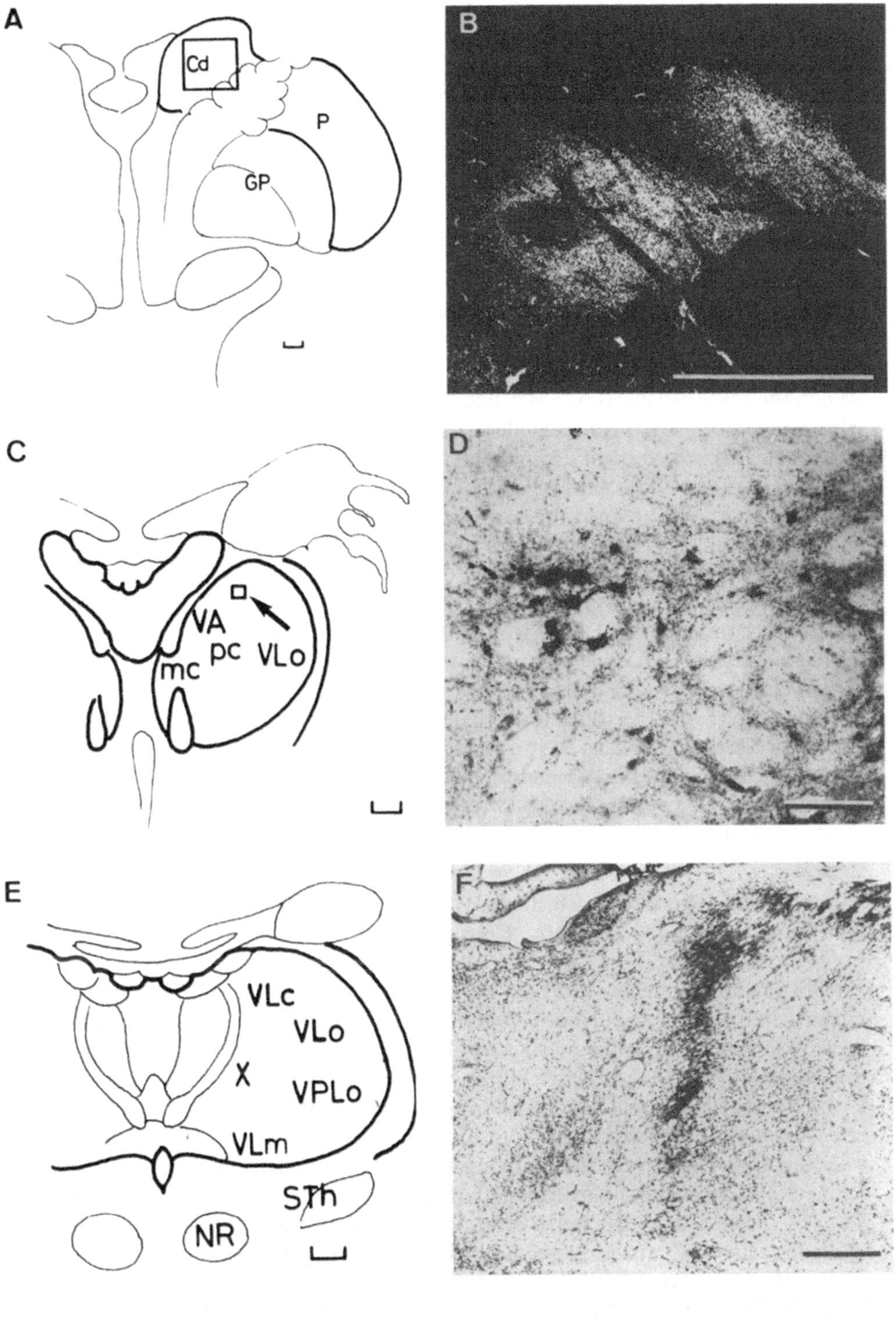
A
Cd
P
GP
B
C
VA
pc
VLo
mc
D
E
VLc
VLo
X
VPLo
VLm
STh
NR
F

in earlier studies (which were produced by strong and long-lasting repetitive stimulation of the cortical surface) were mediated partly via the motor cortex or via other complex routes. In summary, our results suggest that the descending spinal projection from the SMA is organized in discrete efferent zones controlling proximal as well as distal muscles of contralateral forelimb and hind limb, and that these zones are not organized in a rostrocaudal order for the forelimb and hind limb respectively (rather, the forelimb zones tended to be "sandwiched" dorsally and ventrally by hind limb zones). As yet, we have no information as to the laminar distribution of corticospinal terminals at segmental level.

Efferent Projection of the SMA to Other Motor Structures

There is also evidence that the SMA has indirect connections with the spinal cord. When the responsiveness of SMA neurons was tested with a series of stimulating electrodes along the presumed descending trajectory, it was found that many more neurons projected to supraspinal structures than to spinal levels (Macpherson et al. 1982b). The supraspinal target structures of these descending SMA neurons may have included the corpus striatum, the motor thalamus, the red nucleus, the tegmental reticular formation and the pontine nuclei. These are all structures known to receive afferents from the SMA (see e.g. DeVito and Smith 1959; Kuypers and Lawrence 1967; Dhanarjan et al. 1977). We have reinvestigated the connections of the SMA by means of the sensitive anterograde and retrograde tracer HRP conjugated to the lectin WGA. We have confirmed and extended the results of earlier studies. Figures 2 and 3 illustrate some typical projection patterns to the caudate nucleus, the motor thalamus, the parvocellular portion of the red nucleus, the pontine nuclei and the inferior olivary nucleus. A summary diagram of the descending SMA projections is provided in Fig. 4. In addition, the SMA projects to ipsilateral cortical areas also involved in motor control: the premotor cortex and the frontal eye field (cf. also Fig. 6A).

Afferent and Efferent Connections of the SMA with Structures Not Directly Related to Motor Control

The following is not meant as an exhaustive account of the connectivity of the SMA (the material is currently prepared for publication elsewhere); rather we here wish to emphasize some structural relations which may shed some light on the assumed "higher functions" of the SMA.

With respect to cortical inputs to the SMA from subcortical structures which are "non-motor" in a traditional sense, our retrograde labelling study revealed ascending projections from the thalamic mediodorsal nucleus, the amygdalar complex, the nucleus basalis of Meynert, the raphe nuclei and the locus coeruleus (Fig. 5). Cortical afferents to the SMA from areas outside the sensorimotor cortex include prefrontal cortex, cingulate cortex, and the depth of the superior temporal and of the intraparietal sulci (Fig. 6B). Relatively weak corticocortical projections from the SMA to the prefrontal, cingulate, parietal and temporal association cortices were also seen (Fig. 6A).

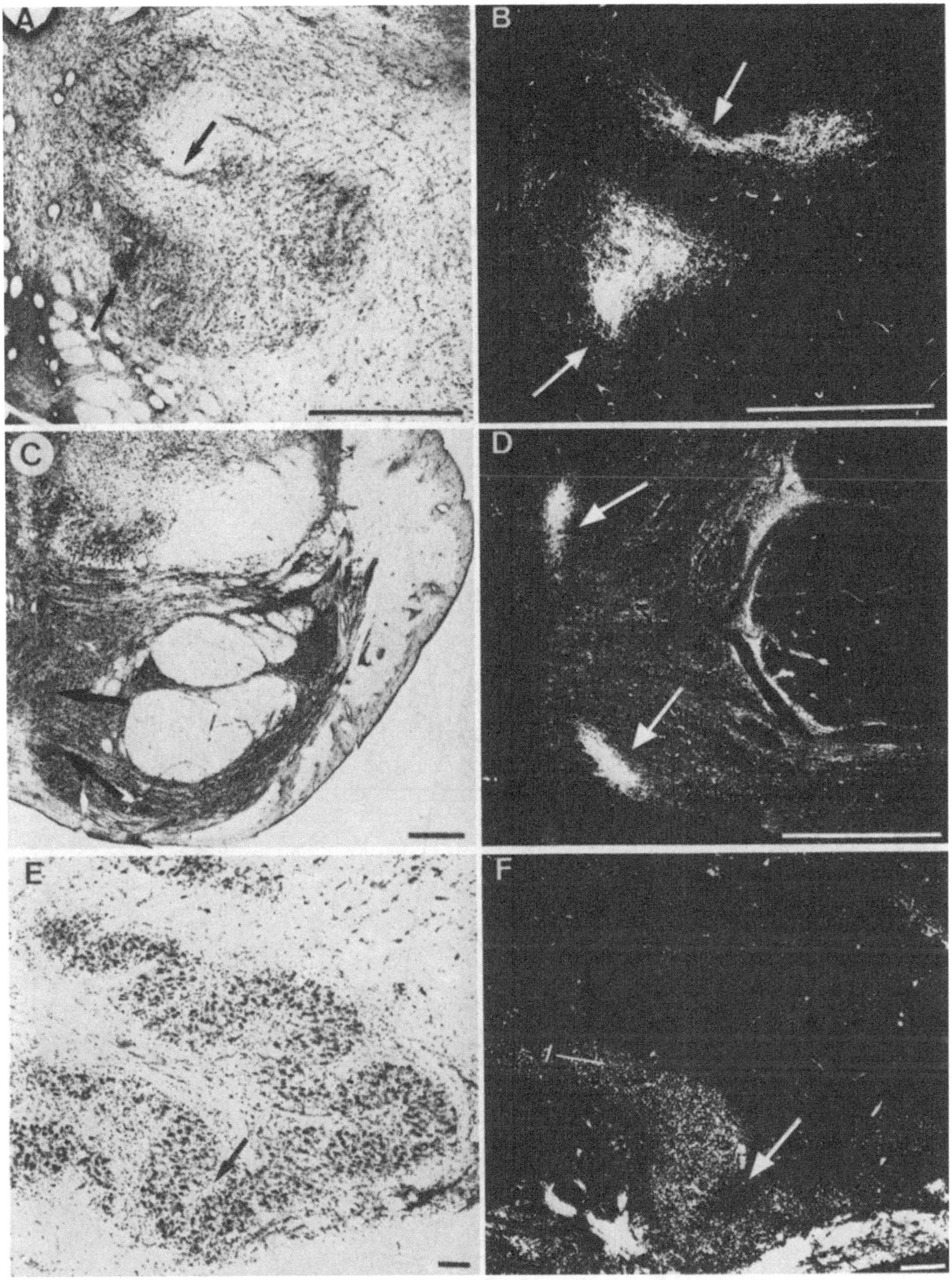

Fig. 3. A Bright field photomicrograph to show anterograde labelling in the ipsilateral red nucleus, pars parvocellularis (TMB reaction, neutral red counterstaining). *B* Dark field micrograph of the same projection area. *C* Nissl-stained transverse section of the pontine nuclei (*PN*) (*left*). *D* Two intense patches of the SMA projection in the paramedian portion of the PN (same region as in *C*). *E* Transverse Nissl section of the inferior olive. *F* Terminal labelling in the ipsilateral principal nucleus, ventral lamella (same region as in *E*). The relatively weak projection extended over more than half of the rostrocaudal extent of the olivary nucleus. Same case as in Fig. 2. *Calibration bars*, 1 mm (*A–D*) and 0.1 mm (*E, F*). Arrows mark patches of anterograde labelling (*right*) and corresponding sites in Nissl sections (*left*).

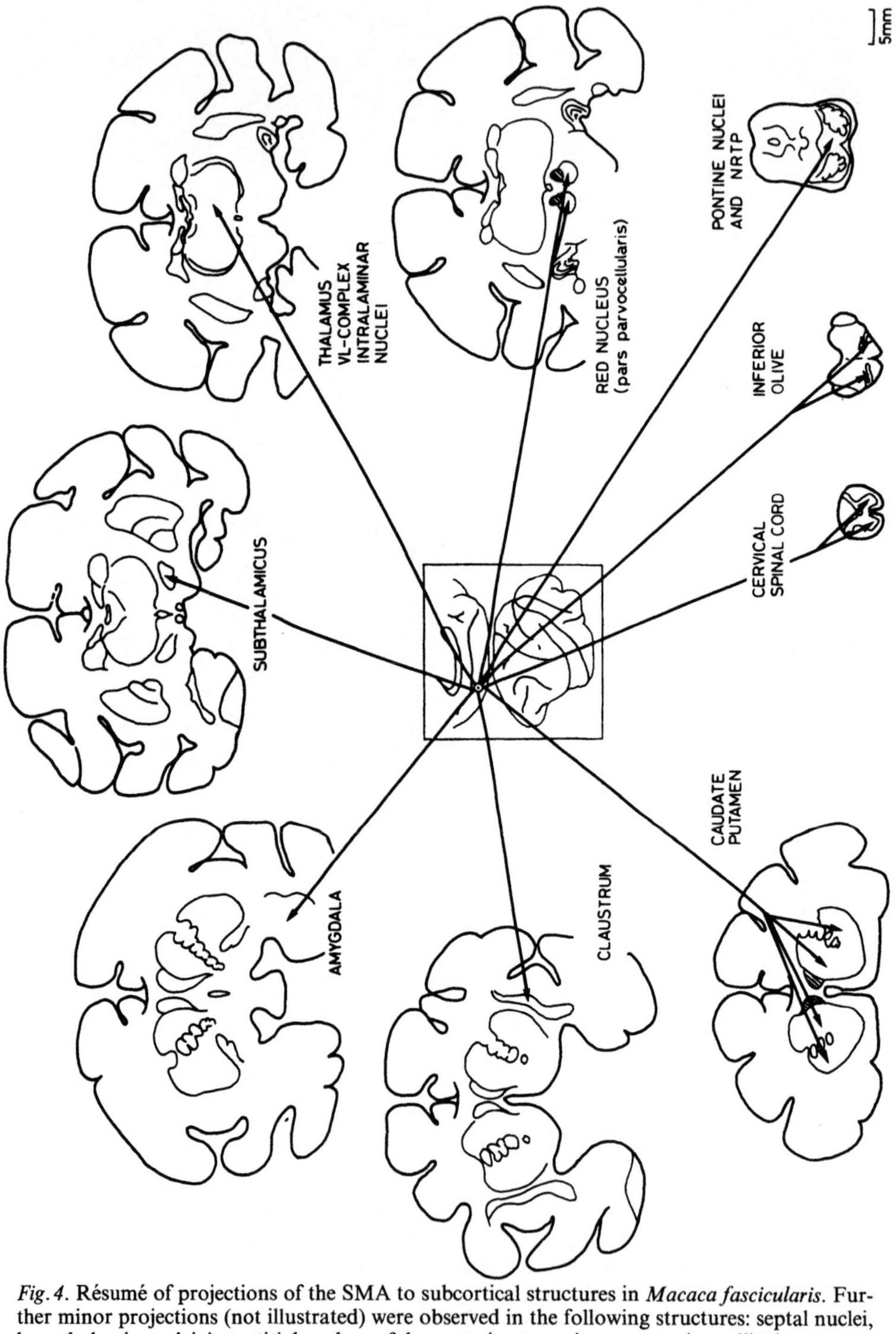

Fig. 4. Résumé of projections of the SMA to subcortical structures in *Macaca fascicularis*. Further minor projections (not illustrated) were observed in the following structures: septal nuclei, hypothalamic nuclei, interstitial nucleus of the posterior commissure, superior colliculus, mesencephalic, pontine and medullary reticular formation. *NRTP*, nucleus reticularis tegmenti pontis; *VL*, nucleus ventralis lateralis.

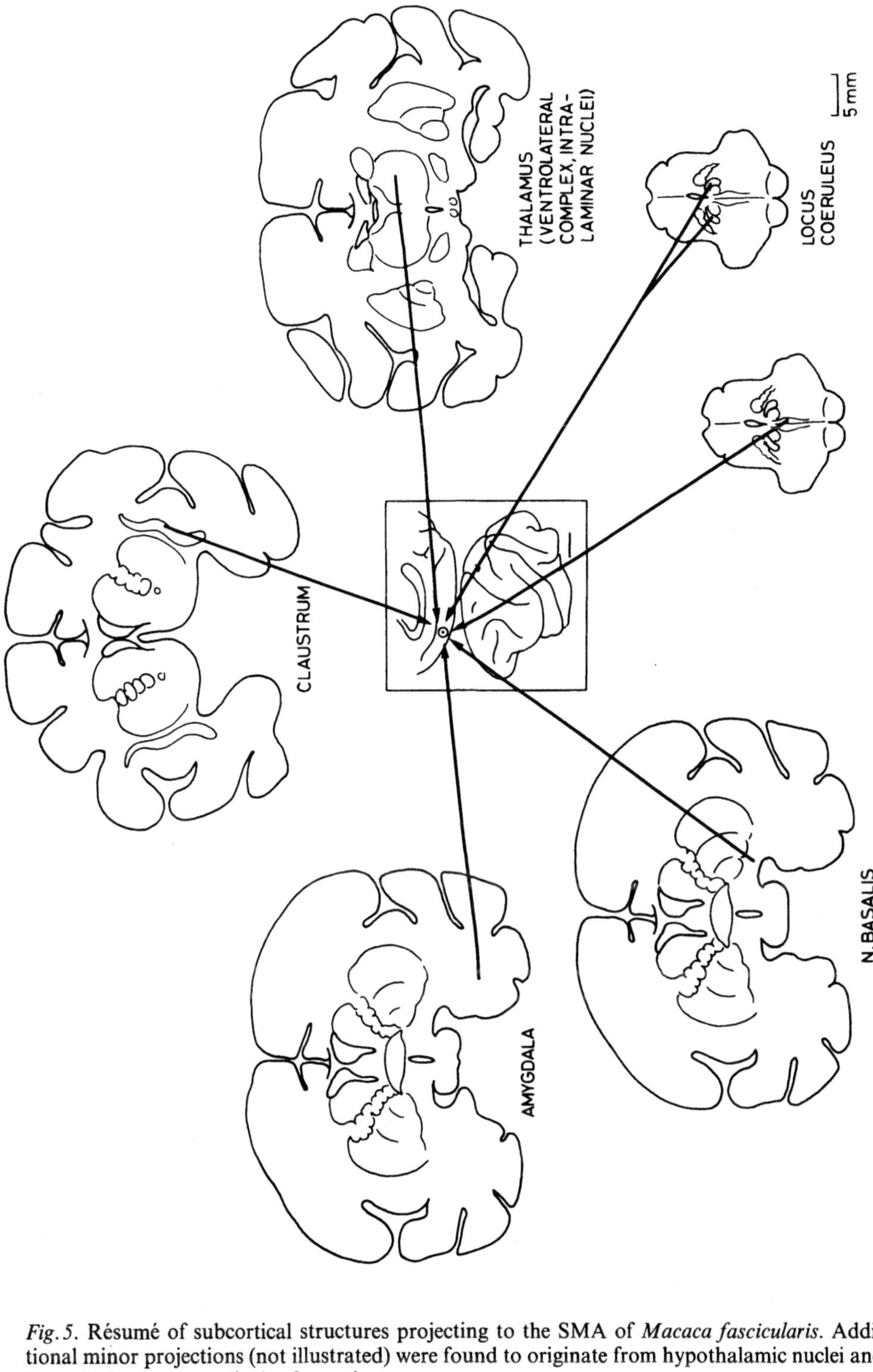

Fig. 5. Résumé of subcortical structures projecting to the SMA of *Macaca fascicularis*. Additional minor projections (not illustrated) were found to originate from hypothalamic nuclei and from the brain stem reticular formation.

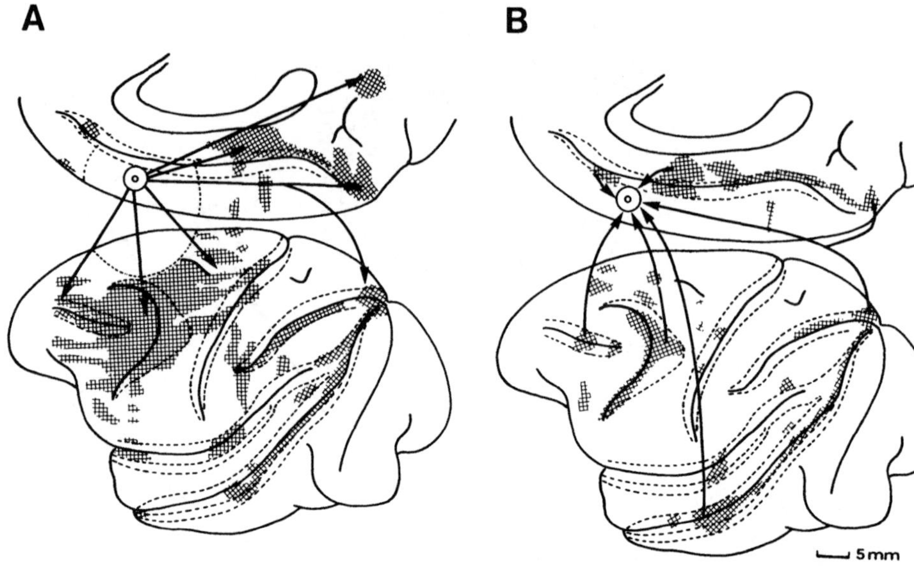

Fig. 6A,B. Schematic representation of afferent and efferent ipsilateral cortical connections of the SMA in *Macaca fascicularis*. *A* Inputs to the SMA. *Arrows* point to areas in which substantial labelling was found (*hatched areas without arrows* were weakly labelled). *B* Outputs of the SMA. The regions marked by *arrows* contained a substantial number of retrogradely labelled cells. The *hatched areas without arrows* contained some scattered cells only.

What Might Be the Significance of a "Higher" As Well As a "Lower" Control Function of the SMA?

Eccles (1982) immediately recognized the important significance of the reported studies on the SMA in humans which indicate that the SMA may play a key role in movement initiation (cf. Introduction). Admittedly, these studies provide indirect evidence only for the alleged higher position of the SMA in the process of movement initiation. Also, they lack precision with regard to localization and (for the blood flow studies) timing. Single-unit recordings in monkeys (Brinkman and Porter 1979; Smith 1979; Tanji et al. 1980; Tanji and Kurata 1982; Wise and Tanji 1981) are promising, but they have so far not provided unique results which had not been seen in other structures. But it is not our intention to review critically those experimental and clinical studies which led to the SMA hypothesis of Eccles (1982). Instead, let us assume that future studies will indeed strengthen the hypothesis. The question posed in this section then is: what might be the advantage of the SMA, as an initiating system for voluntary, especially self-paced movements, also exerting a direct control on the motor apparatus? Our proposition is that the direct influence of the SMA on the motor apparatus is used to "set" the appropriate excitability of spinal circuits and to allow for anticipatory postural adjustments. In addition to the direct corticospinal action, the SMA may also influence spinal neurons via the brain-stem reticular formation. Reticulospinal neurons exert a distributed influence on the spinal cord by means of an extensive collateralization and may thus well serve in the process of rapid and early postural adjustments. Anticipatory adjustments indeed appear to be a common feature of all voluntary movements (e.g. Hess 1943; Gurfinkel et al. 1971; Gahéry and Massion 1981; Hugon et al. 1982; Jung 1982). It has been reported that patients with frontal lobe lesions without paresis may lack these anticipatory postural adjustments (Elner and Gabibov 1977). Evidently, optimal and smooth performance of voluntary movements also requires optimization in postural support. The commands for complex postural adjustments involving widely distributed muscles of the trunk and extremities must be issued at an early stage of movement programming. It is not inferred that the SMA is unique in this respect. Other "premotor areas", i.e. cortical areas projecting to the motor cortex (Muakassa and Strick 1979) may also be engaged in the programming process, such as the anterior cingulate cortex of area 24B, the postarcuate region and parietal area 5. Each of these areas has also a contingent of corticospinal neurons (Murray and Coulter 1981; Macpherson et al. 1982a, b; Toyoshima and Sakai 1982). The various areas may play their specific role (according to their specific connectivity) in initiating movements which are triggered or guided by various sorts of sensory signals or which are "internally" generated (cf. also Wise 1983). The possible role of the SMA in generating anticipatory postural adjustments during volitional movements remains to be investigated.

Acknowledgement. The work of the authors was supported by the Swiss National Science Foundation (Grant No. 3.752.80). Secretarial assistance by Mrs. Rossier is gratefully acknowledged.

References

Biber MP, Kneisley LW, LaVail JH (1978) Cortical neurons projecting to cervical and lumbar enlargements of the spinal cord in young and adult rhesus monkeys. Exp Neurol 59: 492–508

Brinkman C, Porter R (1979) Supplementary motor area in the monkey: activity of neurons during performance of a learned motor task. J Neurophysiol 42: 681–709

Deecke L, Kornhuber HH (1978) An electrical sign of participation of the mesial "supplementary" motor cortex in human voluntary finger movement. Brain Res 159: 473–476

De Vito J, Smith OA (1959) Projections from the mesial frontal cortex (supplementary motor area) to the cerebral hemispheres and brain stem of the *Macaca mulatta*. J Comp Neurol 111: 261–278

Dhanarajan P, Rüegg DG, Wiesendanger M (1977) An anatomical investigation of the corticopontine projection in the primate (*Saimiri sciureus*). The projection from motor and somatosensory areas. Neuroscience 2: 913–922

Eccles JC (1982) The initiation of volunatry movements by the supplementary motor area. Arch Psychiatr Nervenkr 231: 423–441

Elner AM, Gabibov GA (1977) Possibility of recovery of the postural synergy after removal of the brain tumour. Agressologie 18: 69–73

Gahéry Y, Massion J (1981) Co-ordination between posture and movement. TINS 4: 199–202

Gurfinkel VS, Kots JM, Paltsev FI, Feldman AG (1971) In: Gelfand IM, Gurfinkel VS, Fomin SSV, Tsetlin ML (eds) Models of the structural functional organization of certain biological systems. MIT Press, Cambridge, pp 382–395

Hess WR (1943) Teleokinetische und ereismatische Kräftesysteme in der Biomotorik. Helv Physiol Acta 1: C62–C63 (English translation in: Akert K (1981) Biological order and brain organization. Springer, Berlin Heidelberg New York, pp 265–268 (Selected works of WR Hess))

Hugon M, Massion J, Wiesendanger M (1982) Anticipatory postural changes induced by active unloading and comparison with passive unloading in man. Pflügers Arch 393: 292–296

Jung R (1982) Postural support of goal-directed movements: the preparation and guidance of voluntary action in man. Acta Biol Acad Sci Hung 33: 201–213

Kuypers HGJM, Lawrence DG (1967) Cortical projections to the red nucleus and the brain stem in the rhesus monkey. Brain Res 4: 151–188

Macpherson JM, Marangoz C, Miles TS, Wiesendanger M (1982a) Microstimulation of the supplementary motor area (SMA) in the awake monkey. Exp Brain Res 45: 410–416

Macpherson JM, Wiesendanger M, Marangoz C, Miles TS (1982b) Corticospinal neurones of the supplementary motor area of monkeys. A single unit study. Exp Brain Res 48: 81–88

Muakassa KF, Strick PL (1979) Frontal lobe inputs to primate motor cortex: evidence for four somatotopically organized "premotor" areas. Brain Res 177: 176–182

Murray EH, Coulter JD (1981) Organization of corticospinal neurons in the monkey. J Comp Neurol 195: 339–365

Penfield W, Welch K (1951) Supplementary motor area of the cerebral cortex. Arch Neurol 66: 289–317

Roland PE, Larsen B, Lassen NA, Skinhoj E (1980) Supplementary motor area and other cortical areas in organization of voluntary movements in man. J Neurophysiol 43: 118–136

Smith AM (1979) The activity of supplementary motor area neurons during a maintained precison grip. Brain Res 172: 315–327

Tanji J, Kurata K (1982) Comparison of movement-related activity in two cortical motor areas of primates. J Neurophysiol 48: 633–653

Tanji J, Taniguchi K, Saga T (1980) Supplementary motor area: neuronal response to motor instructions. J Neurophysiol 43: 60–68

Toyoshima K, Sakai H (1982) Exact cortical extent of the origin of the corticospinal tract (CST) and the quantitative contribution to the CST in different cytoarchitectonic areas. A study with horseradish peroxidase in the monkey. J Hirnforsch 23: 257–269

Wiesendanger M (1981) Organization of secondary motor areas of cerebral cortex. In: Brooks VB (ed) Handbook of physiology, vol II, part 2. Am Physiol Soc, Bethesda, pp 1121–1147

Wiesendanger M, Séguin JJ, Künzle H (1973) The supplementary motor area – a control system for posture? In: Stein RB, Pearson KC, Smith RS, Redford JB (eds) Control of posture and locomotion. Plenum, New York, pp 33–346

Wise SP (1983) The non-primary motor cortex and its role in the cerebral control of movement (to be published)

Wise SP, Tanji J (1981) Supplementary and precentral motor cortex: contrast in responsiveness to peripheral input in the hindlimb area of the unanesthetized monkey. J Comp Neurol 195: 433–451

Woolsey CN, Settlage PH, Meyer DR, Spencer W, Hamuy TP, Travis AM (1952) Patterns of localization in precentral and "supplementary" motor areas and their relation to the concept of a premotor area. Res Publ Assoc Res Nerv Ment Dis 30: 238–264

Sensorimotor Integration in an Insect's Brain: The Neurophysiological Basis for the Flashing Dialog in the Firefly *Luciola lusitanica* (Charp.)

M. BRUNELLI and F. MAGNI, Pisa and Perugia

Summary. The reflex and direct excitability of the photomotor center, tested by applying a brief photic stimulus to the eyes or single electric stimuli to the protocerebrum of *Luciola lusitanica* at different time intervals from a spontaneous or a photically driven flash, was found to be depressed for 1200 ms after a spontaneous flash and for 900–1000 ms after a photically driven one. Direct excitability testing performed at various intervals from a subliminal photic stimulus shows a short-lasting (100-ms) increase followed by a long-lasting (900–1000 ms) depression. No depression of the peripheral response elicited by electrical stimulation of the lantern and of the electroretinographic responses was observed.

The excitability changes induced by photic stimulation of the eyes and those arising in the wake of a spontaneous flash summate at the level of the photomotor center: the excitatory effect of a photic stimulus is suppressed, and the summated inhibitory influences produce by postinhibitory rebound a long-latency flashing response when the photic stimulus is timed to occur 250–300 ms after a spontaneous flash.

This interval coincides with the critical one required for the entrainment of a flashing dialog between male and female. A flashing dialog between a male and an "artificial female" can be set up only if the artificial female's flash occurs within this critical delay range.

It is concluded that the flashing dialog between the two sexes of *Luciola lusitanica* is due to the central interaction of an "intrinsic" inhibitory mechanism associated with spontaneous flashing in the male and an "extrinsic" inhibition produced by the female response.

Key words: *Luciola lusitanica – Protocerebrum – Photogenic volleys – Flash – Flashing dialog*

Introduction

Flashing of luminous insects has long been recognized as a mating signal, enabling sex recognition and providing directional guidance. This involves a specific communication system, which has evolved in different genera and species along different strategies (see Lloyd 1971).

In *Luciola lusitanica* the specificity resides in the exact timing of the flash emitted by the female in response to a male's spontaneous flash. When this occurs, the flash of the female entrains in turn a male's flashing response followed by a

Experimental Brain Research, Suppl. 9
© Springer-Verlag Berlin · Heidelberg 1984

female flash, and so on. A "flashing dialog" is thus established, which lasts until the two partners meet and copulation takes place (Papi 1969). Although these phenomena have been througly investigated from the ethological point of view (Papi 1969; see Lloyd 1971), no attempt has yet been made to elucidate the neurophysiological mechanism underlying the mating communication.

Previous investigations (Bagnoli et al. 1976) have shown that the spontaneous flashing of the male of *L. lusitanica* is due to the discharge of impulses from a protocerebral structure (photomotor center), whose activity, tonic in nature, is made rhythmic by the inhibitory influence of an oscillator located in the lobula. It has also been demonstrated that several optic lobe structures (lamina and medulla) exert modulatory (tonic and phasic) influences on the basic flashing rhythm (Bagnoli et al. 1976).

Evidence has been produced that a prolonged inhibition of the photomotor center is associated with each spontaneous flash in males; moreover, brief photic stimuli exert a dual action on the photomotor centers of both males and females of *L. lusitanica*, namely a short-lasting excitation (which may elicit a reflex flash in the female and in the nonflashing male) followed by a prolonged inhibition, the duration of which is directly related to the intensity of the photic stimulus. These inhibitory events set the upper limit to the repetition rate of both spontaneous and photically driven flashing and strongly suggest that the so-called late responses occasionally observed in the male following brief photic stimulation are the result of postinhibitory rebound of the photomotor neurons (Brunelli et al. 1977).

Although the male's photomotor center is spontaneously active, no fundamental differences could be found in this species in the basic organization of the photomotor centers in the two sexes, as far as the response to a photic stimulus is concerned (Brunelli et al. 1977). Clearly, then, the mechanisms underlying the flashing dialog must be sought in the mutual interplay of excitatory and inhibitory events elicited in the photomotor center of either sex by the flashing of the partner.

The present experiments were performed with the aim of determining the time course of these events in both sexes, in relation to those occurring in the male during spontaneous flashing. The information gathered by our experiments allows a neurophysiological interpretation of the flashing dialog to be attempted.

Materials and Methods

The experiments were performed on male specimens of *L. lusitanica*. The animals were collected in the field at dusk and used the same night. The techniques used have been fully described in previous papers (Buonamici and Magni 1967). Briefly, the animals were attached ventral side up on a small Lucite board; flashes were recorded by means of a photomultiplier; the "photogenic volleys" (Case and Buck 1963) were recorded by means of a fine platinum electrode inserted into the lantern. After suitable amplification and filtering, both signals were displayed on the screen of an oscilloscope and photographed on film. In some experiments the electroretinogram was recorded by means of an Ag-AgCl electrode placed under the cornea. Electrical stimulation of the photomotor centers was performed monopolarly by means of a tungsten electrode (tip diameter 10 μm) inserted into the exposed protocerebrum. Electrical stimulation of the lantern was performed bipolarly through two Ag-AgCl electrodes inserted in the lanternal tissue (Buonamici and Magni 1967). A neon bulb driven by a square pulse of variable duration provided the photic stimulus. The light was focused on the animal's eyes by means of a glass fiber light guide and lenses. The stimulus intensity was controlled by means of neutral density filters. The photic stimulus could be timed to occur at variable delays from a trigger signal; the latter was either a preceding photic or electric stimulus, or the signal from a Schmidt-Trigger driven by spontaneous flash. All experiments were performed at night, in darkness, and at room temperature (22°–24°C).

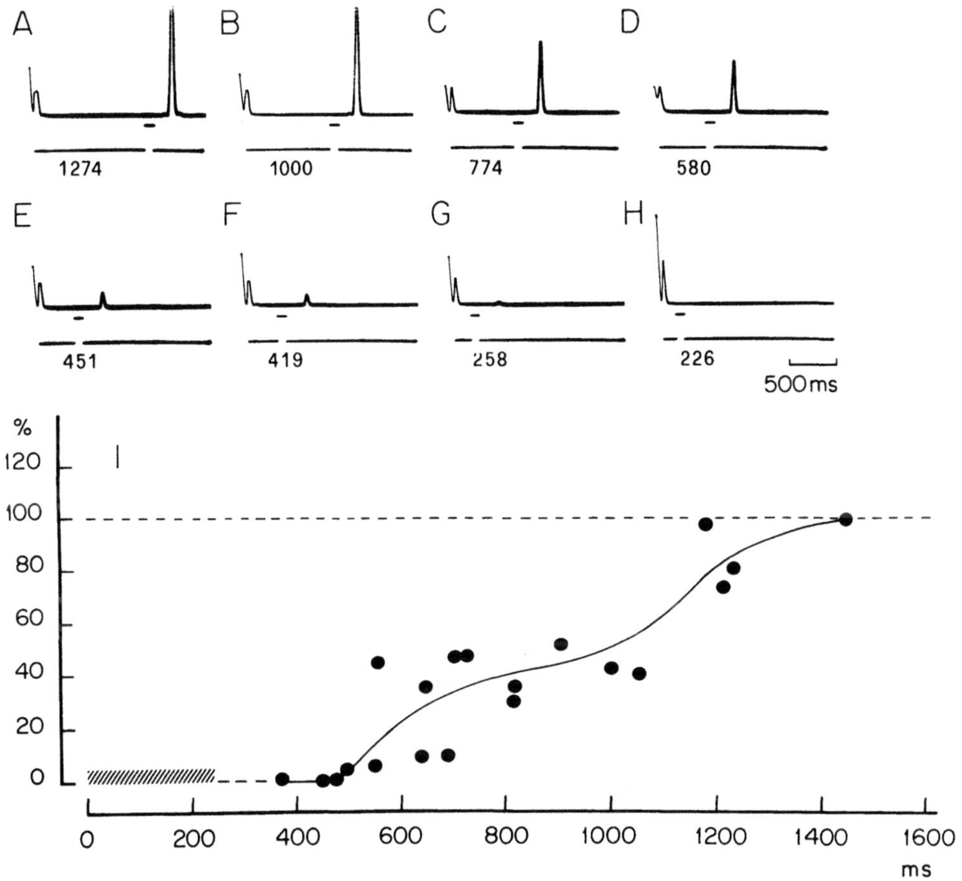

Fig. 1. A–H Reflex flashes (*upper traces*) evoked in a spontaneously flashing male of *L. lusitanica* by 100-ms light pulses (*lower traces*) applied at different delays (indicated in milliseconds by the figures below each panel) after the peak of a spontaneous flash. Same amplification throughout. *I* Plot of the amplitude of the reflex flash (expressed as percent of the maximum, *ordinate*) against delay (*abscissa*). The *shaded bar* covers the latency between the spontaneous photogenic volley and the peak of the spontaneous flash.

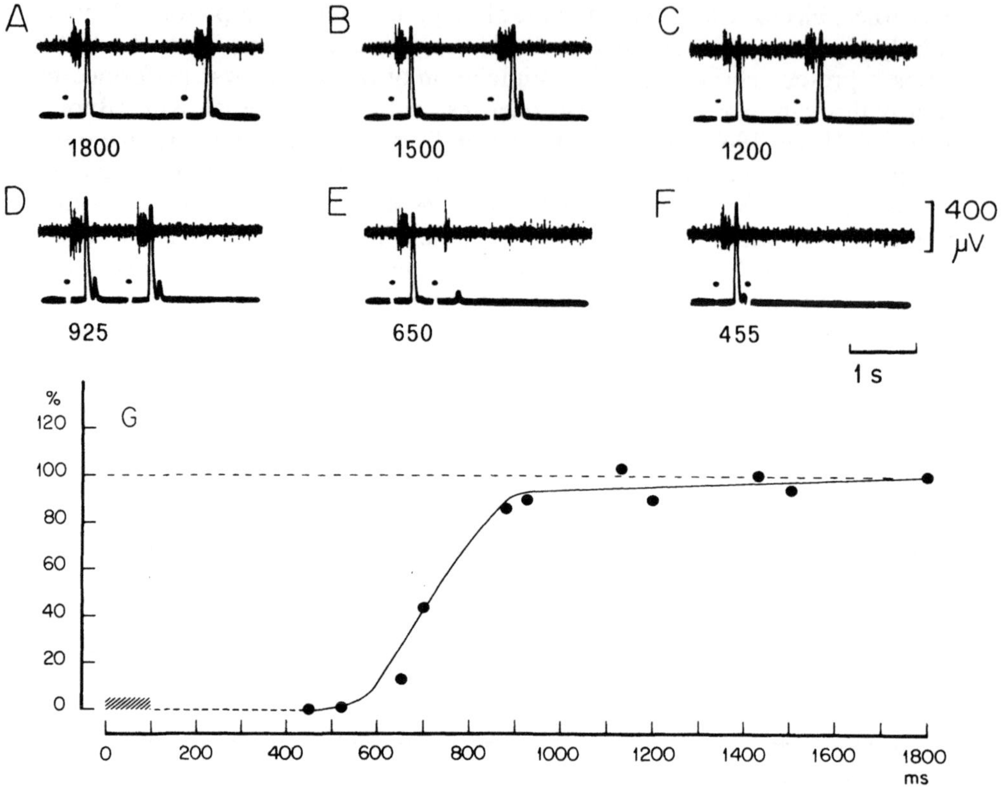

Fig. 2. A–F Light flashes (*lower traces*) and photogenic volleys (*upper traces*) evoked in a non-flashing male of *L. lusitanica* by paired 100-ms light pulses (*interruptions in the lower traces*) separated by the intervals indicated in milliseconds by the figure below each panel. Same amplification throughout; voltage calibration applies to *upper traces* only. *G* Plot of the amplitude of the response to the second stimulus of the pair (percent of maximum, *ordinate*) against inter-stimulus separation (*abscissa*). The *shaded bar* covers the duration of the first light pulse.

Results

Reflex Excitability Changes of the Photomotor Centers After Spontaneous and Photically Driven Flashing

The excitability of the photomotor centers of a spontaneously flashing male was tested by measuring the amplitude of the reflex response to a bright photic stimulus (100 ms) of constant intensity applied at various times after a spontaneous flash. The test stimulus is ineffective in eliciting a response when applied within a period of 250 ms from the peak of a spontaneous flash (Fig. 1H). Increasing the interval leads to the reappearance of the reflex response (Fig. 1G) which grows in amplitude as the interval is made progressively longer to reach the maximal (100%) amplitude at a delay of about 1200 ms (Fig. 1A–F).

Since each flash is initiated by a "photogenic volley" which starts about 150 ms before the beginning of light emission (Buonamici and Magni 1967; Brunelli et al. 1977) and the time to peak of a flash is about 100 ms (Buonamici and

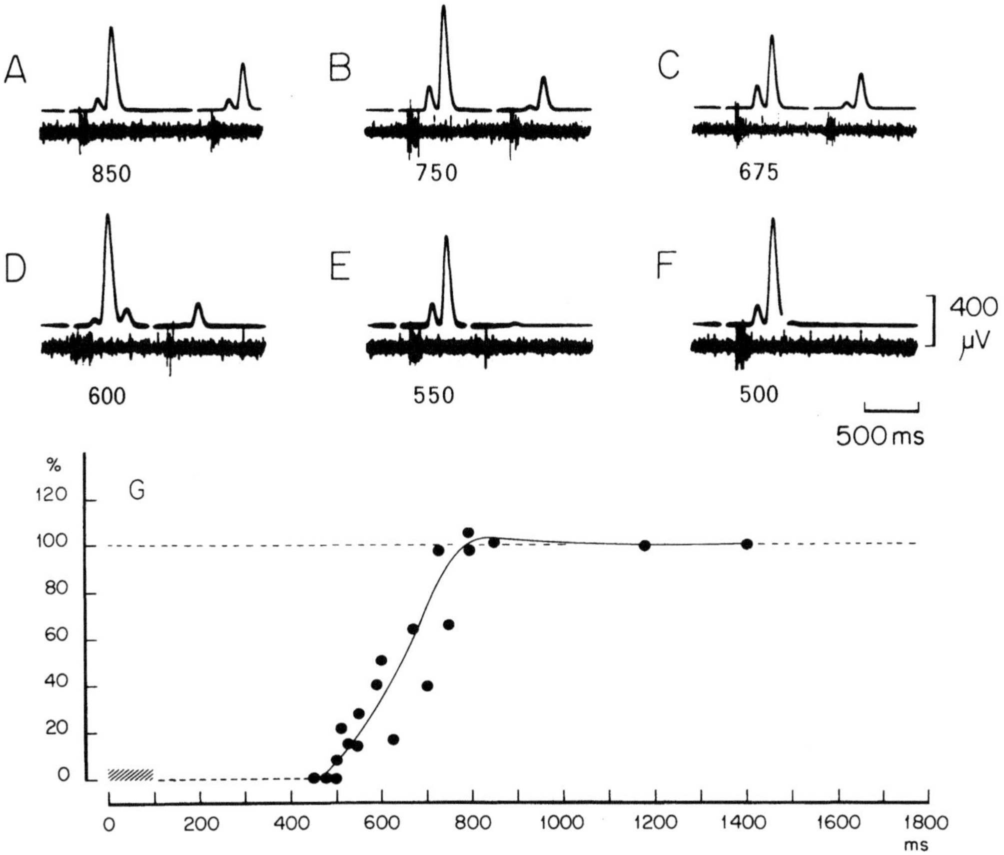

Fig. 3A–G. Same as Fig. 2, but in a female of *L. lusitanica. Upper traces:* photomultiplier output; *lower traces*: photogenic volleys. Stimulus marker: *interruption of the upper trace*. Voltage calibration applies to lower traces; same amplification throughout.

Magni 1967), the effective time course of the excitability of the photomotor centers can be obtained by adding 250 ms to each measured delay. The results of such calculations are displayed in graphical form in Fig. 1I, which shows the full time course of the photomotor center reflex excitability.

In nonflashing males the presentation of a brief photic stimulus elicits a reflex response at short latency ("early response"; Brunelli et al. 1977). The excitability changes following the early response were investigated by measuring the response amplitude to a second identical photic stimulus applied at various intervals after the first. The results are illustrated in Fig. 2. The test stimulus is ineffective when applied within about 450 ms after the first one (Fig. 2F); for longer intervals the response to the test stimulus reappears (Fig. 2E) and increases sharply as the interval is made longer (Fig. 2A–D), reaching the control value at an interval of about 1000 ms. The steep slope of the recovery is shown in the graph of Fig. 2G. A similar result was obtained when the excitability changes following a reflex response were tested in a female, as shown in Fig. 3.

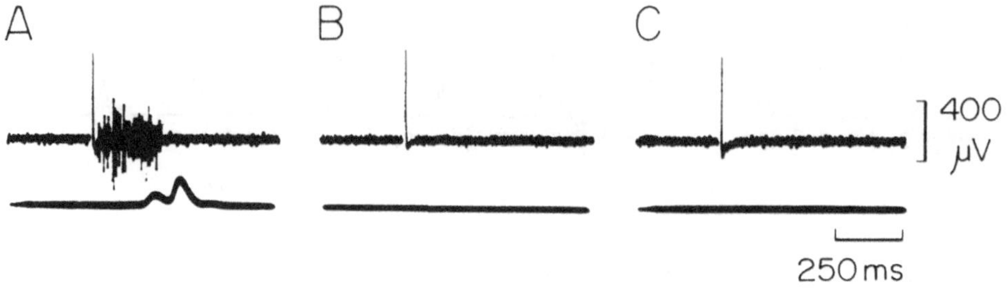

Fig. 4. A Photogenic volley (*upper trace*) and flash (*lower trace*) elicited by single-shock stimulation of the protocerebrum in a male of *L. lusitanica*. *B, C* Lack of effect of an electric shock of the same intensity as that in *A* after the stimulating electrode has been displaced by 100 μm in central and anterior directions respectively. Voltage calibration applies to *upper traces* only; same amplification throughout.

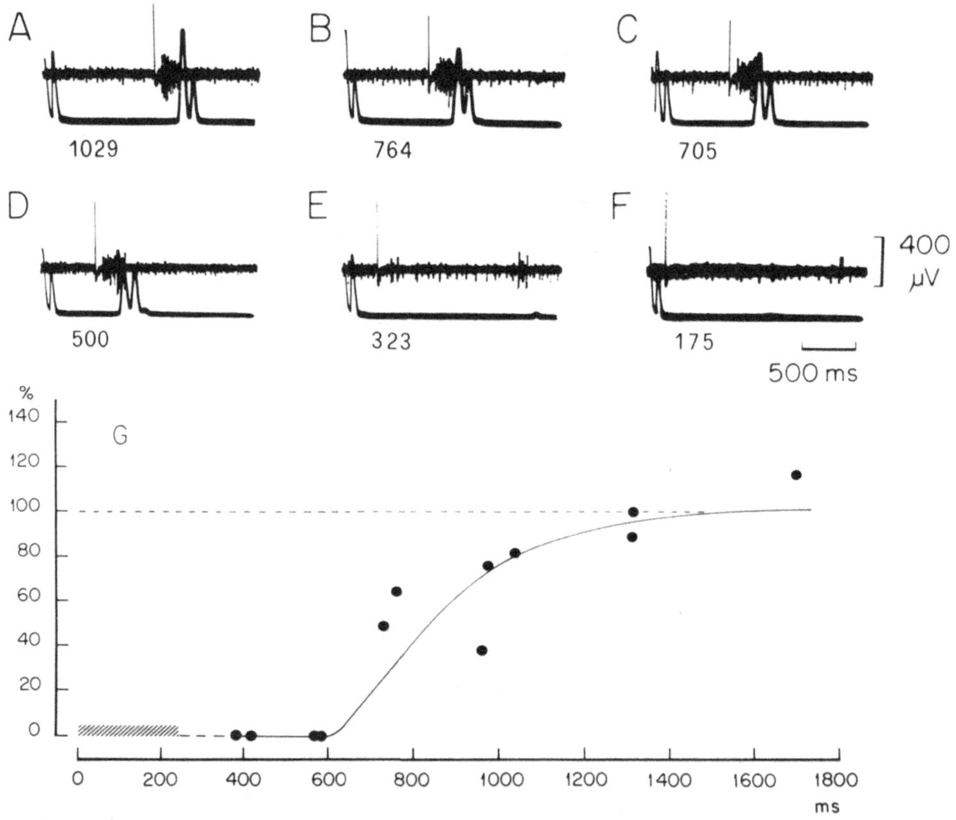

Fig. 5. A–F Photogenic volleys (*upper traces*) and flashes (*lower traces*) evoked by single electric shocks to the protocerebrum (*sharp upward deflections of the upper traces*) at different intervals (indicated in milliseconds by the figures below each panel) from a spontaneous flash in a male of *L. lusitanica*. Voltage calibration applies to *upper traces* only; same amplification throughout. *G* Plot of the amplitude (percent of the maximum) of the electrically evoked flash (*ordinate*) against the delay (*abscissa*). The *shaded bar* covers the latency between spontaneous photogenic volleys and the peak of the spontaneous flash.

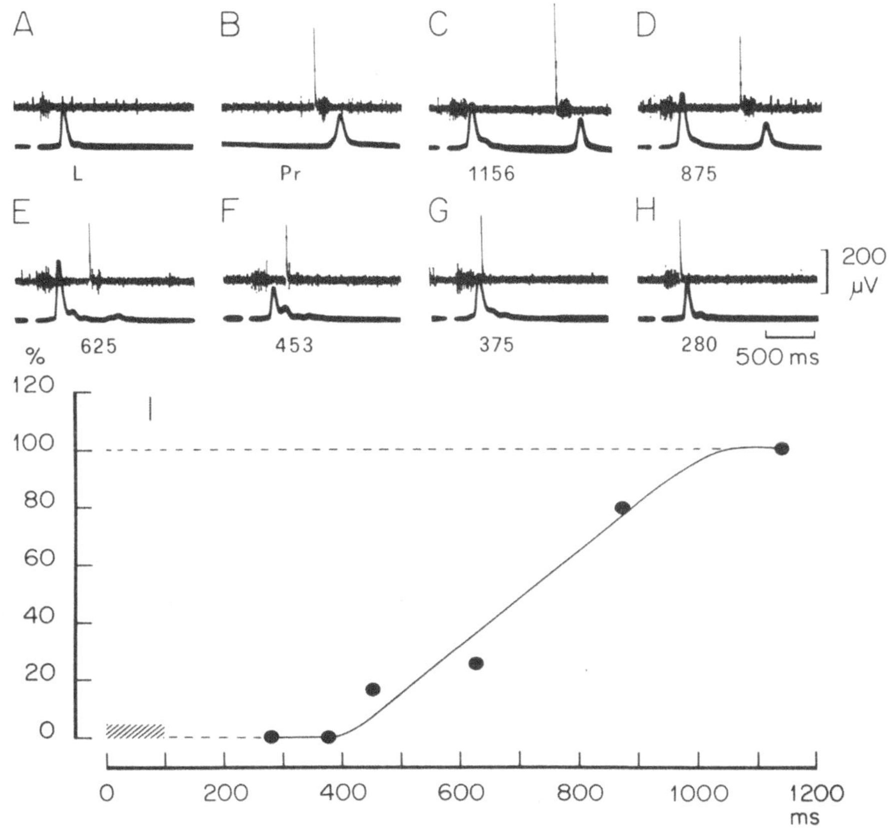

Fig. 6. A,B Photogenic volley (*upper trace*) and flash (*lower trace*) elicited by a 100-ms light pulse (*L, interruption of lower trace in A, arrow*) and by a single shock to protocerebrum (*Pr, upward deflections of upper trace, arrow*) in a nonflashing male of *L. lusitanica. C–H* Paired photic and electrical stimuli delivered to the same animals at the time intervals indicated in milliseconds by the figures below each panel. Voltage calibration applies to *upper traces* only; same amplification throughout. *I* Plot of the amplitude of the response to the electric shock (percent of maximum, *ordinate*) against interstimulus separation (*abscissa*). The *shaded bar* covers the duration of the light pulse.

Changes in the Direct Excitability of the Photomotor Centers Following Spontaneous Flashes and Photic Stimulation

In this series of experiments the excitability of the photomotor centers was tested by means of localized electrical stimulation of the deep protocerebral structures (Bagnoli et al. 1976). Single-shock stimulation applied to these structures through a tungsten microelectrode elicits a typical photogenic volley and a flash response from the lantern (Fig. 4A); displacement of the microelectrode tip by 100–150 μm in dorsoventral (Fig. 4B) and anteroposterior (Fig. 4C) directions leads to the disappearance of the response.

The response to localized electrical stimulation of the protocerebrum was tested at various intervals from the peak of the flashes in a spontaneously flashing

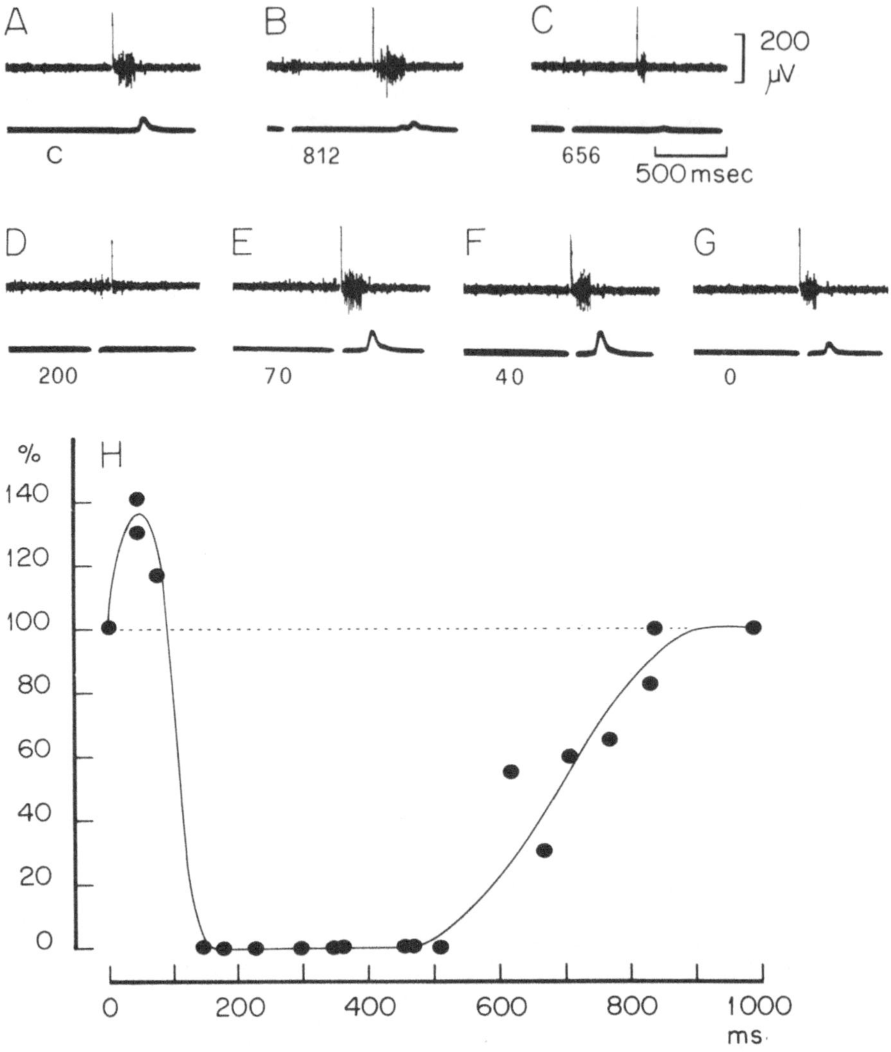

Fig. 7. A Control photogenic volley (*upper trace*) and flash (*lower trace*) elicited by a single electric shock to the protocerebrum (*upward deflection in upper trace*) of a nonflashing male at 05.00 a.m. *B–G* Responses to single electric shock stimulation of the protocerebrum in the same animal preceded by a subliminal photic stimulus (*interruption of lower trace*) at the time intervals indicated by the figures below each panel. Voltage calibration applies to *upper traces* only; same amplification throughout. *H* Plot of the response amplitude (percent of control, *ordinate*) against interstimulus separation (*abscissa*).

male (Fig. 5). Complete suppression of the test response (both photogenic volley and flash) was observed at intervals up to about 350 ms from the peak of the spontaneous flash (Fig. 5E,F); at longer intervals the response quickly recovered in amplitude, recovery being complete at an interval of about 1200 ms (Fig. 5A,D). The time course of the central excitability changes is plotted in Fig. 5G, after cor-

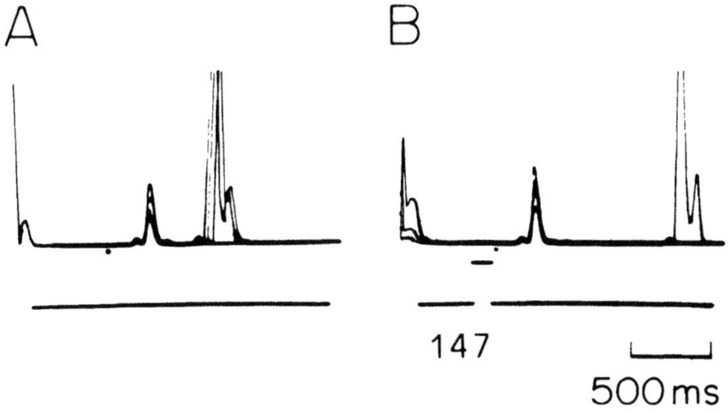

Fig. 8. A "Direct" and "reflex" response (*upper trace*) elicited by single-shock electrical stimulation (*lower trace*) of the lantern in a spontaneously flashing *L. lusitanica* male. *B* The reflex response is abolished and the direct response is unaffected by a photic stimulus applied to the eyes 147 ms in advance of the electric lanternal shock. Superimposed records; same amplification throughout.

rection for the time lag between the beginning of the photogenic volley and the peak of the ensuing spontaneous flash (see above).

The same direct excitability testing was performed in nonflashing males by delivering a test electric shock to the protocerebrum at various time intervals from a photic stimulus, which evoked a reflex early response (Brunelli et al. 1977). The results of this experiment are illustrated in Fig. 6, which shows that both the photogenic volleys and the light emission evoked by the test electric shock (Fig. 6B) are completely suppressed when the latter is timed to occur within about 400 ms from a flash-eliciting photic stimulus (Fig. 6G,H). Recovery of the test response is observed at longer intervals, being complete at an interstimulus interval of about 1000 ms (Fig. 6C–F). The full time course of the phenomenon is shown in the graph of Fig. 6I. Similar results were also obtained in females.

It has been reported that starting from midnight and throughout the day the excitability of the photomotor neurons undergoes a decrease, which has been attributed to the operation of a circadian control (Papi 1969; Brunelli et al. 1970; Bagnoli et al. 1972a, b), so that reflex flashing in response to photic stimulation cannot be elicited (Brunelli et al. 1977). In these conditions the excitability changes induced by a subliminal photic stimulus were investigated by applying a test electric shock to the protocerebrum. A dual effect was consistently observed. As shown in Fig. 7, a short-lasting increase in excitability of the photomotor neurons to direct electrical stimulation (signaled by both the increased duration and frequency of the photogenic volley and the increased flash intensity, Fig. 7E–F) is produced by the subliminal photic stimulus at short delays (up to about 100 ms). At longer intervals the facilitation gives way to a marked depression, which leads to the suppression of the test response up to an interstimulus interval of 500 ms. The test response recovers gradually as the delay is further increased, reaching control value at a delay of about 1000 ms. The full time course of these events is illustrated in the graph of Fig. 7H.

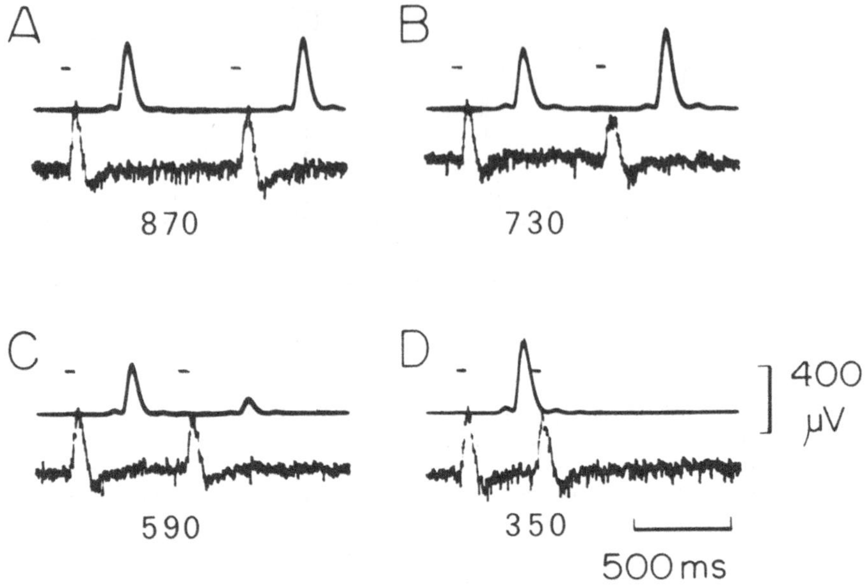

Fig. 9A–D. Flashes (*upper traces*) and electroretinographic responses (*lower traces*) elicited by paired photic stimuli (*βαϱσ αβθφε φππεϱ τϱαςεσΔ*) delivered at different intervals (indicated in milliseconds by the figures below each panel) to the eyes of a nonflashing male of *L. lusitanica*. Voltage calibration applies to *lower traces* only; same amplification throughout.

Central Nature of the Depression Following Spontaneous and Light-Evoked Flashing

The possibilities that peripheral inhibition (Brunelli et al. 1968a) and/or sensory adaptation could be responsible, in part at least, for the depression of the test responses described in the previous sections were tested in two series of experiments.

In a first series a single electric shock was applied to the lantern 600 ms after a flash in a spontaneously flashing male. As illustrated in Fig. 8A, this elicited a direct response at the typical latency, followed by a centrally mediated response at longer and more variable latency (Buonamici and Magni 1967). A photic stimulus was then applied to the animal's eyes about 400 ms after the spontaneous flash. The photic stimulus, albeit itself incapable in these conditions of eliciting a reflex flash, caused the disappearance of the centrally mediated component of the response to the lanternal shock, while the direct component remained completely unaffected (Fig. 8B).

In a second series of experiments, performed in nonflashing males, the reflex flashes elicited by brief photic stimuli were recorded together with the electroretinographic responses (Fig. 9). It was found that when the interval between the paired photic stimuli was made progressively shorter than 600 ms, the response to the second stimulus was at first considerably depressed (Fig. 9C) and finally completely suppressed (Fig. 9D), whereas the electroretinographic responses remained unchanged.

Interactions Between Photically Induced and Spontaneous Excitability Changes

The interaction between the excitability changes of the photomotor center of a spontaneously flashing male and those induced by a photic stimulus is illustrated in Fig. 10. The direct electrical stimulation of the photomotor center at an interval of about 900 ms after a spontaneous flash elicited a response (test response), the amplitude of which was approximately 80% of the maximum (Fig. 10A). This response was completely abolished when the electric shock was preceded by a photic stimulus applied 500 ms in advance (Fig. 10B). The photic stimulus was incapable of eliciting a response by itself, since it occurred at a delay from the spontaneous flash (400 ms) at which complete suppression of the photically driven response has been demonstrated (see Fig. 1). On the other hand, the photic stimulus-electric shock delay was such as to be expected to produce a 80% depression of the test response (see Fig. 6). The complete suppression of the test response observed in this experiment is indicative of summation of the depressant action of light and of that associated with the spontaneous flash. The test response reappeared as the photic stimulus was withdrawn (Fig. 10C).

The summation of the effects of a photic stimulus and the depression occurring in the wake of a spontaneous flash is further substantiated by the observations illustrated in Fig. 11. When bright (more than 100 lux) photic stimuli are applied to a male at short intervals after a spontaneous flash, a late response can be obtained (Fig. 11A), which has been attributed to a postinhibitory rebound (Brunelli et al. 1977). This late response persists as the interval is made progressively longer (Fig. 11B), until at a delay of about 400 ms an early response (Brunelli et al. 1977), signaled by a short photogenic volley, unable to trigger a flash, begins to appear (Fig. 11B, arrowhead). As the interval is further increased, the early response grows progressively larger (Fig. 11C, arrow), finally developing as a full-sized photogenic volley which triggers a flash (Fig. 11D).

The rebound nature of the late response elicited at short intervals after a spontaneous flash by a strong photic stimulus is demonstrated by the experiment illus-

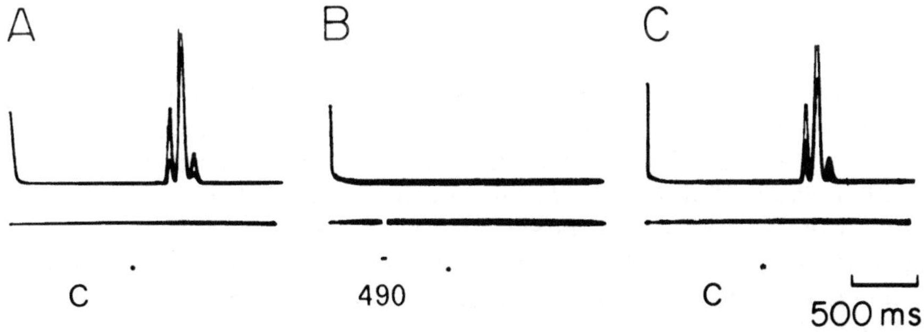

Fig. 10. A Flashing (*upper trace*) elicited by a single electric shock (*lower trace*) applied to the protocerebrum of a *L. lusitanica* male 900 ms after a spontaneous flash. *B* Suppression of the response by a photic stimulus delivered 490 ms before the electric shock. *C* Reapparence of the response upon withdrawal of the photic stimulus. Superimposed traces; same amplification throughout.

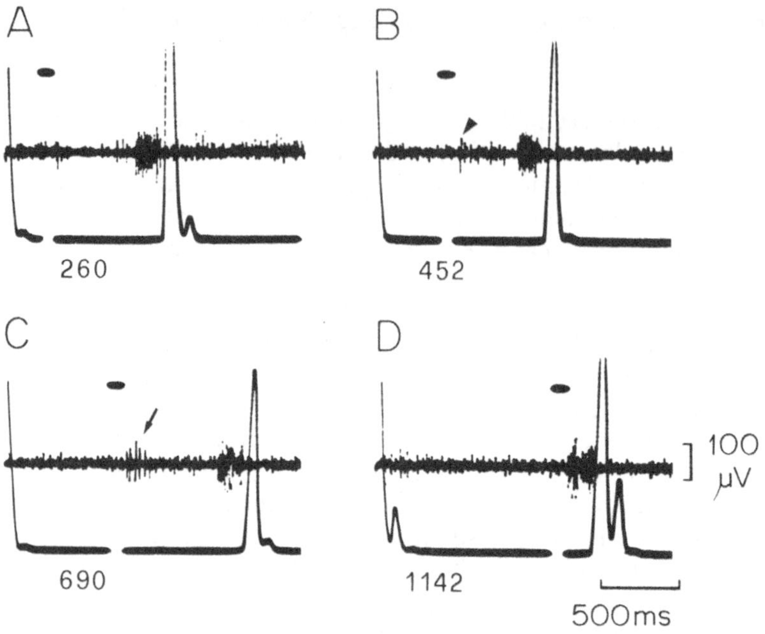

Fig. 11. A Late response of *L. lusitanica* male to a bright photic stimulus (more than 100 lux) applied 260 ms after a spontaneous flash. *B* As the delay is increased a subliminal photogenic volley at short latency is elicited by the photic stimulus (*arrowhead*). *C* At longer delay the early photogenic volley becomes more conspicuous (*arrow*). *D* At a still longer delay the photic stimulus elicits a typical early response. *In each panel upper trace*: photogenic volleys; *lower trace*: photomultiplier output; stimulus signal: *square pulse in lower traces*. Voltage calibration applies to *upper traces* only; same amplification throughout.

trated in Fig. 12, which shows that an increase in the duration of the photic stimulus shifts the late response by the same amount, i.e., the latency of the late response remains constant to the "off," in full agreement with the results of previous investigations (Brunelli et al. 1977).

Entrainment of a Flashing Dialog by Photic Stimulation

Application of a dim photic stimulus at an interval of 250–320 ms after a male's spontaneous flash is invariably followed by a flashing late response occurring at a latency of about 500–600 ms (Fig. 13B). This leads to a sequence of two flashes with a period of about 800–900 ms. As shown in Fig. 13B, a third flash tends to follow the late response at the same period; however, in the absence of successive photic stimuli this tendency is not maintained for more than two cycles. Photic stimuli applied at shorter or longer delays fail to induce this phenomenon (Fig. 13C).

This pattern is fully compatible with the time sequences observed in the flashing of male and female fireflies engaged in a dialog (Fig. 14). The interflash period of the male (b in Fig. 14) is 900 ms; the delay between the peak of the male's flash and the beginning of the female's response is 250 ms (a in Fig. 14) and the latency

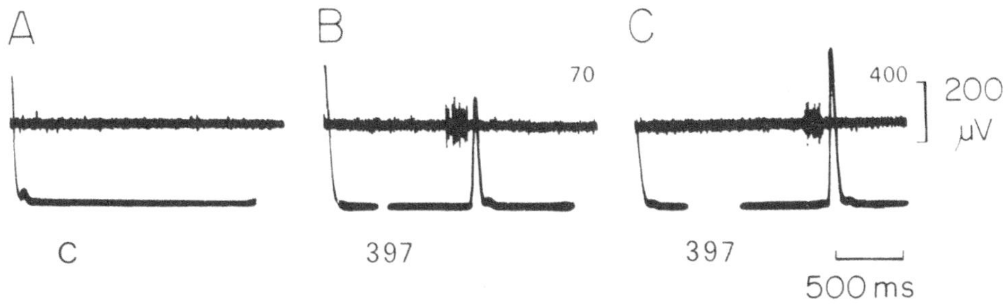

Fig. 12. A Control (*C*) sweep triggered by a spontaneous flash of a *L. lusitanica* male. *B, C* Late responses elicited by bright photic stimuli (70 and 400 ms in duration respectively) delivered to the eyes 397 ms after a spontaneous flash. *In each panel upper trace*: photogenic volleys; *lower trace*: photomultiplier output. Photic stimulus indicated by *interruption in lower traces*. Voltage calibration applies to *upper traces* only; same amplification throughout.

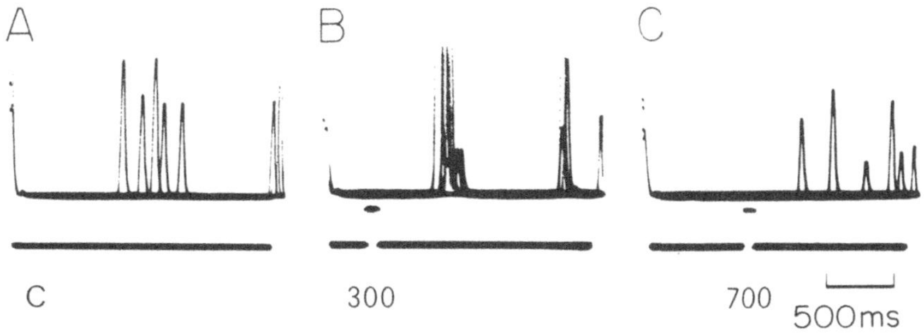

Fig. 13. A Control record (*C*) of spontaneous flashing in a *L. lusitanica* male. Each sweep is triggered by a spontaneous flash. *B* Late responses (*upper traces*) elicited by photic stimuli (*lower trace*) applied 300 ms after a spontaneous flash. Note the tendency of the following flashes to group at the same interval. *C* Lack of effect of a photic stimulus applied 700 ms after a spontaneous flash. Superimposed traces; same amplification throughout

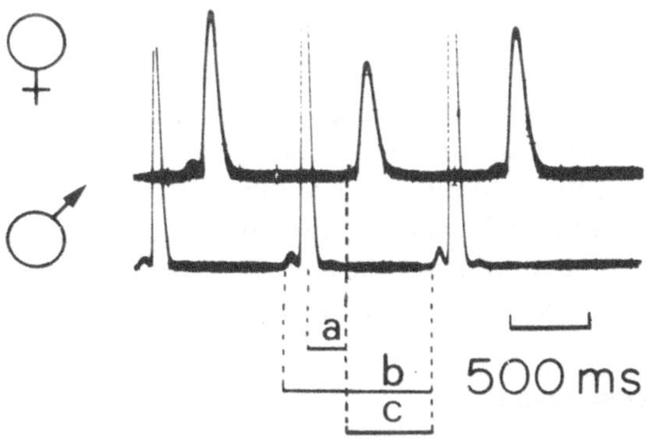

Fig. 14. Flashing of male (♂) and female (♀) *L. lusitanica* recorded simultaneously during a flashing dialog. *a*, latency between the peak of the male's flash and female response; *b*, male's flashing period; *c*, latency between female's flash and male's response.

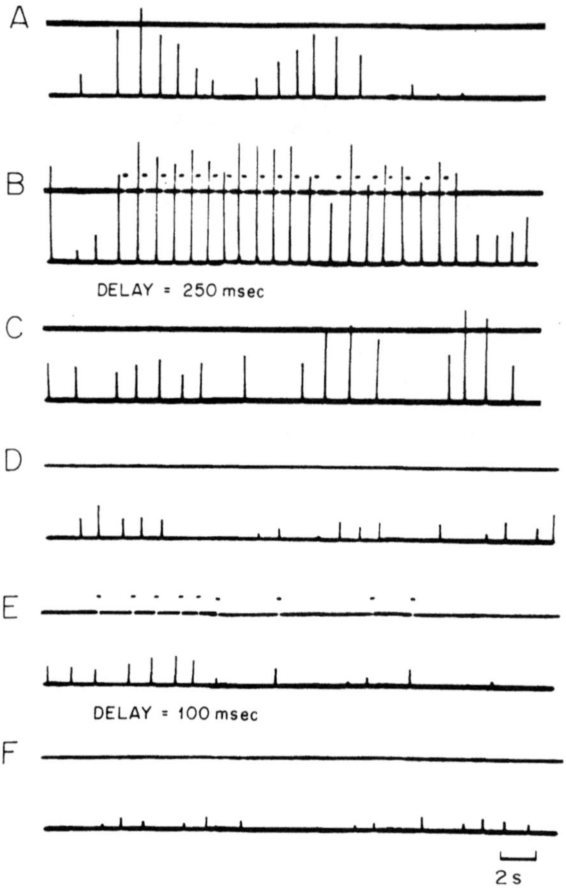

Fig. 15. A–C Continuous records of a dialog between a spontaneously flashing male (*lower trace*) and an artificial female flashing with a delay of 250 ms from each male's flash (*B*). *D–F* Lack of dialog when the delay is shortened to 100 ms (*E*). Same amplification for *A–C* and *D–E*.

between the start of the female's flash and that of the following male's response is 550 ms (c in Fig. 14). In fact, in full agreement with previous findings (Papi 1969) a dialog can be set up between a spontaneously flashing male (Fig. 15A) and an artificial source of pulsed photic stimuli, by setting the delay between the flash and the light pulse at a value of 250–300 ms (Fig. 15B); the artificial dialog stops as soon as the photic pulses are no longer delivered (Fig. 15C). No artificial dialog can be obtained if the delay is shortened to 100 ms (Fig. 15D–F) or made longer than 300 ms (not illustrated).

Discussion

The results of the present experiments confirm and extend those of a previous investigation (Brunelli et al. 1977) by showing that: (a) each spontaneous flash is followed by a prolonged inhibition of flashing, and (b) each flash triggered by brief photic stimulation of the eyes is also followed by an inhibition of shorter duration.

The finding that both reflex and direct excitabilities of the photomotor center are equally depressed for about 1200 ms after each spontaneous flash substantiates the hypothesis put forward in a preceding paper (Brunelli et al. 1977) that the depression is due to an active inhibitory action exerted on the central structure responsible for flashing. This conclusion is compatible with the interplay between photomotor structures of the protocerebrum and an oscillator, which previous lesion and stimulation experiments have shown to be located in the lobula (Bagnoli et al. 1976). Since this intrinsic rhythmicity is independent of photic input, as shown by the fact that the effects of eye illumination are still present after destruction of the lobula or when spontaneous flashing is absent (Bagnoli et al. 1976), it can be proposed that the inhibition of the photomotor centers we have observed in the wake of each spontaneous flash is the expression of a negative feedback control exerted by the lobula oscillator.

Our experiments show that also the flashes triggered by brief photic stimulations both in nonflashing males and in females are followed by a period of reduced direct and reflex excitability, albeit of shorter duration (900 ms) than that observed after a spontaneous flash. This result indicates that photic stimulation exerts an inhibitory action on the photomotor centers, which is mediated by mechanisms different from those responsible for the inhibition associated with spontaneous flashing, in full agreement with the conclusions of a previous investigation (Brunelli et al. 1977).

Concerning the mechanisms responsible for the inhibition associated with a photically triggered flash, the possibility that it may be due to the hormonally mediated peripheral inhibition demonstrated by previous investigations (Brunelli et al. 1968a, 1968b, 1970; Bagnoli et al. 1970, 1972a) is made unlikely by its fast onset and short duration, difficult to reconcile with a humoral mechanism, and is ruled out by our finding that the "direct" flash (Buonamici and Magni 1967) elicited by electrical stimulation of the lantern is unaffected by a preceding photic stimulus, whereas the "reflex" response is suppressed. Nor can the depression be attributed to sensory adaptation, since the amplitude and shape of the electroretinographic responses to paired photic stimuli are still identical at interstimulus separations at which the flash triggered by the second stimulus is suppressed. It thus appears that the photically activated inhibitory mechanism is central in nature.

The fact that a period of depressed excitability follows a photic stimulus subliminal in intensity for triggering a flash rules out the possibility that it may be produced by a recurrent inhibitory circuit, and clearly shows that the photically induced inhibitory action is exerted directly on the photomotor centers by the visual input.

Our finding that a subliminal photic stimulus exerts a dual effect on the excitability of the photomotor centers, namely a short-lasting increase followed by a prolonged depression, fully confirms the conclusion of a previous paper (Brunelli et al. 1977). Although a precise identification is not possible at this stage of the investigations, the mechanisms responsible for the dual effect of brief photic stimuli are likely to be found in the organization of the visual centers, as suggested

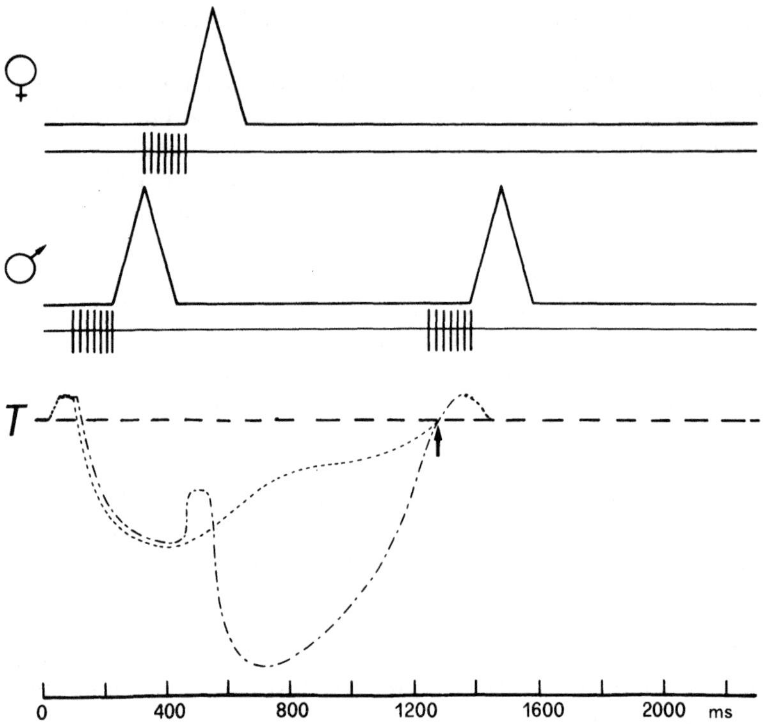

Fig. 16. The proposed mechanism for flashing dialog. ♀: female's flash and photogenic volley. ♂: male's flash and photogenic volley. *Dotted line*, inhibition following each spontaneous flash; *dash-dot line*, algebraic summation of photically evoked inhibition to that associated with the spontaneous flash; *T*, threshold; *arrow*, postinhibitory rebound excitation.

by the excitatory and inhibitory effects elicited by lesions and by electrical stimulations of various visual structures (Bagnoli et al. 1978).

In a previous paper (Brunelli et al. 1977) it was shown that a photic stimulus may, under appropriate conditions, evoke in the male an "off" response (i.e., a flash of long latency which remains constant with respect to the cessation of the photic stimulus), which was interpreted as the result of postinhibitory rebound. The results of the present experiments are in full agreement with this interpretation. In fact, we have observed that the excitatory effect of a photic stimulus does not appear if the latter is timed to occur during the inhibitory period following a spontaneous flash, whereas the inhibitory effects are still observed (and are even stronger, see below), an observation suggesting that the intrinsic (lobula) and extrinsic (photic) influences impinge upon the same photomotor structures. Moreover, we have shown that the response to an electric stimulus applied to the protocerebrum at such an interval from a spontaneous flash that it is partially depressed is completely suppressed when it is preceded by a photic stimulus which by itself is unable to produce a complete suppression of the flash. Finally, a bright photic stimulus applied at a short interval after a spontaneous flash elicits a "late" flashing response, the latency of which is constant with respect to the cessation of the photic stimulus, whereas the photic stimulus applied at longer delay elicits a typical early response. These findings can be best explained on the basis of the interaction of opposing synaptic influences on the neurons of the photomotor center: at short delay after a spontaneous flash the excitatory effect of the photic stimulus is effectively counteracted by the inhibition associated with the spontaneous flash, whereas the inhibitory actions summate, leading to a degree of hyperpolarization adequate for postinhibitory rebound. This interpretation is fully supported by our observation that, if the photic stimulus is timed to occur at a delay from a spontaneous flash at which the intrinsic inhibition has declined from its maximum, the photically induced excitatory effect may appear in the form of a subthreshold photogenic volley, which eventually grows larger and larger as the delay is progressively increased, until it reaches the threshold for triggering a flash and the late response is replaced by an early one.

Along this line of interpretation, the time courses of the two inhibitory actions lead to the predictions that the highest degree of summation and the steepest slope of delay of the inhibition should be obtained at an interval of about 300 ms between spontaneous flash and photic stimulus. This prediction is fully verified by experimental tests, since we have shown that a late response is consistently elicited at a delay of 300 ms, whereas no response is obtained with longer or shorter delays. This delay is coincident with that found to be critical for the entrainment of a flashing dialog in behavioral studies (Papi 1969). In fact, comparison of Figs. 13 and 14 shows that the delay between male flash and female response and that between female response and the following male's flash fall in the same range in the natural dialog and in the laboratory situation, in which the female response is mimicked by a properly timed light stimulus (artificial female). Furthermore, we have been able to entrain a flashing dialog between a male and an artificial female by timing the artificial female to flash 250–300 ms after a male's flash.

Summing up, our results allow the following conclusions:

1. In male fireflies the spontaneous rhythmic flash is secured by the interplay (through inhibitory feedback circuits) between photomotor structures of the protocerebrum and an oscillator located in the lobula (Bagnoli et al. 1976).
2. In both male and female visual nervous structures exert a dual action on the photomotor centers, namely a short-latency excitation and a long-lasting inhibition.
3. The photomotor center appears as the final common pathway for both intrinsic and extrinsic (visual) synaptic inputs, the integration of which determines at any given instant its excitability level.

The quantitative evaluation of the time courses of the excitatory and inhibitory events we have performed in the present study allows a tentative model of the neural events underlying the flashing dialog to be proposed, as diagramatically sketched in Fig. 16. In the female, which does not flash spontaneously, a spontaneous flash of the male elicits a response at short latency (250–300 ms) as a consequence of the excitatory action of the photic stimulus, followed by an inhibition lasting 900–1000 ms. The female flash elicits in turn an identical sequence of excitation and inhibition (dash-dot line) on the photomotor centers of the male which, at this time, are maximally inhibited by the intrinsic rhythmogenic mechanism (dotted line). Hence the excitatory action of the female flash remains well below threshold for the male early response, whereas the inhibitory action is added to the intrinsic one, thus leading to a marked hyperpolarization of the photomotor neurons (dash-dot line). As both inhibitions come to an end at the same time, the steep slope of recovery following the sudden withdrawal of the inhibitory pressures leads to rebound excitation of the male's photomotor centers and to flashing (male's late response, arrow). Since by the time the male's late response occurs the extrinsic inhibition of the female's photomotor center has vanished, the male's late response triggers a new early response in the female and so on.

We are fully aware that the proposed model is incomplete and largely oversimplified; however, it appears to be justified as a plausible explanation in neural terms of the critical timing of the light signals which is required for the entrainment of a dialog. In fact, on its basis the strict requirement for a critical range of delays between males' and females' flashes can be understood in terms of degree of summation of inhibitory actions which must be attained to elicit a rebound response in the male. Any shift outside the proper time range would lead to a summation inadequate to elicit the rebound response and hence to entrain a dialog.

The organization of the neuronal structures and circuitry underlying the phenomena we have described remains to be worked out in detail, a task that can only be accomplished by means of direct investigations at the cellular level.

Acknowledgments. The technical assistance of Mrs. Bruna Margheritti and Mr. Alberto Bertini and the secretarial help of Mr. Sergio Passerini are gratefully acknowledged.

References

Bagnoli P, Brunelli M, D'Aiello V, Magni F (1972a) Further evidence for peripheral inhibition of flashing and for the rôle of the male gonads in *Luciola lusitanica* (Charp.). Arch Ital Biol 108: 181–206

Bagnoli P, Brunelli M, Magni F, Viola M (1972b) The identification of a flash inhibiting substance from the male gonads of *Luciola lusitanica* (Charp.). Arch Ital Biol 110: 16–34

Bagnoli P, Brunelli M, Magni F, Musumeci D (1976) Neural mechanisms underlying spontaneous flashing and its modulation in the firefly *Luciola lusitanica*. J Comp Physiol 108: 133–156

Brunelli M, Buonamici M, Magni F (1968a) Mechanisms for photic inhibition of flashing in fireflies. Arch Ital Biol 106: 85–156

Brunelli M, Buonamici M, Magni F (1968b) Effects of castration on the inhibition of flashing in fireflies. Arch Ital Biol 106: 100–112

Brunelli M, Buonamici M, Magni F, Viola M (1970) The rôle of the male gonades in the peripheral inhibition of flashing in *Luciola lusitanica*. Arch Ital Biol 108: 1–20

Brunelli M, Magni F, Pellegrino M (1977) Excitatory and inhibitory events elicited by a brief photic stimuli on flashing of the firefly *Luciola lusitanica* (Charp.). J Comp Physiol 119: 15–35

Buonamici M, Magni F (1967) Nervous control of the flashing in the firefly *Luciola lusitanica*. Arch Ital Biol 105: 323–338

Case J, Buck JB (1963) Control of flashing in fireflies. II Rôle of central nervous system. Biol Bull 125: 234–250

Lloyd JE (1971) Bioluminescent communications in insects. Annu Rev Entomol 16: 97–122

Papi F (1969) Light emission, sex attraction and male flash dialogues in a firefly *Luciola lusitanica* (Charp.). Mon Zool It (N.S.) 3: 135–184

VII. The Neurophysiological Basis of Mind

John Eccles preparing an experiment in Canberra 1953/54

Anomalous Contours: A Tool in Studying the Neurophysiology of Vision

G. Baumgartner, R. von der Heydt, and E. Peterhans, Zürich

Summary. The responses of neurons in area 17 and 18 to stimuli inducing anomalous contours were recorded. Neurons in area 17 do not respond to stimuli inducing anomalous contours, neurons in area 18 do; the latter also respond to dark or bright bars. It is suggested that neurons in area 18 are reached by convergent inputs from simple or complex field neurons and from stopped-end ("hypercomplex") neurons with orthogonal orientation.

Introduction

The organization of the receptive fields in the retina, the corpus geniculatum laterale, and area 17 has been studied by using elementary visual stimuli (light spots, grids, bars) (Kuffler 1953; Baumgartner 1961a, b; Hubel and Wiesel 1977). Leaving details aside, these studies have revealed a relatively uniform functional pattern of concentric receptive fields in the retina and the corpus geniculatum laterale. In the cortex a number of oriented field types with more specific features organized in columns or bands of common orientation and ocular dominance have been found (Hubel and Wiesel 1977). The circuitry within one level seems to be basically the same throughout the visual field, the difference in resolution depending on the density of receptors and the number of cortical columns per unit area of visual field as expressed in the magnification factor (Daniel and Whitteridge 1961). Some psychophysical parameters such as intensity, flicker fusion, brightness constancy, resolution and border contrast are reflected in the neuronal activity (Jung 1973). But little is yet known about the contribution of intrinsic cortical processing in the construction of perceived visual surrounding; a wide gap therefore exists between psychophysical and physiological approaches. Since perception must be correlated to the distribution of neuronal activity in larger areas of the cortex and the deep nuclei in a defined but yet unknown way, a better understanding of the mechanisms operating within these neuronal subsystems needs to be established. In analogy to the periphery, where the concentric receptive fields define the activity of different neurons and therefore function as organizers of visual information, we may assume that in cortical areas receptive fields are of a more complex nature. Here the term "receptive field" no longer seems meaningful and ought to be replaced by one which captures the functional characteristics. Support for this point comes from studies in areas V1, V2, V3, V4 and V5 (Hubel and Wiesel 1977; Zeki 1978, 1981). However, the interpretation of results obtained with rather elementary stimuli on the input side rests on the as-

Experimental Brain Research, Suppl. 9
© Springer-Verlag Berlin · Heidelberg 1984

Fig. 1. A Discontinuity stimulus inducing an anomalous contour. *B* Stimulus pattern to induce an anomalous bar. Both stimuli could be given in any orientation and the number and spacing of lines and the bar width could be varied. In Figs. 1–4 we reversed the stimulus contrast to simplify reproduction, i.e., the parts shown in black in this paper were bright and appeared superimposed on a homogeneous background with about 50% contrast.

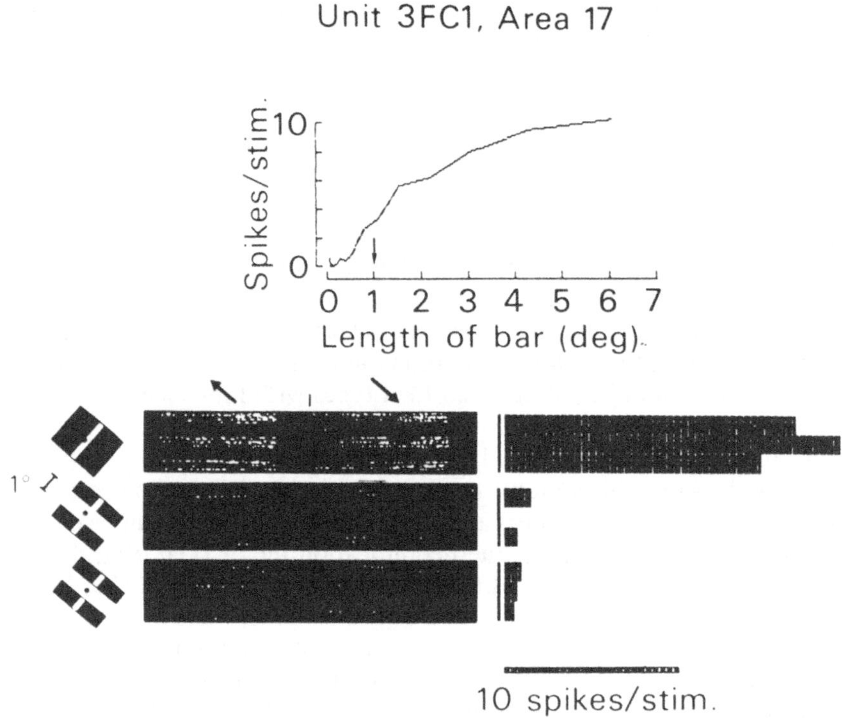

Fig. 2. Neuron in area 17 responding to movement of a dark bar of 45° orientation in both directions. No response to movement of the anomalous contour. The action potentials are represented as *white dots* for three times eight cycles. Saturation is reached with a bar length of about 6°. *Black points* indicate center of receptive field.

sumption that the cortical activity produced by a more complex stimulus can be inferred from the known responses to a number of elementary stimuli, thereby neglecting possible interactions or distortions produced in the cortical machinery itself.

Since there are configurations of stimuli which structure the visual surroundings by producing brightness differences, depths, and patterns without a true

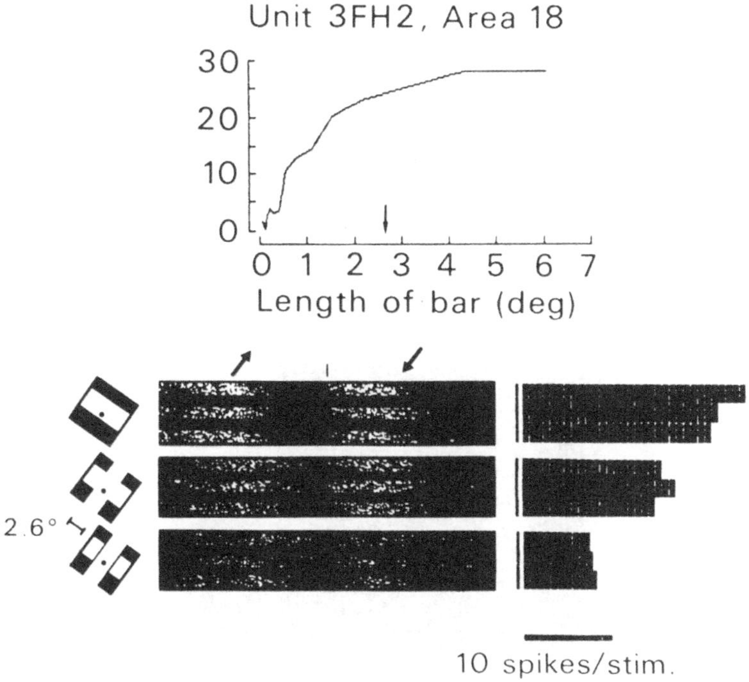

Fig. 3. Neuron of area 18 responding to a dark bar of -35° orientation in both directions. The real bar is most effective, but there is a clear response to the stimulus which produces an anomalous bar. The response is reduced by closing the notches by fine lines. Saturation with a bar length of about 4°.

physical correlate, we have studied the neuronal responses to such "illusory" or "anomalous" contours (Kanizsa 1979). These contours can also be seen dichoptically, exposing either eye only to parts of the stimulus. They must therefore be of cortical origin.

Methods

Rhesus monkeys (*Macaca mulatta*) were trained to fixate a small target and to indicate the period of fixation by pulling a key. They each received a microelectrode holder, implanted above V1 and V2. Single-neuron recordings were then made via glass-insulated platinum-iridium microelectrodes. Stimuli were presented on an oscilloscope screen driven by function generators which allowed patterns of various orientations, movements, and sizes to be produced. First, orientation and receptive field localization were determined with conventional bars and edges that were moved across the receptive field. Then we introduced stimuli with anomalous contours (see Fig. 1) corresponding perceptively to the effective stimulus. Stimulus I, the discontinuity stimulus, consisted of two abutting gratings of lines, slightly displaced, which produced an illusory contour orthogonal to the lines. Stimulus II was composed of two dark (light) bars separated by a gap,

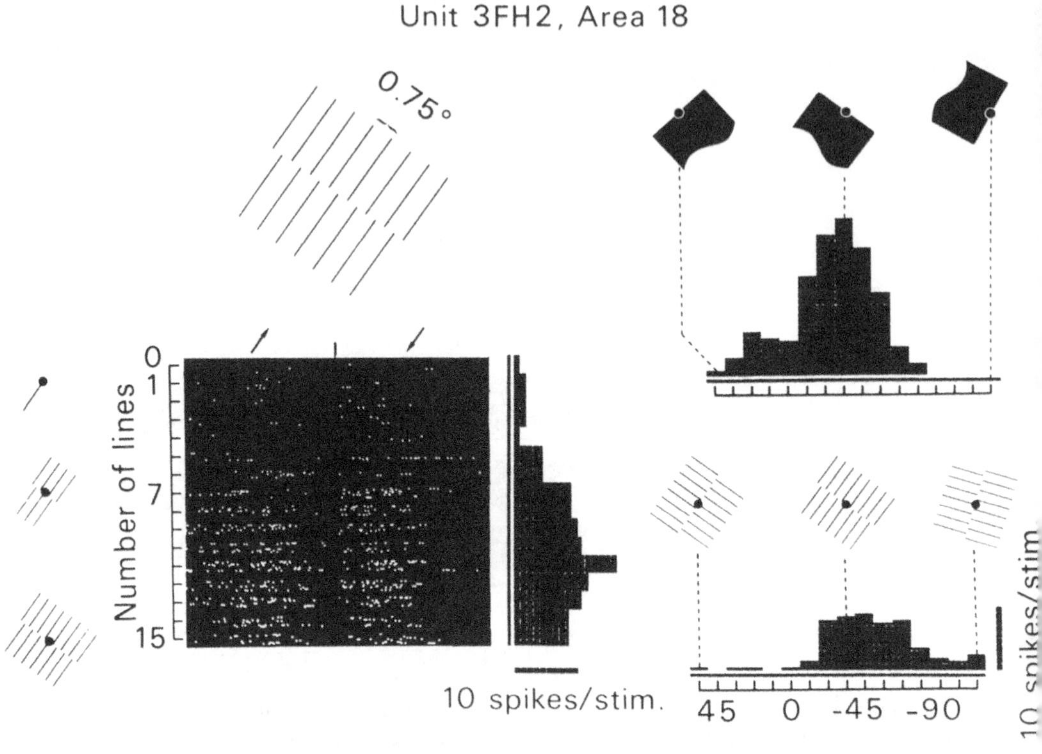

Fig. 4. Response of a neuron in area 18 to a discontinuity contour as a function of the number of lines generating the anomalous contour (*left*). Comparison of orientation tuning with an edge and the discontinuity stimulus (*right*).

with a pair of notches on both sides of the gap. This produces the illusion of a light (dark) bar overlapping the two dark (light) ones, and continuing across the gap. The illusory lines or bars were moved back and forth, and the responses of neurons measured. With stimulus I, the number of lines could be varied; with stimulus II, the gap and the width of the notches could be varied. The notches could be closed by a line in order to reduce the appearance of illusory contours. Recordings were made from 70 cells in area 17 and 80 in area 18.

Results

Out of 70 cells recorded in area 17 and tested with stimulus I, some responded to the grating when the lines were oriented parallel to the orientation of the receptive fields. Many cells did not respond at all. Stimulus II also produced negative results in area 17. Figure 2 shows an example of a neuron of area 17 which responded to a dark bar of 45° orientation, reaching saturation with a bar length of about 6°. In spite of this apparently large receptive field, there was no clear response to an illusory contour of the same orientation when the gap was 1° wide. Figure 3 is an example of a neuron in area 18 which was strongly activated by the

A

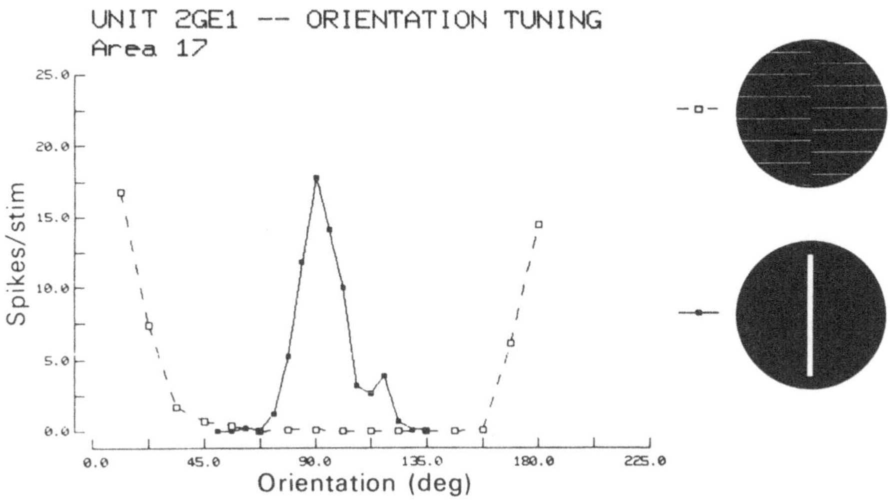

B

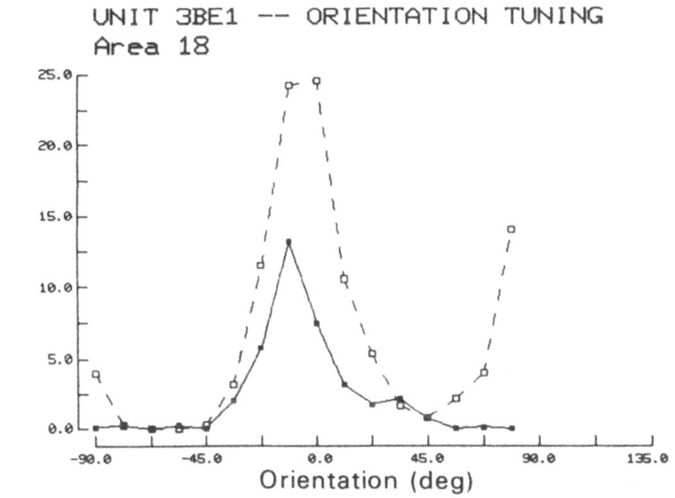

Fig. 5. A Orientation tuning curves of a neuron in area 17. No response to the discontinuity stimulus. *B* Orientation tuning curves of a neuron in area 18. Better response to the discontinuity stimulus inducing an anomalous contour than to the bar.

stimulus with illusory contours bridging a gap of 2.6°. There was a reduction in response when the notches were closed by thin lines. Saturation is reached at a bar length of about 4°. Figure 4 demonstrates the reaction to a grating with illusory contours as a function of the number of lines and the responses to an edge and to discontinuity at various orientations. It can be seen that the orientation

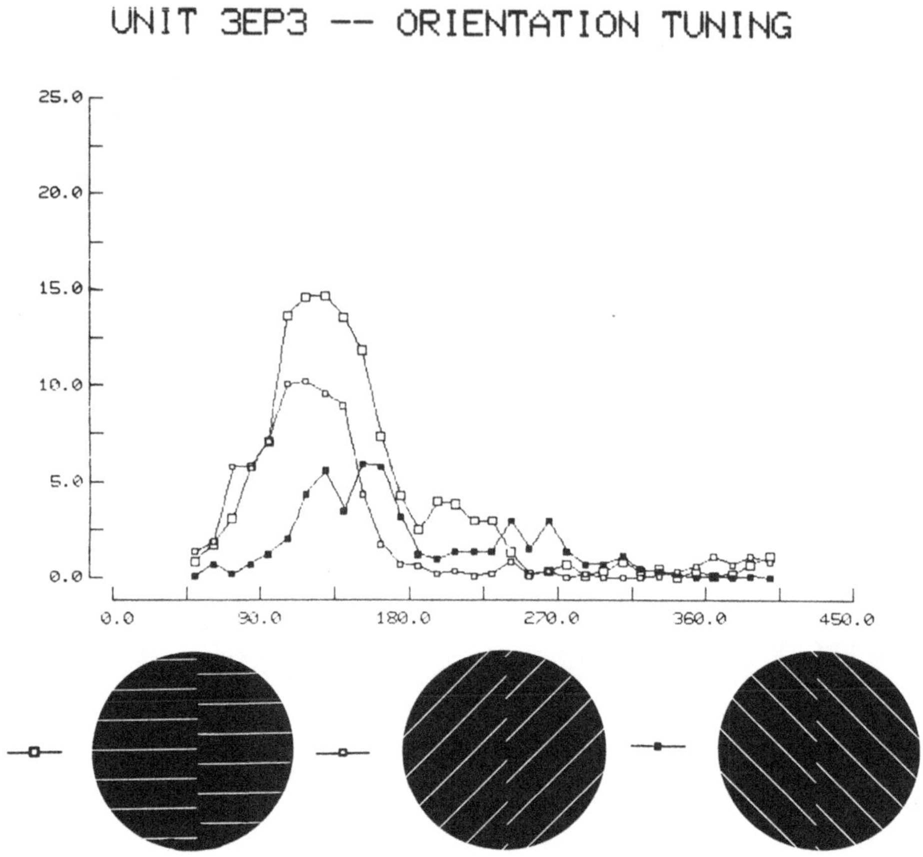

Fig. 6. Anomalous contours as a function of the orientation of inducing lines. The anomalous contour is best perceived and the neuronal response strongest with the lines orthogonal to the contour.

of the edge is nearly the same as the orientation of the illusory contour, which produces a clear activation.

In Fig. 5A and B the orientation tuning of neurons in area 17 and 18 is compared. The response to the illusory contours of the area 17 neuron is zero, while the neuron of 18 shows stronger activation by the discontinuity stimulus producing an illusory contour than by a bright bar. The anomalous contours are best induced by discontinuous lines orthogonal to the illusion. This is valid also for the neuronal response as shown in Fig. 6.

Discussion

The results indicate that only neurons in area 18 begin to respond to stimuli producing anomalous contours. Neurons in area 17 do not. The activation parallels the effectiveness of the stimulus to generate anomalous contours, i.e., the response is weaker under conditions which reduce the appearance of the contours, for example by reducing the number of lines or changing the orientation in the discontinuity figures and bridging the notches by a line in the anomalous bar stimulus. It is somewhat paradoxical that the neurons stimulated by the discontinuity pattern require lines whose orientation is orthogonal to the appearing contour and at the same time are activated by real edges and bars of corresponding orientation, for both stimuli exclude each other in neurons of area 17. One must therefore assume a convergence of signals of simple or complex field neurons of the same orientation as the anomalous contour and some stopped-end ("hypercomplex") neurons with orthogonal orientation in area 17 on neurons in area 18. This indicates that the activity of hypercomplex neurons may not or not only signal the orientations at which they respond best, but may be used also to produce the perception of a nonexistent line orthogonal to the orientation of the stimulus.

The explanations of illusory contours in psychophysics are still a matter of discussion and the data presented here are far from sufficient to offer a neuronal explanation. But they indicate that it should be possible to work out a neuronal solution and, at the same time, gain information about the extension of neuronal cooperation under some defined conditions. So far we do not have sufficient data about the details of convergence and are not sure to which layers these neurons belong. It is also not yet clear whether they are organized in a column-like fashion. But we expect that by improving the stimuli used and correlating the physical conditions with the appearance of anomalous contours we can improve our understanding of visual data processing and cooperativity in visual cortical areas.

References

Baumgartner G (1961a) Kontrastlichteffekte an retinalen Ganglienzellen. In: Jung R, Kornhuber H (eds) Neurophysiologie und Psychophysik des visuellen Systems. Springer, Berlin Göttingen Heidelberg, pp 45–53

Baumgartner G (1961b) Die Reaktionen der Neurone des zentralen visuellen Systems der Katze im simultanen Helligkeitskontrast. In: Jung R, Kornhuber H (eds) Neurophysiologie und Psychophysik des visuellen Systems. Springer, Berlin Göttingen Heidelberg, pp 296–311

Daniel PM, Whitterigde D (1961) The representation of the visual field on the cerebral cortex in monkeys. J Physiol 159: 203–221

Hubel DH, Wiesel TN (1977) Ferrier lecture: functional architecture of macaque monkey visual cortex. Proc R Soc Lond [Biol] 198: 1–59

Jung R (1973) Visual perception and neurophysiology. In: Jung R (ed) Handbook of sensory Physiology, vol VII, 3A. Springer, Berlin Heidelberg New York, pp 1–152

Kanizsa G (1979) Organization in vision. Essays on gestalt perception. Praeger, New York

Kuffler SW (1953) Discharge patterns and functional organization of mammalian retina. J Neurophysiol 16: 37–168

Zeki SM (1978) Uniformity and diversity of structure and function in rhesus monkey prestriate visual cortex. J Physiol 277: 273–290

Zeki SM (1981) The mapping of visual functions in the cerebral cortex. In: Katsuki Y, Norgren R, Sato M (eds) Brain mechanisms of sensation. Wiley, New York, pp 105–128

Attention, Readiness for Action, and the Stages of Voluntary Decision – Some Electrophysiological Correlates in Man

H.H. KORNHUBER, Ulm

Summary. Preceding all kinds of voluntary movement the maximum *Bereitschaftspotential* is over the supplementary motor area. In hand tracking the frontal and medial cortex (including supplementary motor area) show anticipatory behavior, the *Bereitschaftspotential* ends 500 ms earlier than the directed attention potential. The directed attention potential has its maximum at the lateral parietal area, contralateral to the external world side of attention.

In visuomotor learning there is a similar difference between the frontal and medial cortex on the one hand and the posterior-lateral cortex on the other hand: only the frontal and medial cortex shows a high correlation between the enhancement of the negative potential and motor learning (as measured by error reduction).

A theory is presented to explain the temporal correlation of the supplementary motor area activity with the onset of all kinds of voluntary action:

1. What to do is supervised mainly by the orbital cortex with its hypothalamic, limbic and amnestic afferents.
2. How to do it is guided by the frontolateral cortex with its sensory and spatial afferents from the temporal and parietal lobe.
3. When to start is decided by the supplementary motor ares with its multiple afferents from motivation, motor and other systems.

For the explanation of the high degree of human freedom of behavior, central nervous organization is underlined in contrast to the current emphasis on chance events.

Key words: *motor system – motivation – learning – attention – freedom – event related potentials*

All information which is received by an organism should be used for the guidance of its behavior. Direct omniconvergence of all senses is, however, impossible. The function of setting priorities in information processing by means of evaluation is attention (Kornhuber 1983). The lowest part of the attention system is the ascending reticular activating system of the midbrain, which is responsible for the waking state. The middle part consists of the medial and intralaminar nuclei of the thalamus, the external pallidum, and the rostral pole of the thalamus (Hassler 1980). It has access to all cortical areas and via the striatum has afferents from all cortical fields. Thus all sensory, motivational, mnestic, and motor systems influence this important function, which is able to suppress irrelevant information

Experimental Brain Research, Suppl. 9
© Springer-Verlag Berlin · Heidelberg 1984

and to concentrate on important signals. Attention is the function which enables us to make optimum use of our sensory capacities despite the limited span of consciousness. Bilateral lesions of the attention system (in the medial thalamus or pallidum) result in vigil coma or akinetic mutism.

When an electrophysiological correlate of the preparatory process for action in the human brain was found it was called *Bereitschaftspotential* or "readiness potential" (Kornhuber and Deecke 1965) because it was felt that this preparatory process for self-paced action in the absence of external stimuli is different from attention. The maximum amplitude of the *Bereitschaftspotential* is over the midline juxtalimbic supplementary motor area preceding finger, hand, eye, and speech actions (Becker et al. 1972; Deecke et al. 1976; Deecke and Kornhuber 1978; Grözinger et al. 1979). It has therefore been concluded that the supplementary motor area is a way by which drive and will are channeled into action (Kornhuber 1980; Eccles 1982). Supporting evidence comes from studies of the regional cerebral blood flow (Roland et al. 1980). The regional cerebral blood flow method has a poor time resolution, so that it cannot distinguish between events before and after movement onset. It has, however, a good spatial resolution.

Our new experiments (Deecke et al. 1984) clearly distinguish between the attention and readiness for action processes. They show a parietal maximum of the attention potential as should be expected from previous clinical and experimental findings (Brain 1941; Semmes et al. 1963; Lynch et al. 1977). They confirm the medial maximum of the *Bereitschaftspotential*, and they reveal a different time course of the readiness and attention process.

In these experiments the human subjects looked straight ahead with the gaze fixed on a point. In the left visual field there was a television screen on which a moved visual stimulus could be displayed. Alternatively (in the tactile experiment), a moving tactile stimulus was applied to the subject's left-hand palm. The subjects held a stylus in their right hand with a trigger contact. They moved their right hand down voluntarily in a self-paced manner to contact a paper. As soon as the pen was in contact with the paper, a light spot on the television screen started to move for exactly 1 s in a random direction, then changing direction and moving for another second in another random direction. The two directions of the moved stimulus were unpredictable for the subject. However, the slow stimulus speed was constant and the two durations (1 s each) were predictable. At the end of the second stimulus movement, the stimulus returned to the initial position with a fast step. The task for the subjects was to track the two smooth movements of the light spot in their left visual field (or, alternatively, the moved tactile stimulus on the left hand) with their right hand, using the ball-point stylus. The tactile stimulus was out of the subject's sight. Hearing was excluded by a masking noise via earphones. The subjects were instructed that the accuracy of their tracking performance would have an effect on the payment. The visual and tactile tasks were in separate runs of 64 trials each. The whole experiment consisted of 128 visual and 128 tactile trials (after rejection of artifact-contaminated recordings). Control experiments were done with the same stimuli without tracking.

As shown by the grand averages across the 16 subjects, the voluntary hand movement (by which the stimulus was started) was preceded by a *Bereitschaftspotential* which in this case can even be seen over the occipital lobe. The maximum

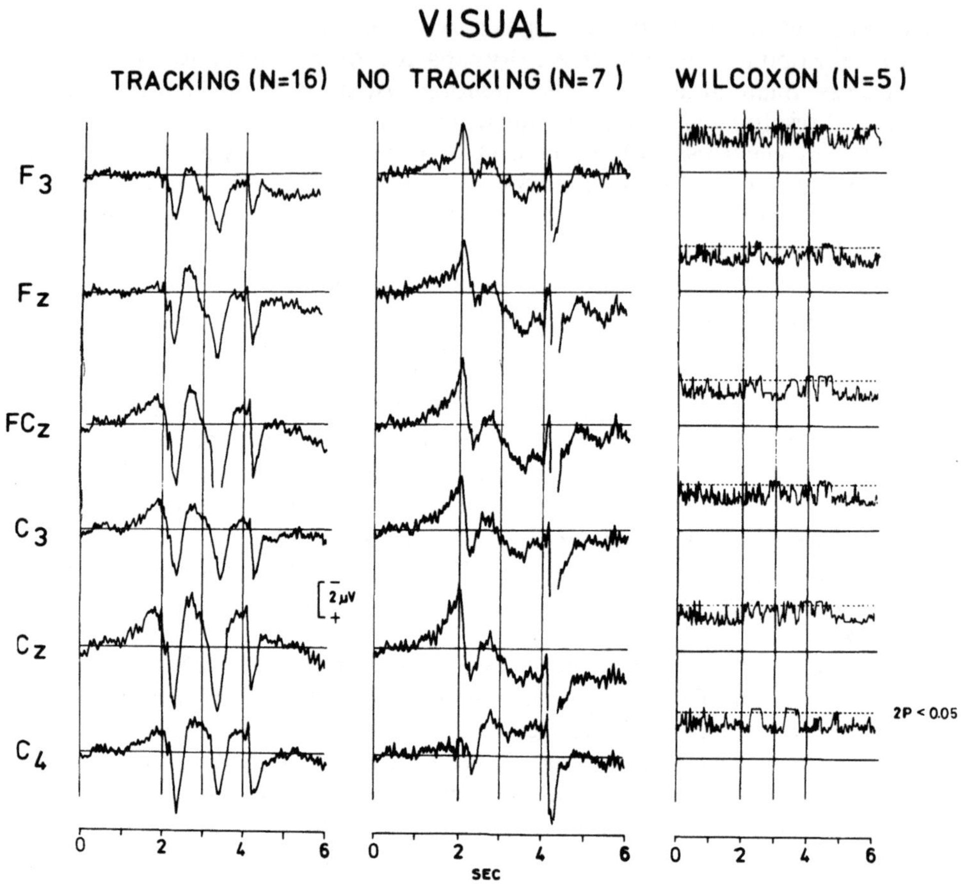

Fig. 1. Cerebral potentials in the visuomotor hand tracking experiment of Deecke et al. (1984) described in the text, comparison between the tracking and no-tracking tasks. Grand averages across 16 subjects for the tracking and seven subjects for the no-tracking task. Monopolar recordings versus linked ears. 10–20 nomenclature for the leads. *Ordinates,* amplitudes in μV, negativity up. *Abscissa,* time in seconds. Movement onset at s 2. Change in direction of the stimulus at s 3, reset of the stimulus at s 4. To the *right,* Wilcoxon test of paired differences. In this figure the frontal and central leads are shown. (Data from Deecke et al. 1984)

Bereitschaftspotential amplitude, however, is over the supplementary motor area at the electrodes Cz and FCz. Then comes a typical premotion positivity (PMP) (Deecke et al. 1969), which again is largest over the supplementary motor area at FCz. It follows over all leads a large positive potential which we call "relaxation potential." There is a sharp negative wave of the sensory evoked potential within the declining slope of the relaxation potential even at frontomedial leads. The fact that this sharp potential is preserved in the grand average is due to the high interindividual consistency. All 16 subjects were similar in their recordings so that even details of the individual potential can be seen in the grand average. However, there is an important difference between the leads: while over the anterior midline and frontal leads the relaxation potential begins with the premotion positivity (about 100 ms before the movement onset), the relaxation begins over the poste-

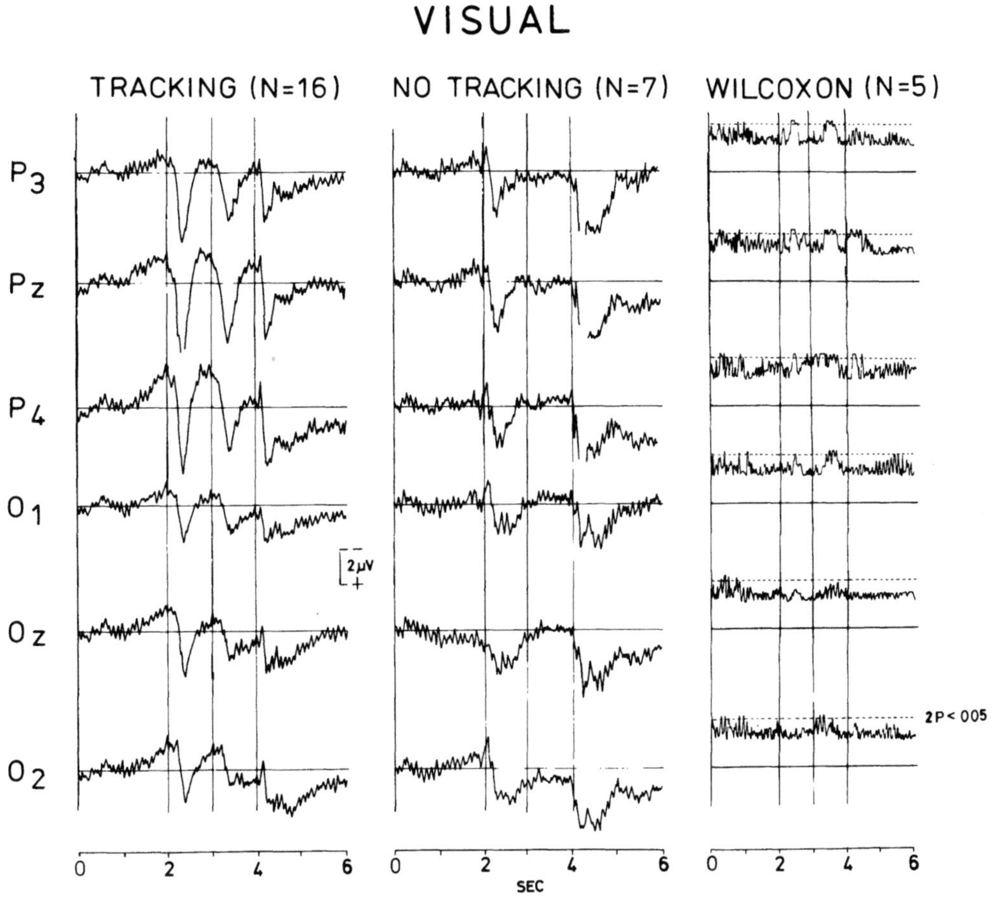

Fig. 2. The same experiment as shown in Fig. 1; here the parietal and occipital leads. (Data from Deecke et al. 1984)

rior and lateral areas of the forebrain about 200 ms after stimulus onset – which corresponds to the fact that the sensory and sensory association areas pay attention to the stimulus direction, while the motivation cortex, after giving its command, can afford to relax earlier.

About 450 ms after stimulus onset the relaxation potential ends and another negative potential begins which again lasts 200–300 ms after stimulus reversal over the parietal and occipital areas. A comparison between the right and left electrodes shows that the potential is larger over the right side, that is to say the side ipsilateral to the moved hand, but contralateral to the external stimulus. This means that the potential is related to attention rather than to the motor process. It has therefore been called "directed attention potential" or DAP. The directed attention potential is largest over the right parietal area.

By contrast, the potential over the anterior midline and over the frontal cortex declines already 300 ms before the stimulus reversal, going into another relaxation phase 500 ms before the posterolateral sensory areas of the cortex. This earlier re-

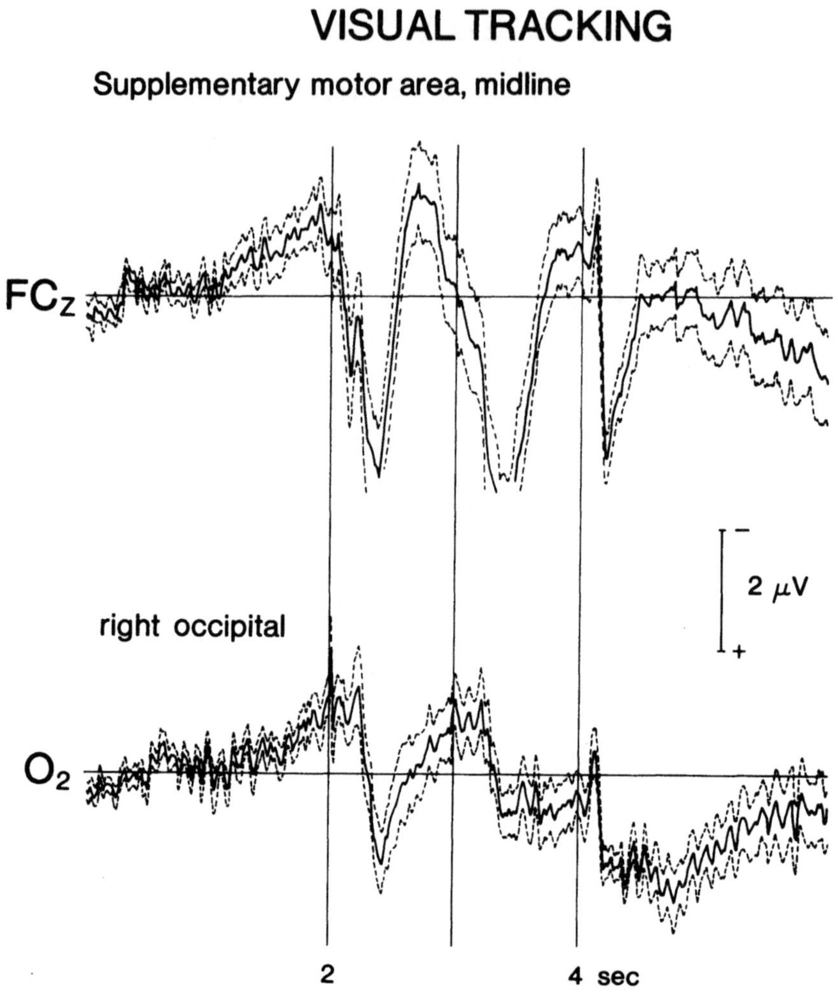

Fig. 3. Same experiment as Fig. 1, comparison of the time course of potentials at FCz (overlying the supplementary motor area) and 02 (right occipital lead). Note the anticipatory behavior of the negative potential at FCz which ends before the stimulus reversal (s 3), about 500 ms prior to the end of the directed attention potential in the occipital lead. Grand average from the 16 subjects, with standard error (*dotted lines*); note the small intersubject variability of the potentials. (Data from Deecke et al. 1984)

laxation is, of course, due to the fact that in our stimulus situation with a fixed time program the motivation system is able to anticipate the change in stimulus direction and to give the necessary commands well in advance. There are, in other words, two clearly different parts of the brain in this situation: the anterior and midline motivational brain with anticipatory behavior and the posterior and lateral attention brain with cognitive behavior.

It is interesting to note that even the motor cortex hand area contralateral to the moved hand (C3) shows a relaxation potential during the tracking. Obviously, the movement is largely executed by subcortical mechanisms (Kornhuber 1974)

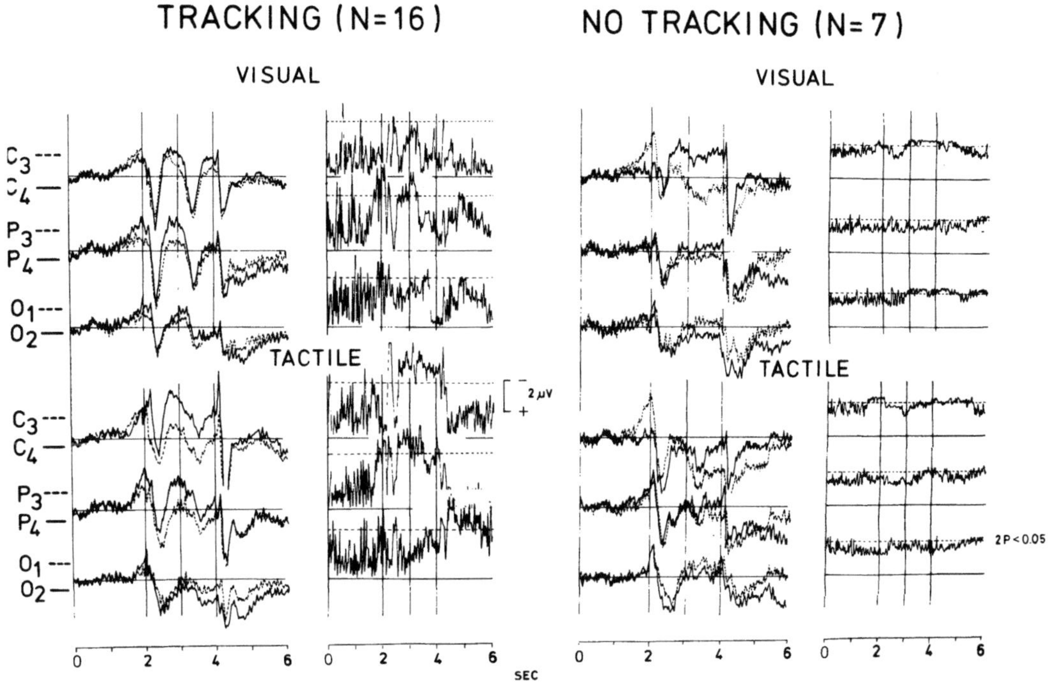

Fig. 4. Same experiment as Fig. 1, comparison to the right and left potentials in the visual and tactile experiments. The directed attention potential is higher over the right hemisphere (right central higher than left central, right parietal higher than left parietal), although the right hand is moving the stylus. (Data from Deecke et al. 1984)

under the guidance of the posterolateral association cortex. This is further evidence for the distributed organization of the motor system (Kornhuber 1984a). This distribution is partly the consequence of the evolution of more advanced systems without destruction of the older systems, but partly also a fundamental principle of organization of the higher nervous systems. The cortex is a parallel computer with many active processors. The superior colliculus of the midbrain is something like the sensorimotor cortex for eye movements (with help from cognitive and motivational mechanisms of the forebrain). The sensorimotor cortex for speech is, in my opinion, Wernicke's speech area of the temporal lobe, because lesions of the motor cortex do not cause aphasia and lesions of Broca's area result only in transient aphasia, whereas lesions of Wernicke's area cause long-lasting aphasia; the most severe kinds of aphasia, however, result from combined lesions of Wernicke's cortex and the basal ganglia (Brunner et al. 1982) – thus a predominantly sensorimotor system, the basal ganglia, is able to compensate partially for Wernicke's speech area. The same tongue movement which during chewing is regulated by the motor cortex with its tactile and proprioceptive afferents (preventing lesions of the tongue by the teeth) is also guided during speech by the temporal lobe with its auditory and mnestic-semantic afferents.

This distributed nature of the motor system, although it is different from the picture presented in the text books, is extremely helpful not only for rapid execution of complex tasks but also for compensation of lesions by learning and finding roundabout ways within the brain (Kornhuber 1984a). The belief in the omnipotency of the motor cortex for all voluntary movements paralyzes the therapeutic will of neurologists.

In the no-tracking control experiments of Deecke et al. (1984) there are, of course, the sensory evoked potentials due to the stimulus onset, reversal, and return events, but there is much less negative potential during the motion of the stimulus. In addition there is no negative *Bereitschaftspotential* at the occipital leads in the no-tracking situation, which shows that the negative occipital potential during the *Bereitschaftspotential* period in the tracking task is an attention potential.

I will not go into further details about these results. It should, however, briefly be discussed why the parietal maximum of the attention activity is not apparent in Grey Walter's expectancy wave or contingent negative variation (CNV), which appears in conditioning experiments with a sound as conditioning and a light flash as indicative stimulus. The CNV has its maximum at frontal leads (Deecke and Kornhuber 1977). There are two reasons for this: firstly, in the CNV situation the attention potential is mixed with the *Bereitschaftspotential*, and there is no condition for anticipatory behavior of the latter. And secondly, there is no sophisticated sensory analysis required in this stimulus situation. What matters is only the temporal relation and the fast response. It is, however, known from single-unit recording in the delayed response task (Fuster 1973) that the convexity of frontal lobe is active in this situation. One of the functions of the frontal cortex is probably the temporal coherence of behavior. Lesions in the lateral surface of the frontal lobe in the monkey result in the well-known defect in the delayed response task. The frontal animal is not able to concentrate on the task for more than a few seconds or to inhibit the effects of interfering signals; as soon as the stimuli are out of sight they are out of the animal's mind (Jacobson 1935).

The distiction between the motivational and attentional brain is not only obvious from the slow DC shifts but also from the frequency content of the EEG. We have now developed a method for the analysis of the frequency content of the EEG from short durations. This was difficult because of the artifacts caused by cutting out the movement-related parts of the traces. With this method it turns out that the anterior medial cortex, corresponding probably to the supplementary motor area, shows an activation of the theta rhythm during the *Bereitschaftspotential* period before onset of voluntary movement. This is in contrast to the posterolateral areas of the brain. A similar theta activation over the supplementary motor area occurs during the evaluation period following aimed hand movements (Foit et al. 1983). The source of the theta activity might be the hippocampus.

The difference between the anterior and medial motivational brain and the posterior-lateral sensory brain also appears clearly in experiments on human visuomotor learning (Lang et al. 1983). When inverted (mirror) hand tracking is compared to direct tracking, the negative cerebral potentials are higher during the inverted tracking task. The motor learning (as measured by error reduction) in the inverted tracking task is highly correlated with the enhancement of the negative potential. This correlation holds only for the frontolateral and frontomedial

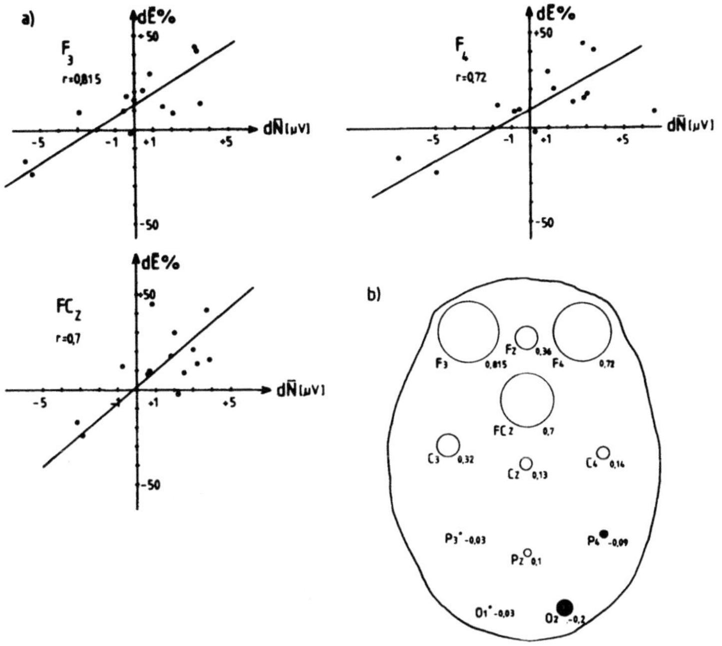

Fig. 5. a Correlations between the reduction of error in motor performance (*dE%*) and the enhancement of cortical negativity (*dN*) when the tracking and inverted tracking tasks in the experiment of Lang et al. (1983) are compared for every subject. There are significant correlations only in frontolateral (*F3* and *F4*) and frontomedial (*FCz*) areas. *b* Topography of the correlations in the experiment of Lang et al. (1983) (inverted "mirror" tracking vs tracking tasks). Frontal areas *up*, occipital *down*. The *diameters* of the circles are proportional to the correlation factors, positive correlations open circles, negative filled circles. (Lang et al. 1983)

electrodes including the supplementary motor area, but not for the posterior sensory and association areas (Fig. 5).

If all of the frontal lobe is devoted to motivational mechanisms, why is only the medial cortex, especially the supplementary motor area, active as shown by a surface-negative potential shift preceding simple finger or hand (Deecke et al. 1969), eye (Becker et al. 1972), or mouth (Grözinger et al. 1979) movements, but not the frontolateral areas? I believe it is because the first two of the three questions that have to be answered by voluntary actions, namely what, how, and when, are already answered beforehand in the experimental situation, so that only the third question (when to start) remains to be decided, and this is, as I now presume, the function of the supplementary motor area. The supervision of the task "what to do" may be a function mainly of the orbital cortex with its hypothalamic, limbic, and mnestic afferents. "How to do it" is in novel situations probably mainly a task for the frontolateral cortex with its afferents from the parietal and temporal teleceptive association areas. This hypothesis is based regarding the orbital and frontolateral cortex on the results of Karl Kleist (1934). Lesions of the orbital cortex disturb the moral self-continuity of the individual, resulting in disinhibition of primitive drives. Lesions of the lateral convexity of the frontal lobe disturb the adaptation of behavior to the external situation, as also shown by the delayed response deficit (Jacobson 1935). On the other hand, the electrical potentials correlate best over the supplementary motor area with the moment immediately preceding a start of voluntary action (Kornhuber and Deecke 1965; Deecke et al. 1969; Deecke and Kornhuber 1978).

An interesting aspect of the results of both Kleist (1934) and Lang et al. (1983) is that a widespread opinion regarding the function of the cerebral cortex is probably erroneous: the assumption that the cerebral cortex subserves cognition while motivation is mainly localized in some small subcortical ganglia. About half of the human cerebral cortex, the frontal and medial areas, subserves motivation and will. This agrees obviously with the importance of motivational and related acts in human consciousness. Since the time of Aristoteles the definition of man has been made by reference to language. If, however, will is properly distinguished from drive and adequately understood as an ability to set priorities for long-term planning and action, one could also define man as the animal with will. It is characteristic of our time that in the *Encyclopedia Britannica* of 1983 there is an entry neither under freedom nor under will (except for will in the sense of testament).

Human will, as human mind, is related to freedom – a most confused subject today. I feel grateful to Eccles for discussing the freedom of will seriously (Eccles and Zeier 1980). It is nowadays a fashion to explain freedom by chance events. However, chance events occur in the earthworm as well as in man, but there is a tremendous difference between them in the degree of freedom. It is higher abilities because of higher nervous organization that create higher freedom (Kornhuber 1978, 1984b), and it is mainly the large association cortex which is the basis of this; a large part of the association cortex, however, is devoted to motivation and will.

Acknowledgment. This work was supported by *Deutsche Forschungsgemeinschaft*. I greatfully appreciate the help of my co-workers Prof. Lüder Deecke, Dr. Michael Lang, Dr. Wilfried Lang and Anselm Kornhuber.

References

Becker W, Hoehne U, Iwase K, Kornhuber HH (1972) Bereitschaftspotential, prämotorische Positivierung und andere Hirnpotentiale bei sakkadischen Augenbewegungen. Vision Res 12: 421–436

Brain WR (1941) Visual disorientation with special reference to lesions of the right cerebral hemisphere. Brain 64: 244–272

Brunner RJ, Kornhuber HH, Seemüller E, Suger G, Wallesch CW (1982) Basal ganglia participation in language pathology. Brain Lang 16: 281–299

Eccles JC (1982) The initiation of voluntary movements by the supplementary motor area. Arch Psychiat Nervenkr 231: 423–441

Eccles JC, Zeier H (1980) Gehirn und Geist. Kindler, Zürich

Deecke L, Scheid P, Kornhuber HH (1969) Distribution of readiness potential, pre-motion positivity and motor potential of the human cerebral cortex preceding voluntary finger movements. Exp Brain Res 7: 158–168

Deecke L, Grözinger B, Kornhuber HH (1976) Voluntary finger movement in man: cerebral potentials and theory. Biol Cybernetics 23: 99–119

Deecke L, Kornhuber HH (1977) Cerebral potentials and the initiation of voluntary movement. In: Desmedt JE (ed) Attention, voluntary contraction and event-related cerebral potentials. Karger, Basel, pp 132–150 (Prog clin neurophysiol, vol 1)

Deecke L, Kornhuber HH (1978) An electrical sign of participation of the mesial "supplementary" motor cortex in human voluntary finger movement. Brain Res 159: 474–476

Deecke L, Heise B, Kornhuber HH, Lang M, Lang W (1984) Brain potentials associated with voluntary manual tracking: Bereitschaftspotential, conditioned premotion positivity, directed attention potential and relaxation potential; anticipatory activity of the limbic and frontal cortex. In: Karrer R, Cohen J, Tueting P (eds) Brain and information: event-related potentials. Ann NY Acad Sci 425

Foit A, Grözinger B, Kornhuber HH, Diekmann V (1983) Motor programming and outcome analysis center: the supplementary motor area. EEG and ERP changes related to aimed movement. Seventh int conf on event-related potentials of the brain EPIC VII, Florence, p 85

Fuster JM (1973) Unit activity in prefrontal cortex during delayed response performance; neuronal correlates of transient memory. J Neurophysiol 36: 61–78

Grözinger B, Kornhuber HH, Kriebel J (1979) Participation of mesial cortex in speech: evidence from cerebral potentials preceding speech production in man. In: Creutzfeldt O, Scheich H, Schreiner C (eds) Hearing mechanisms and speech. Springer, Berlin Heidelberg New York, pp 189–192 (Exp Brain Res Suppl 2)

Hassler R (1980) Brain mechanisms of intention and attention with introductory remarks on other volitional processes. In: Kornhuber HH, Deecke L (eds) Motivation motor and sensory processes of the brain: electrical potentials, behaviour and clinical use. Elsevier North-Holland, Amsterdam, pp 585–614 (Prog in brain res, vol 54)

Jacobson CF (1935) Functions of the frontal association area in primates. Arch Neurol Psychiat 33: 558–569

Kleist K (1934) Gehirnpathologie. JA Barth, Leipzig

Kornhuber HH (1974) Cerebral cortex, cerebellum and basal ganglia: an introduction to their motor functions. In: Schmitt FO, Worden FG (eds) The neurosciences, Third Study Program. MIT Press, Cambridge, pp 267–280

Kornhuber HH (1978) Geist und Freiheit als biologische Probleme. In: Die Psychologie des 20. Jahrhunderts, vol VI. Kindler, Zürich, pp 1122–1130

Kornhuber HH (1980) Introduction. In: Kornhuber HH, Deecke L (eds) Motivation motor and sensory processes of the brain: electrical potentials, behaviour and clinical use. Elsevier North-Holland, Amsterdam, pp IX–XII (Prog in brain res, vol 54)

Kornhuber HH (1983) Functional interpretation of the multimodal convergence in the central nervous system of vertebrates. In: Horn (ed) Fortschr Zool, vol 28. Fischer, Stuttgart

Kornhuber HH (1984a) Mechanisms of voluntary movement. In: Prinz W, Sanders AF (eds) Cognition and motor processes. Springer, Berlin Heidelberg New York Tokyo, pp 163–173

Kornhuber HH (1984b) Von der Freiheit. In: Lindauer M, Schöpf A (eds) Wie erkennt der Mensch die Welt? Grunlagen des Erkennens, Fühlens und Handelns. Klett, Stuttgart

Kornhuber HH, Deecke L (1965) Hirnpotentialänderungen bei Willkürbewegungen des Menschen: Bereitschaftpotential und reafferente Potentiale. Pflügers Arch ges Physiol 284: 1–17

Lang W, Lang M, Kornhuber A, Deecke L, Kornhuber HH (1983) Human cerebral potentials and visumotor learning. Pfügers Arch Eur J Physiol 399: 342–344

Lynch JC, Mountcastle VB, Talbot WH, Yin TCT (1977) Parietal lobe mechanisms for directed visual attention. J Neurophysiol 40: 362–389

Roland PE, Larsen B, Lassen NA, Skinhoj (1980) Supplementary motor area and other cortical areas in organization of voluntary movements in man. J Neurophysiol 43: 118–136

Semmes S, Weinstein S, Genth L, Teuber HL (1963) Correlates of impaired orientation in personal and extrapersonal space. Brain 86: 747–772

Electrophysiological Cues of the Language-Dominant Hemisphere in Man: Slow Brain Potentials During Language Processing and Writing*

RICHARD JUNG, Freiburg/i.Br.

Summary. Slow brain potentials during language processing and writing were recorded using averaging methods. The lateral preponderance of negative potentials over the left and right cerebral hemispheres was compared in 24 right-handers and 12 left-handers. The left or right preponderance appeared best in transverse leads between corresponding brain regions.

Language tasks and writing tasks were separated in time with forward averaging of two subsequent periods to distinguish their cerebral correlates. Language processing was tested without vocalization in order to exclude artifacts caused by speech. The influence of the writing hand was investigated by comparing the potentials during writing with either right or left hand.

Slow brain potentials during language and writing tasks showed large interindividual variations of form and polarity but fairly constant forms in the same individual. Selected samples of repeated tests were evaluated in right-handers and in two groups of left-handers writing with the left or the right hand respectively.

Language processing caused larger negativities in left than in right parieto-temporal, precentral, and frontal regions in most right- and left-handers. This is thought to indicate speech dominance of the left hemisphere. During writing, the preponderant lateral negativity was mainly determined by the writing hand, which caused larger contralateral negativity mainly in precentral leads.

During the two test periods of language processing and writing, different polarities of slow potentials appeared, with both earlobes serving as reference. In some individuals negativity during the language task reversed to positivity during the writing task.

Most left-handers showed a left-hemispheric preponderance during language tasks in spite of the right-sided cerebral location of hand dominance. None of the five left-handers writing with the right hand and only one of the seven left-handers writing with the left hand showed right-sided preponderance during language processing. Hence hand dominance and language dominance may be independent hemispheric qualities.

Handedness and speech dominance appear to develop rather independently in the cerebral hemispheres. They are later coordinated in writing. Language dominance and hand dominance coincide in the left hemisphere in right-handers, but are found in opposite hemispheres in most left-handers. The hand preference

* Supported by the Deutsche Forschungsgemeinschaft in Sonderforschungsbereich 70 (Hirnforschung und Sinnesphysiologie). I am very grateful to Dr. A. Hufschmidt, B. Natsch, and H. Kapp for effective methodological help and to Dr. M. Greenlee for language correction.

Experimental Brain Research, Suppl. 9
© Springer-Verlag Berlin · Heidelberg 1984

for writing is acquired by learning and does not necessarily follow innate hand dominance.

Introduction

During the last decade Eccles has become interested in higher functions of the cerebral cortex and the speech centers (Eccles 1980, 1981), but has not been able to explore these problems experimentally. It may, therefore, be appropriate to report some of our electrophysiological results on the language-dominant hemisphere in honor of his 80th birthday.

A previous study of slow potentials elicited during writing in right-handers described the interaction of the writing hand and of the language dominance (Jung et al. 1982). The readiness potentials appearing before movements (Kornhuber and Deecke 1965, Deecke and Kornhuber 1978) and the aiming potentials during goal directed movements were larger contralateral to the moving hand in right- and left-handers (Grünewald et al. 1979). In a new series of experiments in right- and left-handers the language processing and the writing act itself were separated in time to avoid the influence of the writing hand (Jung and Hufschmidt 1983). All language tasks were performed without vocalization in order to exclude the many artifacts caused by speaking (Grözinger et al. 1975). In the following some results are described that may help to identify the language-dominant hemisphere in man.

Methods

The slow brain potentials were recorded from the scalp with both ears as reference using a Mingograph (Siemens) of ten or 16 channels and a time constant of 5 s and stored on magnetic tape. Averaging was achieved backward in the first series of experiments, A, partly published in 1982 (Jung et al. 1982), and forward in the second series, B. Backward averaging summated the slow potential shifts before, during, and after writing in the A series. The forward-averaging B yielded summations in two periods: (I) during language tasks without writing, and (II) during the writing down of the results of the language problem solved during period I (Jung and Hufschmidt 1983). The duration of periods I and II was about 5 s in each case and could be varied according to the writing time for a total duration of 12–18 s.

The EEG potential shifts were led from nonpolarizable surface electrodes over symmetrical brain regions of the left and the right hemispheres, using both ears as common reference: bilateral precentral, parietal, frontal, parietotemporal, and occipital electrodes C_3 and C_4, P_3, P_4, F_3, F_4, PT_3, PT_4, O_3, O_4 of the international 10/20 electrode system of Jasper (1958). Midline electrodes were positioned at the vertex (C_z), parietal (P_z) and occipital (O_z) and also recorded with ear reference.

In addition, transverse bipolar leads recorded between corresponding left and right brain regions were used to record unilateral preponderance of negativity: leads between C_3 and C_4, P_3 and P_4, PT_3 and PT_4, and F_3 and F_4, as shown in Figs. 4–8.

Eye movements, writing pressure, and electromyogram (EMG) of hand muscles were recorded in separate channels of the Mingograph and, in part, were also summated by averaging. The writing pressure varied between 150 and 400 g in different subjects. In all records the pressure amplitudes were usually made 5–10 mm in height and no further quantification was attempted.

Averaging

A. The writing potentials were collected by retrograde summation of 32–64 repeated writing acts, each starting backward from 2.5 s after the end of the writing movements, recorded as writing pressure by a specially built pin. In addition, EMG potentials from the first interosseus muscle were recorded with a short time constant of 0.1 s.

B. Comparison of language processing and writing with forward averaging was conducted in two periods. Following an acoustic stimulus (pip 1), the subject was requested to solve a verbally presented language task, (a), in about 5 s without speaking and write it down, (b), after pip 2 in the subsequent 5–7 s. The brain potentials were summated for 12–18 s after the first pip, together with writing pressure and signals of the periods (a) and (b). Averaging was carried out off-line from magnetic tape recordings or/and on-line with a specially adapted computer.

Experimental Procedure

A. Writing tasks with backward averaging were either repeated writing of the same word or writing of different dictated works. Various control experiments were described in an earlier paper (Jung et al. 1982). Writing with the experienced hand was compared to that with the opposite inexperienced hand (Figs. 2,6,7).

B. Language tasks were: completion of sentences, translation of German words into English, translation of English words into German, finding common nouns for two words, and memorizing name and address. These language tests (I) preceded a 5 s writing period, (II) in which the results were written down. Subjects were asked to refrain from speaking or making eye movements during recording.

The total averaging time was usually 8–13 s in the A experiments and 12–18 s in the B experiments. The time was varied according to the length of writing, since writing with the inexperienced hand was usually slower (see Fig. 2).

Eye movements were suppressed by fixating a point next to the writing line. Subjects were asked not to follow their script with the eyes. Eye movements were recorded as controls, but not averaged. Respiration was recorded as control in some A experiments and in all B experiments, partly with averaging.

The polarity of the averaged EEG was designed so that negativity points upward and positivity downward at the scalp electrode over the brain region in relation to ear reference. In transverse leads higher negativities at the left electrode were poled upward, higher right negativities downward. In the transverse leads of Figs. 4–8, left preponderance was marked by horizontal, right by vertical hatching.

The peaks of negativity in the averaged unipolar leads were measured against an average level between the tasks when no eye movements occurred. Records contaminated by eye movement artifacts were eliminated by marking "good" recordings for summation in averaging data from A (Jung et al. 1982). In averaging data from B, some contamination by eye movements sometimes occurred in the last 2 or 3 s of the summation period of 15 s. These late artifacts did not, however, interfere with the previous periods of language processing and writing.

Subjects

Thirty-three healthy adults aged 20–35 years, who had had several years of writing experience after school, and three school-aged children participated in the experiments: 21 males and 15 females. Twenty-four were right-handers and 12 left-handers (nine adults, two boys aged 12 and 15, and one girl aged 16). Seven left-handers wrote with the left and five with the right hand.

Methodological Limits

Local and lateral differences of various brain regions were recorded in precentral, parietal, and parietotemporal areas and the midline in most experiments. The restricted number of simultaneous recordings, eight in the first and 15 in the second series, allowed only few bilateral recordings from frontal and occipital areas. Therefore, the frontal and occipital potential changes are not described in this paper and will be studied in further experiments.

Another limit is the time constant of 5 s and the lack of DC recordings in our experiments. We can assume, however, that within the averaged time periods of 5 s for language processing and writing the slow potentials were not too much deformed by the time constant.

A third limitation is the comparative averaging of transverse leads and "unipolar" leads using ear reference. When right and left precentral and parietal leads were recorded with ear reference, the corresponding transverse leads were computed by substraction of the right and left unipolar records from the tape. Such precentral transverse records, obtained by subtraction to demonstrate the lateral predominance, are shown in Figs. 4 and 5 together with the original recordings with ear reference.

Results

A. Brain Potentials During Writing in Right-Handers

Averaging Slow Potentials Before and During Writing with the Right Hand. Twenty-three right-handers were studied by retrograde summation of slow potentials. All experiments that showed disturbed baselines caused by eye movements and galvanic skin reflexes were discarded. The results of a detailed study of different writing tasks in four selected subjects were described and illustrated in an earlier paper (Jung et al. 1982). Figures 1–3 show some essential findings that may be summarized as follows.

Slow cerebral potential shifts were recorded from scalp electrodes at symmetrical positions in the left and right cerebral hemispheres and midline with both earlobes serving as reference. In addition, bipolar leads between the scalp elec-

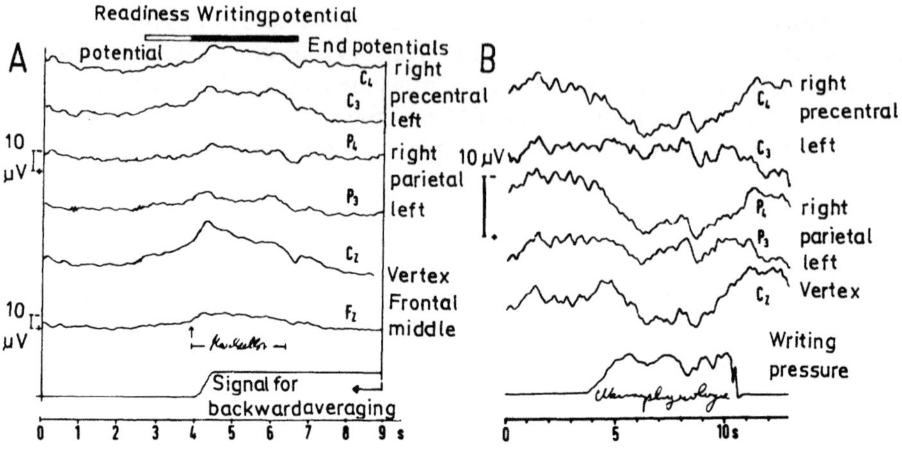

Fig. 1A,B. Slow potentials accompanying writing of repeated words in two right-handers recorded by backward averaging. The predominant polarity with ear reference was surface negative in *A* and surface positive in *B* after averaging 64 words written with the right hand. *A* Repeated writing of the name elicits negative potentials over both hemispheres, largest at the vertex and in the left precentral region. Right-hander, male, aged 31. *B* Repeated writing of "*Neurophysiologie*" elicits mainly positive potentials except in the left precentral region contralateral to the writing hand. Right-hander, male, aged 29. (From Jung et al. 1982)

trodes in sagittal and transverse planes were analyzed. All recordings were summated by backward averaging. Repeated writing of the same word was compared to writing different dictated words, to drawing, and to other control experiments. Surface-negative readiness potentials appeared 1–3 s before writing, similar to those preceding other voluntary movements.

During writing, the writing potentials in different cortical regions began with a negative increase in the readiness potential, usually followed by a plateau or positive waves. One or several biphasic potentials persisted for another 2 s after writing had ceased.

The writing potential showed large interindividual variations of form and polarity but had fairly constant forms in the same individual. The largest initial negativity occurred at the vertex and in the left motor region contralateral to the writing hand.

The characteristic intraindividual constancy of the pattern of writing potentials facilitated quantification in the same subject. Even with different writing tasks, such as writing the same word repeatedly or writing different dictated words, the characteristic polarity pattern retained its approximate form in the same individual (Jung et al. 1982). By comparing the amplitude over symmetrical left and right cortical regions during the writing action, possible signs of language dominance of one hemisphere were explored.

Left-Hemispheric Preponderance in Writing with the Right and Left Hand. In all subjects a larger negativity of the writing potentials occurred in the precentral region contralateral to the writing hand. This also occurred when the potentials were predominantly positive. The cerebral projection of the writing hand evi-

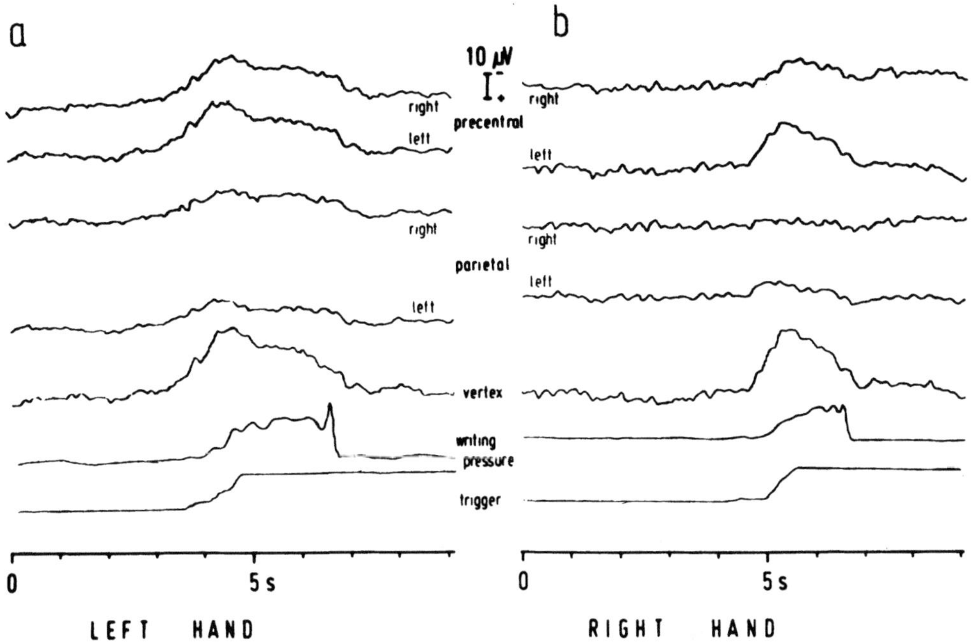

Fig. 2. Left-sided preponderance of slow potentials of a right-hander during writing of dictated words with either hand. *a* Writing with the left hand: the left precentral and parietal regions (C_3, P_3) show consistently larger amplitudes than the symmetrical leads from the right (C_4, P_4) with ear reference. Although the left hand effect increases the amplitudes over the contralateral right precentral and parietal regions in comparison to those in *b*, they remain smaller than on the symmetrical left side. *b* Writing with the right hand results in large amplitudes in the contralateral left precentral and parietal regions and small potentials in the right regions.
Writing with the inexperienced left hand is slower (3.5 s) than with the right hand (1.8 s). Right-hander, male, aged 31. Average of 32 different words four letters in length. (From Jung et al. 1982)

dently complicated the effect of the language-dominant hemisphere: the left-sided preponderance in right-handers appeared partly as a motor effect of the writing action. This interaction between the writing hand and speech dominance could be analyzed by asking the subject to write with the nonexperienced left hand. A lateral preponderance independent of the writing hand was regarded as a sign of language dominance.

The larger negative amplitude over the left hemisphere during language processing and writing were most obvious in subjects who showed a high negative plateau in the writing potential (Jung et al. 1982). Therefore, the lateral differences in the writing potentials were studied by employing different tasks in these subjects who wrote with either hand. Some lateral differences during writing with the right and left hand are shown in Figs. 2 and 3, which give a survey of various writing tasks.

Figure 2 compares writing with the left and with the right hand following dictation. All dictated words were four letters in length. Writing with the inexperienced left hand was considerably slower, as evident from the duration of writing and of the vertex potentials in Fig. 2a and b. A left-sided laterality

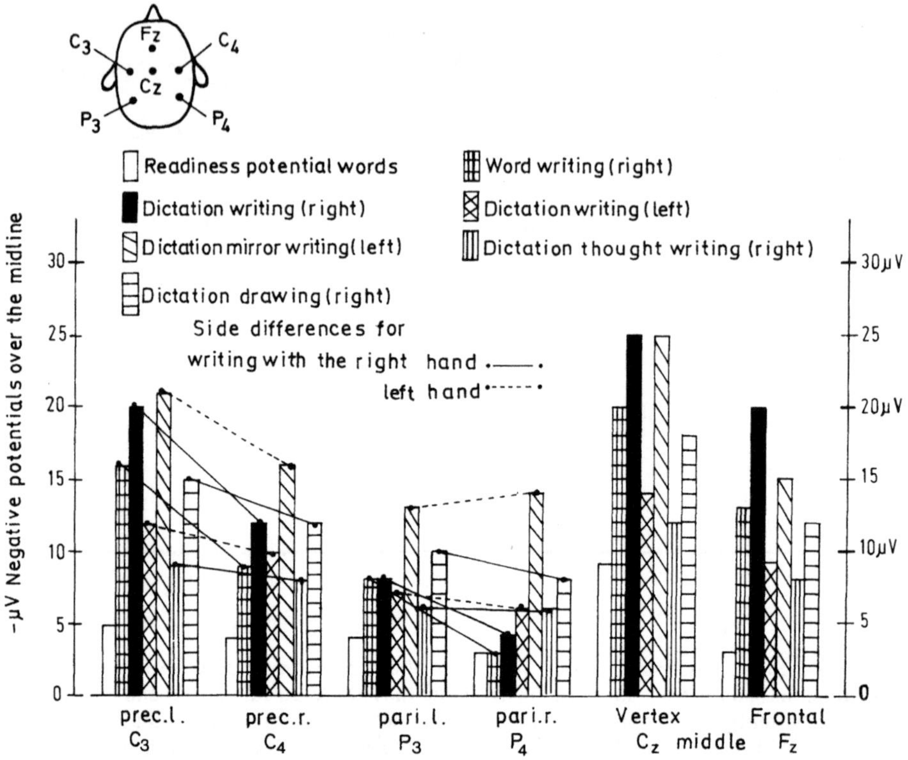

Fig. 3. Maximal negative left-sided preponderance of slow potentials during writing and drawing in a right-hander. This subject, also shown in Figs. 1A, 2, and 4 (W.M., aged 31), had very large negative amplitudes of writing potentials with both ears serving as reference; the largest peaks were compared in left and right precentral (C_3, C_4), parietal (P_3, P_4), vertex (C_z), and midfrontal regions (F_z) during writing and drawing. The *slopes of the lines* connecting the corresponding columnar peaks from left to right indicate the left-sided preponderance. *Continuous lines* designate writing with the right hand and *interrupted lines* left-handed writing. Right-sided preponderance is only observed in the parietal cortex during the writing of mirror script with the left hand. (From Jung et al. 1982)

with larger negative potentials over the left precentral and parietal regions was still present when the left hand was used, although the right hemisphere, contralateral to the left hand, now showed higher amplitudes than when the right hand was used. Because of this hand effect the left-sided preponderance is less marked while subjects write with the left hand. However, because the left-hemispheric preponderance is still evident, one may suppose that cerebral language processing occurs after word dictation to convert the acoustically perceived word into the motor pattern required for writing. We assumed that such persistent left-sided preponderance could be an accompaniment of language processing necessary for the writing act in the left speech-dominant hemisphere (Jung et al. 1982).

Writing dictated words that require language processing caused larger potentials in the left hemisphere than did repeated word writing. The lateral preponderance of negative peaks in both precentral and parietal areas during various writing tasks is shown in Fig. 3.

Figure 3 demonstrates the larger negativity of the left cortex in the difference between the slopes of the amplitudes in symmetrical recordings falling off from the left toward the right. The left-sided preponderance of the precentral area was preserved even when dictated words were written with the left hand. All writing tasks using the left hand are marked by hatched lines between the potential amplitudes of symmetrical left and right cerebral regions. In the parietal cortex an opposite right-sided preponderance appeared when the left hand was used to write mirror script. This action may need more motor programming and more spatial construction than normal writing. In this subject the right parietal cortex seemed to be more active during this unusual task of mirror writing, but in others we found no significant right parietal preponderance. When the same subject wrote mirror script with the right hand, he also showed higher amplitudes over the right parietal region.

In this subject mirror script with the left hand was the only exception to rule that greater negative amplitudes are exhibited in writing potentials over the language-dominant hemisphere. Other individuals, however, showed more exceptions to the normal left-sided preponderance when exhibiting large positive deflections in the writing potentials. In these subjects the measurement of negative preponderance was more difficult because effects of the writing hand and language dominance sometimes could not be distinguished clearly.

The observation that negative potentials with larger left-hemispheric amplitudes appeared after verbal stimuli also when the left hand was used was the first electrophysiological sign of language dominance. To prove that language information is processed in the speech-dominant hemisphere before and during writing, the effect of the writing hand had to be excluded by more detailed experiments.

B. Language Tasks and Writing in Right-Handers

It was not possible to distinguish the writing hand effect and the language dominance by recording the writing potentials using our methods published in 1982 (Jung et al. 1982). Furthermore, we wanted to avoid the rather time-consuming and complicated method of backward averaging. On the other hand, spoken language also had to be avoided because all slow potentials before and during speech are confounded by large respiration artifacts and other disturbances (Grözinger et al. 1975). These serious impairments of brain potential recordings during spoken language and the interaction of language processing and writing hand were excluded by language tasks that do not require speech and by use of forward averaging. The effect of the writing hand was excluded by a temporal separation of language processing and writing (Jung and Hufschmidt 1983).

Two Periods of Language Task (I) and Writing (II). Triggered by a first signal, the language tasks preceded the writing of the answer by about 5 s. The following tasks yielded similar results of left-sided preponderance of negative potentials: completion of sentences, categorizing of two words, translation of German words into English and vice versa, rehearsing and associating words preceding the writing act. During these tasks we measured the averaged slow brain potentials with 12–24 electrodes. When language processing and script performance were aver-

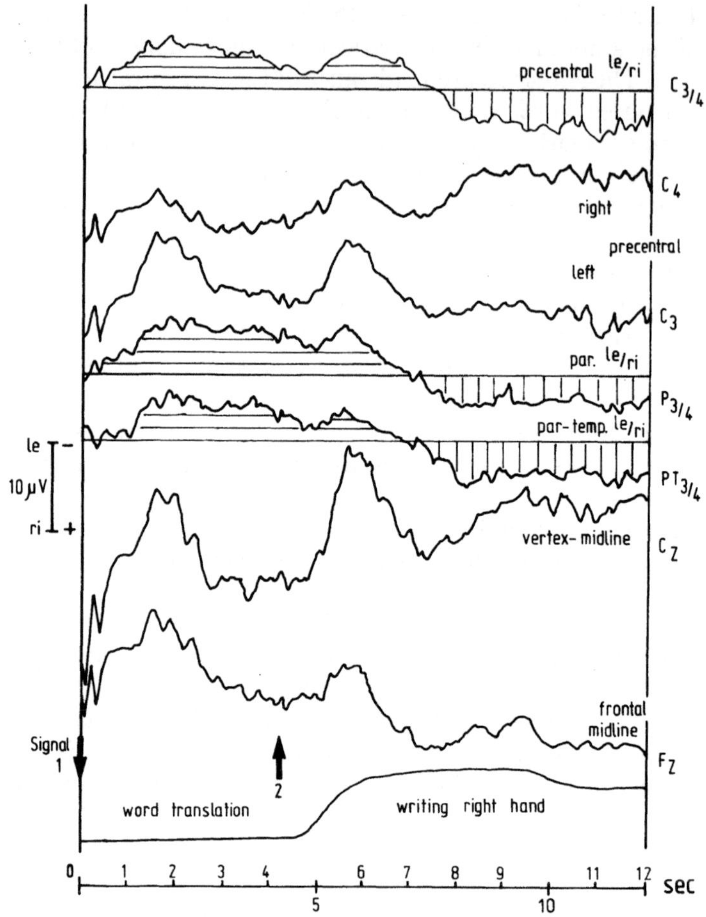

Fig. 4. Left-sided preponderance of slow potentials during language tasks and writing in a male right-hander, aged 31. During translation of German words after signal 1 (↓), all leads show surface-negative potentials with higher amplitudes over the left hemisphere. In the transverse leads between symmetrical left and right areas ($C_{3/4}$, $P_{3/4}$, $PT_{3/4}$), left-sided preponderance is marked by *horizontal hatching*. After signal 2 (↑), the subject writes down the translated English words with his right hand. Subsequently, another peak of negativity and left-sided preponderance appears, the writing potential. During the last phase of writing and following the writing act all transverse leads reverse into a period of right-sided preponderance (*vertical hatching*). After the first signal evoked potentials appear that are larger in "unipolar" leads with ear reference (C_3, C_4, C_z, F_z). The negative peaks show highest amplitudes in the vertex midline for slow potentials during translation and writing releated to ear reference. The left preponderance is somewhat larger in parietal than in precentral and parietotemporal areas. Average of 65 tasks.

aged in two distinct periods of about 5 s, their EEG accompaniments were different. By comparing the first time period (a) with the second (b), we were able to analyze the effects of pure language processing over both hemispheres and the midline and relate them to the writing act with the left and right hand.

In using two trigger signals, the first signaling the subject to start the language task, the second to start writing the answer, our experimental procedure is comparable to Walter's method of recording the expectancy wave, now called CNV (contingent negative variation), between the conditional and the "imperative" signals (Walter 1964/65). Of course, the language task involves more than simply expecting the second signal and being ready to act.

In 24 right-handers the negative potentials during these language tasks showed higher amplitudes over several regions of the left cerebral hemisphere than over the right (Fig. 4). The amplitude differences of transverse leads were often larger between the left and right precentral ($C_{3/4}$) and parietal ($P_{3/4}$) areas than between the temporal parietal areas ($PT_{3/4}$) located at the sensory speech center of Wernicke (1874). In some subjects the frontal areas showed a larger left preponderance in negative potentials during the first period of language processing.

In contrast to case with undelayed writing of dictated words (in the A experiments), the delayed writing potentials following the period of language processing in the B experiments were mostly brief peaks. In the transverse leads small peaks of left-sided preponderance were usually reversing into longer periods of right preponderance when writing stopped (Fig. 4).

In the subject W.M. the mean durations of the precentral writing potentials for words with four letters were as follows: negativity with ear reference 2.5–4 s in the A series and 1.5–2.5 s in the B series, left-sided preponderance 2–2.5 s in A and 1.5–2 s in B. Similar brief durations of left-sided writing potentials with a following reversal into right preponderance also occurred in some left-handers when they wrote with their right hand (Fig. 6b).

These experiments show that both duration and lateral preponderance of the writing potentials are shorter when they follow a language task (B series) than when they appear in a combined task of language processing and writing of dictated words (A series).

Another method of investigating the electrophysiology of language tasks was the comparision of verbal and visual stimuli that elicited the drawing of geometrical figures (Jung et al. 1982). Verbal evocation of figure drawings also showed larger negativities over left precentral, partietal, and parietotemporal regions when compared with mere visual copying of the same figures. Our experiments with this method, however, have been limited to a few subjects. Therefore, the results in a larger population using this and similar methods will be reported later.

C. Language Tasks and Writing in Left-Handers

Only 12 left-handers, seven writing with their left hand and five with their right hand, were tested, four of them in two to three repeated experiments that yielded consistent results, although the amplitudes of lateral preponderance varied slightly in different experiments. Ten of these left-handers showed a larger negativity in the left hemisphere during language tasks and only one in the right hemisphere; one had no clear lateral predominance. The larger negativity appeared in the left precentral (C_3), parietal (P_3), and parietotemporal (PT_3) electrodes with ear reference. Recordings from frontal lateral electrodes (F_3, F_4) were only obtained in six subjects and also showed left-sided preponderance. The preponderance is clearly demonstrated in transverse leads, as in Figs. 5–8. In all figures up-

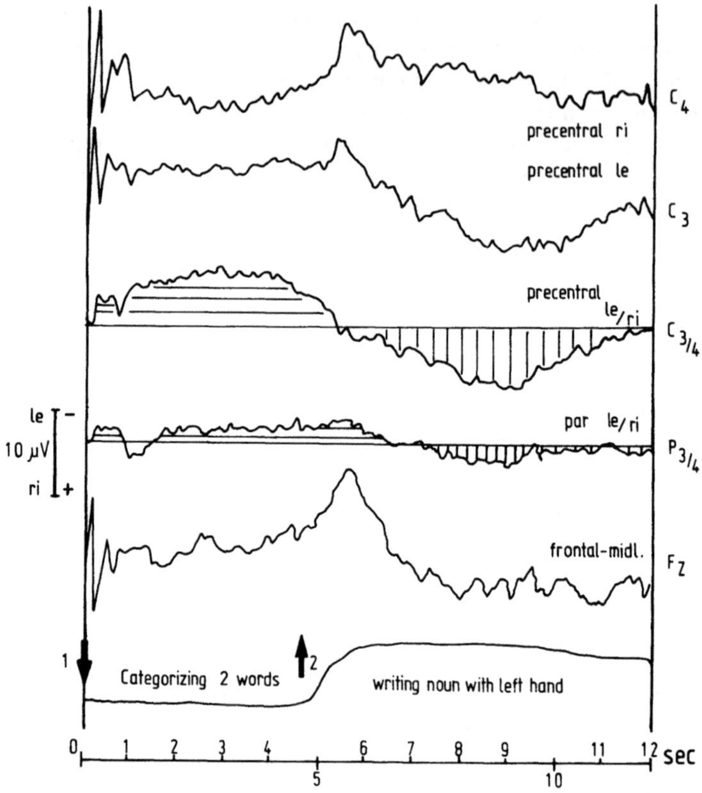

Fig. 5. Left-sided preponderance in a left-hander during language processing changes into right-sided preponderance during writing with the left hand. The language processing consisted of 30 categorizations of two words. After this the resulting common noun was written down 30 times with the left hand. During language processing, the left precentral and parietal regions show predominant negativity. Writing with the left hand in the second period results in higher contralateral right-sided negativity and reversal of the transverse leads from precentral and parietal regions. The frontal midline lead with ear reference also reverses from negativity during the language task to positivity during writing after a brief negative peak at the onset of writing. *Horizontal hatching* indicates left-sided preponderance, *vertical hatching* right-sided preponderance in transverse leads. Left-hander aged 31 writing with the left hand; during the first 4 school years he had written with the right hand. Average of 30 tasks.

Fig. 6a,b. Left-sided preponderance in a left-hander during language processing and late reversal to right preponderance during writing with left hand. In all transverse leads the negativity is larger at the left side (*horizontal hatching*) during rehearsal of name and address after signal 1 (↓). In *a* the left-sided preponderance reverses to right preponderance (*vertical hatching*) during the writing of the name in mirror script with left hand following a small biphasic wave elicited by signal 2 (↑). In *b* the left preponderance is maintained during writing with right hand. All midline electrodes with ear reference show a negative polarity during language processing, but reverse to positivity during writing after an initial negative peak. The subject, a male aged 27, used to write with his right hand. Averages of 30 tasks using either hand.

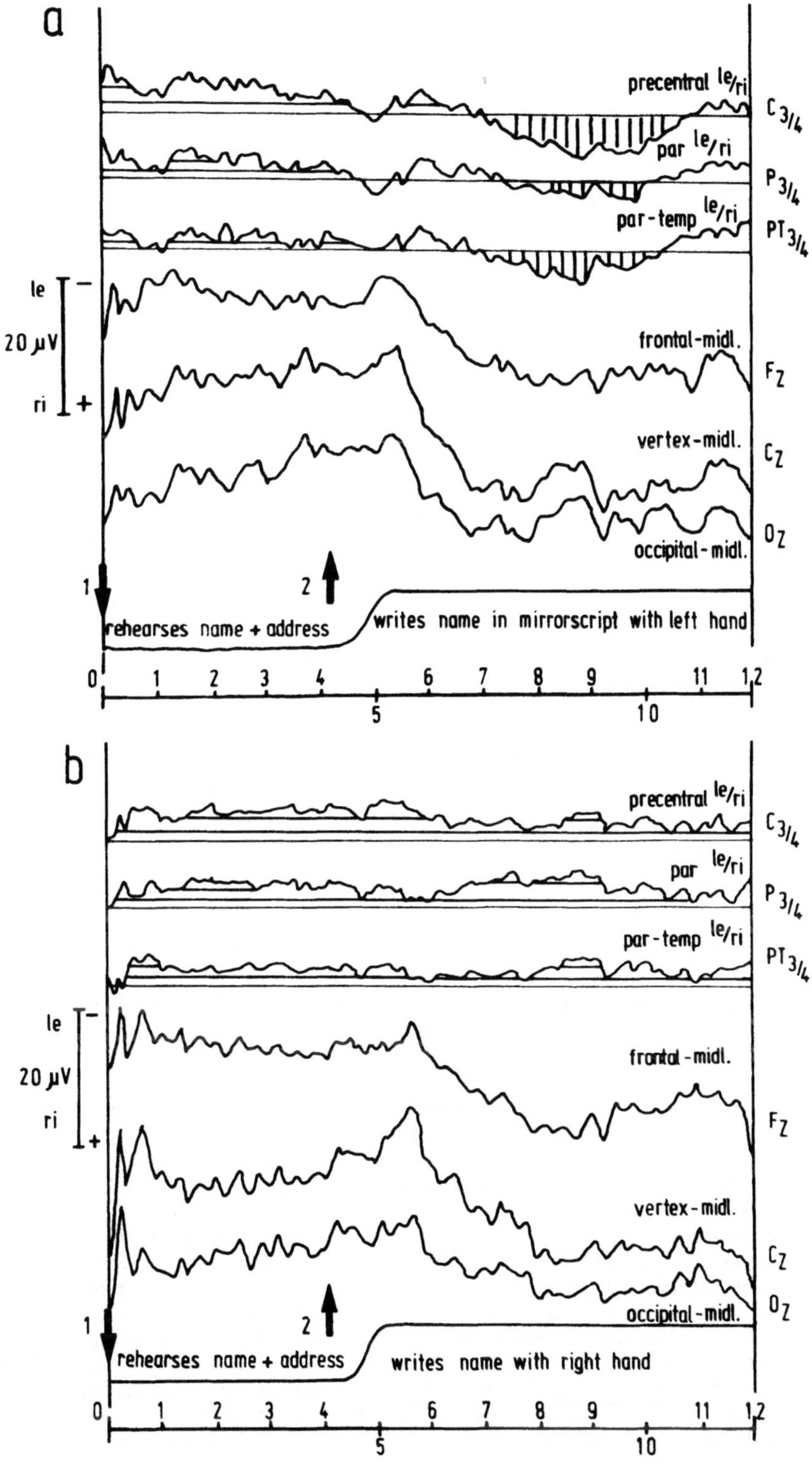
a
precentral le/ri
C3/4
par le/ri
P3/4
par-temp le/ri
PT3/4
le
20 µV
ri
frontal-midl.
Fz
vertex-midl.
Cz
occipital-midl.
Oz
1
2
rehearses name + address
writes name in mirrorscript with left hand
0 1 2 3 4 5 6 7 8 9 10 11 12
b
precentral le/ri
C3/4
par le/ri
P3/4
par-temp le/ri
PT3/4
le
20 µV
ri
frontal-midl.
Fz
vertex-midl.
Cz
occipital-midl.
Oz
1
2
rehearses name + address
writes name with right hand
0 1 2 3 4 5 6 7 8 9 10 11 12

ward deviation from the midline and horizontal hatchings indicate that negativity is larger at the left electrode. Downward deviation and vertical hatching means right preponderance.

During language processing the maximal amplitude of lateral preponderance in precentral, parietal, and parietotemporal regions showed small intraindividual and large interindividual variations. The peaks of left preponderance during language tasks ranged between 3 and 12 μV for amplitudes larger in left than in corresponding right brain regions, measured in transverse leads of nine left-handers with left preponderance. The left-sided preponderance during the subsequent period of writing with the right hand had similar amplitudes but was often of shorter duration. The short peaks were followed by a return to middle levels or a reversal to right preponderance during the later writing act. The reversal to the right usually had smaller amplitudes than left-sided preponderance during language tasks. It often began earlier in parietal and parietotemporal than in precentral leads, as shown in Fig. 7b.

As a rule, the left preponderance during the language-processing period I was maintained in the initial period II of the test when writing with the right hand began. During the following writing the left preponderance often reversed slightly into right preponderance (Figs. 6b and 7b). Writing with the left hand usually reversed the transverse leads from left preponderance during language processing to right preponderance. This reversal with the writing hand was more pronounced in precentral than in parietal and parietotemporal leads (Figs. 6a and 7a).

Both left-hander groups, those writing with the right and those writing with the left hand, showed larger potentials over the left hemisphere during language tasks. Nine adult left-handers, among them five who had written with the left hand for more than 20 years, showed left-sided preponderance during language processing. It appeared that the learned hand preference for writing with the left or right hand had little influence on the lateral preponderance of the negativity.

One exception of larger negativity over the right hemisphere was found in a 12-year-old boy who had written with his left hand for 6 years. In the group of 12 left-handers, among whom seven wrote with their left hand, he alone showed a large right-sided preponderance during language tasks, indicating that his right parietal, parietotemporal, and precentral regions were more involved in language processing than his corresponding left areas (Fig. 8).

Fig. 7a,b. Left-sided precentral preponderance in a left-hander during language tasks changed in opposite directions during writing with *a* left and *b* right hand.
The left-sided negativity in precentral leads appears during translation from German into English. This left-sided preponderance reverses to the right side while the left hand writes, but increases during right-handed writing. All other transverse leads show insignificant laterality before and during left-handed writing. A slight parietal left preponderance appears in *b* during translation and increases before and during writing with the less experienced right hand followed by a late reversal. Large evoked potentials appear in all leads with ear reference after the first signal but not after the second. The midline electrodes show slight biphasic negative-positive waves during language processing and large negative peaks during and after the writing of the translated English words of four letters in length. The writing times of left and right hand show smaller differences (1.9 s left, 2.4 s right) than in the right-hander of Fig. 2, who was inexperienced in writing with the left hand. The subject was a male left-hander aged 31; during the first 4 school years he had written with the right hand. Averages of 30 tasks using either hand.

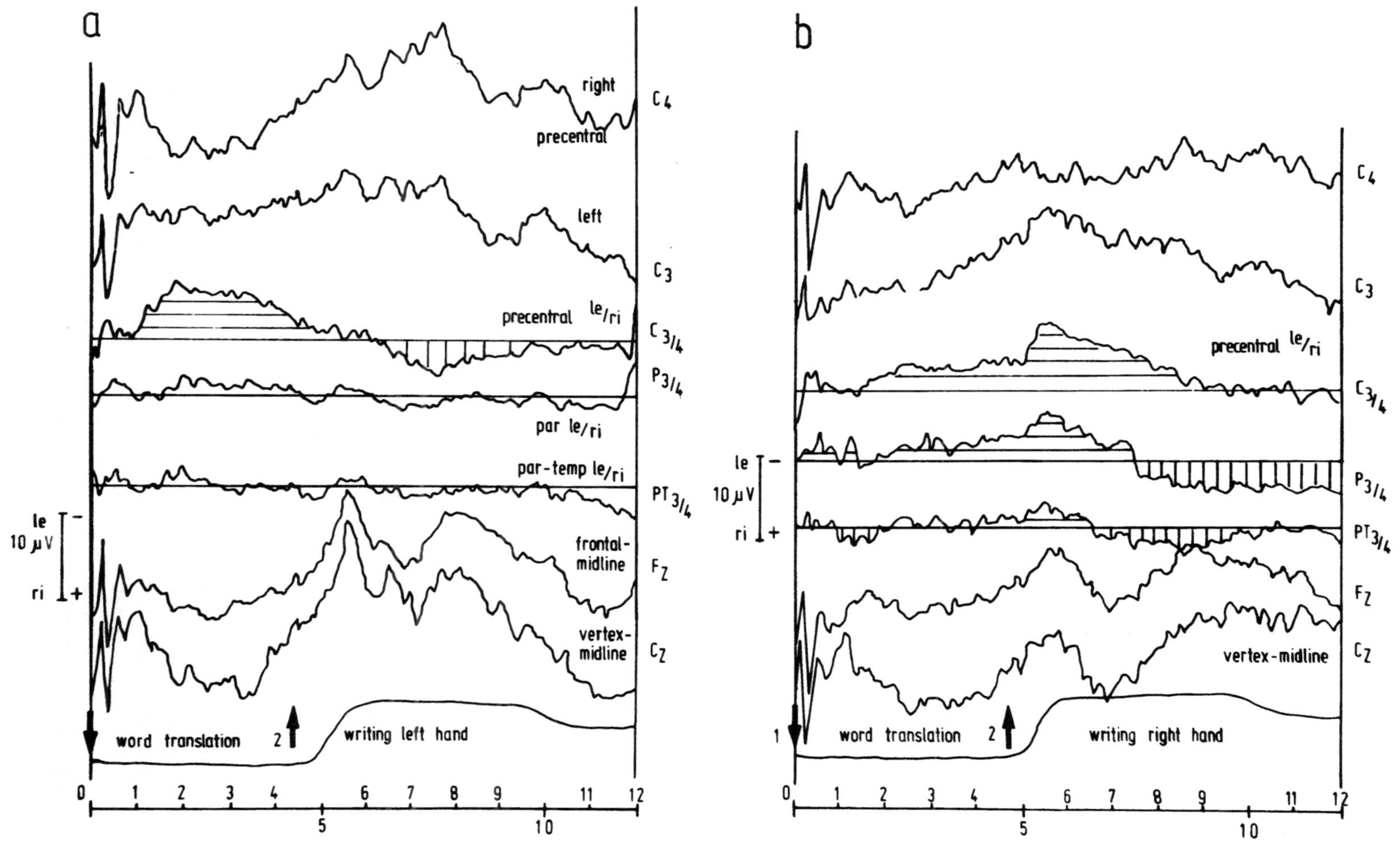

a
right
precentral
C4
left
C3
precentral le/ri
C3/4
P3/4
par le/ri
par-temp le/ri
PT3/4
le
10 µV
ri
frontal-midline
FZ
vertex-midline
CZ
word translation
2
writing left hand
b
vertex-midline
1
writing right hand

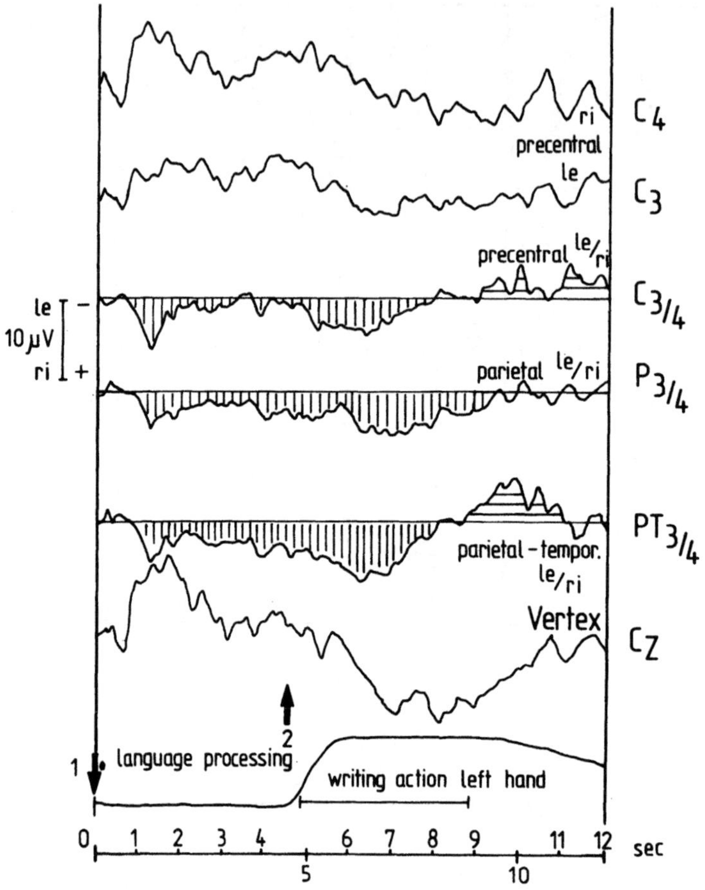

Fig. 8. Language dominance in the right hemisphere of a 12-year-old male left-hander writing with his left hand. During language processing (categorizing two words) a larger negativity of the right-sided electrode is shown in all transverse leads, indicated by a *hatched downward deviation.* After signal 2 (↑), writing with the left hand further prolongs the right preponderance of negativity, although the writing potentials reverse from negativity during language processing to positivity during writing in the precentral and vertex leads with ear reference. Signal 1 (↓) elicits small evoked potentials preceding the slow potentials. In the *upper records* the larger surface-negative potentials of the right precentral electrode using ear reference can be compared to that of the left. Average of 30 tasks.

Another left-hander showed no clear lateral preponderance during the language tasks. In the other ten a left-sided preponderance appeared with varying amplitudes during the various language tasks, such as categorizing of two words (Fig. 5), rehearsal of name and address (Fig. 6), and translation (Fig. 7). Figures 5–8 were selected to demonstrate clear lateral preponderance in four left-handers. The other eight subjects had smaller amplitudes of negative laterality, in seven with left-sided preponderance, in one of doubtful laterality.

One may conclude that the majority of left-handers exhibit electrophysiological signs of left-hemispheric language processing, although somewhat less often than right-handers. The left-sided preponderance is found in left-handers writing with the left hand and in those writing with the right hand. Only one of 12 left-handers showed signs of language processing in his right hemisphere.

Whether our electrophysiological criteria will be able to demonstrate right-sided language dominance in a larger percentage of left-handers, as has been estimated on the basis of neurological experience (Hécaen et al. 1981; Hécaen and Sauguet 1971; Hécaen and DeAjuriaguerra 1963; Milner et al. 1966) and carotid injection tests (Wada and Rasmussen 1960), can only be answered when a larger number have been examined using our method.

Local and Amplitude Limits of Lateral Preponderance. The lateral difference of negativity during language processing appeared with relatively high amplitudes of 6–12 μV in some right-handers (Fig. 4) and left-handers (Figs. 5 and 7). In others, however, it was rather small or restricted to one or two transverse leads: the lateral difference measured only 2–5 μV in amplitude, oscillating around the equipotential level. When such small lateral differences occured in several leads these records were difficult to evaluate for language dominance. In one left-hander with small and varying lateral differences the side of preponderance was uncertain. Figure 7a shows similar small amplitudes only in transverse leads from parietal and parietotemporal areas of a left-handed man who had changed from writing with the right hand to writing with the left at the age of 10. A larger left-sided preponderance during translation appeared in his precentral leads. Why some subjects showed larger lateral preponderance in precentral than in parietal or parietotemporal leads during language processing cannot yet be explained. A possible relation to early motor programming of the subsequent writing appears improbable, because the left-sided laterality appeared before writing with either hand: the left-sided preponderance of precentral leads occurred during various modes of language processing before either the ipsilateral (Figs. 5, 6a, 7a) or the contralateral hand (Figs. 6b, 7b) wrote down the result.

Discussion

The present experiments describing electrophysiological signs of language dominance in one cerebral hemisphere were based on small samples taken from a limited number of right- and left-handers, as shown in Figs. 1–7. These preliminary results may be used to demonstrate how language dominance can be investigated by use of electrophysiological techniques.

The results supplement earlier investigations of writing potentials (Jung et al. 1982) and were surprising in three ways: (a) the high frequency of left-sided language dominance in left-handers, (b) the wide distribution of the left-sided preponderance over many brain regions during language processing, and (c) the different polarities of the potentials accompanying language tasks and writing. Before discussing these points I would like to offer reasons for the preference in the selected samples.

Sampling Individual Recordings Versus Interindividual Statistics. The records of slow potentials during language processing and writing are less disturbed by artifacts than those during spoken language (Grözinger et al. 1975). Our results present, however, other methodological problems that limit data processing and statistics. The large interindividual variations of form and polarity preclude the statistical evaluation of many subjects. If one were to pool 20 records of writing potentials with surface-positive polarities and 20 with negative, one would obtain an average potential near zero. Taking grand averages from such groups would only show such crude effects as the influence of the writing hand. On the other hand, the intraindividual constancy of potential forms allows significant comparisons of tested actions and different tasks evaluated in the same subject, who serves as a sample of a group. Therefore, we have selected representative samples of right- and left-handers who write with either their left or right hand. Backward averaging (Figs. 1–3 (Jung et al. 1982)) and, for this paper, forward averaging (Figs. 4–8) were used to demonstrate electrophysiological cues of language dominance.

Slow potentials from many cortical regions are easy to record but difficult to evaluate. They have the general limitations evident in the analysis of polygraphic recordings, which may be overcome in further experiments with computer averaging in larger subject samples.

Constancy of Individual Patterns. A similar intraindividual constancy – though large interindividual variations of form and polarity – was also found for slow potentials during language tasks in both right-handers and left-handers, as decribed earlier for the writing potentials in right-handers (Jung et al. 1982). The local scalp potentials when recorded with ear reference showed characteristic and rather fixed patterns for each individual during language processing and writing. These patterns were also maintained during different language and writing tasks.

Language Dominance and Handedness. The demonstration of left-hemispheric dominance of language processing in right-handers may be considered simply an electrophysiological confirmation of well-established facts. The speech dominance of right-handers in the left hemisphere has been known since Broca's discovery in 1865 (Broca 1865) and Wernicke's work in 1874 (Wernicke 1874). It would be more interesting and useful for clinical applications if we could demonstrate electrophysiological criteria for the language dominance in left-handers. The results, however, showing left-sided predominance during language tasks in nearly all left-handers, were a surprise to us.

We expected that at least left-handers writing with the left hand would show a larger right-sided negativity because they might process their language in the right hemisphere contralateral to their writing hand. In this group, however, only one subject of seven showed right-sided preponderance (Fig. 8). Five exhibited the electrophysiological indications of left-sided language processing, as did the five left-handers writing with their right hand.

Left-Sided Language Dominance in Left-Handers. If we accept a lateral preponderance of negative slow potentials during language processing as a sign of speech dominance, this appeared to be located in the left hemisphere in the vast majority of our subjects. All right-handers and ten of 12 left-handers showed a left-sided predominance of negativity. Only one left-hander had a clear right-sided predominance, and one had equivocal laterality.

Broca's original concept, which localized the speech centers of left-handers in the right frontal cortex (Broca 1865) opposite to that of right-handers, was doubted following neurological studies in aphasia. Also, after Wernicke (1874) discovered that language was processed mainly in the left temporal areas in right-handers cases of persistent aphasia caused by right-sided temporal lesions in left-handed patients remained rare. The problem is still unsolved with respect to the degree to which language is processed in left-handers in the right, the left, or both hemispheres. Our electrophysiological test showed right-sided language dominance only in one of 12 left-handers. Clearly, a larger number must be examined until the distribution of laterality may be ascertained.

Language and hand dominance are both located in the left hemisphere in right-handers, but are often dissociated in left-handers (Goodglass and Quadfasel 1954; Graves et al. 1982; Hécaen and DeAjuriaguerra 1963; Levy and Gur 1980; Penfield and Roberts 1959). Whether more bilateral language processing occurs in left-handers appears uncertain (Hécaen and Sauguet 1971; Hécaen et al. 1981; Milner et al. 1966). Although our experiments seem to show smaller left-sided preponderance of negativity during language processing in some left-handers, this is insufficient evidence for bilateral language processing and needs confirmation by further experiments.

Possible anatomic correlates of speech dominance are limited to Geschwind's finding of the larger left planum temporale (Galaburda et al. 1978; Geschwind 1974), but the structural basis of hand dominance remains obscure (Witelson 1980). The planum temporale is hidden deeply in the sylvian fissure and, therefore, is not easily accessible to EEG recordings from the surface. This may be a serious limitation of our electrophysiological criteria for language dominance.

Extension of Language Processing Areas. The unilateral predominance of negativity during language processing comprises the parietal, temporal, precentral, and frontal regions (Figs. 4–8). Hence it appears to transcend the speech centers of the cortex. Whether or not this is simply physical spread in extracortical tissues remains uncertain without a detailed localization of potential fields.

In our first series of experiments we neglected the transverse leads between frontal electrodes because of large artifacts caused by eye movements. Frontal electrodes of midline (Figs. 4, 5, 7) and bilateral location may, however, show rather large potential shifts as also recorded in conditioning experiments (Walter 1964/65) and during writing (Jung et al. 1982). In more recent recordings we were able to record a higher lateral predominance of transverse leads from frontal electrodes than from precentral, parietal, and temporal electrodes. The parietal-temporal electrodes are located nearer to sensory, the frontal nearer to motor language centers. Whether the larger left-sided frontal predominance may indicate a participation of the frontal speech area in language processing without vocalization remains equivocal. This may be decided by more detailed analysis of the frontal potential fields.

For localization one must consider the uncertain correlation of the potential fields recorded from the scalp electrodes with the complex cortical structures of both hemispheres in man. Even more detailed analyses of surface potential distributions may be insufficient to show clear correlates of cortical functions and brain potentials. The modular function of the cortex proposed by Szentágothai (1975), Eccles (1980), and Mountcastle (1978) also makes it unlikely that we would find sharp localized areas of electrical activity during complex achievements such as language processing. This may be the reason why rather large areas of the cortical surface participate in language tasks and writing.

Hand Dominance and Hand Preference

Although the term "left hand preference" is often used in the literature for left-handers, innate hand dominance should be distinguished from acquired hand preference (Jung 1977a). Hand preference develops after experience and exercise of certain special actions. In practically all right-handers, the right hand is both dominant and preferred. In many left-handers, however, hand dominance and hand preference may differ according to the subject's education and individual habits. Most German adult left-handers learned writing with the right hand and therefore prefer right-handed writing, although they use their left hand to operate a screwdriver or throw a ball and for sewing, drawing, and strenuous unilateral athletics. In these individuals, therefore, the left hand is dominant for many actions but they have opposite hand preference for writing with the nondominant right hand. This is also true for left-handed artists. They often draw with the left hand but write with the right hand, as did Holbein, Menzel, Klee and others listed in an earlier paper (Jung 1977b). Among modern artists Escher may be added (Witelson 1980), although most younger left-handed painters use the left hand for both drawing and writing. The motor features of dominant and preferent activity and their relation to brain potentials deserve a detailed study. Since Goodglass and co-workers (Graves et al. 1982) described a correlation between mouth asymmetry in speech and the language-dominant hemisphere similar motor patterns and lateral preponderance of slow potentials may give additional clues for research in cerebral dominance.

Lateral Preponderance and Polarities with Ear Reference. The negative or positive polarity of the individual writing potentials recorded with ear reference had surprisingly little influence upon the laterality of negative preponderance in transverse leads during language processing and writing. Even subjects who had mainly surface-positive writing potentials showed a left-sided preponderance of negativity during language tasks and writing with the right hand. This is shown clearly in Figs. 5 and 6 when the positive writing potentials of the records using ear reference are compared to the left-sided preponderance in transverse leads of corresponding brain regions. The marked right-hemispheric preponderance of Fig. 8 was also associated with negative polarity during the language tasks and with positive polarity in the subsequent writing period. Hence the polarities during language processing and writing may vary, but the lateral preponderance of negativity appears more stable.

The lateral preponderance of negativity in transverse leads exhibited during language processing, although somewhat independent of interindividual variations in form and polarity of the slow potentials, ranges between 3 and 10 μV in amplitude. The small amplitudes obtained by averaging in some subjects limit the validity of our technique. Further experiments using large subject samples of right- and left-handers may show whether this technique will develop into a practical and useful method of recognizing language dominance in man.

Possible Clinical Application. A useful method of recognizing language dominance in patients would be an important advance in neurology and neurosurgery. Until now a neurosurgeon has only been able to apply the invasive technique of Wada and Rasmussen (1960) when he wanted to know language dominance before an operation. Our noninvasive method, however, is still too difficult to use in patients because it demands long concentration with steady fixation to exclude the artifacts of eye movements. A similar and less strenuous method would be more desirable, in which the ocular artifacts could be excluded by computer compensation so that the subjects could write normally, that is, visually following the hand movement. Such a method would be better suited for clinical use.

Conclusions

The technique of slow potential recordings, employed in these experiments, poses four problems which have not yet been solved:

1. The small amplitudes of the lateral preponderance of negative potentials
2. The uncertainty of the neutral potential level in averaging
3. The large interindividual variations of surface negativity or positive polarity
4. The influence of eye position and the necessity of steady fixation. Even slight eye movements toward the left or right influence the polarity of transverse leads between symmetrical brain regions.

The middle level should be defined more clearly for both unipolar recordings with ear reference and transverse leads between left and right brain regions. With these problems in mind we have begun to improve our methods of data analysis.

In developing special computer programs to analyze the DC potential fields of the scalp over both hemispheres, we hope to find more reliable electrophysiological criteria for language dominance than the rather crude averaging of local recordings described in this paper.

Note Added in Proof: Since this paper was presented in April 1983 we have made further experiments. In 30 left-handers, only 2 showed cerebral correlates of right-sided language dominance. Two others had uncertain laterality. Right-sided lateralities appeared only in left-handers below the age of 15. All adult left-handers, showed left-sided laterality during language tasks. Group differences appeared only during the writing task between right- and left-handers who usually wrote with the right hand and those who usually wrote with the left.

References

Broca P (1865) Sur le siège de la faculté du language articulé. Bull Soc d'Anthropol 6: 377–393

Deecke L, Kornhuber HH (1978) An electrical sign of participation of the mesial supplementary motor cortex in human voluntary finger movement. Brain Res 159: 473–476

Eccles JC (1980) The human psyche. Springer, Berlin Heidelberg New York
Eccles JC (1981) Language, thought and brain. Epistemologia 4: 97–126
Galaburda AM, LeMay M, Kemper TL, Geschwind N (1978) Right-left asymmetries in the brain: structural differences between the hemispheres that underly cerebral dominance. Science 199: 852–856
Geschwind N (1974) The anatomical basis of hemispheric differentiation. In: Dimond SJ, Beaumont JG (eds) Hemisphere function in the human brain. Elek, London
Goodglass H, Quadfasel FA (1954) Language laterality in left-handed aphasics. Brain 77: 521–548
Graves R, Goodglass H, Landis T (1982) Mouth asymmetry during spontaneous speech. Neuropsychologia 20 [4]: 371–381
Grözinger B, Kornhuber HH, Kriebel J (1975) Methodological problems in the investigation of cerebral potentials preceding speech: determinating the onset and suppressing artifacts caused by speech. Neuropsychologia 13: 263–270
Grünewald G, Grünewald-Zuberbier E, Hömberg V, Netz J (1979) Cerebral potentials during smooth goal-directed hand movements in right-handed and left-handed subjects. Pflügers Arch 381: 39–46
Hécaen H, DeAgostini M, Monzon-Montes A (1981) Cerebral organization in left-handers. Brain Lang 12: 261–284
Hécaen H, Sauguet J (1971) Cerebral dominance in left-handed subjects. Cortex 7: 19–48
Hécaen H, DeAjuriaguerra J (1963) Les Gauchers, prévalence manuelle et dominance cérébrale. P.U.F., Paris
Jasper HH (1958) The ten twenty electrode system of the International Federation. Electroencephalogr Clin Neurophysiol 10: 371–375
Jung R (1977a) Motorisches Lernen, Seitendominanz und Seitenpräferenz. EEG EMG 8: 218
Jung R (1977b) Über Zeichnungen linkshändiger Künstler von Leonardo bis Klee: Linkshändermerkmale als Zuschreibungskriterien. In: Semper attentus. Beiträge für Heinz Götze zum 8. Aug. 1977. Springer, Berlin Heidelberg New York, pp 189–218
Jung R, Hufschmidt A (1983) Can the language dominant hemisphere be recognized during language processing by slow brain potentials in man? Naunyn-Schmiedebergs Arch Pharmacol [Suppl] 322: R76, 302
Jung R, Hufschmidt A, Moschallski W (1982) Langsame Hirnpotentiale beim Schreiben: die Wechselwirkung von Schreibhand und Sprachdominanz bei Rechtshändern. Arch Psychiatr Nervenkr 232: 305–324
Kornhuber HH, Deecke L (1965) Hirnpotentialänderungen bei Willkürbewegungen und passiven Bewegungen des Menschen: Bereitschaftspotential und reafferente Potentiale. Pflügers Arch ges Physiol 284: 1–17
Levy J, Gur RC (1980) Individual differences in psychoneurological organization. In: Herron J (ed) Neuropsychology of left-handedness. Academic, New York, pp 199–210
Milner B, Branch C, Rasmussen T (1966) Evidence for bilateral speech representation in some non-right handers. Trans Am Neurol Assoc 91: 306–308
Mountcastle VB (1978) An organizing principle for cerebral function: the unit module and the distributed system. In: Edelman GM, Mountcastle VB, Schmitt VB (eds) The mindful brain. MIT Press, Cambridge, pp 7–50
Penfield W, Roberts L (1959) Speech and brain mechanisms. Princeton University Press, Princeton
Szentágothai J (1975) The "module concept" in cerebral architecture. Brain Res 95: 475–496
Wada J, Rasmussen T (1960) Intracarotid injection of sodium amytal for the lateralization of cerebral speech dominance. J Neurosurg 17: 266–282
Walter WG (1964/65) Slow potential waves in the human brain associated with expectancy, attention and decision. Arch Psychiatr Nervenkr 206: 309–322
Wernicke C (1874) Der aphasische Symptomencomplex. Eine psychologische Studie auf anatomischer Grundlage. Cohn and Weigert, Breslau, p 72
Witelson SF (1980) Neuroanatomical asymmetry in left-handers: a review and implications for functional asymmetry. In: Herron J (ed) Neuropsychology of left-handedness. Academic, New York, pp 79–113

John and Helena Eccles during the symposium
"Central Correlates of Conscious Experience" in Senanque (Provence) 1978

Subject Index

GPSR Compliance

The European Union's (EU) General Product Safety Regulation (GPSR) is a set of rules that requires consumer products to be safe and our obligations to ensure this.

If you have any concerns about our products, you can contact us on ProductSafety@springernature.com

In case Publisher is established outside the EU, the EU authorized representative is:

Springer Nature Customer Service Center GmbH
Europaplatz 3
69115 Heidelberg, Germany

Batch number: 10838062

Printed by Printforce, the Netherlands